CYCLES OF CONTINGENCY

Life and Mind: Philosophical Issues in Biology and Psychology
Kim Sterelny and Rob Wilson, editors

Cycles of Contingency: Developmental Systems and Evolution, Susan Oyama, Paul E. Griffiths, and Russell D. Gray, editors, 2001

Coherence in Thought and Action, Paul Thagard, 2001

Norms of Nature: Naturalism and the Nature of Functions, Paul Sheldon Davies, 2001

CYCLES OF CONTINGENCY
Developmental Systems and Evolution

edited by Susan Oyama, Paul E. Griffiths, and Russell D. Gray

A Bradford Book
The MIT Press
Cambridge, Massachusetts
London, England

This book was set in Times New Roman by Asco Typesetters, Hong Kong in QuarkXpress, and was printed and bound in the United States of America.

Library of Congress Cataloging-in-Publication Data
Cycles of contingency: developmental systems and evolution / Susan Oyama, Paul E. Griffiths, and Russell D. Gray, editors.
        p.   cm.
    "A Bradford book."
    Includes bibliographical references and index.
    ISBN 0-262-15053-0 (alk. paper)
    1. Developmental psychology. 2. Nature and nuture. 3. Genetic psychology. I. Oyama, Susan. II. Griffiths, Paul E. III. Gray, Russell D.
    BF713 .C93   2000
    155.7—dc21   00-026951

Contents

Preface

The idea for this book emerged from sessions at the International Society for History, Philosophy and Social Studies of Biology, first at Brandeis University in 1993, then at the University of Leuven, Belgium, in 1995 and the University of Washington in Seattle in 1997. The topic of these sessions was the developmental systems approach to development and evolution. Like the papers in those sessions, the papers collected here do not conform to any "party line." Instead, the volume aims to explore the implications, potential and limitations of a group of ideas in which there has been a great deal of interest across a range of disciplines. It aims to locate points of contact and points of disagreement between developmental systems theory and a number of other attempts to improve the conceptual framework of the life sciences. Our main hope for the volume is that it will act as a resource for other people intrigued by these ideas, and who wish to make use of them and build upon them in their own disciplines.

Our thanks as editors must go, first and foremost, to the twenty-two other contributors to the volume. These authors have been exemplary in their ability to meet deadlines and to engage in the productive discussion with the editors and with one another which was made possible by the timely submission of draft manuscripts. The volume was initially developed on a website, to allow authors to read and respond to one another's work, as well as to receive comments from the editors. This has helped the chapters to talk to one another and, we hope, given a volume a sense of unity. To all these people, we say many thanks.

Particular thanks are also due to Ross West of the University of Sydney, who worked as a research assistant through the development of the volume. In addition to maintaining the website and managing the physical production of the manuscript, Ross drew or redrew many of the illustrations and was always on hand to respond to the authors' queries. Russell Gray would like to thank Nicola Gavey for her patience and support during the production of this book.

Acknowledgments

"A Critique of Konrad Lorenz's Theory of Instinctive Behavior" by D. S. Lehrmann first appeared in *Quarterly Review of Biology* 28(4): 337–363 and is reprinted by permission of the University of Chicago Press.

"Gene, Organism, and Environment" by R. C. Lewontin first appeared in D. S. Bendall (Ed.), *Evolution: From Molecules to Men* (Cambridge: Cambridge University Press, 1983), pp. 273–285.

"From Complementarity to Obviation: On Dissolving the Boundaries Between Social and Biological Anthropology, Archaeology, and Psychology" by T. Ingold first appeared in *Zeitschrift für Ethnologie* 123: 21–52.

Contributors

Patrick Bateson
Sub-Department of Animal Behaviour
University of Cambridge
Cambridge, England

David Depew
Communication Studies/POROI
University of Iowa
Iowa City, Iowa

Marcus W. Feldman
Department of Biological Sciences
Stanford University
Stanford, California

Deborah M. Gordon
Department of Biological Sciences
Stanford University
Stanford, California

Peter Godfrey-Smith
Department of Philosophy
Stanford University
Stanford, California

Gilbert Gottlieb
Center for Developmental Science
University of North Carolina
Chapel Hill, North Carolina

Paul E. Griffiths
Department of History & Philosophy of Science
University of Pittsburgh
Pittsburgh, Pennsylvania

Russell Gray
Department of Psychology
University of Auckland
Auckland, New Zealand

Tim Ingold
Department of Sociology
University of Aberdeen
Aberdeen, Scotland

Eve Jablonka
Cohn Institute for the History and Philosophy
of Science and Ideas
Tel Aviv University
Tel Aviv, Israel

Timothy D. Johnston
Department of Psychology
University of North Carolina at Greensboro
Greensboro, North Carolina

Evelyn Fox Keller
Program in Science, Technology, and Society
Massachusetts Institute of Technology
Cambridge, Massachusetts

Peter H. Klopfer
Biology Department
Duke University
Durham, North Carolina

Kevin N. Laland
Sub-Department of Animal Behaviour
University of Cambridge
Cambridge, England

Daniel S. Lehrman
†Deceased

Richard C. Lewontin
Museum of Comparative Zoology
Harvard University
Cambridge, Massachusetts

Lenny Moss
Department of Philosophy
University of Notre Dame
Notre Dame, Indiana

Eva M. Neumann-Held
European Academy for the Study of
Consequences of Scientific and Technological
Advance
Bad Neuenahr-Ahrweiler, Germany

H. Frederik Nijhout
Department of Zoology
Duke University
Durham, North Carolina

F. John Odling-Smee
Institute of Biological Anthropology
Oxford University
Oxford, England

Susan Oyama
John Jay College of Criminal Justice and
The Graduate Center
City University of New York
New York, New York

Kim Sterelny
Department of Philosophy
Research School of the Social Sciences
Australian National University
Australia, and
Department of Philosophy
Victoria University
Wellington, New Zealand

Peter Taylor
Critical and Creative Thinking Program
University of Massachusetts
Boston, Massachusetts

Cor van der Weele
Centre for Bioethics and Health Law
Utrecht University
Utrecht, The Netherlands

Bruce H. Weber
Department of Chemistry and Biochemistry
California State University, Fullerton
Fullerton, California

William C. Wimsatt
Department of Philosophy
University of Chicago
Chicago, Illinois

CYCLES OF CONTINGENCY

1 Introduction: What Is Developmental Systems Theory?

Susan Oyama, Paul E. Griffiths, and Russell D. Gray

The nature/nurture debate is not dead. Open a book, read a newspaper, turn on the TV, read *Science* or *Nature* and you will find yourself bombarded with claims and counterclaims. Are there "genius" genes? If not those, then surely "gay" ones? Is aggression the consequence of social and economic conditions, or is it a product of evolution? Are cognitive differences between men and women due to genetics or upbringing? Can we shape our destiny, or are we robots programmed by our selfish genes? These are not esoteric questions, of concern only to a few academic specialists. Their answers can have social and political consequences. People are quick to draw conclusions about the possibility, or even the *rightness*, of trying to subvert nature's plans. If intelligence is innate, then some would say that compensatory programs are a waste of effort and money. If sexual orientation is a biological given rather than a free choice, then, so it is argued, the language of morality in that context is both inappropriate and useless.

Underlying these vexed questions are a number of oppositions: nature or nurture, genes or environment, biology or culture. Developmental systems theory (hereafter DST) is an attempt to do biology without these dichotomies. This is a more difficult task and requires a greater theoretical reworking of biological concepts than has so far been realized. The standard response to nature/nurture oppositions is the homily that nowadays everyone is an interactionist: All phenotypes are the joint product of genes and environment. According to one version of this conventional "interactionist"[1] position, the real debate should not be about whether a particular trait is due to nature or nurture, but rather how much each "influences" the trait. The nature/nurture debate is thus allegedly resolved in a quantitative fashion. The question is no longer whether intelligence is innate or acquired, but instead whether intelligence is 50 percent or 70 percent genetic. DST rejects the attempt to partition causal responsibility for the formation of organisms into additive components. Such maneuvers do not resolve the nature/nurture debate; they continue it. This is typical of the way in which the traditional view of development morphs into new forms when challenged and returns to plague current academic and social debates. If it is no longer acceptable to ask whether something is instinctive, then we ask whether it has a large genetic component. If that, too, becomes unacceptable, then we ask if there is a genetic predisposition toward it. What we need is the "stake-in-the-heart move" (Oyama 1985: 27)—a way of thinking about development that does not rely on a distinction between privileged, essential causes and merely supporting or interfering causes.

Other concepts in the heartland of contemporary biology, such as inheritance and evolution, need substantial reformulation as well. The reliability of many aspects of development has encouraged biologists, psychologists, and social scientists to postulate some central directing agency or "master molecule." Inheritance and evolution are defined as the passing on and alteration of such master molecules. Other inputs to development tend to be lumped together as "environment" and treated as a standard background that is not itself in need of explanation. In contrast, DST views both development and evolution as processes of *construction* and *reconstruction* in which heterogeneous resources are contingently but more or less reliably reassembled for each life cycle. It is these cycles of contingency that we need to unpack, and it is these cycles that are the subject of this book.

So What Exactly Is DST?

What we have come to term *developmental systems theory* is not a theory in the sense of a specific model that produces predictions to be tested against rival models. Instead, it is a general theoretical perspective on development, heredity and evolution, a framework both for conducting scientific research and for understanding the broader significance of research findings.[2] Many other biological theories play this dual role, perhaps most noticeably in recent years the idea of the "selfish gene" (see chapter 20).

Developmental systems theory is not attributable to one person or group. It draws on insights from researchers in a wide range of areas who have been dissatisfied with crude dichotomous accounts of development and have attempted to formulate an alternative.[3] Table 1.1 draws on a number of past attempts to specify a list of tenets of DST (Gray 1992, 1997; Schaffner 1998; Griffiths and Knight 1998; Oyama 2000b, Introduction). Programs of scientific research are not easily reduced to a set of precisely stated axioms (Kuhn 1970), so these tenets are more like what Schaffner (1998) has called "themes" of developmental systems research. In the rest of this section we expand and comment on these six themes.

Joint Determination by Multiple Causes

It is a truism that all traits are influenced by both genetic and nongenetic factors. According to DST, however, this "interactionist consensus" is little better than the nature/nurture dispute it is supposed to have dissolved. There are many kinds of influences on development, and there are many ways to group these interactants together. DST does not claim that all these sources of causal influence play the same role, nor that all are equally important (whatever that might mean). Rather, different groupings of developmental factors are valuable when addressing different questions. The distinction between genes and every other causal factor in development ("environment") is just one more grouping, possibly helpful for some purposes, much less so for many others. Many developmentally constructive interactions do not fit traditional categories, and for this reason have largely been overlooked or marginalized. Oppositions between genes (or biology) and learning, or between genes (or biology) and culture, are endemic to many fields but are miserably inadequate for capturing the multitude of causal factors needed for any reasonable treatment of ontogeny or phylogeny. DST emphasizes crucial but often overlooked similarities among resources that are usually contrasted. Phenocopying, for instance, occurs when genetic mutations, as well as changes in the outside world, can bring about similar alterations in the organism (Markert and Ursprung 1971; Waddington 1975; see also Oyama 1981 on the significance of this causal parity or symmetry in phenocopying from the perspective of DST). There are bithorax mutants in *Drosophila*, but the

Table 1.1
Major themes in developmental systems theory

1. Joint determination by multiple causes—every trait is produced by the interaction of many developmental resources. The gene/environment dichotomy is only one of many ways to divide up these interactants.

2. Context sensitivity and contingency—the significance of any one cause is contingent upon the state of the rest of the system.

3. Extended inheritance—an organism inherits a wide range of resources that interact to construct that organism's life cycle.

4. Development as construction—neither traits nor representations of traits are transmitted to offspring. Instead, traits are made—reconstructed—in development.

5. Distributed control—no one type of interactant controls development.

6. Evolution as construction—evolution is not a matter of organisms or populations being molded by their environments, but of organism-environment systems changing over time.

bithorax phenotype can also be induced by ether. Genes and ether shocks turn out to be developmentally equivalent in this respect. Phenomena that are usually contrasted to one another can be equivalent in evolution, too. Developmental influences may follow a lineage equally closely through evolution, even though one is genetic and the other "environmental"—genes and dietary cues, for example (see chapters 16 and 23). These often overlooked similarities form part of the evidence for DST's claim of causal parity between genes and other factors of development. The "parity thesis" (Griffiths and Knight 1998) does not imply that there is no difference between the *particulars* of the causal roles of genes and factors such as endosymbionts or imprinting events. It does assert that such differences do not justify building theories of development and evolution around a distinction between what genes do and what every other causal factor does.

Context Sensitivity and Contingency

The demands of interactionism in its conventional form can often be satisfied by merely admitting that every organism must have some genes and some environment. With that out of the way, the real business of settling what is due to nature and what is due to nurture can continue. Typically, this work proceeds by inferring more or less directly the extent of the correlation between genes and phenotype in one or more populations. These methods can be as direct as molecular screening or as indirect as a study of monozygotic twins raised apart. The underlying logic is very similar across this whole range of methods: The stronger the correlation, the more the genes are said to be responsible for the trait. However, as Richard Lewontin has argued, heritability estimates are not measures of global causal importance, nor do they indicate how much a trait can be modified by environmental changes (Lewontin 1974). Heritability estimates measure the proportion of variation in a specific population that is attributable to genetic differences. Be-

havior geneticists these days are quick to admit that high heritability scores do not show that it is hard to alter the trait by nongenetic means. Whenever a number of causal factors interact to produce an outcome, we should expect that the effect of changing one factor will depend on what is happening to the others. However, current theoretical frameworks encourage much more research on "genes for" traits than on statistical interactions among developmental factors in natural populations (see Schlichting and Pigliucci 1998 for a refreshing change).

The persistent tendency to minimize context sensitivity and developmental contingency when studying genetic factors in development is connected to the prevalence of information metaphors in contemporary biology (Keller 1985; Oyama 1985). As long as the DNA is thought of as containing information about developmental outcomes, it will seem sensible to inquire whether outcomes occur because they are represented in the chromosomes. Once an outcome is seen as an expression of the genetic information that controls development, it acquires a special status. It represents what the organism is "meant to be," and deviations from it are misrepresentations of the true nature of the organism—its inner essence, which was conferred on it at the moment of conception (at least in those organisms that have such moments). In such an intellectual framework, context sensitivity is often treated as interference with the basic pattern of biological causation. For DST, contingency is basic, whether the results are expected or surprising.

Extended Inheritance

A traditional way to privilege genes over other causes in development is to argue that genes are the only things organisms inherit from their ancestors. Hence the biological nature of organisms must be in the genes. DST insists on a definition of inheritance that explicitly recognizes the wide range of resources that are "passed on" and are thus available to reconstruct the organism's life cycle. Some of these resources are familiar—chromosomes, nutrients, ambient temperatures, childcare. Others are less familiar, despite the recent explosion of work on "epigenetic inheritance" (see chapter 9). These include chromatin marks that regulate gene expression, cytoplasmic chemical gradients and gut- and other endosymbionts. Another important topic in recent biology is the participation of the organism in the construction of its niche. Hence a further aspect of inheritance is the local physical environment, altered by past generations of the same species and other species as well as the organism's own activities (see chapters 10 and 12). Many of these inherited resources have distinctive roles. DNA is unique in acting as templates for protein synthesis. Membranes are unique in acting as templates for the assembly of proteins into more membrane. Chemical traces from foraging play a characteristic role in diet choice in many rodents. A written text, interpreted in an enormously complex personal and cultural setting, is distinctive in yet other ways. DST explores these diverse roles, acknowledging differences but resisting any attempt to divide them into, say, one set contributing to the organism's "nature" and another that influences its "nurture." Oyama has argued that if these vexed terms are to be retained, then "nature" should refer simply to the outcomes of development and "nurture" to the processes that produce, maintain, and alter those outcomes (Oyama 1985).

Development as Construction

Developmental systems theory is a thoroughly epigenetic account of development. Current use of term "epigenetic" is ambiguous. It is often used to mean "in addition to the genes," as in the phrase "epigenetic inheritance" (see chapter 9). Its basic meaning, however, derives from a contrast between preformationist and epigenetic theories of development. Classical preformationism held that the egg contains a tiny organism, so that embryologists only had to explain increases

in size, not the generation of biological order (Pinto-Correia 1997). Modern preformationism is subtler: The organism is not preformed in the egg, but the information that programs its development is preformed in the genes. By contrast, an epigenetic account of development is one that never sidesteps the task of explaining how a developmental outcome is produced. The claim that development occurs because it is programmed to occur or because it has been selected by evolution is merely a promissory note redeemable against future developmental biology. Similar views about biological explanation have been described elsewhere as "constructivist interactionism" (Callebaut and Stotz, forthcoming; Oyama 2000b; see also chapter 15, this volume). Despite its clumsiness, this phrase succinctly expresses two major themes of developmental systems theory. The life cycle of an organism is developmentally constructed, not programmed or preformed. It comes into being through interactions between the organism and its surroundings as well as interactions within the organism.

The conviction that development involves multiple, interdependent causes is entirely compatible with the practical requirements of research. Practicing researchers have to draw boundaries around the system to be studied, placing certain factors in the foreground while taking others for granted. In this book the contributions of Gilbert Gottlieb, Frederik Nijhout, Deborah Gordon, Pat Bateson, and Peter Klopfer show that a strategic concern with complexity and context dependence need not interfere with the process of identifying individual causal contributions by controlling other variables. The practice of changing one variable at a time while holding others constant is important, but it is incomplete. Additional investigation is required, both to show how a causal factor is coupled in a system of causes and to reveal the ways in which these links change over time. It does not require considering everything at once, as some seem to fear, but can be done by coordinating diverse investigations. In fact, having a richer strategic vision should allow researchers to make more intelligent and flexible "tactical" use of reductionistic research strategies (Wimsatt 1980; see also chapter 17).

Distributed Control

Taking a systems perspective on developmental processes means, among other things, attending to the ways in which the developing organism functions as a resource for its own further development. The organism helps determine which other resources will contribute to that development, as well as the impact they will have. The roles played by the vast and heterogeneous assembly of interactants that contribute to a life-course are system-dependent and change over time. So DST creates an inhospitable context for moves that preempt the investigation of actual processes by identifying one type of resource as controlling or directing the process, leaving other interactants to function as background conditions, raw materials, or sources of disturbance. We believe that despite the widespread talk of genetic blueprints and programs in contemporary biology, there is no scientifically defensible sense in which a subset of developmental resources contains a program or set of instructions for development.[4]

The most obvious way to defend talk of genetic programs and genetic information is to argue that it is in some unproblematic sense related to information theory. "Information" in this statistical sense is the systematic dependence of a signal on a source, a dependence that is created by a set of channel conditions. In the case of development, the genes are typically taken to be the source, so the channel conditions are all the other resources needed for development to occur. But in information theory the role of source and channel condition can be reversed. A source is simply one channel condition whose current state the signal is being used to investigate. If all other resources

are held constant, the outcomes of development can give us information about the genes, but if the genes are held constant, outcomes give us information about whichever other resource we have decided to let vary. Every resource whose state affects development could thus be considered a "source of developmental information."

Maynard Smith (2000) has recently suggested that this does not capture what is really meant when biologists talk about genetic information. According to him, biologists use information in an intentional or semantic sense—a gene has an intended meaning as well as causal consequences, and this intentional sense of information reveals the true asymmetry between genes and other developmental resources. Sterelny and Griffiths say: "A distinctive test of intentional or semantic information is that talk of error or misrepresentation make sense. A map of Sydney carries semantic information about the layout of Sydney. Hence it makes sense to say of any putative map that it is wrong or that it has been misread" (1999: 104).

Maynard Smith draws on Ruth Millikan's (1984) attempt to explain intentional information in evolutionary terms. This "teleosemantic" theory says that things carry intentional information about whatever evolution has selected them to represent; a gene contains intentional information about a phenotype that it has been selected to produce. But the same can be said of any developmental resource whose presence can be given an evolutionary explanation. As we argued in the section on extended inheritance, many kinds of developmental resources are inherited and evolve. The teleosemantic account does not show that only genes carry developmental information.

We believe that the heuristic value of the idea of developmental information in certain contexts is more than outweighed by its misleading connotations. Locating information in a single type of developmental resource obscures the context-dependency of causation by localizing control.

Some contributors to the present volume, however, are happier to deploy information concepts than we are, and do so in their chapters.

Evolution as Construction

The idea of construction through the interaction of many different factors is applicable to evolution as well as development, and it highlights striking similarities between the two processes (Oyama 1992). Just as there are no preexisting representations or instructions that shape organisms from within, there are no preexisting niches or environmental problems that shape populations from without (see chapter 6). Evolutionary change is the result of interactions in which outcomes are codetermined, or co-constructed, by populations and environments with their own, often intricately interrelated, histories and characteristics; outcomes are not imposed by or prefigured in only certain interactants. Extended inheritance both increases the range of developmental outcomes that can be given evolutionary explanations and alters our view of evolutionary dynamics (see chapter 16). If evolution is change in developmental systems, then, as just noted, it is no longer possible to think of evolution as the shaping of the organism to fit an environmental niche. Rather, the various elements of the developmental systems coevolve. Organisms construct their niches both straightforwardly by physically transforming their surroundings and, equally importantly, by changing which elements of the external environment are part of the developmental system and thus able to influence the evolutionary process in that lineage.

Aim of This Book

A common response to the confusion and political mischief caused by the nature/nurture debate is to scold the participants for their empirical sloppiness and technical errors (Kitcher 1985, 2001). Though this is often a valuable exercise,

we believe that it does not go far enough (Oyama 1987). Like Levins and Lewontin (1985), we think that the fundamental problem lies in the way causation is viewed in biological systems. This book does not therefore focus on debates about the (mis)application of biological concepts and techniques in arguments about intelligence, aggression, and gender, but rather on the concepts themselves: on the kinds of understandings of biology, development, and evolution that make poor practice likely and render critiques of that practice less than optimally effective. The purpose of the present collection is thus to provide a forum for exploring DST's alternative conceptions of development, heredity, and evolution. The contributors did not necessarily have to identify themselves as proponents of DST, nor were they required to adhere to any party line. In fact, some were selected because they could offer informed criticism. We selected contributors whose writing promised to extend, illustrate, challenge, parallel, or contrast with issues raised in DST. Many were invited to present their own empirical work, but they were asked to confront developmental systems ideas in a substantive way as well. In addition, authors were sometimes encouraged to address key concepts such as genetic information, program, and transmission.

Many of the papers collected here aim at nothing less than a root-and-branch transformation of contemporary biological thought. We do not expect such large hopes to be fulfilled in the immediate future, but we will be gratified if we can convince some readers that there are some real objections to ideas that have previously seemed unproblematic—ideas such as the gene/environment dichotomy and the genetic program for development. We also hope that by drawing attention to alternative ways of thinking about biological systems, and to some programs of research that embody them, we will persuade others to try out some of these alternatives in their own work.

Part I deals with key influences on developmental systems thinking, starting with Daniel S.

Lehrman's classic 1953 critique of the concept of innateness. As Timothy Johnston shows in his introduction to this essay, aspects of what we would now call a systems view of development can be discerned in Lehrman's work and in that of other animal behavior researchers such as Kuo (1967) and Schneirla (1966). The work of these early critics of traditional understandings of instinct and maturation remains relevant because those understandings are still with us. Eschewing pseudo-explanatory notions such as instinct and innateness allows us to ask a host of questions about the actual causal influences at each stage of development. Gilbert Gottlieb has devoted his career to elucidating these influences. In chapter 4 he looks back on this long and fruitful history of research and outlines the conceptual framework he has helped build. The ongoing research tradition stemming from Lehrman, Kuo, and Schneirla and ably represented by Gottlieb (see Michel and Moore 1995 for a recent text in this tradition) provides a standing reply to the challenge that developmental dichotomies are the only way to render complex developmental systems amenable to empirical study.

In the remainder of part I, Richard Lewontin examines the implications of his constructionist view of development and evolution for research in genetics in a new introduction to his classic paper "Gene, Organism and Environment" (1983) which is reproduced here as chapter 6. In this and other papers published around the same time, Lewontin questioned the traditional model of adaptation in which the environment acts on a passive organism and fits it to a preexisting ecological niche. Lewontin's discussion of the role of organism in constructing its own niche is elaborated elsewhere by Kevin Laland, John Odling-Smee, and Marcus Feldman (chapter 10) and by Paul Griffiths and Russell Gray (chapter 16).

Part II looks at attempts to reformulate the idea of heredity so as to do justice to the facts of development. The essays by Eva Neumann-Held and Lenny Moss focus on the concept of the gene itself. Tracking the changing definition of "gene"

and describing the relationship between the genes of classical transmission genetics and those of molecular biology has proved extremely difficult (Falk 1984, 1986; Kitcher 1982, 1984; Sarkar 1998). Neumann-Held and Moss show how a developmental perspective can move this debate forward, in part by revealing how the neglect of development has distorted theoretical conceptions of genes and gene action and marginalized other developmental factors. Despite these similarities in approach, Moss and Neumann-Held propose quite different reconceptualizations of the gene. Both are grounded in the practice of contemporary genetics, and both provide ways to integrate developmental systems thinking into genetics. The future dialectic between these two views should prove illuminating.

The remaining chapters in part II reflect two exciting innovations in recent evolutionary biology. Eva Jablonka's book with Martha Lamb, *Epigenetic Inheritance and Evolution* (Jablonka and Lamb 1995), makes it difficult to go on minimizing the theoretical and empirical significance of extragenetic mechanisms of cellular heredity. Until now, phenomena such as DNA imprinting have been assimilated by conventional neo-Darwinism by invoking developmental programs in the cell or by arguing that their impact on evolution is small when compared to genetic inheritance. Needless to say, Jablonka herself does not accept such maneuvers, but rather traces out the full implications of these phenomena for theories of development and evolution.

Kevin Laland, John Odling-Smee, and Marcus Feldman introduce their work on the significance of niche construction, giving many examples of the reciprocal influence of organisms and their surroundings that Lewontin described in the early 1980s. Laland and his coauthors show how models of gene-culture coevolution can be adapted to explore such phenomena. The theme of niche construction runs through much of this volume, and it is treated in a variety of ways (compare chapters 10, 16, and 19, for instance). There is still some uncertainty about the best way to integrate this important concept into our picture of evolutionary change.

Even readers who are sympathetic to the concept of a developmental system and aware of the problems it helps to resolve sometimes ask what difference it makes in practice. The chapters in part III, like Gottlieb's chapter in part I, show how the ideas of DST are deployed in actual research. The papers in this section make it clear that similar ideas have been generating important work in a number of fields for many years. One aim of the present volume is to bring these research traditions into closer contact with one another and to show how they complement one another. H. Frederik Nijhout provides a developmental perspective on the genotype/phenotype relationship. He shows what happens when a simple but realistic developmental model of how genes influence phenotypic traits is added to a conventional population genetic treatment of evolution. The results of this modeling exercise challenge the conventional idea that the selection of phenotypes results in the selection of "genes for" those phenotypes. Deborah Gordon gives a DST-style treatment of an entity above the level of the individual organism (an ant colony and its nest). She shows how behavioral patterns that are both complex and flexible can be regulated without a central locus of control. Transindividual units also figure in Peter Klopfer's contribution. Klopfer examines relations between infants and parents, and evaluates several popular metaphors of development. Patrick Bateson reviews some pervasive confusions over heritability and development, and gives examples of the kind of adaptive developmental flexibility that has been a focus of his long-term interest in fusing the developmental and functional perspectives.

Part IV deals with the overall impact that developmental systems ideas can have on biological theory. Susan Oyama's "Terms in Tension" considers the difficulties of employing terms such as *interaction*, *system*, and *construction*, which have complex histories and conflicting theoretical im-

plications. She suggests that although such histories complicate the theoretician's (and reader's) task, they can also be put to good use in elaborating an account of life processes that is adequate to the phenomena of developmental and evolutionary biology and the social sciences.

In "Darwinism and Developmental Systems," Paul Griffiths and Russell Gray systematically redefine in developmental systems terms the key concepts of evolutionary theory: inheritance, natural selection, adaptation, and lineage. They aim to show that the DST formulation of evolution can do all the explanatory work of the conventional ones and can actually extend the range of phenomena that can be given adaptive/historical explanations. Continuing the focus on evolution, William Wimsatt's chapter considers the implications of the fact that evolution must operate by producing systems capable of reliably reconstructing themselves, but also capable of evolutionary change. He argues that this has immediate implications for the sorts of systems that can evolve. Like Wimsatt, Bruce Weber and David Depew are longstanding advocates of the need for a resynthesis of evolution and development. In their chapter they explore the prospects for DST as a source of that new synthesis and also as a way to forge links between biology and the physical sciences.

Part IV concludes with Tim Ingold's examination of biology/culture oppositions in anthropology. He shows how much of our understanding of human life relies on ignoring our role in the construction of our environment, and hence in the construction of ourselves. Ingold also explores links between the gene/environment and biology/culture oppositions and, perhaps the most famous of all these dichotomies, that of mind and body.

Part V contains essays more concerned to explore and evaluate DST than to add to it or provide arguments in its support. Peter Godfrey-Smith confronts the question raised earlier: Is DST a contribution to biological science or something more like a "philosophy of nature"? In doing so, he provides valuable insights into the variety of functions performed by scientific approaches in general. Kim Sterelny evaluates DST's attempt to extend the concept of inheritance, taking up the challenge implicit in DST's claim to provide a principled definition of inheritance. He argues that conventional neo-Darwinism has principled reasons for excluding some of what DST wants to include. Central DST themes of historical contingency, distributed control, and heterogeneity of developmental interactants are taken up in Peter Taylor's chapter. Like several other authors, Taylor discusses the problem of allocating causal responsibility in complex systems, and, in addition, he addresses the crucial questions of agency and the limitations of trying to bring about change solely through the medium of ideas.

The work of Evelyn Fox Keller also resonates strongly with many DST themes. This is surely the case in her analysis of the metaphors that have sustained dichotomous thinking in modern biology (Keller 1985). In her contribution to this book, however, Keller argues that contrary to DST's emphasis on the multiplicity of possible boundaries, some boundaries really are special. In particular, she argues that DST is in danger of neglecting the importance of the cell membrane and of the skin that comes to define the body.

In the final chapter of part V, Cor van der Weele takes up the issue of the relationship between scientific approaches and ethical systems, making explicit some of the concerns about science/society relations that are implicit in other chapters. She also discusses the "critical science" tradition to which some of the authors collected here have contributed. Her concept of an ethics of attention is relevant to the use of a systems perspective to counteract the hierarchies of importance built into prevailing ways of thinking. In raising these issues, van der Weele brings us back to the social, economic, and political problems that were alluded to in the beginning of this introduction, problems that have motivated many of the authors collected here to pay closer attention to the ways in which we conceptualize our natures.

Notes

1. The term *interactionism* is widely used but has such a broad spectrum of meanings as to be almost useless. In chapter 15, Oyama examines some of these multiple senses and the problems caused by their coexistence.

2. "Approach" or "perspective" might thus be preferable, but the DST label seems to have stuck. Fausto-Sterling's (2000) excellent treatment of research on sexuality shows this perspective at work in a field especially laden with just the sorts of social freight alluded to earlier.

3. In the areas of animal behavior and psychobiology these include Bateson (1984, 1991), Gottlieb (1997), Hinde (1968), Johnston (1982, 1987), Klopfer (1973), Kuo (1967), Lehrman (1953, 1970), and Schneirla (1966). In genetics, Lewontin (1974) and Waddington (1975); in developmental biology, Nijhout (1990); and in molecular biology, Stent (1981).

4. Argument and evidence for this view, which will strike many people as surprising, can be found in Godfrey-Smith (1999); Gray (2001); Griffiths and Gray (1994); Johnston (1987); Moss (1992); Nijhout (1990); Oyama (1985); Sarkar (1996); Sterelny and Griffiths (1999); Strohman (1997).

References

Bateson, P. P. G. (1984). Genes, evolution, and learning. In P. Marler and H. S. Terrace (Eds.), *The Biology of Learning,* pp. 75–88. Berlin: Springer-Verlag.

Bateson, P. P. G. (1991). Are there principles of behavioural development? In P. Bateson (Ed.), *The Development and Integration of Behaviour,* pp. 19–39. Cambridge: Cambridge University Press.

Callebaut, W., and K. Stotz. (forthcoming). *Bioepistemology and the Challenge of Development and Sociality.* Cambridge, MA: MIT Press.

Falk, R. (1984). The gene in search of an identity. *Human Genetics* 68: 195–204.

Falk, R. (1986). What is a gene? *Studies in History and Philosophy of Science* 17: 133–173.

Fausto-Sterling, A. (2000). *Sexing the Body: Gender Politics and the Construction of Sexuality.* New York: Basic Books.

Godfrey-Smith, P. (1999). Genes and codes: Lessons from the philosophy of mind? In V. Hardcastle (Ed.), *Biology Meets Psychology: Constraints, Conjectures, Connections,* pp. 305–331. Cambridge, MA: MIT Press.

Gottlieb, G. (1997). *Synthesizing Nature-Nurture: Prenatal Roots of Instinctive Behavior.* Mahwah, NJ: Lawrence Erlbaum.

Gray, R. D. (1992). Death of the gene: Developmental systems strike back. In P. E. Griffiths (Ed.), *Trees of Life: Essays in Philosophy of Biology,* pp. 165–209. Dordrecht: Kluwer Academic.

Gray, R. D. (1997). "In the Belly of the Monster": Feminism, developmental systems, and evolutionary explanations. In P. A. Gowaty (Ed.), *Evolutionary Biology and Feminism,* pp. 385–413. New York: Chapman and Hall.

Gray, R. D. (2001). Selfish genes or developmental systems? Evolution without replicators and vehicles. In R. Singh, C. Krimbas, J. Beatty, and D. Paul (Eds.), *Thinking about Evolution: Historical, Philosophical and Political Perspectives,* pp. 184–207. Cambridge, England: Cambridge University Press.

Griffiths, P. E., and R. D. Gray (1994). Developmental systems and evolutionary explanation. *Journal of Philosophy* 91: 277–304.

Griffiths, P. E., and R. D. Knight (1998). What is the developmentalist challenge? *Philosophy of Science* 65: 253–258.

Hinde, R. A. (1968). Dichotomies in the study of development. In J. M. Thoday and A. S. Parkes (Eds.), *Genetic and Environmental Influences on Behavior,* pp. 3–14. New York: Plenum.

Jablonka, E., and M. J. Lamb (1995). *Epigenetic Inheritance and Evolution: The Lamarkian Dimension.* Oxford: Oxford University Press.

Johnston, T. D. (1982). Learning and the evolution of developmental systems. In H. C. Plotkin (Ed.), *Learning, Development, and Culture,* pp. 411–442. New York: Wiley.

Johnston, T. D. (1987). The persistence of dichotomies in the study of behavioral development. *Developmental Review* 7: 149–182.

Keller, E. F. (1985). *Refiguring Life: Metaphors of Twentieth Century Biology.* New York: Columbia University Press.

Kitcher, P. (1982). Genes. *British Journal of Philosophy of Science* 33: 337–359.

Kitcher, P. (1984). 1953 and all that: A tale of two sciences. *Philosophical Review* 93: 335–373.

Kitcher, P. (1985). *Vaulting Ambition: Sociobiology and the Quest for Human Nature.* Cambridge, MA: MIT Press.

Kitcher, P. (2001). Battling the undead: How (and how not) to resist genetic determinism. In R. Singh, K. Krimbas, J. Beatty, and D. Paul (Eds.), *Thinking about Evolution: Historical, Philosophical and Political Perspectives,* pp. 396–414. Cambridge: Cambridge University Press.

Klopfer, P. H. (1973). *On Behavior: Instinct Is a Cheshire Cat.* Philadelphia: Lippincott.

Kuhn, T. (1970). *The Structure of Scientific Revolutions.* 2d ed. Chicago: University of Chicago Press.

Kuo, Z.-Y. (1967). *Dynamics of Behavior Development.* (2d ed. 1976.) New York: Plenum.

Lehrman, D. S. (1953). A critique of Konrad Lorenz's theory of instinctive behavior. *Quarterly Review of Biology* 28: 337–363.

Lehrman, D. S. (1970). Semantic and conceptual issues in the nature-nurture problem. In L. R. Aronson, E. Tobach, D. S. Lehrman, and J. S. Rosenblatt (Eds.), *Development and Evolution of Behavior: Essays in Memory of T. C. Schneirla,* pp. 17–52. San Francisco: W. H. Freeman.

Levins, R., and R. Lewontin (1985). *The Dialectical Biologist.* Cambridge, MA: Harvard University Press.

Lewontin, R. C. (1974). The analysis of variance and the analysis of cause. *American Journal of Human Genetics* 26: 400–411.

Lewontin, R. C. (1983). Gene, Organism and Environment. In D. S. Bendall (Ed.), *Evolution: From Molecules to Men,* pp. 273–285. Cambridge: Cambridge University Press.

Lewontin, R. C., S. Rose, and L. J. Kamin (1984). *Not in Our Genes.* New York: Pantheon.

Markert, C. L., and H. Ursprung (1971). *Developmental Genetics.* Englewood Cliffs, NJ: Prentice-Hall.

Maynard Smith, J. (2000). The concept of information in biology. *Philosophy of Science* 67: 177–194.

Michel, G. F., and C. L. Moore. (1995). *Developmental Psychobiology: An Interdisciplinary Science.* Cambridge, MA: MIT Press.

Millikan, R. G. (1984). *Language, Thought, and Other Biological Categories.* Cambridge, MA: MIT Press.

Moss, L. (1992). A kernel of truth? On the reality of the genetic program. In D. Hull, M. Forbes, and K. Okruhlik (Eds.), *Proceedings of the Philosophy of Science Association, 1992,* vol. 1, pp. 335–348. East Lansing, MI: Philosophy of Science Association.

Nijhout, H. F. (1990). Metaphors and the roles of genes in development. *Bioessays* 12: 441–446.

Oyama, S. (1981). What does the phenocopy copy? *Psychological Reports* 48: 571–581.

Oyama, S. (1985). *The Ontogeny of Information: Developmental Systems and Evolution.* (2d ed., 2000.) Cambridge: Cambridge University Press. Durham, NC: Duke University Press.

Oyama, S. (1987). Review of P. Kitcher, "Vaulting Ambition." *Canadian Philosophical Review* 7: 203–205.

Oyama, S. (1992). Ontogeny and phylogeny: A case of metarecapitulation? In P. E. Griffiths (Ed.), *Trees of Life: Essays in Philosophy of Biology,* pp. 211–240. Dordrecht: Kluwer.

Oyama, S. (2000a). Causal democracy and causal contributions in DST. *Philosophy of Science,* 67 (proceedings): 332–347.

Oyama, S. (2000b). *Evolution's Eye: A Systems View of the Biology-Culture Divide.* Durham, NC: Duke University Press.

Pinto-Correia, C. (1997). *The Ovary of Eve.* Chicago: University of Chicago Press.

Sarkar, S. (1996). Biological information: A sceptical look at some central dogmas of molecular biology. In S. Sarkar (Ed.), *The Philosophy and History of Molecular Biology: New Perspectives,* pp. 187–232. Dordrecht: Kluwer.

Sarkar, S. (1998). *Genetics and Reductionism.* Cambridge: Cambridge University Press.

Schaffner, K. F. (1998). Genes, behavior, and developmental emergentism: One process, indivisible? *Philosophy of Science* 65: 209–252.

Schlichting, C. D., and M. Pigliucci. (1998). *Phenotypic Evolution: A Reaction Norm Perspective.* Sunderland, MA: Sinauer.

Schneirla, T. C. (1966). Behavioral development and comparative psychology. *Quarterly Review of Biology* 41: 283–302.

Stent, G. (1981). Strength and weakness of the genetic approach to the development of the nervous system. In W. M. Cowan (Ed.), *Studies in Developmental Neurobiology,* pp. 288–320. Oxford: Oxford University Press.

Strohman, R. C. (1997). The coming Kuhnian revolution. *Nature Biotechnology* 15: 194–200.

Sterelny, K., and P. E. Griffiths. (1999). *Sex and Death: An Introduction to the Philosophy of Biology.* Chicago: University of Chicago Press.

Waddington, C. H. (1975). *The Evolution of an Evolutionist.* Ithaca, NY: Cornell University Press.

Wimsatt, W. C. (1980). Reductionistic research strategies and their biases in the units of selection controversy. In T. Nickles (Ed.), *Scientific Discovery: Case Studies,* pp. 213–259. Dordrecht: Reidel.

I INFLUENCES

2 Toward a Systems View of Development: An Appraisal of Lehrman's Critique of Lorenz

Timothy D. Johnston

The work of Daniel S. Lehrman provides the conceptual foundation for a great deal of the empirical and theoretical work on behavioral development that has been undertaken in the last half century. Most of his publications are empirical contributions to the literature on behavioral endocrinology, and many of those are among the most widely cited papers in that field. His broader impact on our thinking about behavioral development and evolution comes primarily from two theoretical publications: his 1953 critique of Konrad Lorenz's theory of instinct, excerpted in the chapter that follows, and a later chapter (Lehrman 1970) that reiterates and extends the ideas presented in the earlier paper. Between them, those two publications contain what is still one of the clearest statements of the systems view of development, modern articulations of which are represented by the chapters in the present volume. Lehrman himself did not use the terminology of systems thinking, but his analysis of development illustrates many of the most significant features of modern developmental systems theory (DST).

As implied by the title of his 1953 paper, Lehrman articulated his developmental views in reaction to Lorenz's presentation of classical ethological instinct theory, which emerged in the writings of Lorenz and his collaborators, especially Nikolaas Tinbergen, during the 1930s and 1940s (see Lehrman's bibliography for citations to this literature). Lorenz's was not, of course, the first theory of instinct to engage the attention of psychologists, nor was Lehrman's the first criticism of the concept of instinct or of the categories of inherited and acquired behavior as useful ways to approach the analysis of behavior.

The Inherited and the Acquired in Behavior, 1890–1953

Before about 1920, use of the categories "inherited" and "acquired" to explain the origins of behavior was relatively uncontroversial in psychology. Inherited behavior (instinct) was understood to be an inherent part of the individual's makeup that resulted from the evolutionary history of the species to which it belonged. Instincts were especially important in nonhuman animals, although they were also thought to account for at least some human behavior. Although the mechanisms of inheritance were not well understood (see Maienschein 1987; Sapp 1983), it was generally, if implicitly, assumed that one could speak of the inheritance of behavior as straightforwardly as one could speak about the inheritance of a morphological or physiological trait, about which no controversy existed.

Evolutionary writers like Charles Darwin (1871), George Romanes (1888), and Conwy Lloyd Morgan (1895) used the concept of instinct to explain the behavior and mental abilities of human and nonhuman animals, but it was William James and William McDougall whose writings were most responsible for stimulating a debate over the explanatory utility of the concept in psychology. James (1890) drew on Darwinian evolutionary theory in establishing functionalism as an important force in American psychology around the turn of the last century. James defined instinct as "the faculty of acting in such a way as to produce certain ends, without foresight of the ends, and without previous education in the performance" (James 1890, vol. 2: 383), emphasizing both the teleological and the nativistic aspects of instinct. In James's wide-ranging and rather diffuse psychology, instinct was just one among many determinants of behavior. In McDougall's system, conversely, instinct became the foundation on which all behavior was based. McDougall (1908) argued that all human behavior has an instinctive core that provides both motivation and direction for behavior. Learning may play an important role in determining what particular objects and behaviors become associated with various instincts, but the instincts themselves are inherited, unalterable, and essential to the adaptive organization of the behavioral repertoire. For about the first two decades of the twentieth century, the concept of instinct was repeatedly invoked as an explanation for behavior, most often not on the basis of careful experimentation and theoretical analysis, but by simply postulating an instinct whenever a type of behavior seemed in need of explanation. James had legitimized this approach (he listed more than twenty human instincts, including instincts of shyness, fear, acquisitiveness, play, and modesty). Although McDougall offered a more fully developed analysis, he too provided lists of instincts that were intended to explain the whole range of human behavior. Indeed, such lists were common in writings about instinct of this period: One writer counted nearly 850 major types of instinct proposed in the psychological literature between 1900 and 1920 (Bernard 1924).

This cavalier use of the concept, coupled with the vitalistic and teleological character of McDougall's instinct theory (which made it anathema to the mechanistic S-R psychology of Watsonian behaviorism; see Boakes 1984, chap. 8) soon provoked a backlash. Beginning with a paper by Dunlap (1919), the "anti-instinct movement" in American psychology criticized the concept of instinct on a number of grounds—that it invoked vitalistic forces, that instinct theorists frequently offered no evidence for the existence of the instincts they proposed, and that attributing instinctive behavior to "inborn dispositions" leaves unanswered the question of how the behavior comes into being. The last of these is most directly relevant to understanding the impact of the anti-instinct movement on developmental thinking.

Perhaps the most persistent and unforgiving of the anti-instinct critics was Zing-Yang Kuo, who wrote five theoretical articles criticizing the concept of instinct between 1921 and 1930, in addition to a series of experimental reports analyzing the prenatal development of behavior in chicks. In his first two papers, Kuo (1921, 1922) argued that there is no evidence supporting the claim that instinctive behavior exists in human or nonhuman animals, and that all supposed instincts can in principle be explained as the outcome of environmental influences of one kind or another. Although some of Kuo's arguments are overblown, he made an important fundamental point: Unless we know in detailed, mechanistic terms what it means to say that instinctive behavior patterns are "inherited," the use of instinct as an explanatory category produces a "finished psychology," or at least a "finished" developmental psychology. That is, it tends to block further investigation into the ontogeny of the behavior by purporting to *explain* when all it really does is to *label*. It may be correct to identify certain elementary movement patterns, what Kuo called "unlearned reaction units," as inherited, but more complex coordinations of behavior (including those usually identified as instincts) offer too many opportunities for the influence of experience to be attributed to heredity.

In his later papers, Kuo (e.g., 1924) went even farther, arguing that "*in a strictly behavioristic psychology, with its emphasis on laboratory procedure and with its insistence on physiological explanation of behavior, there is practically no room for the concept of heredity*" at all (p. 428, original italics). In this paper, he withdrew his earlier concession that some simple reactions may be inherited because "so long as there are inherited reactions, simple as they may be, there is justification for the use of the term instinct" (p. 439). "The traditional sharp distinction between inherited and acquired responses," he wrote, "should be abolished. All responses must be looked upon as the direct result of stimulation, as interactions between the animal and its environment" (p. 439). Kuo insisted that the idea of inherited behavior was simply too vague and nonspecific to do more than obscure questions about the origins of the behavior in question.

Kuo's arguments were based more on theoretical principle than on empirical evidence, of which very little pertaining to the early development of behavior was then available. However,

Carmichael (1925) bolstered Kuo's position by pointing out that even anatomical structure does not develop independently of environmental influences. He summarized embryological research showing that many presumably "inherited" structures in fish and amphibians require particular environmental conditions to develop normally and concluded: "Heredity and environment are not antithetical, nor can they expediently be separated; for in all maturation there is learning: in all learning there is hereditary maturation" (p. 260).

Debate over the utility of the instinct concept, and over the related concept of maturation to explain the development of instinctive behavior, continued throughout the next twenty-five years. From the outset, contributors to the debate had recognized that the term might be used in more than one way. Indeed, Dunlap (1919) opened the debate by arguing that it was important "to distinguish between the instinct as a group of activities *teleologically* defined, and the instinct as a *physiological* group" (p. 307; original emphasis). Most of the anti-instinct writers aimed their criticisms at the former use of the term, most closely associated with McDougall's theory, although, as we have seen, others (like Kuo) wished to reject any kind of inherited behavior at all. But some authors, especially those trained in comparative and physiological psychology, were much more sympathetic toward Dunlap's second use, "instinct as a *physiological* group," by which he meant "a certain definite group of muscular and glandular performances ... resulting from a definite stimulus or complex of stimuli" (p. 308). For example, Lashley (1938) rejected the use of instinct in the teleological sense (which he referred to as "a dynamics of imaginary forces" [p. 447]) but accepted its value when referring to specific, identifiable behavior and its underlying physiology—Dunlap's "physiological" instincts. Lashley made no apologies for using the concept of instinct in this sense, asserting that "the distinction between genetic and environmental influences ... is of real significance for problems of the physiological basis of behavior" (p. 447). Other authors used behavioral and physiological data to support their claims that visual perceptual organization (Hebb 1937) and sexual behavior in rats (Beach 1951; Stone 1951) are innate, although Hebb and Beach later modified their views (Beach 1955; Hebb 1953).

In 1951, Tinbergen published his classic and influential account of ethological theory, *The Study of Instinct*. Although the ethological approach to behavior had been a prominent feature of the European scientific scene since the 1920s, it only became widely known to American psychologists starting around 1950 (Dewsbury 1984: 131ff). Much of the earlier literature was written in German or Dutch, and what was published in English had appeared in biological journals or conference proceedings not usually read by American psychologists (e.g., Tinbergen 1942; Lorenz 1950). Nonetheless, the European literature, and especially the work of Lorenz, described a theory of instinct that avoided McDougall's teleology, while providing an account that was more fully developed theoretically than any available in the work of American comparative and physiological psychologists. Furthermore, the theory was based on extensive observations, and some experiments, on a large number of species under natural conditions. These features of Lorenz's theory made it possible for Lehrman to articulate a critique of the concept of instinct that revealed its deficiencies for the analysis and understanding of behavior more clearly than did earlier criticisms.

Systems Thinking in Lehrman's 1953 Critique of Lorenz's Instinct Theory

Lehrman's paper was only one of several criticisms of the concept of instinct that appeared around 1953, although it was by far the most comprehensive. Lehrman's fluency in German allowed him to work with the original papers on which Lorenz's ideas were based (Silver and

Rosenblatt 1987), rather than relying on summaries and secondhand accounts written in English, and this enhanced the credibility and impact of his criticism. Drawing on the earlier work of Kuo and Schneirla (e.g., 1949), Lehrman provided a framework for thinking about development that he and others could build into a coherent alternative to instinct theory. Although Lehrman did not identify his account as a systems approach to development, it embodies some of the most significant elements of current systems thinking.

The most readily evident "systems" features of Lehrman's critique were his repudiation of the distinction between innate and acquired behaviors, or elements of behavior, and his recognition that behavior cannot be isolated from the rest of the organism's physiological and anatomical makeup. He provided both theoretical arguments and empirical evidence to show that behavior cannot be neatly divided into the categories of learned and innate. Instead, he argued, we must analyze the development of every pattern of behavior in terms of a continuing interaction between the organism and its environment (*not* between the genotype and the environment, as is sometimes proposed). Although the mechanisms of learning sometimes are involved in this interaction, other, less obvious contributions of experience play a part as well. In particular, he rejected the isolation (or deprivation) experiment as an adequate tool for analyzing development, pointing out that self-stimulation is still possible for an isolated animal. Later work, especially that of Gottlieb (e.g., 1971, 1991) has revealed the importance of self-stimulation for normal behavioral development. Lehrman did not go so far as to consider organism and environment as part of a single system, but he clearly understood that we cannot partition behavior into elements and give a separate account of the development of each of them, an insight that provides the core of the systems view of development.

Lehrman also criticized the ethologists for treating each innate behavior as a neatly sepa-

rable element of the organism, dependent on its own specific physiological substrate. For example, he suggested that the gradual postnatal improvement of pecking in chicks "is very probably due in part to an increase in strength of the leg muscles and to an increase in balance and stability of the standing chick, which results partly from this strengthening of the legs and partly from the development of equilibrium responses" (Lehrman 1953: 344), rather than to the maturation of specific underlying neural circuits. This is quite similar to the dynamical systems view of human locomotor development presented by Thelen (e.g., Thelen, Kelso, and Fogel 1987; Thelen 1995), which explains development not just in terms of increasing behavioral coordination, but also in terms of the growing strength and mass of the infant's limbs.

Although the term *system*, in the sense in which it is employed by modern developmental systems theorists, does not appear in Lehrman's paper, there is no doubt that he would have been entirely sympathetic to the aims of systems theory as it is represented in this volume. The ideas that we should seek an explanation of development in the interactions that occur within the developing organism and between organism and its environment, and that the organism is more than just a nervous system that processes information, are central to the account of development that Lehrman advanced against Lorenz and the classical ethological theory of which he was the chief architect and foremost advocate.

The Response to Lehrman's Critique

Lehrman's paper was one of three published at about the same time that criticized the concepts of instinct and innate behavior as useful bases for developmental thinking. Both Hebb (1953) and Beach (1955) marshaled similar arguments to those presented by Lehrman, in the process retreating from earlier positions in which they had defended the utility of the concept. Hebb (1953:

44) explicitly noted that he had earlier accepted the existence of innate perceptual organization in part because he had overlooked the significance of some of his own data and he took issue with Tinbergen's (1951) proposal that we should study the organization of innate behavior before studying learning, an approach that Hebb concluded would be logically impossible. All three papers were influential in shaping subsequent discussion about the concept of instinct.

As might be expected, many ethologists rejected the arguments of Lehrman and his psychological colleagues, although others were more receptive. For example, speaking at a conference that brought ethologists and their critics together soon after publication of Lehrman's critique, Tinbergen (1955) conceded: "Let me say right at the beginning, that I admit that we must drop this use of the word 'innate' for the reasons already given, namely that the word can be applied only to differences, not to characters, and also because tests such as ours [i.e., the deprivation experiment] exclude only part of all possible environmental influences [p. 102] … instead of drawing a positive conclusion or a pseudo-conclusion by saying this response is innate, I want to specify the little bit we have found out about the ontogeny of this response, saying that, at the moment when we studied it, it was not yet conditioned" [p. 106]. Similarly, in his influential textbook on animal behavior, Hinde (1966) adopted Lehrman's developmental approach and largely dispensed with any attempt to draw sharp distinctions between learned and innate behavior. Other ethologists, however, refused to accept the developmentalists' critique.

One of their responses was to read Lehrman's position as claiming that all behavior is the result of trial-and-error learning and that heredity has no effect on behavior (e.g., Eibl-Eibesfeldt 1961; Eibl-Eibesfeldt and Kramer 1958; Hess 1962; Klinghammer and Hess 1964; Lorenz 1956). This is manifestly *not* what Lehrman and others were saying, but it seems to have been a widely accepted paraphrase of their position among at least some ethologists in the 1950s and 1960s. Lehrman did argue that the development of all behavior involves the influence of experience, but, like his mentor Schneirla, he urged a much broader interpretation of "experience" than the very narrow set of circumstances generally subsumed under the term "learning." In several papers, ethologists attempted to counter Lehrman's assertion that some particular behavior is not innate by showing that it could not be explained as the outcome of standard conditioning procedures ("trial-and-error learning") and thus concluding that it must, in fact, be innate. It seemed difficult for these ethologists to recognize that Lehrman was proposing to abandon the learned-innate distinction in toto, not reclassifying purportedly innate behaviors as learned (Lehrman 1957).

The most important and influential response to Lehrman's criticisms came from Lorenz himself in the form of a long paper in German (Lorenz 1961), later translated into English and expanded into a book (Lorenz 1965). In this book there is an important shift in Lorenz's position regarding the defining feature of instincts. Originally, Lorenz (e.g., 1937) had drawn the learned/innate distinction on the basis of whether the development of a behavior is determined by genes or environment. (This was a very strict distinction; indeed, where a behavior appeared to be a blend of learning and instinct, Lorenz argued that it would always be possible to identify purely learned and purely innate elements, interlocked or intercalated to form the behavior pattern.) But he now proposed that the real difference lies in the source of the *information* that determines its adaptiveness. If adaptiveness is determined by information acquired during an individual's development, then the behavior is learned; if it is determined by information acquired during the evolutionary history of the species, then the behavior is innate. This information metaphor formed the core of Lorenz's long reply to Lehrman (1965) and echoed points made by other ethologists who wished to preserve the learned-innate distinction (e.g., Hess 1962;

Thorpe 1961, 1963; see Beer 1973). It is a seriously flawed metaphor (Oyama 1985; Johnston 1987) but its attractiveness for Lorenz helps to clarify a large part of the disagreement between him and Lehrman: Whereas Lehrman's primary interest was in understanding the *development* of behavior, Lorenz's interest was in understanding its *adaptiveness*. The deprivation experiment, criticized so effectively by Lehrman, is for Lorenz only incidentally concerned with explicating development; its primary goal is to reveal the source of the information that makes behavior adaptive (Hess 1962: 223; Lorenz 1965: 83ff).

The Legacy of Lehrman's Critique

In the decade or so immediately after the publication of Lehrman's critique, discussion of his ideas appeared in various theoretical papers devoted to the analysis and reanalysis of the concept of instinct and the relation between learned and innate behavior. Many of these (cited earlier) defended the concept of instinct, but others built on Lehrman's arguments and developed his position further (Hinde 1966; Jensen 1961; Lehrman 1956; Ross and Denenberg 1960; Schneirla 1956, 1966). The debate seemed, however, to engage a rather limited group of writers, and the impact of his ideas was slow to be felt in the mainstream of research on comparative psychology and behavioral development. For example, none of the eight chapters on comparative psychology that appeared in the *Annual Review of Psychology* from 1953 to 1963 dealt seriously with the theoretical issues he raised. There are also few empirical papers from this period that use Lehrman's ideas as a framework for designing experiments and interpreting data. An examination of the four journals that published most of the research on animal behavior between 1954 and 1964 (*Animal Behaviour*, *Behaviour*, the *Journal of Comparative and Physiological Psychology*, and *Zeitschrift für Tierpsychologie*) found almost no papers that draw on Lehrman's analysis in any substantive way. There are frequent citations of

Lorenz's and Tinbergen's work (especially in the three European journals), a few direct rebuttals of Lehrman's criticisms of Lorenz (mainly in the *Zeitschrift*), and frequent references to Lehrman's empirical studies of behavioral endocrinology by those working in that field. Many papers that seem to fall squarely into the domain of Lehrman's analysis ignore his 1953 paper entirely, and others mention it only in passing.

After 1953, most of Lehrman's own publications dealt with his empirical research on the relations among hormones, behavior, and environment in the ring dove, exemplifying the kind of behavioral research that he had advocated, but not elaborating his theoretical position very much (see Lehrman 1962). In 1970, he published another major theoretical statement in a volume prepared in memory of his mentor, T. C. Schneirla, who had died in 1968. This chapter (Lehrman 1970) both expanded on the position articulated in 1953 and also responded directly to Lorenz's (1965) defense of instinct theory. I do not think that there was any fundamental change in Lehrman's theoretical position between 1953 and 1970, but he provided additional examples to support his view and emphasized some points that were only implicit in his earlier paper. For example, he discussed the problems that arise when questions about developmental mechanisms are confused with questions about the phenomena of adaptation to the environment. Both are entirely legitimate kinds of questions about behavior, but they require different methods for their investigation; most importantly, a particular answer to a question in one domain does not imply anything about answers to questions in the other domain. "As I hope the discussion so far has made clear, I have not been trying to avoid the concepts of survival value and phylogenetic adaptation, but only to prevent them from being merged with the concepts of the *causal* analysis of *development*" (Lehrman 1970: 37; original emphasis). The same distinction had also been made by Tinbergen (1963), who made it clear that these two kinds of questions are entirely complementary in the study of behavior.

Lorenz (1965) had claimed that the adaptive information specifying an animal's innate behavior is in the form of a blueprint coded in the genes that controls the unfolding ("strictly determined maturation") of those patterns that are innate. Lehrman wrote about the genetic blueprint in terms that are strikingly reminiscent of Kuo's (1921) when he objected to the use of instinct on the grounds that it tends to produce a finished psychology: "I believe that the comfort and satisfaction gained from disposing of the problems of ontogenetic development by the use of such concepts are misleading, and are based upon the evasion or dismissal of the most difficult and interesting problems of development" (Lehrman 1970: 34). Almost thirty years later the "genetic blueprint" is an idea whose use has become increasingly widespread and that continues to have a strong grip on the imaginations of scientists and lay persons alike (see Nelkin and Lindee 1995). The rapid growth of molecular and human behavioral genetics and the advent of the Human Genome Project have made it more persuasive than ever to speak of behavior as being innate or genetically encoded, although Lehrman's arguments against such claims are as cogent now as they were in 1953. However, as shown by the contributions to this volume, the progress of DST during the same period has also been impressive, and it is to be hoped that more scientists and educated laypersons will be encouraged to use this perspective for thinking about the development of behavior.

References

Beach, F. A. (1951). Instinctive behavior: Reproductive activities. In S. S. Stevens (Ed.), *Handbook of Experimental Psychology*, pp. 387–434. New York: John Wiley & Sons.

Beach, F. A. (1955). The descent of instinct. *Psychological Review* 62: 401–410.

Beer, C. G. (1973). Species-typical behavior and ethology. In D. A. Dewsbury and D. A. Rethlingshaver (Eds.), *Comparative Psychology: A Modern Survey,* pp. 21–77. New York: McGraw-Hill.

Bernard, L. L. (1924). *Instinct: A Study in Social Psychology.* New York: Henry Holt.

Boakes, R. (1984). *From Darwin to Behaviorism: Psychology and the Minds of Animals.* Cambridge: Cambridge University Press.

Carmichael, L. (1925). Heredity and environment: Are they antithetical? *Journal of Abnormal and Social Psychology* 20: 245–260.

Darwin, C. R. (1871). *The Descent of Man, and Selection in Relation to Sex.* London: John Murray.

Dewsbury, D. A. (1984). *Comparative Psychology in the Twentieth Century.* Stroudsburg, PA: Hutchinson Ross Co.

Dunlap, K. (1919). Are there any instincts? *Journal of Abnormal Psychology* 14: 307–311.

Eibl-Eibesfeldt, I. (1961). The interactions of unlearned behaviour patterns and learning in mammals. In J. M. Delafresnaye (Ed.), *Brain Mechanisms and Learning,* pp. 53–73. Oxford: Blackwell.

Eibl-Eibesfeldt, I., and S. Kramer. (1958). Ethology, the comparative study of animal behavior. *Quarterly Review of Biology* 33: 181–211.

Gottlieb, G. (1971). *Development of Species Identification in Birds.* Chicago: University of Chicago Press.

Gottlieb, G. (1991). Experiential canalization of behavioral development: Theory. *Developmental Psychology* 27: 4–13.

Hebb, D. O. (1937). The innate organization of visual activity: I. Perception of figures by rats reared in total darkness. *Journal of Genetic Psychology* 51: 101–126.

Hebb, D. O. (1953). Heredity and environment in mammalian behaviour. *British Journal of Animal Behaviour* 1: 43–47.

Hess, E. H. (1962). Ethology: An approach toward the complete analysis of behavior. *New Directions in Psychology* 1: 157–266.

Hinde, R. A. (1966). *Animal Behavior: A Synthesis of Ethology and Comparative Psychology* (1st ed.). New York: McGraw Hill.

James, W. (1890). *Principles of Psychology.* New York: Henry Holt.

Jensen, D. D. (1961). Operationism and the question "Is this behavior learned or innate?" *Behaviour* 17: 1–8.

Johnston, T. D. (1987). The persistence of dichotomies in the study of behavioral development. *Developmental Review* 7: 149–182.

Klinghammer, E., and E. H. Hess. (1964). Parental feeding in ring doves (*Streptopelia roseogrisea*): innate or learned? *Zeitschrift für Tierpsychologie* 21: 338–347.

Kuo, Z. Y. (1921). Giving up instincts in psychology. *Journal of Philosophy* 18: 645–664.

Kuo, Z. Y. (1922). How are our instincts acquired? *Psychological Review* 29: 344–365.

Kuo, Z. Y. (1924). A psychology without heredity. *Psychological Review* 31: 427–448.

Lashley, K. A. (1938). Experimental analysis of instinctive behavior. *Psychological Review* 45: 445–471.

Lehrman, D. S. (1953). A critique of Konrad Lorenz's theory of instinctive behavior. *Quarterly Review of Biology* 28: 337–363.

Lehrman, D. S. (1956). On the organization of maternal behavior and the problem of instinct. In P. P. Grassé (Ed.), *L'Instinct dans le comportement des animaux et de l'homme*, pp. 475–520. Paris: Masson.

Lehrman, D. S. (1957). Nurture, nature, and ethology (Review of W. H. Thorpe, "Learning and instinct in animals," Cambridge University Press, 1956). *Contemporary Psychology* 2: 103–104.

Lehrman, D. S. (1962). Interaction of hormonal and experiential influences on development of behavior. In E. L. Bliss (Ed.), *Roots of Behavior: Genetics, Instinct, and Socialization in Animal Behavior*, pp. 142–156. New York: Hoeber.

Lehrman, D. S. (1970). Semantic and conceptual issues in the nature-nurture problem. In L. R. Aronson, E. Tobach, D. S. Lehrman, and J. S. Rosenblatt (Eds.), *Development and Evolution of Behavior*, pp. 17–50. San Francisco: W. H. Freeman.

Lorenz, K. Z. (1937). Über die Bildung des Instinktbegriffes. *Naturwissenschaften* 25: 289–300, 307–318, 324–331. [Translated as Lorenz (1957). The nature of instinct. In C. H. Schiller (Ed.), *Instinctive Behavior: Development of a Modern Concept*, pp. 129–175. New York: International Universities Press.]

Lorenz, K. Z. (1950). The comparative method in studying innate behaviour patterns. *Symposia of the Society for Experimental Biology* 4: 221–268.

Lorenz, K. Z. (1956). The objectivistic theory of instinct. In P. P. Grassé (Ed.), *L'Instinct dans le comportement des animaux et de l'homme*, pp. 51–76. Paris: Masson.

Lorenz, K. Z. (1961). Phylogenetische Anpassung und adaptive Modifikation des Verhaltens. *Zeitschrift für Tierpsychologie* 18: 139–187.

Lorenz, K. Z. (1965). *Evolution and Modification of Behavior*. Chicago: University of Chicago Press.

Maienschein, J. (1987). Heredity/development in the United States, circa 1900. *History and Philosophy of the Life Sciences* 9: 79–93.

McDougall, W. (1908). *Introduction to Social Psychology*. London: Methuen & Co.

Morgan, C. L. (1895). *An Introduction to Comparative Psychology*. London: Walter Scott.

Nelkin, D., and M. S. Lindee. (1995). *The DNA Mystique: The Gene as a Cultural Icon*. New York: W. H. Freeman.

Oyama, S. (1985). *The Ontogeny of Information: Developmental Systems and Evolution*. Cambridge: Cambridge University Press. (2d rev. ed., Durham, NC: Duke University Press, 2000.)

Romanes, G. J. (1888). *Mental Evolution in Man*. London: Kegan Paul & Trench.

Ross, S., and V. H. Denenberg. (1960). Innate behavior: The organism in its environment. In R. H. Waters, D. A. Rethlingshaver, and W. E. Caldwell (Eds.), *Principles of Comparative Psychology*, pp. 43–73. New York: McGraw-Hill.

Sapp, J. (1983). The struggle for authority in the field of heredity, 1900–1932: New perspectives on the rise of genetics. *Journal of the History of Biology* 16: 311–342.

Schneirla, T. C. (1949). Levels in the psychological capacities of animals. In R. W. Sellars, V. J. McGill, and M. Farber (Eds.), *Philosophy for the Future*, pp. 243–286. New York: Macmillan.

Schneirla, T. C. (1956). Interrelationships of the "innate" and the "acquired" in instinctive behavior. In P. P. Grassé (Ed.), *L'Instinct dans le comportement des animaux et de l'homme*, pp. 387–452. Paris: Masson.

Schneirla, T. C. (1966). Behavioral development and comparative psychology. *Quarterly Review of Biology* 41: 283–302.

Silver, R., and J. S. Rosenblatt. (1987). The development of a developmentalist: Daniel S. Lehrman. *Developmental Psychobiology* 20: 563–570.

Stone, C. P. (1951). Maturation and "instinctive" functions. In C. P. Stone (Ed.), *Comparative Psychology* (3d ed.), pp. 30–61. New York: Prentice-Hall.

Thelen, E. (1995). Motor development: A new synthesis. *American Psychologist* 50: 79–95.

Thelen, E., J. A. S. Kelso, and A. Fogel. (1987). Self-organizing systems and infant motor development. *Developmental Review* 7: 39–65.

Thorpe, W. H. (1961). Comparative psychology. *Annual Review of Psychology* 12: 27–50.

Thorpe, W. H. (1963). Ethology and the coding problem in germ cell and brain. *Zeitschrift für Tierpsychologie* 20: 529–551.

Tinbergen, N. (1942). An objectivistic study on the innate behavior of animals. *Bibliotheca Biotheoretica* 1: 39–98.

Tinbergen, N. (1951). *The Study of Instinct.* Oxford: Oxford University Press.

Tinbergen, N. (1955). Psychology and ethology as supplementary parts of a science of behavior. In B. Schaffner (Ed.), *Group Processes,* pp. 75–167. New York: Josiah Macy Jr. Foundation.

Tinbergen, N. (1963). On aims and methods of ethology. *Zeitschrift für Tierpsychologie* 20: 404–433.

3 A Critique of Konrad Lorenz's Theory of Instinctive Behavior

Daniel S. Lehrman

Beginning about 1931, Konrad Lorenz, with his students and collaborators (notably N. Tinbergen), has published numerous behavioral and theoretical papers on problems of instinct and innate behavior which have had a widespread influence on many groups of scientific workers (Lorenz 1931, 1932, 1935, 1937; Lorenz & Tinbergen 1938; Lorenz 1939; Tinbergen 1939; Lorenz 1940, 1941; Tinbergen 1942, 1948, 1950; Lorenz 1950; Tinbergen 1951). Lorenz's influence is indicated in the founding of the *Zeitschrift für Tierpsychologie* in 1937 and in its subsequent development, and also in the journal *Behaviour*, established in 1948 under the editorship of an international board headed by Tinbergen.

Lorenz's theory of instinctive and innate behavior has attracted the interest of many investigators, partly because of its diagrammatic simplicity, partly because of its extensive use of neurophysiological concepts, and partly because Lorenz deals with behavior patterns drawn from the life cycle of the animals discussed, rather than with the laboratory situations most often found in American comparative psychology. These factors go far toward accounting for the great attention paid to the theory in Europe, where most students of animal behavior are zoologists, physiologists, zoo curators or naturalists, unlike the psychologists who constitute the majority of American students of animal behavior (Schneirla 1946a).

In recent years Lorenz's theories have attracted more and more attention in the United States as well, partly because of a developing interest in animal behavior among American zoologists

and ecologists, and partly through the receptive audience provided for Lorenz and his colleague, Tinbergen, by American ornithologists. The ornithologists were interested from the start, especially because a great part of the material on which Lorenz based his system came from studies of bird behavior, but the range of interest in America has widened considerably. Lorenz and his theories were recently the subject of some discussion at a conference in New York at which zoologists and comparative psychologists were both represented (Riess 1949), and are prominently represented in the recent symposium on animal behavior of the Society of Experimental Biologists (Armstrong 1950; Baerends 1950; Hartley 1950; Koehler 1950; Lorenz 1950; Tinbergen 1950), and extensively used in several chapters of a recent American handbook of experimental psychology which will be a standard sourcebook for some years to come (Beach 1951; Miller 1951; Nissen 1951).

Because Lorenz's ideas have gained wide attention, and in particular because a critical discussion [337–338] of these matters should bring usefully into review Lorenz's manner of dealing with basic problems in the comparative study of behavior, a consideration of Lorenz's system and school seems very desirable at this time.

[*Editor's Note:* In the section omitted here (pp. 338–340) Lehrman uses egg-rolling behavior in the gray goose to illustrate some of the features of a classically defined "instinctive" behavior. When a goose sees an egg that has rolled out of its nest, it stands up, reaches out with its neck so that its bill is hooked over the far side of the egg, and pulls the egg back into the nest, using side-to-side movements of its bill to prevent the egg from rolling away. This example illustrates *appetitive behavior*—stretching the neck out toward the egg; the *instinctive act* or *consummatory behavior*—the highly stereotyped movement of the bill that pulls

This paper originally appeared in the *Quarterly Review of Biology* 28: 337–363 (1953). Page breaks and omissions are indicated in the text as (e.g.) [337–338]. Only those references cited in the excerpts reprinted here are included in the references. The references have been reformatted in the style of other contributions to this volume, and a few errors have been corrected.

the egg back toward the nest; the *innate releasing pattern* (or simply *releaser*)—the appearance of the egg outside the nest and the hard feeling of the egg against the bill, which together elicit the instinctive movement by triggering an *innate releasing mechanism*; and the *taxis*, or orienting movement—the side-to-side adjustment of the position of the bill that prevents the egg from rolling away.]

... [338–340]

Problems Raised by Instinct Theories

Even this brief summary brings to light several questions which ought to be critically examined with reference to the theory. These are questions, furthermore, which apply to instinct theories in general. Among them are: (1) the problem of "innateness" and the maturation of behavior; (2) the problem of levels of organization in an organism; (3) the nature of evolutionary levels of behavioral organization, and the use of the comparative method in studying them; and (4) the manner in which physiological concepts may be properly used in behavior analysis. There follows [340–341] an evaluation of Lorenz's theory in terms of these general problems.

"Innateness" of Behavior

The Problem

Lorenz and Tinbergen consistently speak of behavior as being "innate" or "inherited" as though these words surely referred to a definable, definite, and delimited category of behavior. It would be impossible to overestimate the heuristic value which they imply for the concepts "innate" and "not-innate." Perhaps the most effective way to throw light on the "instinct" problem is to consider carefully just what it means to say that a mode of behavior is innate, and how much insight this kind of statement gives into the origin and nature of the behavior.

Tinbergen (1942), closely following Lorenz, speaks of instinctive acts as "highly stereotyped, coordinated movements, the neuromotor apparatus of which belongs, in its complete form, to the hereditary constitution of the animal." Lorenz (1939) speaks of characteristics of behavior which are "hereditary, individually fixed, and thus open to evolutionary analysis." Lorenz (1935) also refers to perceptual patterns ("releasers") which are presumed to be innate because they elicit "instinctive" behavior the *first time* they are presented to the animal. He also refers to those motor patterns as innate which occur for the first time when the proper stimuli are presented. Lorenz's student Grohmann (1938), as well as Tinbergen and Kuenen (1939), speak of behavior as being innately determined because it matures instead of developing through learning.

It is thus apparent that Lorenz and Tinbergen, by "innate" behavior, mean behavior which is hereditarily determined, which is part of the original constitution of the animal, which arises quite independently of the animal's experience and environment, and which is distinct from acquired or learned behavior.

It is also apparent, explicitly or implicitly, that Lorenz and Tinbergen regard as the major *criteria* of innateness that: (1) the behavior be stereotyped and constant in form; (2) it be characteristic of the species; (3) it appear in animals which have been raised in isolation from others; and (4) it develop fully formed in animals which have been prevented from practicing it.

Undoubtedly, there are behavior patterns which meet these criteria. Even so, this does not necessarily imply that Lorenz's *interpretation* of these behavior patterns as "innate" offers genuine aid to a scientific understanding of their origin and of the mechanisms underlying them.

In order to examine the soundness of the concept of "innateness" in the analysis of behavior, it will be instructive to start with a consideration of one or two behavior patterns which have already been analyzed to some extent.

Pecking in the Chick

Domestic chicks characteristically begin to peck at objects, including food grains, soon after hatching (Shepard and Breed 1913; Bird 1925; Cruze 1935; and others). The pecking behavior consists of at least three highly stereotyped components: head lunging, bill opening and closing, and swallowing. They are ordinarily coordinated into a single resultant act of lunging at the grain while opening the bill, followed by swallowing when the grain is picked up. This coordination is present to some extent soon after hatching, and improves later (even, to a slight extent, if the chick is prevented from practicing).

This pecking is stereotyped, characteristic of the species, appears in isolated chicks, is present at the time of hatching, and shows some improvement in the absence of specific practice. Obviously, it qualifies as an "innate" behavior, in the sense used by Lorenz and Tinbergen.

Kuo (1932a–d) has studied the embryonic development of the chick in a way which throws considerable light on the origin of this "innate" behavior. As early as three days of embryonic age, the neck is passively bent when the heartbeat causes the head (which rests on the thorax) to rise and fall. The head is stimulated tactually by the yolk sac, which is moved mechanically by amnion contractions synchronized with the heartbeats which cause head movement. Beginning about one day later, the head first bends *actively* in response to tactual stimulation. At about this time, too, the bill begins to open and close when the bird nods—according to Kuo, apparently through nervous excitation furnished by the head movements through irradiation in the still-incomplete nervous system. Bill-opening and closing become independent of head-activity only somewhat later. After about 8 or 9 days, fluid forced into the throat by the bill and head movements causes swallowing. On the twelfth day, bill-opening always follows head-movement. [341–342]

In the light of Kuo's studies the "innateness" of the chick's pecking takes on a different character from that suggested by the concept of a unitary, innate item of behavior. Kuo's observations strongly suggest several interpretations of the *development* of pecking (which, of course, are subject to further clarification). For example, the head-lunge arises from the passive head-bending which occurs contiguously with tactual stimulation of the head while the nervous control of the muscles is being established. By the time of hatching, head-lunging in response to tactual stimulation is very well established (in fact, it plays a major role in the hatching process).

The genesis of head-lunging to visual stimulation in the chick has not been analyzed. In *Amblystoma*, however, Coghill (1929) has shown that a closely analogous shift from tactual to visual control is a consequence of the establishment of certain anatomical relationships between the optic nerve and the brain region which earlier mediated the lunging response to tactual stimulation, so that visual stimuli come to elicit responses established during a period of purely tactual sensitivity. If a similar situation obtains in the chick, we would be dealing with a case of intersensory equivalence, in which visual stimuli, because of the anatomical relationships between the visual and tactual regions of the brain, became equivalent to tactual stimuli, which in turn became effective through an already analyzed process of development, which involved conditioning at a very early age (Maier and Schneirla 1935).

The originally diffuse connection between head-lunge and bill-opening appears to be strengthened by the repeated elicitation of lunging and billing by tactual stimulation by the yolk sac. The repeated elicitation of swallowing by the pressure of amniotic fluid following bill-opening probably is important in the establishment of the post-hatching integration of bill-opening and swallowing.

Maternal Behavior in the Rat

Another example of behavior appearing to fulfill the criteria of "innateness" may be found in the maternal behavior of the rat.

Pregnant female rats build nests by piling up strips of paper or other material. Mother rats will "retrieve" their pups to the nest by picking them up in the mouth and carrying them back to the nest. Nest-building and retrieving both occur in all normal rats; they occur in rats which have been raised in isolation; and they occur with no evidence of previous practice, since both are performed well by primiparous rats (retrieving may take place for the first time only a few minutes after the birth of the first litter of a rat raised in isolation). Both behavior patterns therefore appear to satisfy the criteria of "innateness" (Wiesner and Sheard 1933).

Riess (pers. com.), however, raised rats in isolation, at the same time preventing them from ever manipulating or carrying any objects. The floor of the living cage was of netting so that feces dropped down out of reach. All food was powdered, so that the rats never carried food pellets. When mature, these rats were placed in regular breeding cages. They bred, but did *not* build normal nests or retrieve their young normally. They scattered nesting material all over the floor of the cage, and similarly moved the young from place to place without collecting them at a nest-place.

Female rats do a great deal of licking of their own genitalia, particularly during pregnancy (Wiesner and Sheard 1933). This increased licking during pregnancy has several probable bases, the relative importance of which is not yet known. The increased need of the pregnant rat for potassium salts (Heppel and Schmidt 1938) probably accounts in part for the increased licking of the salty body fluids as does the increased irritability of the genital organs themselves. Birch (pers. com.) has suggested that this genital licking may play an important role in the development of licking and retrieving of the young. He is raising female rats fitted from an early age with collars made of rubber discs, so worn that the rat is effectively prevented from licking its genitalia. Present indications, based on limited data, are that rats so raised eat a high percentage of their young, that the young in the nest may be found under

any part of the female instead of concentrated posteriorly as with normal mother rats, and that retrieving does not occur.

These considerations raise some questions concerning nativistic interpretations of nest-building and retrieving in the rat, and concerning the meaning of the criteria of "innateness." To begin with, it is apparent that practice in carrying food pellets is partly equivalent, for the development of nest-building and retrieving, to practice in carrying nesting-material, and in carrying the young. Kinder (1927) has shown that nest-building activity [342–343] is inversely correlated with environmental temperature, and that it can be stopped by raising the temperature sufficiently. This finding, together with Riess's experiment, suggests that the nest-building activity arises from ordinary food (and other object) manipulation and collection under conditions where the accumulation of certain types of manipulated material leads to immediate satisfaction of one of the animal's needs (warmth). The fact that the rat is generally more active at lower temperatures (Browman 1943; Morgan 1947) also contributes to the probability that nest-building activity will develop. In addition, the rat normally tends to stay close to the walls of its cage, and thus to spend much time in corners. This facilitates the collection of nesting material into one corner of the cage, and the later retrieving of the young to that corner. Patrick and Laughlin (1934) have shown that rats raised in an environment without opaque walls do not develop this "universal" tendency of rats to walk close to the wall. Birch's experiment suggests that the rat's experience in licking its own genitalia helps to establish retrieving as a response to the young, as does its experience in carrying food and nesting material.

Maturation-vs.-Learning, or Development?

The Isolation Experiment
These studies suggest some second thoughts on the nature of the "isolation experiment." It is obvious that by the criteria used by Lorenz and other instinct theorists, pecking in the chick and

nest-building and retrieving in the rat are not "learned" behavior. They fulfil all criteria of "innateness," i.e., of behavior which develops without opportunity for practice or imitation. Yet, in each case, analysis of the developmental process involved shows that the behavior patterns concerned are not unitary, autonomously developing things, but rather that they emerge ontogenetically in complex ways from the previously developed organization of the organism in a given setting.

What, then is wrong with the implication of the "isolation experiment," that behavior developed in isolation may be considered "innate" if the animal did not practice it specifically?

Lorenz repeatedly refers to behavior as being innate because it is displayed by animals raised in isolation. The raising of rats in isolation, and their subsequent testing for nesting behavior, is typical of isolation experiments. The development of the chick inside the egg might be regarded as the ideal isolation experiment.

It must be realized that an animal raised in isolation from fellow-members of his species *is not necessarily isolated from the effect of processes and events which contribute to the development of any particular behavior pattern.* The important question is not "Is the animal isolated?" but *"From what* is the animal isolated?" The isolation experiment, if the conditions are well analyzed, provides at best a negative indication that certain specified environmental factors probably are not directly involved in the genesis of a particular behavior. However, the isolation experiment by its very nature does not give a positive indication that behavior is "innate" or indeed any information at all about what the process of development of the behavior really consisted of. The example of the nest-building and retrieving by rats which are isolated from other rats but not from their food pellets or from their own genitalia illustrates the danger of assuming "innateness" merely because a *particular* hypothesis about learning seems to be disproved. This is what is consistently done by Tinbergen, as, for

example, when he says (1942) of certain behavior patterns of the three-spined stickleback: "The releasing mechanisms of these reactions are all innate. A male that was reared in isolation ... was tested with models before it had ever seen another stickleback. The ... [stimuli] ... had the same releaser functions as in the experiments with normal males." Such isolation is by no means a final or complete control on possible effects from experience. For example, is the "isolated" fish uninfluenced by its own reflection from a water film or glass wall? Is the animal's experience with human handlers, food objects, etc., really irrelevant?

Similarly, Howells and Vine (1940) have reported that chicks raised in mixed flocks of two varieties, when tested in a Y-maze, learn to go to chicks of their own variety more readily than to those of the other variety. They concluded that the "learning is accelerated or retarded ... because of the directive influence of innate factors." In this case, Schneirla (1946b) suggests that the effect of the chick's experience with its own chirping during feeding has not been adequately considered as a source of differential learning previous to the experiment. This criticism may also be made of a similar study by Schoolland (1942) using chicks and ducklings.

Even more fundamental is the question of what [343–344] is meant by "maturation." We may ask whether experiments based on the assumption of an absolute dichotomy between maturation and learning ever really tell us *what* is maturing, or how it is maturing? When the question is examined in terms of *developmental* processes and relationships, rather than in terms of preconceived categories, the maturation-versus-learning formulation of the problem is more or less dissipated. For example, in the rat nest-building probably does not mature autonomously—and it is *not* learned. It is not "nest-building" which is learned. Nest-building develops in certain situations through a developmental process in which at each stage there is an identifiable interaction between the environment and organic processes, and within the organism; this interaction is based on the preceding stage of development and gives rise to the succeeding stage. These interactions are present from the earliest (zygote) stage. Learning may emerge as a factor in the animal's behavior even at early embryonic stages, as pointed out by Carmichael (1936).

Pecking in the chick is also an emergent—an integration of head, bill, and throat components, each of which has its own developmental history. This integration is already partially established by the time of hatching, providing a clear example of "innate" behavior in which the statement "It is innate" adds nothing to an understanding of the developmental process involved. The statement that "pecking" is innate, or that it "matures," leads us *away* from any attempt to analyze its specific origins. The assumption that pecking grows *as* a pecking *pattern* discourages examination of the embryological processes leading to pecking. The elements out of whose interaction pecking emerges are not originally a unitary pattern; they *become* related as a consequence of their positions in the organization of the embryonic chick. The understanding provided by Kuo's observations owes nothing to the "maturation-versus-learning" formulation.

Observations such as these suggest many new problems the relevance of which is not apparent when the patterns are nativistically interpreted. For example, what is the nature of the rat's temperature-sensitivity which enables its nest-building to vary with temperature? How does the animal develop its ability to handle food in specific ways? What are the physiological conditions which promote licking of the genitalia, etc.? We want to know much more about the course of establishment of the connections between the chick's head-lunge and bill-opening, and between bill-opening and swallowing. This does *not* mean that we expect to establish which of the components is learned and which matured, or "how much" each is learned and how much matured. The effects of learning and of structural factors

differ, not only from component to component of the pattern, but also from developmental stage to developmental stage. What is required is a continuation of the careful analysis of the characteristics of each developmental stage, and of the transition from each stage to the next.

Our scepticism regarding the heuristic value of the concept of "maturation" should not be interpreted as ignorance or denial of the fact that the physical growth of varied structures plays an important role in the development of most of the kinds of behavior patterns under discussion in the present paper. Our objection is to the *interpretation* of the role of this growth that is implied in the notion that the *behavior* (or a *specific* physiological substrate for it) is "maturing." For example, the post-hatching improvement in pecking ability of chicks is very probably due in part to an increase in strength of the leg muscles and to an increase in balance and stability of the standing chick, which results partly from this strengthening of the legs and partly from the development of equilibrium responses (Cruze 1935). Now, isolation or prevention-of-practice experiments would lead to the conclusion that this part of the improvement was due to "maturation." Of course it is partly due to growth processes, *but what is growing is not pecking ability,* just as, when the skin temperature receptors of the rat develop, what is growing is not nest-building activity, *or anything isomorphic with it.* The use of the categories "maturation-vs.-learning" as explanatory aids usually gives a false impression of unity and directedness in the growth of the behavior pattern, when actually the behavior pattern is not primarily unitary, nor does development proceed in a straight line toward the completion of the pattern.

It is apparent that the use of the concept of "maturation" by Lorenz and Tinbergen as well as by many other workers is not, as it at first appears, a reference to a process of development but rather to ignoring the process of development. To say of a behavior that it develops by maturation is tantamount to saying that the obvious forms of learning do not influence it, and that we therefore do not [344–345] consider it necessary to investigate its ontogeny further.

Heredity-vs.-Environment, or Development?

Much the same kind of problem arises when we consider the question of what is "inherited." It is characteristic of Lorenz, as of instinct theorists in general, that "instinctive acts" are regarded by him as "inherited." Furthermore, inherited behavior is regarded as sharply distinct from behavior acquired through "experience." Lorenz (1937) refers to behavior which develops "entirely independent of all experience."

It has become customary, in recent discussions of the "heredity-environment" problem, to state that the "hereditary" and "environmental" contributions are both essential to the development of the organism; that the organism could not develop in the absence of either; and that the dichotomy is more or less artificial. (This formulation, however, frequently serves as an introduction to elaborate attempts to evaluate what part, or even what percentage, of behavior is genetically determined and what part acquired [Howells 1945; Beach 1947; Carmichael 1947; Stone 1947].) Lorenz does not make even this much of a concession to the necessity of developmental analysis. He simply states that some behavior patterns are "inherited," others "acquired by individual experience." I do not know of any statement of either Lorenz or Tinbergen which would allow the reader to conclude that they have any doubts about the correctness of referring to behavior as simply "inherited" or "genically controlled."

Now, what exactly is meant by the statement that a behavior pattern is "inherited" or "genically controlled"? Lorenz undoubtedly does not think that the zygote contains the instinctive act in miniature, or that the gene is the equivalent of an entelechy which purposefully and continuously tries to push the organism's development in a particular direction. Yet one or both of these preformistic assumptions, or their equivalents,

must underlie the notion that some behavior patterns are "inherited" as such.

The "instinct" is obviously not present in the zygote. Just as obviously, it is present in the behavior of the animal after the appropriate age. The problem for the investigator who wishes to make a causal analysis of behavior is: How did this behavior come about? The use of "explanatory" categories such as "innate" and "genically fixed" obscures the necessity of investigating developmental *processes* in order to gain insight into the actual mechanisms of behavior and their inter-relations. The problem of development is the problem of the development of new structures and activity patterns from the resolution of the interaction of *existing* structures and patterns, within the organism and its internal environment, and between the organism and its outer environment. At any stage of development, the new features emerge from the interactions within the *current* stage and between the *current* stage and the environment. The interaction out of which the organism develops is *not* one, as is so often said, between heredity and environment. It is between *organism* and environment! And the organism is different at each different stage of its development.

Modern physiological and biochemical genetics is fast destroying the conception of a straight-line relationship between gene and somatic characteristic. For example, certain strains of mice contain a mutant gene called "dwarf." Mice homozygous for "dwarf" are smaller than normal mice. It has been shown (Smith and MacDowell 1930; Keeler 1931) that the cause of this dwarfism is a deficiency of pituitary growth hormone secretion. Now what are we to regard as "inherited"? Shall we change the name of the mutation from "dwarf" to "pituitary dysfunction" and say that dwarfism is not inherited as such—that what is inherited is a hypoactive pituitary gland? This would merely push the problem back to an earlier stage of development. We now have a better understanding of the origin of the dwarfism than we did when we could only say it is "genically determined." However, the pituitary function developed, in turn, in the context of the mouse as it was when the gland was developing. The problem is: What was that context and how did the gland develop out of it?

What, then, is inherited? From a somewhat similar argument, Jennings (1930) and Chein (1936) concluded that only the zygote is inherited, or that heredity is only a stage of development. There is no point here in involving ourselves in tautological arguments over the definition of heredity. It is clear, however, that to say a behavior pattern is "inherited" throws no light on its *development* except for the purely negative implication that *certain types* of learning are not directly involved. Dwarfism in the mouse, nest-building in the rat, pecking in the chick, and the "zig-zag [345–346] dance" of the stickleback's courtship (Tinbergen 1942) are all "inherited" in the sense and by the criteria used by Lorenz. But they are not by any means phenomena of a common type, nor do they arise through the same kinds of developmental processes. To lump them together under the rubric of "inherited" or "innate" characteristics serves to block the investigation of their origin just at the point where it should leap forward in meaningfulness. (Anastasi and Foley 1948, considering data from the field of human differential psychology, have been led to somewhat the same formulation of the "heredity-environment" problem as is presented here.)

Taxonomy and Ontogeny

Lorenz (1939) has very ably pointed out the potential importance of behavior elements as taxonomic characteristics. He has stressed the fact that evolutionary relationships are expressed just as clearly (in many cases more clearly) by similarities and differences in behavior as by the more commonly used physical characteristics. Lorenz himself has made a taxonomic analysis of a family of birds in these terms (Lorenz 1941), and others have been made by investigators in-

fluenced by him (Delacour and Mayr 1945; Adriaanse 1947; Baerends and Baerends-van Roon 1950). This type of analysis derives from earlier work on the taxonomic relations of behavior patterns by Whitman (1898, 1919), Heinroth (1910, 1930), Petrunkevitsch (1926), and others.

Lorenz's brilliant approach to the taxonomic analysis of behavior characteristics has had wide influence since it provides a very stimulating framework in which to study species differences and the specific characteristics of behavior. However, it does not necessarily follow from the fact that behavior patterns are species-specific that they are "innate" as patterns. We may emphasize again that the systematic stability of a characteristic does not indicate anything about its mode of development. The fact that a characteristic is a good taxonomic character does not mean that it developed autonomously. The shape of the skull bones in rodents, which is a good taxonomic character (Romer 1945), depends in part upon the presence of attached muscles (Washburn 1947). We cannot conclude that because a behavior pattern is taxonomically stable it must develop in a unitary, independent way.

In addition it would be well to keep in mind that the species-characteristic nature of many behavior patterns may result partly from the fact that all members of the species grow in the same environment. Smith and Guthrie (1921) call such behavior elements "coenotropes." Further, it is not at all necessary that these common features of the environment be those which seem a priori to be relevant to the behavior pattern under study. Lorenz's frequent assumption (e.g., 1935) that the effectiveness of a given stimulus on first presentation demonstrates an innate sensory mechanism specific for that stimulus is not based on analysis of the origin of the stimulus-effectiveness, but merely on the fact that Lorenz has eliminated the major alternative *he* sees to the nativistic explanation.

Thorpe and Jones (1937) have shown that the apparently innate choice of the larvae of the flour moth by the ichneumon fly *Nemerites* as an object in which to deposit its eggs is actually a consequence of the fact that the fly larva was *fed* on the larvae of the flour moth while it was developing. By raising *Nemerites* larvae upon the larvae of *other* kinds of moth Thorpe and Jones caused them, when adult, to choose preponderantly these other moths on which to lay their eggs. The choice of flour-moth larvae for oviposition is quite characteristic of *Nemerites* in nature. In view of Thorpe and Jones' work, it would obviously be improper to conclude from this fact that the choice is based on innately-determined stimuli. Yet, before their paper was published, the species-specific character of the behavior would have been just as impressive evidence for "innateness" as species-specificity *ever* is.

Taxonomic analysis, while very important, is not a substitute for concrete analysis of the ontogeny of the given behavior, as a source of information about its origin and organization.

Levels of Organization

Levels of "Innateness"

Animals at different evolutionary levels show characteristic differences in the extent and manner of learning. In addition, within the same animal's behavior different activities may be more or less susceptible to the influence of learning, and may be affected in different ways by learning (Schneirla 1948, 1949).

Lorenz explains these facts in terms of the richness of the animal's instinctive equipment. As described above, his conception is that instinctive [346–347] behavior is sharply different from all behavior leading up to the performance of the instinct. This "appetitive" behavior is conceived of as the sole evolutionary source of all learned and intelligent behavior. Thus he says:

appetitive behavior, as the sole root of all 'variable' behavior, not only is physiologically something fundamentally different from the automatism of instinctive behavior, but ... the two different processes appear as 'substitutes' *(vikariierend)* for each other, in that the

higher (phylogenetic) development of the one makes the other superfluous and stops its development. The reaching of a higher psychic performance goes hand-in-hand with a reduction of the automatisms that take part in the action, leaving a behavior pattern with the same function as the one originally existing. (Lorenz 1937)

Again:

It is a peculiarity of many behavior patterns of higher animals, that *innate instinctive elements and individually-acquired elements immediately follow each other*, within a functionally unitary chain of acts ... I have characterized this phenomenon as instinct-training interlacement. Similar interlacements occur between instinctive acts and intelligent or insightful behavior.... The essence of such an interlacement is that, within a chain of innate instinctive acts there is a definite point, which point is innately determined, where a learned act is inserted. This learned act must be acquired by each individual in the course of its ontogenetic development. In such a case, the chain of innate acts has a *gap*, in which, instead of an instinctive act, there is a '*capacity to acquire*'. (Lorenz 1937) [All emphases are Lorenz's.]

It is apparent that Lorenz regards differences in the extent to which learning occurs as representing differences in the size of the gaps in the chain of innate behavior. He considers any given "component" of behavior as "innate" or not "innate." This is entirely consistent with his virtual identification of "innate" with "autonomously developing."

However, we have already tried to make it clear that behavior patterns classified as "innate" by *any* criterion do not all fall into the same category with respect to embryonic origin, developmental history, or level of organization. Lorenz notes that more or fewer of the components of behavior may be "innate." But nowhere does he recognize that *one* component may be more or less "innate," or "innate" in one or another *manner*. We may call attention to an important difference between the pecking of the chick and the nest-building of the rat, both behavior patterns which develop without specific practice of the patterns: A major part of the learning which appears to be antecedent to the emergence of pecking in the chick occurs before hatching, while much of the learning which is antecedent to the emergence of nest-building in the rat occurs after birth.

Shall we call those behavior patterns "innate" which develop before birth and not those which develop after? This would be fruitless in view of the demonstrated existence of prenatal conditioning (Ray 1932; Gos 1933; Spelt 1948; Hunt 1949), and unsatisfactory in view of the problem of the so-called postnatal "maturation" of various "innate" behavior patterns (Grohmann 1938). But we must recognize that different behavior patterns may involve learning at different ontogenetic stages to different extents, and in different ways. For example, much less of the behavior of the rat is *directly* a consequence of the specific characteristics of its structure than in the case of the earthworm (Maier and Schneirla 1935). The involvement of learning in the development of the rat's behavior is different from and occurs at different developmental stages from that of the chick. Further, some responses of the rat (such as licking of a painful spot) are very much less subject to change by learning than others, such as care of young (Sperry 1945; Uyldert 1946). These are not differences in the *number* of behavioral elements which are "innate," but rather in the *way* in which the structures are involved in the development of behavior at different evolutionary levels and for different behavior patterns.

Lorenz does not fully utilize the idea of levels of organization of behavior, apparently because his concept of "innateness" is not the result of analysis of the development of behavior; it is in part the result of a preconception that "innate" and "not-innate" are the two categories into which behavior logically falls. Consequently Lorenz and his school have classified behavior as "innate" and "not-innate" on the basis of criteria which when carefully examined appear to be arbitrary. Their category of "innate" therefore includes very different kinds of behavior, which involve learning in many different ways. Lorenz's

concept of "innate" behavior represents a lumping-together of many different kinds and levels of behavior on the basis of an essentially phenotypic classification, and the imposition of preconceived categories upon that classification.

Evolutionary Levels

Since Lorenz does not discuss the existence of qualitative differences with respect to modes of [347–348] development within his category of "innate" behavior it is not surprising that his conception of the evolution of behavior lacks any notion of qualitative change. Lorenz maintains at all levels a sharp distinction between "instinctive acts" and "appetitive behavior" (which includes all oriented, goal-directed, and variable types of behavior at all levels). He says:

> If we consider the unbroken series of forms of corresponding modes of behavior, which extends in a smooth progression from protozoa to man, we must determine that we cannot distinguish between taxis, on the one hand, and, on the other, behavior guided by the simplest intelligence (Einsicht). We cannot here distinguish between taxis and, in the case of our frog, an intelligence which might (anthropomorphically speaking) be limited to the knowledge: 'There sits the fly'. (Lorenz 1937)

This is restated in a later paper (Lorenz 1939): "No sharp line can be drawn between the simplest orienting-reaction and the highest 'insightful' behavior."

It might be pointed out that whether we can distinguish various levels of behavioral organization depends in part on our assiduity in *attempting* to distinguish them. Preconceptions about the number and kind of categories into which behavior ought to fall naturally has an important effect on the kind of examination we make of behavior patterns and the kinds of distinctions we find ourselves able to make among them.

In the quotation above we have translated as "smooth" (progression) Lorenz's word "stufenlöse," which might be more literally translated "without steps" or "without levels." This is a gratuitous and very misleading oversimplification on Lorenz's part. The transition from protozoa to man is *not* "stepless." There are *characteristic* structural differences between phyletic levels, and these differences are responsible for *characteristic* differences in the organization of behavior. A protozoan is not like a simpler man. It is a different kind of organism, with behavior which depends in different *ways* on its structure. The analysis of behavior mechanisms at different levels (Schneirla 1946b) shows that it is frequently misleading to speak of *behavior* patterns or elements as homologous when they seem to serve similar (or the "same") functions and have superficially similar characteristics. Analysis of structural organizations out of which the specific behavior patterns emerge shows that similar behaviors at different phyletic levels often are end-products of evolutionary selection leading to the similar behavior, but deriving from different structures so that the underlying processes and mechanisms are not the same.

Lorenz's application of the concept of evolutionary change does not consist of analyzing the different ways in which behavior patterns at different evolutionary levels depend on the structure and life of the organism. It consists rather of abstracting aspects of behavior, reifying them as specific autonomous mechanisms, and then citing them as demonstrations of "evolution" in a purely descriptive taxonomic sense. Taxonomically, this procedure is often extremely valuable, but by its implicit assumption that "elements" of behavior maintain their nature regardless of change in the organization in which they are embedded (more properly, we should say from which they emerge), it hinders rather than helps analysis of the behavior patterns themselves.

... [348–358]

"Vacuum Activities"

The so-called "vacuum activities" or Leerlauf-reaktionen are regarded by Lorenz and Tinber-

gen as evidence of the accumulation of reaction-specific energy in the instinctive center until it "forces" its way through the inhibiting innate releasing mechanism and "goes off " without any detectable external stimulus.

Lees (1949) has cited the example of the cyclical colony activities of the ant *Eciton hamatum* (Schneirla 1938) as an example of "something akin to 'vacuum activity.'" Colonies of this army ant pass regularly through *statary and nomadic* phases, each lasting about 20 days. As Lees points out [based on Schneirla's (1944) description]:

During the statary phase the bivouac, to which the single queen is confined, remains *in situ* and raiding activities are minimal. During the nomadic phase the position of the bivouac is changed each nightfall and strong raiding parties emerge from the colony. This activity is in no way related to the abundance or scarcity of food in the neighborhood.

This cyclic behavior thus appears to Lees to have the character of a "vacuum activity," in that it occurs periodically without any noticeable change in the external stimulus-conditions. This is very misleading, for Schneirla's (1938, 1944) analysis of this behavior has shown that the change from statary to nomadic behavior is a consequence of the growth of a great new brood of ants. When the callow workers emerge from their cocoons, their movements stimulate the adult workers to great activity. As the callows mature and cease to be dependent on the adults, their energizing effect is lessened. At this point, the emergence of wriggling larvae from the eggs supplements the diminishing activating effect of the callows on the adults. When the larvae pupate, and become inactive, the adults are no longer subject to trophallactic (Wheeler 1928) stimulation, and the colony changes to its statary period.

The point that is relevant to our discussion is that Schneirla's analysis leads to a conception that is the *opposite* of that implied by the notion of "vacuum activity." The periodic recurrences are *not* the result of the building up of energy in any animal's nervous system. They are the result of the periodic recurrences of inter-individual stimulating effects. The behavior is not represented "in advance" in *any* of the animals in the colony; it emerges in the course of the ants' relationships with one another and with the environment. There is no "reaction-specific energy" being built up. The periodicity is a result of the periodicity of the queen's egg-laying, which is *not* a "center" having *any* characteristics corresponding to the behavior. And even this is not a *direct* relationship. If the number of larvae in a colony is experimentally reduced by 50 per cent, thus reducing their total stimulating effect, a normal nomadic phase cannot occur. Recent findings (Schneirla and Brown 1950) have in fact confirmed the hypothesis that each of the regular large-scale egg-delivering episodes in the queen's function basic to the cycle is a specific outcome of her over-feeding, due to a maximal stimulation of the colony by the brood. This event, occurring inevitably at the end of each nomadic phase, is a "feed-back" type of function, not at all related to the implications of "vacuum activity."

The restrictive nature of such categorical theories as that of Lorenz is very well illustrated by Lees' remarks on *Eciton*. The actual development process leading to the periodic performances of this ant are well understood, and are *known* to have no essential relationship to any "reaction-specific energy" in any nervous system; further they are *known* not to be "innate" as such (Schneirla 1938). The processes leading to this behavior surely have nothing to do with the processes leading to "vacuum activities" in a fish. Yet the superficial similarity is sufficient to cause Lees to cite the ant's behavior as an example of a type of behavior described for vertebrates. This is a good example of the tendency encouraged by such theories to look for cases fitting the theoretical categories in many types of behavior, rather than analysis of the processes involved in the development of any one behavior pattern.

Conclusion

We have summarized the main points of Lorenz's instinct theory, and have subjected it to a critical examination. We find the following serious flaws:

1. It is rigidly canalized by the merging of widely different kinds of organization under inappropriate and gratuitous categories.

2. It involves preconceived and rigid ideas of innateness and the nature of maturation.

3. It habitually depends on the transference of concepts from one level to another, solely on the basis of analogical reasoning.

4. It is limited by preconceptions of isomorphic [358–359] resemblances between neural and behavioral phenomena.

5. It depends on finalistic, preformationist conceptions of the development of behavior itself.

6. As indicated by its applications to human psychology and sociology, it leads to, or depends on (or both), a rigid, preformationist, categorical conception of development and organization.

Any instinct theory which regards "instinct" as immanent, preformed, inherited, or based on specific neural structures is bound to divert the investigation of behavior development from fundamental analysis and the study of developmental problems. Any such theory of "instinct" inevitably tends to short-circuit the scientist's investigation of intraorganic and organism-environment developmental relationships which underlie the development of "instinctive" behavior.

Acknowledgments

I am greatly indebted to Dr. T. C. Schneirla (who originally suggested the writing of this paper) and to Dr. J. Rosenblatt for many stimulating and helpful discussions of the problems discussed here. Dr. Schneirla in particular has devoted much attention to criticism of the paper at various stages.

The following people also have read the paper, in part and at various stages, and have made many helpful suggestions and comments: Drs. H. G. Birch, K. S. Lashley, D. Hebb, H. Klüver, L. Aronson, J. E. Barmack, L. H. Hyman, L. H. Lanier, and G. Murphy. Since these scientists differ widely in the extent of their agreement or disagreement with various points of my discussion, I must emphasize that none of them is in any way responsible for any errors of omission or commission that may appear.

References

Adriaanse, M. S. C. (1947). *Ammophila campestris* Latr. und *Ammophila adriaansei* Wilcke. Ein Beitrag zur vergleichenden Verhaltensforschung. *Behaviour* 1: 1–34.

Anastasi, A., and J. P. Foley, Jr. (1948). A proposed reorientation in the heredity-environment controversy. *Psychological Review* 55: 239–249.

Armstrong, E. A. (1950). The nature and function of displacement activities. *Symposia of the Society of Experimental Biology* 4: 361–394.

Baerends, G. P. (1950). Specializations in organs and movements with a releasing function. *Symposia of the Society of Experimental Biology* 4: 337–360.

Baerends, G. P., and J. M. Baerends-van Roon. (1950). An introduction to the study of the ethology of cichlid fishes. *Behaviour (Supplement)* 1: 1–242.

Beach, F. A. (1947). Evolutionary changes in the physiological control of mating behavior in mammals. *Psychological Review* 54: 297–315.

Beach, F. A. (1951). Instinctive behavior, reproductive activities. In S. S. Stevens (Ed.), *Handbook of Experimental Psychology*, pp. 387–434. New York: John Wiley & Sons.

Bird, C. (1925). The relative importance of maturation and habit in the development of an instinct. *Pedagogical Seminar* 32: 68–91.

Browman, L. G. (1943). The effect of controlled temperatures upon the spontaneous activity rhythms of the albino rat. *Journal of Experimental Zoology* 94: 477–489.

Carmichael, L. (1936). A re-evaluation of the concepts of maturation and learning as applied to the early development of behavior. *Psychological Review* 43: 450–470.

Carmichael, L. (1947). The growth of the sensory control of behavior before birth. *Psychological Review* 54: 316–324.

Chein, I. (1936). The problems of heredity and environment. *Journal of Psychology* 2: 229–244.

Coghill, G. E. (1929). *Anatomy and the Problem of Behavior.* London: Cambridge University Press.

Cuze, W. W. (1935). Maturation and learning in chicks. *Journal of Comparative Psychology* 19: 371–409.

Delacour, J., and E. Mayr. (1945). The family Anatidae. *Wilson Bulletin* 57: 3–55.

Gos, E. (1933). Les reflexes conditionnels chez l'embryon d'oiseau. *Bulletin du Société Royal des Sciences de Liège* 4–5: 194–199; 6–7: 246–250.

Grohman, J. (1938). Modifikation oder Funktionsreifung? Ein Beitrag zur Klärung der wechselseitigen Beziehungen zwischen Instinkhandlung und Erfahrung. *Zeitschrift für Tiersychologie* 2: 132–144.

Hartley, P. H. T. (1950). An experimental analysis of interspecific recognition. *Symposia of the Society of Experimental Biology* 4: 313–336.

Heinroth, O. (1910). Beiträge zur Biologie, namentlich Ethologie und Psychologie der Anatiden. *International Ornithological Congress (Berlin)* 5: 589–702.

Heinroth, O. (1930). Ueber bestimmte Bewegungsweisen bei Wirbeltieren. *Gesellschaft Naturforschender Freunde zu Berlin, 1929,* pp. 333–342.

Heppel, L. A., and C. L. A. Schmidt. (1938). Studies on the potassium metabolism of the rat during pregnancy, lactation and growth. *University of California Publications in Physiology* 8: 189–205.

Howells, T. R. (1945). The obsolete dogmas of heredity. *Psychological Review* 52: 23–34.

Howells, T. R., and D. O. Vine. (1940). The innate differential in social learning. *Journal of Abnormal and Social Psychology* 35: 537–548.

Hunt, E. L. (1949). Establishment of conditioned responses in chick embryos. *Journal of Comparative and Physiological Psychology* 42: 107–117.

Jennings, H. S. (1930). *The Biological Basis of Human Nature.* New York: Norton & Co.

Keeler, C. (1931). *The Laboratory Mouse.* Cambridge, MA: Harvard University Press.

Kinder, E. F. (1927). A study of the nest-building activity of the albino rat. *Journal of Experimental Zoology* 47: 117–161.

Koehler, O. (1950). Die Analyse der Taxisanteile instinktartigen Verhaltens. *Symposia of the Society of Experimental Biology* 4: 269–303.

Kuo, Z. Y. (1932a). Ontogeny of embryonic behavior in Aves. I. The chronology and general nature of the behavior of the chick embryo. *Journal of Experimental Zoology* 61: 395–430.

Kuo, Z. Y. (1932b). Ontogeny of embryonic behavior in Aves. II. The mechanical factors in the various stages leading to hatching. *Journal of Experimental Zoology* 62: 453–489.

Kuo, Z. Y. (1932c). Ontogeny of embryonic behavior in Aves. III. The structure and environmental factors in embryonic behavior. *Journal of Comparative Psychology* 13: 245–272.

Kuo, Z. Y. (1932d). Ontogeny of embryonic behavior in Aves, IV. The influence of embryonic movements upon the behavior after hatching. *Journal of Comparative Psychology* 14: 109–122.

Lees, A. D. (1949). Modern concepts of instinctive behaviour. *Science Progress in the Twentieth Century* 37: 318–321.

Lorenz, K. (1931). Beiträge zur Ethologie sozialer Corviden. *Journal für Ornithologie* 79: 67–127.

Lorenz, K. (1932). Betrachtungen über das Erkennen der arteigenen Triebhandlungen der Vögel. *Journal für Ornithologie* 80: 50–98.

Lorenz, K. (1935). Der Kumpan in der Umwelt des Vogels. *Journal für Ornithologie* 83: 137–213, 289–313.

Lorenz, K. (1937). Ueber den Begriff der Instinkthandlung. *Folia Biotheoretica* 2: 17–50.

Lorenz, K. (1939). Vergleichende Verhaltensforschung. *Zoologische Anzeitung* 12 (Suppl. band): 69–102.

Lorenz, K. (1940). Durch Domestikation verursachte Störungen arteigenen Verhaltens. *Zeitschrift für Angewandte Psychologie und Charaktarkunde* 59: 2–81.

Lorenz, K. (1941). Vergleichende Bewegungsstudien an Anatinen. *Journal für Ornithologie* 89 (Sonderheft): 194–294.

Lorenz, K. (1950). The comparative method in studying innate behavior patterns. *Symposia of the Society of Experimental Biology* 4: 221–268.

Lorenz, K., and N. Tinbergen. (1938). Taxis und Instinkthandlung in der Eirollbewegung der Graugans. I. *Zeitschrift für Tierpsychologie* 2: 1–29.

Maier, N. R. F., and T. C. Schnierla. (1935). *Principles of Animal Psychology*. New York: McGraw-Hill Co.

Miller, N. E. (1951). Learnable drives and rewards. In S. S. Stevens (Ed.), *Handbook of Experimental Psychology*, pp. 435–472. New York: John Wiley & Sons.

Morgan, C. T. (1947). The hoarding instinct. *Psychological Review* 54: 335–341.

Nissen, H. W. (1951). Phylogenetic comparison. In S. S. Stevens (Ed.), *Handbook of Experimental Psychology*, pp. 347–386. New York: John Wiley & Sons.

Patrick, J. R., and R. M. Laughlin. (1934). Is the wall-seeking tendency in the white rat an instinct? *Journal of Genetic Psychology* 44: 378–389.

Petrunkevitsch, A. (1926). The value of instinct as a taxonomic character in spiders. *Biological Bulletin* 50: 427–432.

Ray, W. S. (1932). A preliminary study of fetal conditioning. *Child Development* 3: 173–177.

Riess, B. F. (1949). The isolation of factors of learning and native behavior in field and laboratory studies. *Annals of the New York Academy of Sciences* 51: 1093–1102.

Romer, A. S. (1945). *Vertebrate Paleontology*. Chicago: University of Chicago Press.

Schneirla, T. C. (1938). A theory of army-ant behavior based upon the analysis of activities in a representative species. *Journal of Comparative Psychology* 25: 51–90.

Schneirla, T. C. (1944). The reproductive functions of the army-ant queen as pacemakers of the group behavior pattern. *Journal of the New York Entomological Society* 52: 153–192.

Schneirla, T. C. (1946a). Contemporary American animal psychology in perspective. In P. L. Harriman (Ed.), *Twentieth Century Psychology*, pp. 306–316. New York: Philosophical Library.

Schneirla, T. C. (1946b). Problems in the biopsychology of social organization. *Journal of Abnormal and Social Psychology* 41: 385–402.

Schneirla, T. C. (1948). Psychology, Comparative. *Encyclopaedia Britannica* 18: 690–708.

Schneirla, T. C. (1949). Levels in the psychological capacities of animals. In R. Sellars, V. J. McGill, and M. Farber (Eds.), *Philosophy for the Future*, pp. 243–286. New York: Macmillan & Co.

Schneirla, T. C., and R. Z. Brown. (1950). Army-ant life and behavior under dry-season conditions. 4. Further investigation of cyclic processes in behavioral and reproductive functions. *Bulletin of the American Museum of Natural History* 95: 263–354.

Schooland, J. B. (1942). Are there any innate behavior tendencies? *Genetic Psychology Monographs* 25: 219–287.

Shepard, J. F., and F. S. Breed. (1913). Maturation and use in the development of an instinct. *Journal of Animal Behaviour* 3: 274–285.

Smith, P. E., and E. C. MacDowell. (1930). An hereditary anterior-pituitary deficiency in the mouse. *Anatomical Record* 46: 249–257.

Smith, S., and E. R. Guthrie. (1921). *General Psychology in Terms of Behavior*. New York: Appleton.

Spelt, D. K. (1948). The conditioning of the human fetus *in utero*. *Journal of Experimental Psychology* 38: 338–346.

Sperry, R. W. (1945). The problem of central nervous reorganization after nerve regeneration and muscle transposition. *Quarterly Review of Biology* 20: 311–369.

Stone, C. P. (1947). Methodological resources for the experimental study of innate behavior as related to environmental factors. *Psychological Review* 54: 342–347.

Thorpe, W. H., and F. G. W. Jones. (1937). Olfactory conditioning in a parasitic insect and its relation to the problem of host selection. *Proceedings of the Royal Society of London, B* 124: 56–81.

Tinbergen, N. (1939). On the analysis of social organization among vertebrates, with special reference to birds. *American Midland Naturalist* 21: 210–234.

Tinbergen, N. (1942). An objectivistic study of the innate behaviour of animals. *Bibliotheca Biotheoretica* 1: 39–98.

Tinbergen, N. (1948). Physiologische Instinktforschung. *Experientia* 4: 121–133.

Tinbergen, N. (1950). The hierarchical organization of nervous mechanisms underlying instinctive behaviour. *Symposia of the Society of Experimental Biology* 4: 305–312.

Tinbergen, N. (1951). *The Study of Instinct*. Oxford: Oxford University Press.

Tinbergen, N., and D. J. Kuenen. (1939). Ueber die auslösenden und die richtunggebenden Reizsituationen der Sperrbewegung von jungen Drosseln (*Turdus*

m. merula L., und *T. e. ericetarum* Turton). *Zeitschrift für Tierpsychologie* 3: 37–60.

Uyldert, I. E. (1946). A conditioned reflex as a factor influencing the lactation of rats. *Acta Brevia Neerlandica de Physiologia, Pharmacologia, Microbiologia* 14: 86–89.

Washburn, S. L. (1947). The relation of the temporal muscle to the form of the skull. *Anatomical Record* 99: 239–248.

Wheeler, W. M. (1928). *The Social Insects*. New York: Harcourt, Brace & Co.

Whitman, C. O. (1899). Animal behavior. *Biological Lectures, 1898*, pp. 285–338.

Whitman, C. O. (1919). The behavior of pigeons. *Publications of the Carnegie Institute* 257: 1–161.

Wiesner, B. P., and N. M. Sheard. (1933). *Maternal Behaviour in the Rat*. London: Oliver & Boyd.

4 A Developmental Psychobiological Systems View: Early Formulation and Current Status

Gilbert Gottlieb

[August Weismann's] proposition, that effective variation is due to the influence of nutrition upon the determinants contained in the reserve germ-plasm, seems to throw too much stress on the nutrition and environment, too little on the inherent activities of living matter. But if it be regarded as an expression of the fact that all effective variation is a joint product of the inherent activities of germinal cells and the conditioning effects of their environment, it is a self-evident proposition which may be cheerfully accepted.
—Morgan 1893–94

It remains inscrutable why we have had to labor so diligently in the recent past to attain a cheerful acceptance of what was self-evident over a hundred years ago. But I suppose that is what this book is about. I have been given the pleasant task of describing the history and currently status of the concept of probabilistic epigenesis (or a developmental psychobiological systems view).

Behavioral Embryology

Two ideas that I held from my graduate student days were (1) the overriding importance of prenatal development and (2) the bidirectional influences of structure and function (construed very broadly to include the psychological as well as the behavioral, neuroanatomical, and neurophysiological levels). When I did my doctoral dissertation on the sensitive period for imprinting in ducklings, I noticed that, although all of the eggs were set in the incubator at the same time, the ducklings hatched over a two-day period. Therefore, in addition to timing the sensitive period from the time of hatching, I timed the ducklings' age from the onset of incubation. My reasoning was that those that hatched early were less mature at the standard times of training than those that hatched later, the latter being more mature as the birds were all trained at one of five posthatch age ranges (either 4–7 hrs, 8–12 hrs, 13–17 hrs, 18–22 hrs, or 23–27 hrs after hatch-

ing). I reasoned that if the prenatal period of development was important, then the two-day span of hatching had to have consequences for the ducklings' neurosensory, neuromotor, perceptual, and behavioral competence at the time they were trained (exposed) to the imprinting object. Indeed, when I plotted the imprintability of the ducklings as a function of posthatch age at training, I did not find a sensitive period, but when I plotted the same behavioral results according to "developmental age" (timed from the onset of incubation to training) I did find a critical period before which they did not imprint and after which they did not imprint. Imprinting was defined as learning the visual characteristics of the imprinting object as judged by a later behavioral preference test with the familiar object versus an unfamiliar object.

Soon after I published these results (Gottlieb 1961), which suggested developmental age as a more precise age baseline than posthatch age for measuring the sensitive period, I realized that others were interpreting developmental age as "maturational age," a too-narrow connotation that I expressly wished to avoid in explicitly choosing the term developmental age over maturational age. So, Peter Klopfer and I immediately embarked on an experiment to show that the best developmental age for imprinting could be moved around, depending on the sensory experience to which the ducklings were exposed prior to the training trial (Gottlieb and Klopfer 1962). This experiment resulted from my implicit belief in the bidirectionality of structure and function. If the ducklings were reared with visual and social experience from hatching till training, their sensitive periods for auditory and visual imprinting were different from ducklings that were reared in darkness and in social isolation from hatching till training. The sensitive period for imprinting was not exclusively a function of maturation but depended also on the nature and extent of the birds' prenatal and postnatal experiences prior to entering into the imprinting situation.

At this point in my career, having had the benefit of assistance from Zing-Yang Kuo, who kindly consented to work in my laboratory in 1963 to teach me how to do behavioral observations of embryos, I embarked on the experimental analysis of the role of normally occurring embryonic auditory experience in structuring the species-specific auditory perceptual capabilities observed after hatching. As a way of furthering my education in the techniques of behavioral embryology, Kuo and I visited Viktor Hamburger's laboratory in 1963. Hamburger, a distinguished neuroembryologist, was now turning his efforts to the study of motor behavior in the chick embryo, with the assistance of two psychologists, postdoctoral associate Martin Balaban and Ronald Oppenheim, who was a graduate student at that time. Hamburger was claiming the spontaneous autonomy of motor behavior in the early chick embryo, whereas Kuo's observations through the "windows" in the egg shell suggested to him that, while the early chick embryo was indeed motorically active, the extent, kind, and frequency of movement was intimately related to the environmental contingencies of the egg and the chick embryo's own body parts at each phase of development. To make the difference in views on this matter even starker, Balaban (not Hamburger) told me it was possible to remove the embryo from the shell and observe its normal behavior unimpeded without the necessity of Kuo's windowing procedure. In my naïve way, I was delighted and began freeing embryos from their prison with abandon, while Kuo stared in silent disbelief. Pretty quickly it became clear that the embryos began dying rather than exhibiting their species-typical motor movements, so with Balaban's assent (he had only done a few embryos before I got there), we agreed this was not a usable experimental procedure and Kuo and I would go back to the old tried and true way of observing the embryos *in situ* through windows made in the egg shell.

Shortly after Kuo and I completed our descriptive study of the behavior of the duck embryo in the fall of 1963 and he had returned to his home in Hong Kong to begin work on his final book, *The Dynamics of Behavior Development*, I received an invitation from Ethel Tobach and her colleagues in the Animal Behavior Department at the American Museum of Natural History, to write a chapter for a book in honor of the career and imminent retirement of T. C. Schneirla. I decided to write a theoretical review of the history and present state of thinking about behavioral embryology. Since Schneirla had been an admirer of Kuo's embryological observations and he, and his student Daniel Lehrman (1953), had used those observations to good advantage in their critiques of instinct theory, it seemed especially fitting to do a review of behavioral embryology for Schneirla's Festschrift.

Predetermined and Probabilistic Epigenesis of Behavior

In my literature review, I found two rather different conceptualizations of behavioral embryology, one I called predetermined epigenesis and the other probabilistic epigenesis. The various adherents to the predetermined view had in common "the assumption that the behavioral sequence is predetermined by factors of neural growth and differentiation (maturation), which have an essentially invariant schedule of appearance. In this view, nonsensory environmental factors merely support development or allow behavioral development to occur and sensory stimulation does not influence or determine the course of behavioral development in any significant way" (Gottlieb 1970: 112). In this view, behavior was an epiphenomenon of neural maturation and did not in itself contribute to development. For example, Coghill (1929: 16) wrote that embryonic behavior patterns become "organized through a regular order of sequence of definite phases in the growth of the nervous system."

Later, Anokhin (1964: 85) wrote, "It is true that the systemogenetic type of the maturation and ... growth is ... evidently inborn ... preformed, and in fact, in the process of the ontogenesis, they correspond demonstrably to the ecological factors of that species of animal." This latter point on the importance of prenatal development to prepare, as it were, the neonate for survival in its infantile environment cannot be gainsaid, but that can also be regularly achieved with the help of normally occurring species-typical behavioral and environmental influences.

I used the term *probabilistic epigenesis* to designate the alternative view to predetermined epigenesis. Probabilistic epigenesis holds that behavioral development of individuals within a given species does not follow an invariant or inevitable course, and, more specifically, that the sequence and outcome of individual behavioral development is probable (with respect to norms) rather than certain.

Further, I wrote:

A degree of uncertainty in behavioral development is demanded by the overwhelming importance ascribed to stimulative factors at almost all stages of prenatal development. In this view, stimulative factors are regarded, in a weak sense, as merely facilitating the development of behavior along certain lines or, in a much stronger sense, as channeling or forcing the development of behavior along certain lines. Stimulative events are broadly construed and include six main factors: (1) presensory or nonsensory mechanical agitation; (2) interoceptive stimulation; (3) proprioceptive stimulation; (4) exteroceptive stimulation; (5) neurochemical (e.g., hormonal) stimulation; and (6) the musculoskeletal effects of use or exercise.

A second way in which probabilistic epigenesis differs from predetermined epigenesis is that certain theorists [I had in mind E. B. Holt, Z.-Y. Kuo, and T. C. Schneirla] implicitly assume that probabilistic epigenesis necessitates a bidirectional structure-function hypothesis. The conventional version of the structure-function hypothesis is unidirectional in the sense that structure is supposed to determine function in an essentially nonreciprocal relationship. The unidirectionality of the structure-function relationship is one of the main assumptions of predetermined epigenesis. The bidirectional version of the structure-function relationship is a logical consequence of the view that the course and outcome of behavioral epigenesis is probabilistic: it entails the assumption of reciprocal effects in the relationship between structure and function whereby function (exposure to stimulation and/or movement or musculoskeletal activity) can significantly modify the development of the peripheral and central structures that are involved in these events. (Gottlieb 1970: 123)

Although the edited book in which my chapter appeared was not published until 1970, the words were penned in 1965. Even at this relatively early date I was able to find some evidence for the influence of stimulative factors (as broadly defined above) on prenatal development, and also for the bidirectionality of the structure-function relationship. Although the concepts of predetermined and probabilistic epigenesis would seem to fall in a general way into the ancient nature/ nurture dichotomy, what I was most pleased with in the formulation of predetermined and probabilistic epigenesis was explicitly avoiding the genes/environment dichotomy. It was clear to me that genetic activity was involved in both formulations, so I just left out that aspect entirely. The questions dividing the two camps were the all-sufficiency of neural maturation versus the contribution of stimulative and behavioral factors in facilitating and channeling prenatal neural maturation, and these were empirically testable questions.

Due to Schneirla's sudden death in 1968, I did not have the opportunity to discuss probabilistic epigenesis and the bidirectional S-F hypothesis with him (the Festschrift was supposed to be a surprise), but I feel confident that he would have welcomed these notions as entirely compatible with his own thinking on these matters, as exemplified, for example, in his 1956 review, in which he extensively discusses maturation and experience interrelationships in prenatal development and in his 1966 paper, in which he mentions "maturation-experience integrations of the embryo" (p. 290) in referring to Kuo's research. Elsewhere, he wrote of the "fusion" of maturation and experience (Schneirla 1965).

I did send a prepublication copy of my chapter to Kuo in 1965, at which time he was writing his book *The Dynamics of Behavior Development*, originally published in 1967. In that book he was generous enough to credit me with the explicit notion of S-F bidirectionality, so I felt that I had indeed made some contribution by explicating what I believed to be inherent in Kuo's (and Holt's and Schneirla's) views on early behavioral development.

Probabilistic epigenesis and S-F bidirectionality rely heavily on prenatal sensory (as well as motor) function as an important contributor to species-typical development, so in the next reviews I undertook to examine the full range of evidence for sensory functioning in bird embryos (Gottlieb 1968) and in a variety of mammalian fetuses including humans (Gottlieb 1971a). What I found was that each sensory system begins to function while still undergoing maturation (i.e., while cell division, migration, growth, and differentiation are still going on), so each system could contribute to its own normal prenatal (as well as postnatal) development. Thus, normally occurring spontaneous and evoked function (experience broadly defined) could play a role in the rate at which a sensory system becomes completely mature as well as contributing to the overall competence of the system by, for example, influencing the number of neurons, as well as the size of neurons and their axonal and dendritic fields. Thirty years later, this is now well established, of course, but it was not at the time (review in Black and Greenough 1998). The positive influence of functional activity on the nervous system is not restricted to the prenatal period but continues into adulthood (Kempermann, Kuhn, and Gage 1997).

I viewed the functional contribution to maturing (but not yet completely mature) systems as an especially important feature of S-F bidirectionality. It is now a common finding in studies of neural plasticity. Even here, however, function (experience) is thought to become important only after, say, all the neural connections are made. Synapses are lost if they are not functionally validated. Recently, this idea has become somewhat more refined to say that there is an early "exuberant" overproduction of synapses which are then pruned down by experience (and the lack thereof). This is different from saying that, in the first instance, the synapses are built in part by function. It is only very recently that the appropriate experiments have been done to show that normally occurring spontaneous neural activity is necessary to get synapses into the correct area of the brain *in the first place* (Catalano and Shatz 1998).

The Developmental Manifold Concept

In 1971, my own research program with duck embryos and hatchlings paid off in what I believed to be a conceptually valuable way. (I hoped others would agree, but this agreement proved to be a rockier and longer road than I anticipated [Gottlieb 1997].) In 1965, I had shown that ducklings and chicks hatched in incubators, and thus deprived of maternal contact, could nonetheless identify the maternal assembly call of their own species after hatching. The only vocal-auditory experience they had was exposure to their own and sib vocalizations prior to entering the test situation. In 1966, I was able to show that enhancing exposure to sib vocalizations lowered the latency and increased the duration of their behavioral response to their own species maternal call. However, it was necessary to devise an embryonic devocalization procedure to truly rule in the critical importance of the embryonic vocalizations in perfecting the perceptual selectivity of the response that was evident after hatching. With the help of John Vandenbergh, I was able to devise an embryonic muting operation that did not otherwise interfere with the health of the embryo and hatchling (Gottlieb and

Vandenbergh 1968). Now, the selectivity of the postnatal response to the species' maternal call could be examined in ducklings that had not experienced their own or sib vocalizations. Lo and behold, the devocalized mallard duckling's usual auditory selectivity was not in place—they could not distinguish the mallard maternal call from the chicken maternal call. The control birds that had been allowed to hear their own embryonic vocalizations for eighteen to twenty-three hours before being devocalized did show the usual preference for the mallard maternal call over the chicken maternal call (Gottlieb 1971b: 141–142).

The outcome of these experiments led to the formulation of the developmental manifold concept, a forerunner to West and King's (1987) notion of "settling nature and nurture into an ontogenetic niche":

The present results indicate that the epigenesis of species-specific auditory perception is a probabilistic phenomenon, the threshold, timing, and ultimate perfection of such perception being regulated jointly by organismic and sensory stimulative factors. In the normal course of development, the manifest changes and improvements in species-specific perception do not represent merely the unfolding of a fixed or predetermined organic substrate independent of normally occurring sensory stimulation. With respect to the evolution of species-specific perception, natural selection would seem to have involved a selection for the entire developmental manifold, including both the organic and normally occurring stimulative features of ontogeny. With respect to the heavy emphasis placed here on the role of normally occurring stimulation in perfecting species-specific perception, it is pertinent to point out that the same condition may also hold for the development of species-typical action patterns. In the past it has been usual in sensory isolation or motor deprivation studies to ask merely whether the deprived animal can subsequently perform the species-typical behavior, not when or how well the animal performs it. As we move into an era of increasingly sophisticated analyses of the development of behavior, it will not be altogether surprising to find that normally occurring sensory stimulation or motor movement is essential to the normal threshold, timing, and perfection of behavior

conventionally regarded as instinctive or innate. If this prediction turns out to be correct, the nature-nurture controversy may all but evaporate, and a consensus will have been reached on the idea that structure only fully realizes itself through function. (Gottlieb 1971b: 156–157)

A full understanding of the development (and evolution) of behavior will necessarily include genetic factors, so I went on to say, "One of the main questions in the ontogeny of behavior concerns the contribution of genetic (molecular) and other biochemical factors. While there is agreement that such factors provide an indispensable impetus to neural maturation, the exact nature or mechanics of the molecular and biochemical control of early neural maturation is not yet known" (Gottlieb 1971b: 157).

From the beginning of my work in the early 1960s I was eager to include the genetic contribution and seized upon an early opportunity to collaborate with two neurologists, one of whom (George Paulson) was adept at brain dissection in birds and the other of whom (Stanley Appel) was running a laboratory in which protein synthesis in the nervous system was being studied. According to what was then already known protein synthesis was the result of messenger RNA activity and mRNA activity was a consequence of DNA activity (DNA→mRNA→protein). So, protein synthesis could be used as an indirect measure of genetic activity. Accordingly, I prepared two groups of duck embryos for Paulson and Appel by exposing them for several days before hatching either to tape-recorded sib vocalizations or to extra visual stimulation by incubating them in a lighted incubator chamber. The control group was incubated in the dark and in acoustic isolation from other embryos. The point of the experiment was to look for enhanced protein synthesis in the auditory and visual parts of the brain in the treated groups.

Subsequently, Appel told us that both experimental groups showed an enhancement of protein synthesis in the synaptic regions of the brain stem, in which the auditory nuclei are located, and in the optic lobes of the brain, which mediate auditory as well as visual stimulation. This was a clear indication that the extra auditory and visual stimulation had enhanced gene expression in the embryo. This, of course, implied a bidirectional S-F relationship all the way to the genetic level during the embryonic period, and it meant that genetic activity could be influenced by normally occurring exteroceptive sensory stimulation and thus result in an enhancement of neural maturation. The experiment was completed in 1965, and Paulson and I pleaded with Appel to complete the data analysis so we could share the results with interested colleagues by publishing them. Alas, overly hectic research, clinical, and other duties took precedence in Appel's life, so Paulson and I were frustrated in our desire to see this work to completion in the form of publication in a refereed journal. At the time there was only one other study in the literature implicating exteroceptive influences on genetic activity, a study involving the influence of vestibular-mediated learning on changes in nuclear RNA base ratios in vestibular nerve cells in rats (Hydén and Egyházi 1962). It was not until 1967 that Rose published a study showing enhanced RNA and protein synthesis in the visual cortex of rats as a consequence of visual stimulation.

Normally Occurring Environmental and Behavioral Influences on Gene Activity

It was in 1976 that I first extended the unidirectionality and bidirectionality of S-F function to the genetic level, and elaborated on this topic in reviews published in 1983 (p. 13) and 1991 (p. 13) as follows:

Predetermined Epigenesis

Unidirectional Structure-Function Development

Genetic activity (DNA→RNA→Protein)→ structural maturation→function, activity, or experience

Probabilistic Epigenesis

Bidirectional Structure-Function Development

Genetic activity (DNA↔RNA↔Protein)↔ structural maturation↔function, activity, or experience

As it applies to the nervous system, structural maturation refers to neurophysiological and neuroanatomical development, principally the structure and function of nerve cells and their synaptic interconnections. The unidirectional S-F view assumes that genetic activity gives rise to structural maturation that then leads to function in a nonreciprocal fashion, whereas the bidirectional view holds that there are reciprocal influences among genetic activity, structural maturation, and function. In the unidirectional view, the activity of genes and the maturation process are pictured as relatively encapsulated or insulated, so that they are uninfluenced by feedback from the maturation process or function, whereas the bidirectional view assumes that genetic activity and maturation are affected by function, activity, or experience. The bidirectional or probabilistic view applied to the usual unidirectional formula calls for arrows going back to genetic activity to indicate feedback serving as signals for the turning on and off of genetic activity. The usual view, as in the central dogma of molecular biology described below, calls for genetic activity to be regulated by the genetic system itself in a strictly feedforward manner.

It was in 1998 that I was able, at last, to write a comprehensive empirical review of bidirectionality at the genetic level, in the context of a critique of the central dogma of molecular biology, a highly influential and outstanding example of predetermined epigenesis.

The Central Dogma

The central dogma asserts that "information" flows in only one direction from the genes to the structure of the proteins that the genes bring

DNA ⟶ DNA $\overset{?}{\underset{\longrightarrow}{\xleftarrow{\,\cdots\,}}}$ RNA ⟶ Protein

Figure 4.1
Central dogma of molecular biology. The right-going arrows represent the central dogma. The discovery of retroviruses (represented by the left-going arrow from RNA to DNA) was not part of the dogma but, after the discovery, was said by Crick (1970) not to be prohibited in the original formulation of the dogma (Crick 1958). (From Gottlieb 1998. Copyright © by the American Psychological Association. Reprinted with permission.)

about through the formula DNA→RNA→ Protein. (Messenger RNA [mRNA] is the intermediary in the process of protein synthesis. In the lingo of molecular biology, the DNA→RNA is called transcription and the RNA→Protein is called translation.) After retroviruses were discovered in the 1960s—in which RNA reversely transcribes DNA through the enzyme reverse transcriptase—Crick wrote a postscript to his 1958 paper in which he congratulated himself for not claiming that reverse transcription was impossible: "In looking back I am struck not only by the brashness which allowed us to venture powerful statements of a very general nature, but also by the rather delicate discrimination used in selecting what statements to make" (Crick 1970: 562). Any ambiguity about the controlling factors in gene expression in the central dogma was removed in a later article by Crick, in which he specifically says that the genes of higher organisms are turned on and off by other genes (Crick 1982: 515). Figure 4.1 shows the central dogma of molecular biology in the form of a diagram.

The Genome According to Central Dogma

The picture of the genome that emerges from the central dogma is (1) one of encapsulation, setting the genome off from supragenetic influences, and (2) a largely feedforward informational process in which the genes contain a blueprint or master

plan for the construction and determination of the organism. In this view, the genome is not seen as part of the holistic, bidirectional developmental-physiological system of the organism, responsive to signals from internal cellular sources such as the cytoplasm of the cell, cellular adhesion molecules (CAMs), or to extracellular influences such as hormones, and certainly not to extraorganismic influences such as stimuli or signals from the external environment.

The main point of my review was to extend the normally occurring influences on genetic activity to the external environment, thereby further demonstrating that the genome is not encapsulated and is in fact a part of the organism's general developmental-physiological adaptation to environmental stresses and signals: Genes express themselves appropriately only in responding to internally and externally generated stimulation. Further, in this holistic view, while genes participate in the making of protein, protein is also subject to other influences, and protein must be further stimulated and elaborated to become part of the nervous system (or other systems) of the organism, so that genes operate at the lowest level of organismic organization and they do not, in and of themselves, produce finished traits or features of the organism. The organism is a product of epigenetic development, which includes the genes as well as many other supragenetic influences. Since this latter point has been the subject of numerous contributions (reviewed in Gottlieb 1992) I shall not deal with it further here, but, rather, restrict myself to documenting that the activity of genes is regulated in just the same way as the rest of the organism, being called forth by signals from the normally occurring external environment, as well as the internal environment. While this fact is not well known in the social and behavioral sciences, it is surprising to find that it is also not widely appreciated in biology proper (Strohman 1997). In biology, the external environment is seen as the agent of natural selection in promoting evolution, not as a crucial feature of individual development (van der Weele 1995). Many biologists subscribe to the notion that "the genes are safely sequestered inside the nucleus of the cell and out of reach of ordinary environmental effects" (Wills 1989: 19).

Normally Occurring Environmental Influences on Gene Activity

As can be seen in table 4.1, a number of different naturally occurring environmental signals can stimulate gene expression in a large variety of organisms from nematodes to humans. To understand the findings summarized in table 4.1, the nongeneticist will need to recall that the sequence of amino acids in proteins is determined by the sequence of nucleotides in the gene that "codes" for it, operating through the intermediary of mRNA. So there are three levels of evidence of genetic activity in table 4.1: protein expression or synthesis, mRNA activity, and genetic activity itself. As noted in table 4.1, there are important neural and behavioral correlations to genetic activity, even though the activity of the genes is quite remote from these effects. The posttranslational expression of genes beyond the initial synthesis of protein involves the intervention of many factors before the end product of gene activity is realized.

The fact that normally occurring environmental events stimulate gene activity during the usual course of development in a variety of organisms means that genes and genetic activity are part of the developmental-physiological system and do not stand outside of that system.

From the Central Dogma of Molecular Biology to Probabilistic Epigenesis

Once again, the main purpose here is to place genes and genetic activity firmly within a holistic developmental-physiological framework, one in which genes not only affect each other and mRNA but are affected by activities at other levels of the system, up to and including the external

environment. This holistic developmental system of bidirectional, coactional influences is captured schematically in figure 4.2. In contrast to the unidirectional and encapsulated genetic predeterminism of the central dogma, a probabilistic view of epigenesis holds that the sequence and outcomes of development are probabilistically determined by the critical operation of various endogenous and exogenous stimulative events.

The probabilistic-epigenetic framework presented in figure 4.2 is based not only on what we now know about mechanisms of individual development at all levels of analysis, but also derives from our understanding of evolution and natural selection. As everyone knows, natural selection serves as a filter and preserves reproductively successful phenotypes. These successful phenotypes are a product of individual development, and thus are a consequence of the adaptability of the organism to its developmental conditions. Therefore, natural selection has preserved (favored) organisms that are adaptably responsive to their developmental conditions, both behaviorally and physiologically. Organisms with the same genes can develop very different phenotypes under different ontogenetic conditions.

Because the probabilistic-epigenetic view presented in figure 4.2 does not portray enough detail at the level of genetic activity, it is useful to flesh that out in comparison to the previously described central dogma of molecular biology. As shown in figure 4.3, the original central dogma explicitly posited one-way traffic from DNA→RNA→Protein, and was silent about any other flows of "information" (Crick 1958). In the bottom of figure 4.3, probabilistic epigenesis, being inherently bidirectional in the horizontal and vertical levels (figure 4.2) has information flowing not only from RNA→DNA but between Protein↔Protein and DNA→DNA. The only relationship that is not yet supported is Protein→RNA, in the sense of reverse translation (Protein altering the structure of RNA), but there are other influences of Protein on RNA *activity* (not its structure) which would support such a direc-

Table 4.1
Normally Occurring Environmental and Behavioral Influences on Gene Activity

Species	Environmental Signal or Stimulus	Result (alteration in)	Author(s)
Nematodes	absence or presence of food	neuronal *daf-7* gene mRNA expression, inhibiting or provoking larval development	Ren et al. 1996
Fruit flies	heat stress during larval development	transient elevated heat shock proteins and thermotolerance	Singh & Lakhotia 1988
Fruit flies	light-dark cycle	PER and TIM protein expression and circadian rhythms	Lee et al. 1996; Myers et al. 1996
Various reptiles	Incubation temperature	sex determination	Reviewed in Bull 1983; van der Weele 1995
Songbirds (canaries, zebra finches)	conspecific song	forebrain mRNA	Mello, Vicario, & Clayton 1992
Hamsters	light-dark cycle	pituitary hormone mRNA and reproductive behavior	Hegarty et al. 1990
Mice	acoustic stimulation	*c-fos* expression, neuronal activity, tonotopy in auditory system	Ehret & Fisher 1991
Mice	light-dark cycle	*c-fos* mRNA expression in suprachiasmatic nucleus of hypothalamus, circadian locomotor activity	Smeyne et al. 1992
Rats	tactile stimulation	*c-fos* expression and number of somatosensory cortical neurons	Mack & Mack 1992
Rats	learning task involving vestibular system	nuclear RNA base ratios in vestibular nerve cells	Hydén & Egyházi 1962
Rats	visual stimulation	RNA and protein synthesis in visual cortex	Rose 1967
Rats	Environmental complexity	brain RNA diversity	Uphouse & Bonner 1975; review in Rosenzweig & Bennett 1978
Rats	prenatal nutrition	cerebral DNA (cerebral cell number)	Zamenhof & van Marthens 1978
Rats	infantile handling; separation from mother	Hypothalamic mRNAs for corticotropin-releasing hormone throughout life	Meaney et al. 1996
Cats	visual stimulation	visual cortex RNA complexity (diversity)	Grouse et al. 1979
Humans	academic examinations taken by medical students (psychological stress)	interleukin 2 receptor mRNA (immune system response)	Glaser et al. 1990

Note: mRNA, messenger RNA; PER and TIM are proteins arising from *per* (period) and *tim* (timeless) gene activity. (From Gottlieb 1998. Copyright © 1998 by the American Psychological Association. Reprinted by permission.)

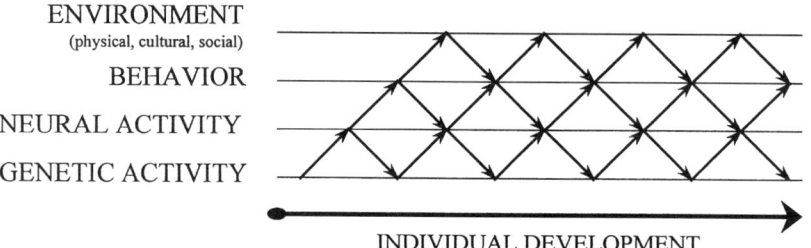

ENVIRONMENT
(physical, cultural, social)

BEHAVIOR

NEURAL ACTIVITY

GENETIC ACTIVITY

INDIVIDUAL DEVELOPMENT

Figure 4.2
Probabilistic-epigenetic framework: Depiction of the completely bidirectional and coactional nature of genetic, neural, behavioral, and environmental influences over the course of individual development. (From *Individual Development and Evolution: The Genesis of Novel Behavior* by Gilbert Gottlieb. Copyright © 1992 by Oxford University Press, Inc. Used by permission of Oxford University Press, Inc.)

tional flow. For example, a process known as *phosphorylation* can modify proteins such that they activate (or inactivate) other proteins (Protein→Protein) which, when activated, trigger rapid association of mRNA (Protein→RNA activity). When mRNAs are transcribed by DNA they do not necessarily become immediately active but require a further signal to do so. The consequences of phosphorylation could provide that signal (Protein→Protein→mRNA activity→ Protein). A process like this appears to be involved in the expression of "fragile X mental retardation protein" under normal conditions and proves disastrous to neural and psychological development when it does not occur.[1]

In summary, the central dogma lies behind the persistent trend in biology and psychology to view genes and environment as making identifiably separate contributions to the phenotypic outcomes of development. Quantitative behavior genetics is based on this erroneous assumption. Although genes no doubt play a constraining role in development, the actual limits of these constraints are quite wide and, most importantly, can not be specified in advance of experimental manipulation or accidents of nature. There is no doubt that development is constrained at all levels of the system (figure 4.2), not only by genes and environments.

Finally, one hopes that the emphasis here on normally occurring environmental influences on gene activity does not raise the spectre of a new, subtle form of "environmentalism." If I were to say organisms are often adaptably responsive to their environments, I do not think that would label me an environmentalist. So, by calling attention to genes being adaptably responsive to their internal and external environments, I am not being an environmentalist but merely including genetic activity within the probabilistic-epigenetic framework that characterizes the organism and all of its constituent parts.

In view of the findings reviewed here, in the future it would be most important to eschew both genetic determinism and environmental determinism, as we now should understand that it is truly correct (not merely a verbalism) to say that environments and genes necessarily cooperate in bringing about any outcome of individual development. (For exceptions see note 1.) The remainder of the book will no doubt enlarge on this theme many times over and in various contexts, both evolutionary and developmental. I appreciate being given the opportunity to recount my contribution to this still-maturing synthesis. In addition to the writings of the psychologists Zing-Yang Kuo, T. C. Schneirla, and Daniel Lehrman, the probabilistic-epigenetic view builds

Genetic Activity According To Central Dogma

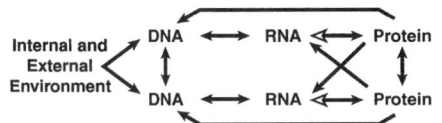

Genetic Activity According To Probabilistic Epigenesis

on the open-systems view of development championed by the biologists Ludwig von Bertalanffy (1962), Paul Weiss (1939), and Sewall Wright (1968), as I have described more fully elsewhere (e.g., Gottlieb 1992, 1997; Gottlieb, Wahlsten, and Lickliter 1998). In this brief overview I have not covered certain features of probabilistic epigenesis: nonobvious experiential contingencies in instinctive behavior, developmental causality, and the significance of coaction for individual development. These topics are covered in Gottlieb (in press).

Figure 4.3
Different views of influences on genetic activity in the central dogma and probabilistic epigenesis. The filled arrows indicate documented sources of influence, while the open arrow from Protein back to RNA remains a theoretical possibility in probabilistic epigenesis and is prohibited in the central dogma (as are Protein↔Protein influences). Protein→Protein influences occur when (1) prions transfer their abnormal conformation to other proteins and (2) when, during normal development, proteins activate or inactivate other proteins as in phosphorylation example described in text. The filled arrows from Protein to RNA represent the activation of mRNA by protein as a consequence of phosphorylation, for example. DNA↔DNA influences are termed "epistatic," referring to the modification of gene expression depending on the genetic background in which they are located. In the central dogma genetic activity is dictated solely by genes (DNA→DNA), whereas in probabilistic epigenesis internal and external environmental events activate genetic expression through proteins (Protein→DNA), hormonal, and other influences. To keep the diagram manageable, the fact that behavior and the external environment exert their effects on DNA through internal mediators (proteins, hormones, etc.) is not shown; nor is it shown that the protein products of some genes regulate the expression of other genes. (Further discussion in text.) (From Gottlieb 1998. Copyright © 1998 by the American Psychological Association. Reprinted with permission.)

Acknowledgment

The author's work is funded in part by NIMH grant P50-MH-52429.

Note

1. The label of "fragile X mental retardation protein" makes it sound as if there is a gene (or genes) that produces a protein that predisposes to mental retardation whereas, in actual fact, it is this protein that is *missing* (absent) in the brain of fragile X mental retardates, and thus represents a failure of gene (or mRNA) expression rather than a positive genetic contribution to mental retardation. The same is likely true for other "genetic" disorders, whether mental or physical: These most often represent biochemical *deficiencies* of one sort or another due to the lack of expression of the requisite genes and mRNAs to produce the appropriate proteins necessary for normal development. Thus, the search for "candidate genes" in psychiatric or other disorders is most often a search for genes that are not being expressed, not for genes that are being expressed and causing the disorder. So-called cystic fibrosis genes and manic-depression genes, among others, are in this category. The instances that I know of in which the presence of genes causes a problem are Edward's syndrome and trisomy 21 (Down syndrome), wherein the presence of an extra, otherwise normal, chromosome 18 and 21, respectively, causes problems because the genetic system is adapted for two, not three, chromosomes at each location. In some cases, it is of course possible that the expression of mutated genes can be involved in a dis-

order, but, in my opinion, it is most often the lack of expression of normal genes that is the culprit.

References

Anokhin, P. K. (1964). Systemogenesis as a general regulator of brain development. *Progress in Brain Research* 9: 54–86.

Black, J. E., and W. T. Greenough. (1998). Developmental approaches to the memory process. In J. L. Martinez and R. P. Kesner (Eds.), *Learning and Memory: A Biological View* (3rd ed.), pp. 55–88. New York: Academic Press.

Bull, J. J. (1983). *Evolution of Sex Determining Mechanisms*. Menlo Park, CA: Benjamin/Cummings.

Catalano, S. M., and C. J. Shatz. (1998). Activity-dependent cortical target selection by thalamic axons. *Science* 281: 559–562.

Coghill, G. E. (1929). *Anatomy and the Problem of Behaviour*. Cambridge: Cambridge University Press.

Crick, F. H. C. (1958). On protein synthesis. *Symposia of the Society for Experimental Biology: No. 12. The Biological Replication of Macromolecules*, pp. 138–163. Cambridge: Cambridge University Press.

Crick, F. (1970). Central dogma of molecular biology. *Nature* 227: 561–563.

Crick, F. (1982). DNA today. *Perspectives in Biology and Medicine* 25: 512–517.

Ehret, G., and R. Fisher. (1991). Neuronal activity and tonotopy in the auditory system visualized by *c-fos* gene expression. *Brain Research* 567: 350–354.

Glaser, R., S. Kennedy, W. P. Lafuse, R. H. Bonneau, C. Speicher, J. Hillhouse, and J. K. Kiecolt-Glaser. (1990). Psychological-stress-induced modulation of interleukin 2 receptor gene expression and interleukin 2 production in peripheral blood leukocytes. *Archives of General Psychiatry* 47: 707–712.

Gottlieb, G. (1961). Developmental age as a baseline for the determination of the critical period in imprinting. *Journal of Comparative and Physiological Psychology* 54: 422–427.

Gottlieb, G. (1965). Imprinting in relation to parental and species identification by avian neonates. *Journal of Comparative and Physiological Psychology* 59: 345–356.

Gottlieb, G. (1966). Species identification by avian neonates: Contributory effect of perinatal auditory stimulation. *Animal Behaviour* 14: 282–290.

Gottlieb, G. (1970). Conceptions of prenatal behavior. In L. R. Aronson, E. Tobach, D. S. Lehrman, and J. S. Rosenblatt (Eds.), *Development and Evolution of Behavior: Essays in Memory of T. C. Schneirla*, pp. 111–137. San Francisco: W. H. Freeman.

Gottlieb, G. (1971a). Ontogenesis of sensory function in birds and mammals. In E. Tobach, L. R. Aronson, and E. Shaw (Eds.), *The Biopsychology of Development*, pp. 111–137. New York: Academic Press.

Gottlieb, G. (1971b). *Development of Species Identification in Birds: An Inquiry into the Prenatal Determinants of Perception*. Chicago: University of Chicago Press.

Gottlieb, G. (1976). Conceptions of prenatal development: Behavioral embryology. *Psychological Review* 83: 215–234.

Gottlieb, G. (1983). The psychobiological approach to developmental issues. In M. M. Haith and J. J. Campos (Eds.), *Handbook of Child Psychology: Infancy and Developmental Psychobiology* (vol. 2, 4th ed.), pp. 1–26. New York: Wiley.

Gottlieb, G. (1991). Experiential canalization of behavioral development: Theory. *Developmental Psychology* 27: 4–13.

Gottlieb, G. (1992). *Individual Development and Evolution: The Genesis of Novel Behavior*. New York: Oxford University Press.

Gottlieb, G. (1997). *Synthesizing Nature-Nurture: Prenatal Roots of Instinctive Behavior*. Mahwah, NJ: Erlbaum.

Gottlieb, G. (1998). Normally occurring environmental and behavioral influences on gene activity: From central dogma to probabilistic epigenesis. *Psychological Review* 105: 792–802.

Gottlieb, G. (in press). Probabilistic epigenesis of development. In J. Valsiner and K. Connolly (Eds.), *Handbook of Developmental Psychology*. London: Sage.

Gottlieb, G., and P. H. Klopfer. (1962). The relation of developmental age to auditory and visual imprinting. *Journal of Comparative and Physiological Psychology* 55: 821–826.

Gottlieb, G., and J. G. Vandenbergh. (1968). Ontogeny of vocalization in duck and chick embryos. *Journal of Experimental Zoology* 168: 307–325.

Gottlieb, G., D. Wahlsten, and R. Lickliter. (1998). The significance of biology for human development: A developmental psychobiological systems perspective. In

R. M. Lerner (Ed.), *Handbook of Child Psychology: Theoretical Models of Human Development* vol. 1, pp. 233–273. (5th ed.) New York: Wiley.

Grouse, L. D., B. K. Schrier, C. H. Letendre, and P. G. Nelson. (1980). RNA sequence complexity in central nervous system development and plasticity. *Current Topics in Developmental Biology* 16: 381–397.

Hegarty, C. M., J. A. Jonassent, and E. L. Bittman. (1990). Pituitary hormone gene expression in male golden hamsters: Interactions between photoperiod and testosterone. *Journal of Neuroendocrinology* 2: 567–573.

Hydén, H., and E. Egyházi. (1962). Nuclear RNA changes of nerve cells during a learning experiment in rats. *Proceedings of the National Academy of Sciences, USA* 48: 1366–1373.

Kempermann, G., H. G. Kuhn, and F. H. Gage. (1997). More hippocampal neurons in adult mice living in an enriched environment. *Nature* 386: 493–495.

Kuo, Z-Y. (1967). *The Dynamics of Behavior Development*. New York: Random House.

Lee, C., V. Parikh, T. Itsukaichi, K. Bae, and I. Edery. (1996). Resetting the *Drosophila* clock by photic regulation of PER and a PER-TIM complex. *Science* 271: 1740–1744.

Lehrman, D. S. (1953). A critique of Konrad Lorenz's theory of instinctive behavior. *Quarterly Review of Biology* 28: 337–363.

Mack, K. J., and P. A. Mack. (1992). Induction of transcription factors in somatosensory cortex after tactile stimulation. *Molecular Brain Research* 12: 141–147.

Meaney, M. J., J. Diorio, D. Francis, J. Widdowson, P. LaPlante, C. Caldji, S. Sharma, J. P. Seckl, and P. M. Plotsky. (1996). Early environmental regulation of forebrain glucocorticoid receptor gene expression: Implications for adrenocortical responses to stress. *Developmental Neuroscience* 18: 49–72.

Mello, C. V., D. S. Vicario, and D. F. Clayton. (1992). Song presentation induces gene expression in the songbird forebrain. *Proceedings of the National Academy of Sciences, USA* 89: 6818–6822.

Morgan, C. L. (1893–94). Dr. Weismann on heredity and progress. *The Monist* 4: 30.

Myers, M. P., K. Wager-Smith, A. Rothenfluh-Hilfiker, and M. W. Young. (1996). Light-induced degradation of TIMELESS and entrainment of the *Drosophila* circadian clock. *Science* 271: 1736–1740.

Ren, P., C-S. Lin, R. Johnson, P. S. Albert, D. Pilgrim, and D. L. Riddle. (1996). Control of *C. elegans* larval development by neuronal expression of a TGFß homolog. *Science* 274: 1389–1391.

Rose, S. P. R. (1967). Changes in visual cortex on first exposure of rats to light: Effect on incorporation of tritiated lysine into protein. *Nature* 215: 253–255.

Rosenweig, M. R., and E. L. Bennett. (1978). Experiential influences on brain anatomy and brain chemistry in rodents. In G. Gottlieb (Ed.), *Early Influences*, pp. 289–327. New York: Academic Press.

Schneirla, T. C. (1956). Interrelationships of the "innate" and the "acquired" in instinctive behavior. In P-P. Grassé (Ed.), *L'Instinct dans le comportement des animaux et de l'homme*, pp. 387–452. Paris: Masson.

Schneirla, T. C. (1965). Aspects of stimulation and organization in approach/withdrawal. Processes underlying vertebrate behavioral development. *Advances in the Study of Behavior* 1: 1–74.

Schneirla, T. C. (1966). Behavioral development and comparative psychology. *Quarterly Review of Psychology* 41: 283–302.

Singh, A. K., and S. C. Lakhotia. (1988). Effect of low-temperature rearing on heat shock protein synthesis and heat sensitivity in *Drosophilia melanogaster*. *Developmental Genetics* 9: 193–201.

Smeyne, R. J., K. Schilling, L. Robertson, D. Luk, J. Oberdick, T. Curran, and J. I. Morgan. (1992). FoslacZ transgenic mice: Mapping sites of gene induction in the central nervous system. *Neuron* 8: 13–23.

Strohman, R. C. (1997). The coming Kuhnian revolution in biology. *Nature Biotechnology* 15: 194–200.

Uphouse, L. L., and J. Bonner. (1975). Preliminary evidence for the effects of environmental complexity on hybridization of rat brain RNA to rat unique DNA. *Developmental Psychobiology* 8: 171–178.

van der Weele, C. (1995). *Images of Development: Environmental Causes in Ontogeny*. Doctoral dissertation, Vrije University.

von Bertalanffy, L. (1962). *Modern Theories of Development: An Introduction to Theoretical Biology*. New York: Harper.

Weiss, P. (1939). *Principles of Development: A Text in Experimental Embryology*. New York: Holt, Rinehart and Winston.

West, M., and A. King. (1987). Settling nature and nurture into an ontogenetic niche. *Developmental Psychobiology* 20: 549–562.

Wills, C. (1989). *The Wisdom of the Genes: New Pathways in Evolution*. New York: Basic Books.

Wright, S. (1968). *Evolution and the Genetics of Population: Vol. 1. Genetic and Biometric Foundations*. Chicago: University of Chicago Press.

Zamenhof, S., and E. van Marthens. (1978). Nutritional influences on prenatal brain development. In G. Gottlieb (Ed.), *Early Influences*, pp. 149–186. New York: Academic Press.

Gene, Organism and Environment: A New Introduction

Richard C. Lewontin

The essay "Gene, Organism and Environment" takes up two fundamental metaphors that inform theoretical, experimental, and natural historical practice in biology. These are the metaphors of development, which carries the implication of an unfolding or unrolling of an internal program that determines the organism's life history from its origin as a fertilized zygote to its death, and the metaphor of adaptation, which asserts that evolution consists in the shaping of species to fit the requirements of an autonomous external environment. That is, both in developmental and in evolutionary biology, the inside and the outside of organisms are regarded as separate spheres of causation with no mutual dependence. The burden of the essay is that these metaphors mislead the biologist because they fail to take account of the interactive processes that link the inside and the outside. The changes that occur in an organism during its life from conception to death depend uniquely on both the cell constituents that are present in the fertilized egg and on the sequence of environments through which the organism passes in its lifetime. But the nature of these environments is not independent of the organism because at every instant the life activities of the organism determine what constitutes the relevant combinations of external physical states and they simultaneously cause changes in those states. The question that remains unaddressed by the essay is, "So what?"

It is easy enough to criticize the metaphors that are used in science. None is perfect and all capture only some parts of reality while falsifying others. The "billiard ball" model of molecules in a gas calls attention to the elastic collisions and the preservation of momentum in the kinetic theory of gases, but real billiard balls do not undergo perfectly elastic collisions, and they have different colors and make a clicking sound when they strike each other. Yet these truths about billiard balls are familiar to every physicist, and no one is led astray by them. Indeed, the only perfect metaphor would be a complete description of the object of interest, in which case it is no longer metaphorical. We are constantly being urged to see the world of living phenomena in a different and better way, using new organizing principles: the organism as a computer program, the organism as a Boolean network, the organism as a self-organizing machine. But the question that must always be asked of such proposals is how the practice of biologists would be changed in a way that would allow us to answer previously unanswerable questions. What new experiment, what new testable prediction, what explanation of a previously mysterious or contradictory observation will flow from a change in point of view? What will the dialectical point of view about organism and environment contribute to the doing of biology? The biological philosopher only interprets the world; the point, however, is to change it.

When we consider how the understanding of genes and environment in "development" (we have no other word for it) has been effective in biology, we see a curiously split history. In applied biology, especially in plant and animal breeding, the understanding that there is a unique interaction between genotype and environment in development has been of fundamental practical importance for nearly one hundred years. The standard method of breeding crops for, say, increase in yield, is to grow the various varieties under test in several years and in several locations in the region of production. The variety chosen for release is not necessarily the one with highest average yield over environments because uniformity of result over years and locations is given strong weight. A seed company that sells a variety with slightly higher average yield, but that causes palpable loss to a quarter of the farmers, will soon go out of business. The result of the breeding practice that takes account of genotype-environment interactions has been the evolution in crop plants of less and less environmental sensitivity, but a significant amount always remains and is taken account of in agricultural genetics. In contrast, nearly every developmental geneticist working on morphogenesis in laboratory model organisms ignores completely the effect of the interaction of genes and environment. One source of the difference in practice is the difference in problematic. For the developmental geneticist using *Drosophila* or *Caenorhabditis*, the question being asked is how the transcription from one gene by the cell influences the transcription of other genes. But this leads to a second difference, which is in the nature of the genetic variation that is used. To study the signaling pathways in the transcription of genes, mutants of major qualitative effect are employed, and there is a strong bias toward gene loci and their mutations whose effects can be seen reliably. The environment enters in only as a set of laboratory conditions necessary for the survival and maximal expression of the abnormal mutations. Every theoretical and experimental agricultural geneticist, on the other hand, is concerned with continuously and subtly varying physiological and morphological traits in which the interaction with the environment is critical. The *Drosophila* geneticist is at leisure to choose the best genotypes and phenotypes for the purposes of a "knockout" experiment, in which the transcription of a gene is prevented and the signaling pathway of which it is a part is blocked. The soybean geneticist is stuck with the trying to find multilocus genotypes that will increase seed yield by a couple of percent.

The standard program of developmental genetics is reinforced by the discovery that the gene signaling pathways that differentiate the front end from the back end of an animal, or that are active in the laying down of body segmentation, are distributed throughout the entire range of animal forms and may even have homologs in plants. Thus the developmental geneticist has the satisfaction of studying a system that has been preserved since the pre-Cambrian. But a great deal of evolution is precisely of the sort studied by

quantitative geneticists and seems not susceptible to knockout gene experiments. If developmental genetics is to be really relevant to understanding the production of form, it cannot constrain itself to the universalities of form, but must also explain the variation, and in particular why an increase in temperature causes an increase in size for one genotype, while it causes a decrease in size for another genotype and is nonmonotonic in its effect on a third. Presumably the way in which this is to be done is, at first, to continue the study of the same signaling pathways, but to create experiments that will cause small perturbations, both genetic and environmental, to those pathways and then to analyze at a molecular level how those perturbations result in small changes in morphogenesis. But such experiments have a greater importance than the explanation of the observed genotype/environment interactions in development. The ultimate purpose of developmental genetics must be to provide an articulated story of the coming into being of particular macromolecular structures at particular places at particular moments in the life of an organism. The existence of nonlinearities and nonmonotonicities in genotype/environment interactions tells us that the correct story cannot be given purely as a diagram of gene-gene signaling. A crucial material mechanism is missing for the establishment of morphogenetic gradients and the localization of qualitatively different macromolecules when only internal autonomous processes are considered. The realization of the program of explanation of developmental biology cannot be accomplished without including the influence of the external.

The effect of taking the dialectic between organism and environment seriously is, if anything, even more profound for evolutionary biology than for studies of development. If it is true that the life activities of an organism enter into the specification of its environment, then different phenotypes will make different environments. Because evolution occurs by a change in the frequencies of heritable types in a population, it follows that during the evolution of a change in the frequencies of heritable types in a population the environments experienced by the population are changing. But this, in turn, means that selective forces, which are a consequence of relations between organisms and their environments, are also changing in a way that is sensitive to the change in population composition. The result is that, realistically, selection coefficients are frequency-dependent, and theoretical modeling of the effect of natural selection must use frequency-dependent formulations. In the history of the development of population genetic theory some small attention has been paid to frequency-dependent models, but for the most part they are regarded as special cases, to be considered in textbooks and monographs under the rubric of "Some Complexities." Nearly every model of natural selection assumes constant fitnesses or, at the most, fitnesses that vary with some autonomous extrinsic force or stochastically. The conclusions from frequency-independent models of selection are, in general, invalid for frequency dependent cases. For example, natural selection does not maximize average fitness and may even minimize it. The first consequence of taking a constructionist view seriously would be the abandonment of frequency independent selection models and their replacement by more realistic ones. There is, however, an immense cost to be paid. Once we admit frequency dependence into our modeling, there are no purely theoretical constraints on what may be predicted. Every frequency dependent model makes its own peculiar predictions. Every biological situation will need to be specifically modeled, and small errors in the claims about biology may lead to radical errors in evolutionary prediction. Moreover, the theoretician becomes inextricably tied to the experimentalist and natural historian, because without the biological details there is no applicable model. It is not clear that theoretical population geneticists and evolutionists are prepared to give up the disciplinary autonomy that they currently possess, an autonomy that licenses them to engage in model-making with no constraints beyond the elements of genetics.

The program of experimental and observational evolutionary genetics does not escape the implications of constructionism. The form of the classical experiment on the determination of the fitnesses of different genotypes is to measure some components of the age specific mortality and fecundity schedules of each genotype in isolation in a physical environment or environments determined by the investigator. The resultant fitness estimates are then used to explain what has happened or to predict what will happen during the evolution of the population. But if the fitness of a genotype is dependent upon the mixture of other genotypes in the population, then nothing has been learned by measuring the "fitnesses" of the genotypes in isolation. On the other hand, it is hopeless to measure the net fitnesses of many genotypes in an immense array of different frequency combinations. A realistic evolutionary genetics that takes context dependence of fitness seriously cannot continue to operate at the purely superficial level of directly measuring fitnesses, of characterizing outcomes without understanding their mediation. What is required is an experimental program of unpacking "fitness." This involves determining experimentally how different genotypes juxtapose different aspects of the external world, how they alter that world and how those different environments that they construct affect their own biological processes and the biological processes of others. The understanding of living systems cannot be achieved merely through a description of their details, but neither can that understanding be obtained by ignoring them. The truth may not be in the details, but the details matter and we must have them in hand for our program of explanation. Ironically, the demand for a dialectical view of the evolution of organisms and their environment is, at first pass, a demand that evolutionary biologists take Descartes's original machine metaphor seriously.

6 Gene, Organism and Environment

Richard C. Lewontin

The modern theory of evolution is, as is so often said, a fusion of the two great insights of nineteenth-century biology: Darwin's realization that the variation among species arises from the conversion of variation between individuals within species, and Mendel's discovery of the segregation of discrete factors as the basis for the inheritance of differences between individuals. We are constantly reminding ourselves and others that the immense progress made in biology in the present century rests firmly on these two major discoveries of a previous time. What is not always appreciated, however, is that the legacies of Darwin and Mendel are also responsible for certain difficulties in biology, difficulties that prevent us from some kinds of further progress and which keep us locked into a rigid framework of thought about the development and evolution of organisms. These difficulties arise, ironically, from the very source of Mendel and Darwin's success as biologists, their separation of internal from external forces acting on organisms. For Mendel the internal "factors" were the causes of the form of the organism and were, in the end, the proper objects of study. The "factors," what we now call "genes," were the subjects and the organisms the objects of developmental forces. From this view of gene as the cause of organism has flowed the entire corpus of modern mechanical and molecular genetics. For Darwin, the external world, the environment, acting on the organism was the cause of the form of organisms. The environment, the external world with its autonomous properties, was the subject and the organism was, again, the object acted upon. In Darwinism, the organism is the interaction of two causal sequences, autonomous in their dynamics. Internal forces produced the variation among organisms, and autonomous external forces molded the species on the basis of these autonomous internally caused variations. The essence of Darwin's account of evolution was the separation of causes of *ontogenetic* variation, as

coming from internal factors, and causes of *phylogenetic* variation, as being imposed from the external environment by way of natural selection. (I gloss over Darwin's later flirtation with Lamarckism, since the proposition that the environment specifically engenders heritable adaptive variation is at total variance with the Darwinian mechanism of evolution.) It is from this view of environment as the cause of organisms that the entire corpus of modern evolutionary biology arises.

We cannot appreciate fully the nature of the change in biology wrought by Mendel and Darwin unless we understand the historical importance of the objectification of the organism. Descartes' metaphor of the organism as machine had virtually no impact on biology for two hundred years. So, for example, even Harvey's mechanical description of the circulation of blood was not really accepted until the beginning of the nineteenth century, and Caspar Friedrich Wolff's epigenetic theory and his remarkably modern distinction between genotype and phenotype had no effect in embryology until the *Entwicklungsmechanik* of the latter part of the last century. Biology lacked clear notions of separable causes and effects and, more important, a systematic commitment to the analysis of biological systems along mechanistic lines. Lamarck's view that information from the external world could become permanently incorporated in organisms and their progeny through the mediation of a living being's needs and will was quintessentially representative of pre-Darwinian biology. As Charles Gillespie (1959) so cogently puts it: "Lamarck's theory of evolution belongs to the contracting and self-defeating history of subjective science, and Darwin's to the expanding and conquering history of objective science."

By making organisms the *objects* of forces whose *subjects* were the internal heritable factors and the external environment, by seeing organisms as the *effects* whose *causes* were internal and external autonomous agents, Mendel and Darwin brought biology at last into conformity with the epistemological meta-structure that already characterized physics since Newton and chemistry since Lavoisier. This change in world view was absolutely essential if biology was to progress by making contact with physical science and by becoming quantitative and predictive. The mechanistic reductionism and the clear separation of internal and external were as necessary in the nineteenth century for the creation of a scientific biology as Newton's ideal bodies and perfect determinism were for the physics of the seventeenth. But we must not confuse the historically determined necessity of a particular epistemological stance at one stage in the development of a science with a perfect model that will guarantee all future progress. On the contrary, the very progress made possible by certain revolutionary formulations may lead eventually to results that are in contradiction with those earlier formulations and which can be resolved only by their reexamination. Yet those reexaminations are themselves rooted in the past formulations. Newtonian mechanics and classical optics were in serious contradiction with the newer observations of physics at the end of the nineteenth century, but anyone familiar with the development of the Special Theory of Relativity immediately sees how that theory, in one aspect a negation of Newtonian principles, is built entirely on a Newtonian framework and could never have been developed in the absence of classical physics.

As in physics, so too in biology. As time has passed, Mendel's view of organisms as the manifestation of autonomous internal "factors" with their own laws, and Darwin's view of organisms as passive objects molded by the external force of natural selection, have become increasingly in contradiction with the known facts of developmental and population biology. But the situation in biology has developed rather differently than in physics. All of physical science seemed blocked by the repeated conflict between classical formu-

lations and the new observations of radioactivity, of light, and of astronomy. Without relativity and quantum theory, physics and theoretical chemistry would have ground to a halt. On the other hand, biology, far more diverse in its subject matter, far more loosely tied together into a coherent science, has undergone a radically uneven development. Some branches such as molecular biology have made extraordinary progress by concentrating on just those questions for which the simple mechanical reductionism of the nineteenth century is the perfect epistemology. Developmental biology, the study of cognition and memory, and evolutionary biology, on the other hand, have profited only marginally from these rapid advances. Rather, they are stalled by their attempt to use outdated concepts to confront a rich phenomenology to which these concepts clearly do not apply. Evolutionary biology suffers particularly because it is the nexus of all other biological sciences, so that a lack of progress in developmental biology, in ecology, in behavioral science, all are fatal to a proper understanding of evolution.

Specifically, evolutionary biology must confront two issues about the forms of organisms. One is the ontogenetic process by which the sequence of forms that comprise an individual's life history come into being. The second is the phylogenetic process by which species as collective entities form and change based on the variations among the individuals that make them up. Classical, post-Darwinian, post-Mendelian biology has settled on two metaphors through which the processes are seen. The first, ontogenetic, process is seen as an *unfolding* of a form, already latent in the genes, requiring only an original triggering at fertilization and an environment adequate to allow "normal" development to continue. The second, phylogenetic, process is seen as *problem* and *solution*. The environment "poses the problem"; the organisms posit "solutions," of which the best is finally "chosen." The organism proposes; the environment disposes. These two metaphors are simply the forms of the original

Mendelian view that internal factors make the individual organism and the Darwinian view that external forces determine the collectivity. In the balance of this chapter, I want to make clear why these two metaphors are wrong. Individual development is not an unfolding, and evolution is not a series of solutions to present problems. Rather, genes, organisms, and environments are in reciprocal interaction with each other in such a way that each is both cause and effect in a quite complex, although perfectly analyzable, way. The known facts of development and of natural history make it patently clear that genes do not determine individuals nor do environments determine species.

Gene, Environment, and Organism in Development

I will begin with the obvious. It is well known that Mendel solved the basic problem of the laws of inheritance by investigating the heredity of very special sorts of differences, those in which there was a determinate correspondence between genotype and phenotype. Given the genotype, the phenotype corresponding to it was unambiguously defined, at least under the condition of Mendel's experimental garden. Indeed, it is the essence of the Mendelian methodology that one can read off the genotype given either the individual's phenotype or, in the case of complete dominance, the phenotypes of the progeny of a single test cross. It is not always explicitly pointed out that all of modern biochemical, molecular, and developmental genetics also has depended for progress on finding genetic differences that make clear-cut, non-overlapping phenotypic classes under easily controlled conditions. It is amusing to contemplate where bacteriophage genetics (and, a fortiori, all of molecular genetics) would be now if Benzer's bacteriophage mutants had only differed from each other by 1% in the number of progeny phage they produced on restrictive hosts. One consequence of this methodological history is that textbooks of

genetics (and of evolution) consistently describe organisms as "determined" by their genes. It is a short step to seeing organisms as "lumbering robots" created by their genes "body and mind." It is not hard to see why such unbiological rubbish is taken seriously. Yet, the vast majority of morphological, behavioral, and physiological differences among individuals do not "Mendelize." It is simply not possible to read off the genotypic differences between tall and short individuals of *Rumex acetosella* from their individual phenotypes or from any number of controlled test crosses involving them or their relatives, not to speak of the genotypic difference between faithful spouses and philanderers. The reaction of many evolutionists (encouraged, again, by textbooks) to this obvious fact is to consign such characters to "polygenic control," invoking the multiple factor hypothesis, and thus saving the basic model that genes determine organisms. But quantitative variation of a character is not prima facie evidence that it is influenced by many genes. Single gene mutations affecting eye shape, wing variation and bristle number in *Drosophila*, or enzyme activity in humans, all show quantitative variation in phenotype and, unless care is taken to control the condition of development, considerable overlap between genotypes in their phenotypic distributions. The fundamental general fact of phenogenetics is that the phenotype of organisms is a consequence of non-trivial interaction between genotype and environment during development. All that genes ever do is to specify a norm of reaction over environments. Moreover, fitness too is a phenotype and varies from environment to environment, both because other aspects of the phenotype develop differently in different environments, and because a given shape or behaviour or physiology will confer different fitnesses in different environments. During the fifteen years between 1950 and 1965, population geneticists, largely under the influence of plant and animal breeders and of Schmalhausen's seminal work, *Factors of Evolution* (1949), devoted considerable attention to the environmental contingency of phenotype and to the effects of selection in different environmental regions (see, for example, Lerner 1954, Dobzhansky and Spassky 1944, Falconer 1960, and Robertson 1960), but this work passed out of fashion with the advent of molecular population genetics. Indeed, except for the famous work of Clausen, Keck and Hiesey (1958), no study of the norms of reaction of naturally occurring heterozygous genotypes appeared until the experiments of Gupta on *Drosophila* (Gupta and Lewontin 1982). Thus, the basic data for judging the effects of selection in particular genotypes are simply lacking. The little that is known shows clearly that the developmental responses of different genotypes to varying environments are non-linear and do not allow the simple ordering of genotypes along a one-dimensional scale of phenotype. Norms of reaction cross each other so that no genotype gives a phenotype unconditionally larger, smaller, faster, slower, more or less different than another. These well-known facts seem, however, to have made no impact on evolutionary theorists who continue to speak about selection for a character and about genes that are selected because they produce that character.

A second, less well-known, feature of development is that phenotype is not given even when the genotype and the environment are completely specified. There is a significant effect of "developmental noise" (Waddington 1957) in producing phenotype. The two sides of a *Drosophila* have the same genotype, and no reasonable definition of environment will allow that the left and right sides of a pupa developing halfway up the side of a glass milk bottle in the laboratory are in different environments. Yet, the number of sternopleural bristles and the number of eye facets differ between the two sides of an individual fly. Small events at the level of thermal noise acting during cell division and differentiation have large effects on the final developmental outcome.

It is only the beginning of understanding to agree that internal and external factors contribute to phenotype. The reaction to that knowledge is usually to attempt to assign the relative weights to genotype, environment, and developmental noise, as for example in a heritability study or some other form of the analysis of variance. The second fact that evolutionary biologists must cope with is that the influence of each factor depends upon the influence of the other factors so that no assignment of fixed weights to genetic, environmental, and noise components of variation is possible (Lewontin 1974). There are no more important experimental results for evolutionary biology than those of Rendel (1967) on canalization, yet the early impact of these findings on evolutionary theory seems now to be forgotten. What Rendel and others have shown is the *reciprocity* of effects of genetic state on environmental sensitivity and of environmental state on genetic sensitivity of the developing organism. Moreover, the genotype is itself hierarchically organized so that the environmental sensitivity of a given gene substitution will depend upon genes at other loci. The consequence is that selection can change the average expression of a trait, independently from the variation of that expression in response to variable environment and to developmental noise. Thus, by selection, an organism can be made developmentally insensitive or highly sensitive to perturbations of its genotype, its environment, developmental accidents, or any combination of these. Internal and external facts not only play a role in development, but each determines the role played by the other.

Thus far, my description of development is still in terms of the organism as the object, the effect of causes both internal and external to it, although the causes themselves interact with each other. The final step in the integration of developmental biology into evolution is to incorporate the organism as itself a *cause* of its own development, as a mediating mechanism by which external and internal factors influence its future. To describe phenotype as the consequence of gene, environment, and accident leaves out of account entirely the element of *temporal order* which is of the essence in a developmental process. The organism's phenotype is in a continual state of change from fertilization to death. The phenotype at any instant is not simply the consequence of its genotype and current environment, but also of its phenotype at the previous instant. That is, development is a first order Markov process in which the next step depends upon the present state. The temporal order of environments is thus a critical, but not a sufficient, prediction of future development. If a *Drosophila* adult hatches from the pupal case as an unusually small fly, it does not matter what the contributing causes were to making it small, it now has an unusually high surface to volume ratio, which will play an important role over the rest of its life history. Small changes in ambient temperature or in its own activity will be felt by it in a way different from its larger sibs and will have different effects on its reproductive rate. The way in which its genotype and its future environmental sequence influence the fly are themselves *effects* of the organism as *cause*. The organism is not simply the object of developmental forces, but is the subject of these forces as well. Organisms as entities are one of the causes of their own development.

Organism and Environment in Evolution

Our usual description of evolution by natural selection is framed in terms of the process of *adaptation*. A species' environment exists and changes as a consequence of some autonomous forces outside the species itself. This outside world poses problems for the species, problems of acquiring space, consumables, light, and individuals of the opposite sex. Those most successful in solving the problems, because, by chance, their morphologies, physiologies, and behaviors make them mechanically the best fit to do so, leave the most offspring and thus the species adapts. This view of evolution, however, has certain paradoxical features (Lewontin 1978). One is that all extant species are said to be already adapted to their environments. A good deal of evolutionary biology is taken up with demonstrating that their features represent optimal solutions to environ-

mental problems. What then is the motive power of further evolution? The solution proposed by Van Valen (1973) is that the environment is constantly moving and that species are simply running to keep up. In that case, it is the autonomous forces of environmental change that govern the rate of evolution, and we would be well advised to study the laws of environmental rather than organismic change if we want to understand what has been happening. More paradoxical is the necessity of defining environments without organisms. To make the metaphor of adaptation work, environments or ecological niches must exist before the organisms that fill them. There must be a preferred, denumerable set of combinations of factors that make "environments" and a non-denumerable infinity of combinations of factors that are not. The history of life is then the history of the coming into being of new forms that fit more and more closely into these preexistent niches. But what laws of the physical universe can be used to pick out the possible environments waiting to be filled? In fact, we only recognize an "environment" when we see the organism whose environment it is. Yet so long as we persist in thinking of evolution as adaptation, we are trapped into an insistence on the autonomous existence of environments independent of living creatures. In fact, we should be able to list the environments on Mars, being careful all the while not to be influenced by our knowledge of earthly life! If, on the other hand, we abandon the metaphor of adaptation, how can we explain what seems the patent "fit" of organisms and their external worlds? Fish have fins, and not only fish but whales, seals, penguins and even sea snakes have some sort of flattened body part that the animal uses for swimming. Moles have long claws as do anteaters; birds and bats have wings, and so on. The marvellous fit of organisms to their environments seems obvious. What is left, then, but some concept of a progressive fitting of organisms to predetermined adaptive peaks?

What is left out of this adaptive description of organism and environment is the fact, clear to all natural historians, that the environments of organisms are made by the organisms themselves as a consequence of their own life activities. How do I know that stones are part of the environment of thrushes? Because thrushes break snails on them. Those same stones are not part of the environment of juncos who will pass by them in their search for dry grass with which to make their nests. Organisms do not adapt to their environments; they construct them out of the bits and pieces of the external world. This construction process has a number of features:

(1) Organisms determine what is relevant. While stones are part of a thrush's environment, tree bark is part of a woodpecker's, and the undersides of leaves part of a warbler's. It is the life activities of these birds that determine which parts of the world, physically accessible to all of them, are actually parts of their environments. Moreover, as organisms evolve, their environments, perforce, change. All animals are covered with a thin boundary layer of warm moist air as a consequence of their metabolism. Small ectoparasites may be completely immersed in that boundary layer which thus determines their environmental temperature and humidity. But if natural selection should increase the body size of these parasites, they may emerge through the layer into the stratosphere of a colder drier world. It is the genes of sea-lions that make the sea part of their environment and the genes of lions that make the savannah part of theirs, yet those genes are descended from a common carnivore ancestor.

(2) Organisms alter the external world as it becomes part of their environments. All organisms consume resources by taking up minerals, by eating. But they may also create the resources for their own consumption, as when ants make fungus farms, or trees spread out leaves to catch sunlight. They may create an environment more hospitable to their own species, as for example when beavers raise the water level of a pond, but, in contrast, white pine in New England creates a dense shade that prevents its own reseeding.

Ecological succession is precisely the history of self-destruction of species by alterations of their own environment.

(3) Organisms transduce the physical signals of the external world. Changes in external temperature are not perceived by my liver as thermal changes but as alterations in the concentrations of certain hormones and ions. The photon energy impinging on my retina and the vibrational energy at my ear drums when I see and hear a rattlesnake are immediately changed through the mediation of my central nervous system into changes in adrenalin concentration. But this change is in part a consequence of my biology, since another rattlesnake would presumably react rather differently.

(4) Organisms create a statistical pattern of environment different from the pattern in the external world. Organisms, by their life activities, can damp oscillations, for example in food supply by storage, or in temperature by changing their orientation or moving. They can, on the contrary, magnify differences by using small changes in abundance of food types as a cue for switching search images. They can also integrate and differentiate. Plants may flower only when a sufficient number of days above a certain temperature have been accumulated. *Cladocera* can change from asexual to sexual reproduction in response to a large alteration in oxygen supply, temperature or food supply, irrespective of the absolute level. Even the period of external oscillation can be modulated, as when cicades count a prime number of seasonal fluctuations. It might be objected that the notion of organisms constructing their environments leads to absurd results. After all, hares do not sit around constructing lynxes! But in the most important sense they do. First, the biological properties of lynxes are presumably in part a consequence of selection for catching prey of a certain size and speed, i.e. hares. Second, lynxes are not part of the environment of moose while they are of hares, because of biological differences between moose and hares. Then what about laws of physical nature? Or-

ganisms did not pass the law of gravity. Yet, whether gravitation is an aspect of the environment of an organism depends upon biological properties of the organism, for example size. Bacteria are "outside" gravity, but their very small size makes them subject to a very different physical phenomenon, Brownian motion of molecules, which larger organisms do not notice.

The metaphor of construction rather than adaptation leads to a different formulation of natural selection and evolution. On the adaptive view evolution can be represented formally as a pair of differential equations. The first, describing the change in organisms, O, as a function of organism and environment, E, $dO/dt = f(O, E)$, and a second law of the autonomous change of environment, $dE/dt = g(E)$. A constructionist view makes this into a pair of coupled differential equations in which organism and environment coevolve, each as a function of the other, $dO/dt = f(O, E)$, and $dE/dt = g(O, E)$.

The parallel in population genetics is between frequency-independent and frequency-dependent selection. In the first case, one postulates a set of adaptive peaks which may or may not change in time as a consequence of external forces, and a set of gene frequencies that change in response to the potential field represented by the adaptive peaks. In the second case, the location and existence of the peaks are themselves functions of the genetic composition of the evolving population. The model of the first process is climbing a mountain peak, of the second walking on a trampoline. While population geneticists usually model selection as a frequency-independent process, adding frequency dependence as an added complication of marginal interest, the actual situation is the reverse. Most selective processes are frequency-dependent, as for example are any processes in which fitness depends on relative position in an ordered series, or in which there is competition for resources in short supply with different relative success of different types.

A problem seems to be posed for the constructionist view by the phenomenon of convergence.

If environments do not preexist, how are we to explain the remarkable similarities that evolved independently in different groups? Not only do animals as unrelated as fish, mammals, and birds all have fin-like appendages when they are aquatic, but a whole marsupial fauna developed with "wolves," "moles," "mice," and "cats" (although no ungulates, marine forms, elephants, bats, or horses). The error is to suppose that because organisms construct their environments they can construct them arbitrarily in the manner of a science fiction writer constructing an imaginary world. The coupled equations of coevolution of organism and environment are not unconstrained. Some pathways through the organism-environment space are more probable than others, precisely because there are real physical relations in the external world that constrain change. The construction of an environment that includes living in a fluid medium of a certain density and viscosity places certain quite loose constraints on morphology. Either the organism will be sessile, as many are, and wait for food, or it will be vagile and chase it. If it is vagile it may propel itself by jet action like the squid or by umbrella sculling like the jelly-fish or by flattened appendages. But even in the last case, there are those who undulate up and down (whales), side to side (sharks), fly in the water (rays), beat tiny wings rapidly (sea horses), and a variety of other fin movements. Where there is strong convergence as in certain marsupial-placental pairs, this should be taken as evidence about the nature of constraints on development and physical relations, rather than as evidence for pre-existing niches.

The nagging problem of adaptation has always been what one imagines the progression of events to be during the process itself, since we always observe only the finished product. If a seal's flippers are an adaptation to water, at what stage in the evolutionary history of seals did swimming in water become the "problem" which seals "solved" by losing their legs? No one imagines that a whole group of terrestrial carnivores simply plunged into the water one day, experiencing a new major adaptive problem, and then proceeded to adapt to it by the usual route of natural selection for small increases in flipper-like morphology. Nor, alternatively, can we say that swimming has always been a major problem for carnivores. There is, in fact, no reason why we should not, *a priori*, reverse the entire scenario. Perhaps an early pinniped ancestor acquired slightly flipper-like appendages for an entirely different reason—genetic drift or some preadaptation. Partly aquatic life then became an opportunity rather than a problem. This is not a very satisfactory theory, at least as a typical one for evolution, but for precisely the same reason that the standard adaptive story is unacceptable. Both uncouple organism and environment in such a way that we must regard one as undergoing significant autonomous change and the other as responding.

Concentrating as we always have on the problem of how the organisms change under natural selection, we have neglected to ask seriously how the environment to which the organism is supposedly responding has come to be a problem in the first place. Presumably the non-aquatic carnivore ancestors of the Pinnipedia slowly incorporated the water as a more and more energetically significant aspect of their environment while their morphologies and physiologies changed to make that appropriation more energetically rewarding. Were a complete reconstruction possible, we would be unable to find the moment at which swimming was for the first time a "problem" to be "solved" by the animal.

Organisms, then, both make and are made by their environment in the course of phylogenetic change, just as organisms are both the causes and consequences of their own ontogenetic development. The alienation of internal and external causes from each other and of both from the organism, seen simply as passive result, does not stand up under even the most casual survey of our knowledge of development and natural history. It is a tribute to the power of long-held ideology that the study of evolution continues

to lean so heavily on an impoverished view of the relation between gene, environment, and organism. If the hundredth anniversary of the death of Darwin is not to mark the death of Darwinism, we need to struggle for its transfiguration.

References

Clausen, J., D. D. Keck, and W. W. Hiesey. (1958). Experimental studies on the nature of species, Vol. 3: Environment responses of climatic races of *Achillea. Carnegie Institution of Washington Publication* 581: 1–129.

Dobzhansky, T., and B. Spassky. (1944). Manifestation of genetic variants in *Drosophila pseudoobscura* in different environments. *Genetics* 29: 270–290.

Falconer, D. S. (1960). Selection of mice for growth on high and low planes of nutrition. *Genetical Research, Cambridge* 1: 91–113.

Gillespie, C. C. (1959). Lamarck and Darwin in the history of science. In B. Glass, O. Temkin, and W. L. Straus (Eds.), *Forerunners of Darwin*, pp. 265–291. Baltimore: Johns Hopkins University Press.

Gupta, A. P., and R. C. Lewontin. (1982). A study of reaction norms in natural populations of *D. pseudoobscura. Evolution* 36: 934–948.

Lerner, T. M. (1954). *Genetic Homeostasis.* Edinburgh: Oliver and Boyd.

Lewontin, R. C. (1974). The analysis of variance and the analysis of causes. *American Journal of Human Genetics* 26: 400–411.

Lewontin, R. C. (1978). Adaptation. *Scientific American* 239(9): 156–169.

Rendel, J. M. (1967). *Canalization and Gene Control.* London: Academic Press.

Robertson, F. W. (1960). The ecological genetics of growth in *Drosophila*. II. Selection for large body size on different diets. *Genetical Research, Cambridge* 1: 305–318.

Schmalhausen, I. I. (1949). *Factors of Evolution: The Theory of Stabilizing Selection.* Philadelphia: Blakeston.

Van Valen, L. (1973). A new evolutionary law. *Evolutionary Theory* 1: 1–30.

Waddington, C. H. (1957). *The Strategy of the Genes.* London: Allen and Unwin.

II RETHINKING HEREDITY

Let's Talk about Genes: The Process Molecular Gene Concept and Its Context

Eva M. Neumann-Held

The Problem with the "Gene"

It is almost common knowledge among biologists and philosophers of biology (e.g: Burian 1986; Carlson 1991; Falk 1984; Keller 2000; Kitcher 1982, 1992; Portin 1993) that the classical molecular gene concept is not sufficient anymore in the face of the complex interactive processes being reported by molecular biology. The classical molecular gene concept defined a gene as that segment of DNA which codes for a polypeptide. Thereby, this gene concept implied a structural and functional unity which allows for a one-to-one relationship between DNA and corresponding polypeptide. That seemed to justify the wording that the "information" for the polypeptide resides in this particular segment of DNA. Today, however, we learn from molecular biology on the one hand about numerous genomic organizational features such as genomic imprinting and "overlapping genes," and on the other hand about complicated gene expression features, such as alternative mRNA splicing and mRNA editing. Both types of features have led to the realization that the assumption of a kind of "static" one-to-one relationship between DNA and polypeptide was an oversimplification.

In the face of this situation, it might seem a contradiction that the talk of biologists and the general public contains as many references to genes as in the good old days when there was no reason to question the classical molecular gene concept. However, a closer look on the usage of the word *gene* in communities of biologists shows that gene is used to mean very different things. Falk (1984: 203), for example, observed: "A casual glance at the current genetic literature would be enough to reveal that although the term 'gene' is very much in use, it means different things for different people.... Today the gene is ... a unit, a segment that corresponds to a unit-function, as defined by the individual experimentalist's needs." Portin states (1993: 208) that it is quite "arguable that the old term gene ... is no longer useful, except as a handy and versatile expression, the meaning of which is determined by the context." Sterelny and Griffiths (1999: 144) identify at least two different usages of the term gene in a molecular genetics textbook. They comment: "This suggests that the *gene* does not really name a unit of molecular biology, but is shorthand for any of several different units.... *Gene* is used in molecular biology as a shifting tag rather than as a name for a specific molecular kind" (p. 145). They add: "Molecular biologists often seem to use *genes* to mean 'sequences of the sort(s) that are of interest in the process that I am working on'" (p. 133).

What do we do with this situation? Three positions can be taken. One is to relinquish the term *gene* altogether. Alternatively, one might allow for several different usages, thereby accepting the status quo. But there is also a third possibility, which is to strive to develop precise definitions for *gene*.

A look at the literature reveals that all three positions have their supporters. Kitcher (1992: 130), for example, writes: "Indeed, it is hard to see what would be lost by dropping talk of genes from molecular biology and simply discussing the properties of various interesting regions of nucleic acid." In contrast to this position, I think that more than a term would be lost by dropping *gene* from the biological talk. In addition, it would no longer be possible to name a particular relationship between DNA segments and the course of events that leads to the production of polypeptides. This point will be discussed in more detail below. Furthermore, it will be argued that for any particular context the important question is precisely, what makes particular regions of nucleic acid interesting? Neither a subsumption of "interesting regions" under the name gene nor a focus on these regions in themselves (promoter, coding regions, etc.) can solve this question. I therefore prefer a clarification of the relationship between DNA and polypeptide production to a policy of (supposedly) benign neglect.

The second approach to the multiple usages of gene might be called the liberal standpoint. It just admits that the concept of the gene is "fuzzy." Within this approach one can distinguish a number of lines of argumentation. The earlier quotations from Falk, Portin, and Sterelny and Griffiths, for example, share with each other the claim that the different meanings of the term *gene* do not pose a problem, because the actual object of reference could be deduced by the context. According to Sterelny and Griffiths (1999: 133) the "usage" of the term *gene* is "perfectly satisfactory" because it is grounded in a "rich background of shared assumptions." Rheinberger (2000) defends the vagueness of the term gene even more emphatically by stressing that the term gene proved historically to be of heuristic value, so that vagueness itself would not be grounds for its rejection. Similarly, Keller (2000) argues that it might be just its "disarray" which today may provide "conceptual value" to the term gene.

I agree that in certain research contexts the different usages of the term gene seem not to pose a problem. However, at least when these contexts are left and assumptions about the nature of genes cross disciplinary boundaries or reach the public sphere, different usages of the term gene can become a problem. One example can be found in the writings of Dawkins (1982, 1989), when so-called evolutionary genes are introduced as arbitrary DNA sequences, and then, in a next step, become confused with polypeptide producing molecular genes (for a critical analysis, see Griffiths and Neumann-Held 1999; Neumann-Held 1998). Although a certain amount of vagueness might have heuristic value, when straightforward ambiguities arise in a single discussion there is a clear need for more precision. However, greater precision is not only needed in popular and interdisciplinary contexts, but also in the context of molecular research.

To explain this point, let us take a brief look at the third position in regard to "fuzzy" (Fischer

1995, Rheinberger 2000) gene concepts. This position is represented by those, who, for different reasons, want to try to find a precise meaning for the term *gene*, which nevertheless incorporates the new findings of molecular biology. These efforts are in the realm of biological sciences as well as in philosophy of biology. The journal *Nature* recently published a suggestion on the gene concept by Epp (1997), and almost any textbook of molecular biology and genetics will include some efforts of trying to define the gene. In the realm of philosophy of biology the requirement for the precise introduction of scientific terms motivated Fogle (1990) to suggest a gene concept, whereas Schaffner's (1993) and Waters's (1994) efforts in this regard are designed to help them to defend the possibility that a Mendelian gene concept could be reduced to a molecular gene concept.[1]

In their efforts to define "genes," all of these approaches focus on the nucleic acid sequences that are involved in the synthesis of a polypeptide (note here the similarity to Kitcher's approach). Within these limits, then, the approaches suggest different conceptualizations by including different, functionally relevant parts of DNA in their definition or by allowing that different stages of mRNA processing can be called a gene, as long as they are involved in the course of events that lead to the synthesis of a polypeptide. One might call these approaches, therefore, "nucleic-acid centered gene concepts." However, although this means a certain lack of precision in terminology, I would prefer to call these suggestions "DNA-centered" gene concepts. If it should occur that a differentiation between "DNA" and "RNA centeredness" is required, for the goal of this article, I would point that out.

Although the suggestions in the framework of this third approach are quite sophisticated, one should keep in mind that all definitions of genes, which only focus on nucleic acid sequences actually focus on the question: "Which DNA sequences causally *participate* in the polypeptide synthesis?" Conceptualizing genes in such a way is particularly interesting when the research focus is on the question, whether or not variations in polypeptide sequences are caused by mutations in particular DNA segments. It should be stressed here that such a question certainly reflects a legitimate research interest, and I make this point not only to express some kind of tolerance. Rather, I intend to point out that any effort to define gene (or other scientific terms) should start with a clarification of the purposes, the contexts of research interests, for which a term or an empirical approach is designed. Thus, instead of asking for the causal influence of DNA sequences on polypeptide production, for example, one might also ask whether there are not other, context specific causal influences in addition to DNA. This leads to the further question of how certain DNA segments become involved in polypeptide production at all. The observation that polypeptide synthesis is time and tissue regulated makes particularly obvious the importance of understanding the regulation mechanisms of transcription and translation. Those who defend the need for precise gene definitions try to include the possibility of regulated expression by including in the definition those segments of DNA (and mRNA) which can be shown to be necessary for these regulation mechanisms. However, the (regulated) production of a polypeptide needs more than only DNA sequences. One might wonder, therefore, whether a definition of gene should not also include *more* than only DNA sequences, if the major interest is to understand the (regulated) production of polypeptides. It is exactly at this point and to address these research interests, that I want to suggest a new gene concept. The research contexts in which this gene concept will be of value are mostly, although not exclusively, in developmental genetics.

Genes for Development

Grounds for suggesting a new gene concept for the context of developmental genetics might be best given by looking in more detail at the shortcomings of recent efforts for defining the gene. The common feature of all these definitions is that a gene is defined as one or more pieces of DNA which code for a polypeptide (for examples, see Levin 1985: 685; Knippers 1997: 287, 283ff). Efforts to recognise the complex molecular mechanisms involved are found in Singer and Berg (1991: 440). After explaining their uneasiness with any gene definition, they nevertheless adopt the following definition of the gene "for the purposes of this book":

A combination of DNA segments that together constitute an expressable unit, expression leading to the formation of one or more specific functional gene products that may be either RNA molecules or polypeptides. The segments of a gene include (1) the transcribed region (the *transcription unit*), which encompasses the coding sequences, intervening sequences, any 5' or 3' trailer sequences that surround the ends of the coding sequences, and any regulatory segments included in the transcription unit, and (2) the regulatory sequences that flank the transcription unit and are required for specific expression.

This definition, like most others, concentrates only on DNA sequences, and only on those DNA sequences that are involved in the synthesis of a polypeptide chain. Although such a definition, as stated above, can be read as reflecting specific research interests in the role of DNA sequences, another reading cannot be excluded. For example, there is no mentioning of all other entities, which are not located on the DNA but still might be necessary to allow expression of DNA segments. Therefore, it is easy to read this definition in such a way that the causal arrow leading to the synthesis of a polypeptide has only one important starting point: "a combination of DNA sequences." This attribution of causal primacy would be appropriate in a system where all other environmental, non-DNA located parameters are kept constant, as could be obtained in, for example, in vitro experiments. However, it seems that this is not the (only) context that Singer and Berg have in mind, and it is clearly not the only context that I am interested in. When Singer and Berg specify these DNA sequences that "are required for specific expression," they obviously want to include in their definition the observation that in experimental systems, derived from living beings, not all DNA sequences are expressed all the time. Rather, it is the time and tissue regulated expression that allows for ordered developmental processes at all. Furthermore, it is a well-known problem in in vitro translation systems that numerous polypeptide fragments are synthesized from a given mRNA, and sophisticated techniques are required to identify the polypeptide of interest. Even if these expression systems work very well in vitro, it is still often a matter of luck for gene-technologists whether such gene constructions would work as anticipated when brought back into an *in vivo* system. I will come back to the "in vivo challenges" later.

What can be learned from this so far is that the regulated synthesis of a polypeptide clearly needs more than a coding region alone. It needs regulatory regions, as Singer and Berg point out rightly, but it also needs something that does the regulation, using those DNA sequences as regulatory regions. We can refer to all these influences and entities that perform regulatory functions by means of DNA (or mRNA) sequences as the non-DNA located, or "environmental" entities, factors, parameters, or influences in gene expression.

The significance of these non-DNA located entities for the synthesis of polypeptide chains becomes most obvious in those cases in which the same DNA sequence is part of very different expression processes, leading to different polypeptides.

For example, activation and termination points, respectively, on the DNA (for transcription) and on the mRNA (for translation) are not just there, but need to be recognized as such. That is shown by the observable fact that those sequences which at some point are recognized as, for example, a promoter (transcription starting sequence), serve at another point as a

coding sequence for a polypeptide. Among the other mechanisms that are important for using mRNAs as templates for polypeptide synthesis are so-called *capping* and *polyadenylation* of the primary mRNA. Although examples for the "under-determination" of polypeptides by DNA sequences are discussed at length in Neumann-Held (1999a, which see for detailed references), two examples from this former paper should be briefly mentioned here because, in my opinion, they are particularly fascinating examples of molecular regulation mechanisms. The first example is alternative mRNA splicing. During the processes of alternative mRNA splicing different parts of the primary mRNA of a particular DNA are spliced together. Therefore, different templates are created as "mature mRNA" for the process of polypeptide synthesis. The obvious result is that different polypeptides are synthesized upon use of the same DNA sequence. Alternative splicing is quite common in eukaryotes.

RNA editing, the second example, edits quite literally the primary mRNA by adding, deleting or exchanging particular nucleotides so that different mRNAs are created than the corresponding DNA would have allowed one to predict. Once again, the same piece of DNA is used here in different processes which lead to different polypeptide products. Furthermore, both alternative mRNA splicing and mRNA editing depend on the developmental state of a particular organism, and both mechanisms are not at all rare. Alternative splicing is quite often found in eukaryotes, and mRNA splicing happens very frequently in organelles. Furthermore, the extent of editing is developmental and tissue dependent. One example of this is the human gene for apolipoprotein B (Strachan and Read 1996: 198). In the liver *apoB* codes for a mRNA of 14.1 kb and a protein of 4536 aminoacids, whereas in the intestines, the mRNA has a length of only 7kb, coding for 2152 amino acids. The shorter length of the resulting amino acid chain is caused by a stop codon that is introduced into the mRNA by mRNA editing, which changes a cytosine into an uracile at position 6666. Although some evidence indicates that editing of cytoplasmic mRNA depends on the environment of the nucleic acid sequence (Knippers 1997: 400, 482), this does not explain the tissue specifity of mRNA editing.

In plant mitochondria mRNA editing is apparently not even dependent on the sequence composition. So far, no conclusive models of the mechanisms have been developed here, and the physiological significance of the sometimes extremely complicated mRNA editing processes, for example in protozoan mitochondria, remain a point of speculation (Knippers 1997: 482ff).

To summarize: In these cases of alternative mRNA splicing and mRNA editing, no one-to-one relationship between DNA and polypeptide can be established, and the same piece of DNA can be used for different functional and structural units. The sequence of nucleotides on the DNA does not even allow prediction of the sequence of amino acids in the polypeptide products. This empirical evidence shows that it is not only the presence of a DNA sequence that determines the course of events that lead to the synthesis of a polypeptide but, in addition, specific non-DNA located factors must act on DNA and derived mRNA to determine the particular processing mechanisms. The fate of the DNA thus depends on the developmental state and tissue. One could say in some metaphoric sense that in the interactions with specific environmental factors "meaning" is attributed to particular nucleic acid segments, which, depending on the context, are then used, for example, on one occasion as a promoter and on another as one of several putative coding regions.[2] Therefore, what counts as a significant structure to perform a particular function is determined by the context in which that segment is being used. Strictly speaking, the consensus sequence (TATAAT) of a promoter is only such a consensus sequence when it is used as a promoter, and not, for example, as part of a coding region.

Recently, Evelyn Fox Keller reached very similar results in her analysis of the molecular gene.

She writes: "Considered as a functional unit, the gene is no longer a static entity, set above and apart from the processes that specify cellular and inter-cellular organization, but itself a part and parcel of these processes, defined and brought into existence by the action of a complex self-regulating dynamical system in which and for which the inherited DNA provides the crucial raw material." And further: "*genes* function in these debates to remind modern geneticists of what it is that makes a region of nucleic acid 'interesting' or of what constitutes 'meaningful structure' in the genome" (Keller 2000).

Therefore, Keller suggests: "One way of posing the particular difficulty that is here made explicit might be to paraphrase Howard Pattee's question . . . and ask: How does a sequence become a gene?" (Keller 2000).

It is by posing exactly this question that Keller suggests a difference which I want to emphasize. If one wants to conceptualize the complex molecular mechanisms of polypeptide synthesis processes, then, so I argue, it is necessary to distinguish between DNA sequences on the one hand, and the embedding of DNA sequences in the course of events of polypeptide synthesis on the other hand.

DNA sequences are undoubtedly necessary in the process of making a polypeptide. But in addition, any process of polypeptide synthesis requires factors other than this DNA. Non-DNA located entities are required to determine how DNA segments are used to make one or more transcriptional units. From there, further interactive processes determine the course of the translation processes. This becomes particularly obvious in cases of tissue and/or time regulated polypeptide synthesis.

Therefore, in summary, the synthesis of a linear polypeptide chain is caused by numerous factors which interact with each other in a particular order. Among these factors are particular DNA sequences. Which DNA segments participate in the synthesis of a polypeptide at a particular point cannot be determined by looking at the nucleotide sequence alone, and a particular DNA sequence can become part of different courses of polypeptide syntheses.

A very important point is, however, that once polypeptide synthesis has occurred, it is possible to clearly identify particular DNA segments which featured in the course of events that resulted in the synthesis of this particular polypeptide. Therefore, in retrospect, a particular DNA-polypeptide relationship can be specified. This sounds very similar to Keller's statement that the definition of a gene as that which is "expressible" allows "for an after-the-fact designation of the particular combination of DNA segments actually employed" (Keller 2000).

However, in this quotation Keller seems to limit the use of the word gene to DNA segments again, whereas I want to emphasize that an "after-the-fact designation" is, in principle, possible for any entity that causally participates in the process. From this then, we arrive at a distinction between DNA and gene, and find a new conceptualization of the gene for developmental processes. This concept can be stated as follows:

"Gene" is the process (i.e., the course of events) that binds together DNA and all other relevant non-DNA entities in the production of a particular polypeptide. The term *gene* in this sense stands for processes which are specified by (1) specific interactions between specific DNA segments and specific non-DNA located entities, (2) specific processing mechanisms of resulting mRNA's in interactions with additional non-DNA located entities. These processes, in their specific temporal order, result (3) in the synthesis of a specific polypeptide. This gene concept is relational, and it always includes interactions between DNA and its (developmental) environment.

At this point one might wonder whether I should not include so-called "RNA genes" in this definition as well, thereby changing point (3). The term "RNA genes" is usually used to designate those parts of DNA sequences which do not lead to an mRNA, which then is translated into a polypeptide, but which lead to other kinds of RNAs, for example rRNAs and tRNAs. I have

to admit that I hesitate to include these RNAs in the suggested gene concept. And why should we want to summarize both kinds of processes under the same heading anyway? One argument would be that one should name everything a gene, which includes a DNA segment that is involved in a transcription process. Maybe it was an argument of such a kind which allowed classical molecular gene definitions to include "RNA producing" DNA sequences as "genes." I would hold against this argument, and therefore against this gene definition, the fact that it emphasizes another point than I have in mind. An important feature of my suggested gene definition is that it binds together processes of transcription *and* translation. Under this perspective the main emphasis is *not* on the question, in which kind of production processes particular DNA sequences play an important role, but the emphasis is on the question, how are DNA sequences used in the processes of *polypeptide* production. The whole purpose of trying to suggest PMG as a gene concept for developmental processes is to gain some clarification of the concept gene for at least one clearly defined application area. If we include processes under the heading *gene*, which are limited to only transcription, it is my feeling that we lose some of the just-gained clarity again, and we might wonder later on, whether other DNA regions, which are involved in regulatory functions, should be called genes as well. Therefore, I tend to exclude what is commonly called "RNA genes" from the suggested developmental gene concept, although I have not arrived at a final stand-point in this matter myself.

Coming back to point (3) in the suggested gene definition, it is this last feature in particular, the emphasis on the interaction between DNA and its environment, which allows this gene concept to be viewed as an extension of the developmental systems approach (DSA), as it was developed by Susan Oyama (1985). This claim might be best supported by briefly explaining how DSA actually motivated me to search for a developmental gene concept. Furthermore, such an excursion

PMG = Process Molecular Gene

will allow for further clarification of the context of scientific inquiry for which this gene concept was developed.

A central point in Oyama's DSA is the claim that, if one wants to understand the *development* of any phenotypic trait, a distinction between "genetic" and "nongenetic" causes is not suitable. Although I found the arguments provided by DSA very inspiring, I nevertheless wondered why DSA seemed to stop short when approaching the molecular level itself. Oyama, for example, conceded that "it makes sense in general to say that the primary structure of a polypeptide is encoded on the chromosomes" (Oyama 1985: 70). But what does it mean to make "sense in general"? It could be interpreted as assuming a (more or less) simple correspondence between a DNA segment and a linear polypeptide chain. Therefore, DSA obviously could be understood as being willing to concede that at least on the level of polypeptide synthesis DNA would have "more causal powers" than non-DNA located influences. One may grant at this point that such an interpretation cannot be deduced analytically from Oyama's statement, because one has to differentiate between the *kind* and the *size* of causal role. But because the concept of causality remains unspecified here, Oyama's statement can lead to misunderstandings, particularly in combination with statements of other DSA representatives. Gray granted, for example, that "the nucleotide sequence does specify the primary structure of a protein" (Gray 1992: 190). Again, one might argue that in a certain sense it does, and again, misunderstandings are possible according to which this statement seems to represent a rather standard view point on a more prominent causal role of DNA. What I saw here was, on the one hand, a gap in DSA that pointed to a need for clarification, a blind spot in DSA when applying the approach to the molecular level of developmental processes. On the other hand, I thought that DSA would provide a suitable framework for the reconceptualization of those molecular contexts that allow the synthesis of linear

polypeptide chains. The need for further clarification in this regard was seen by Oyama (1988: 261) herself, when she wrote, for example: "But even leaving aside the complexities of overlapping and discontinuous genes and the problems they pose for the concept of coding, the information-as-sequence must be reconstituted as well." Therefore, I made it my goal to work out a gene concept that extends DSA for application to *developmental processes* on those *molecular* levels of interactions, which have to do with *DNA* and end with the synthesis of *linear polypeptide* chains. It has been pointed out before, but should be stressed again, that all the *italics* in the last sentence indicate points of emphasis and at the same time points of exclusion, which define the purposes of the search for a gene concept for development. The suggested gene concept is *not* for an evolutionary context and *not* about other kinds of molecular interactions than those mentioned. The limits of this gene concept will be discussed further in the next paragraph. But first I need to say a word about naming this gene concept.

When I first presented my gene concept I spoke of conceptualizing the gene in "constructionist ways."[3] The term *constructionism* was used by Gray (1992: 175 and 203, note 23), who applied it to the new viewpoint on development and evolution as it was exemplified in the writings of Lewontin and Oyama. Today, I do not favor the term *constructionist* because, for several reasons, it is likely to lead to misunderstandings. One reason is its phonetic closeness to the term *constructivism*, which is being used in the naming of such different approaches as "radical-constructivism" (for example, Maturana 1985; Maturana and Varela 1990; Schmidt 1987), "social-constructivism" (for example, Berger and Luckmann 1980; Luhmann 1988) and "methodical constructivism" (Janich et al. 1972–73; Lorenzen 1987; Mittelstrass 1973). Another downside of the use of the term *constructionism* is that it suggests a programmatic unity, where there really are a diversity of very different approaches, which have (only?) an "antireduction-ist consensus" in common (Sterelny and Griffiths 1999: 137ff). Developmental systems theory (DST) or approach (DSA) (for a discussion of the distinction between *theory* and *approach* in this context, see chapter 20 of this volume) is one formulation in the antireductionist community, and I understand the gene concept I suggest as an addition or extension to DSA for conceptualizing the molecular course of events in the synthesis of polypeptides. However, because I am no longer using the name *constructionist*, I decided that a more appropriate name for this gene concept would be the *process molecular gene* concept (PMG).

The Limits of PMG

PMG has relevance only for particular contexts of research interests. Some distinctions have been discussed in the preceding paragraph. For example, PMG was not designed as an evolutionary gene concept, and it remains arguable whether it is possible to find a unified gene concept for both developmental and evolutionary research contexts.[4]

PMG is designed for contexts that focus on the relationship between DNA and a polypeptide (or an RNA), because this relationship indicates a process of production that is central for developmental processes on the molecular level. This might be the reason that Singer and Berg (1991) want to preserve this relationship in their suggestion of a molecular gene concept. Furthermore (this is rather a minor point), it was my purpose to extend DSA by the conceptualization of the relationship between DNA and polypeptide synthesis, and therefore it was my purpose to focus on these rather than on other molecular processes.

But if these processes between DNA and synthesis of a linear polypeptide are viewed as important for understanding development on the molecular level, why then, one might ask, stop at the level of linear polypeptide synthesis?

I have not set these limits to PMG because I regard the following processes as unimportant in the production of phenotypic traits (or trait differences). On the contrary, I find them so important that they should be considered independently, particularly independently from the level of DNA sequences. To make this point clear, I need to present my argument in more detail.

I have shown that the process of polypeptide production needs numerous other factors besides DNA. A linear polypeptide chain, however, is then used for numerous further processes, in which this polypeptide is folded and interacts with cofactors, other polypeptides and other cell components. Here, this polypeptide, and possibly its functional loss, becomes a factor in the development of a phenotypic trait or a so-called genetic disease. These levels of process, producing a linear polypeptide chain on one hand and the further interactions of this chain with other components on the other, are often not differentiated. Instead, it is assumed (!) that a mutation on a DNA sequence that influences the functional effect of the polypeptide is sufficient to explain variations in phenotypic traits. Thus, the human geneticist and anthropologist Ulrich Wolf states:

... In particular cases, ontogenetic modification appears to be of minor significance, so that the phenotype of a mutation can be predicted with considerably accuracy. This is no surprise if, depending on the nature of the mutation and the physiological function of the gene affected, the genotype-phenotype relationship is direct.... It is assumed that the total of the non-genetic influences (epigenetic, environmental) are usually so similar or are compensated by the organism to such an extent that the respective mutations act as the major variable during ontogenetic development. (Wolf 1995: 127)

However, it is Wolf's point that very often the genotype-phenotype relationship is more complex and does not necessarily allow for the simplifying *assumption* to which he refers. Wolf stresses that "the phenotype is ... the product of ontogenetic development rather than the mere consequence of the genetic constitution of the

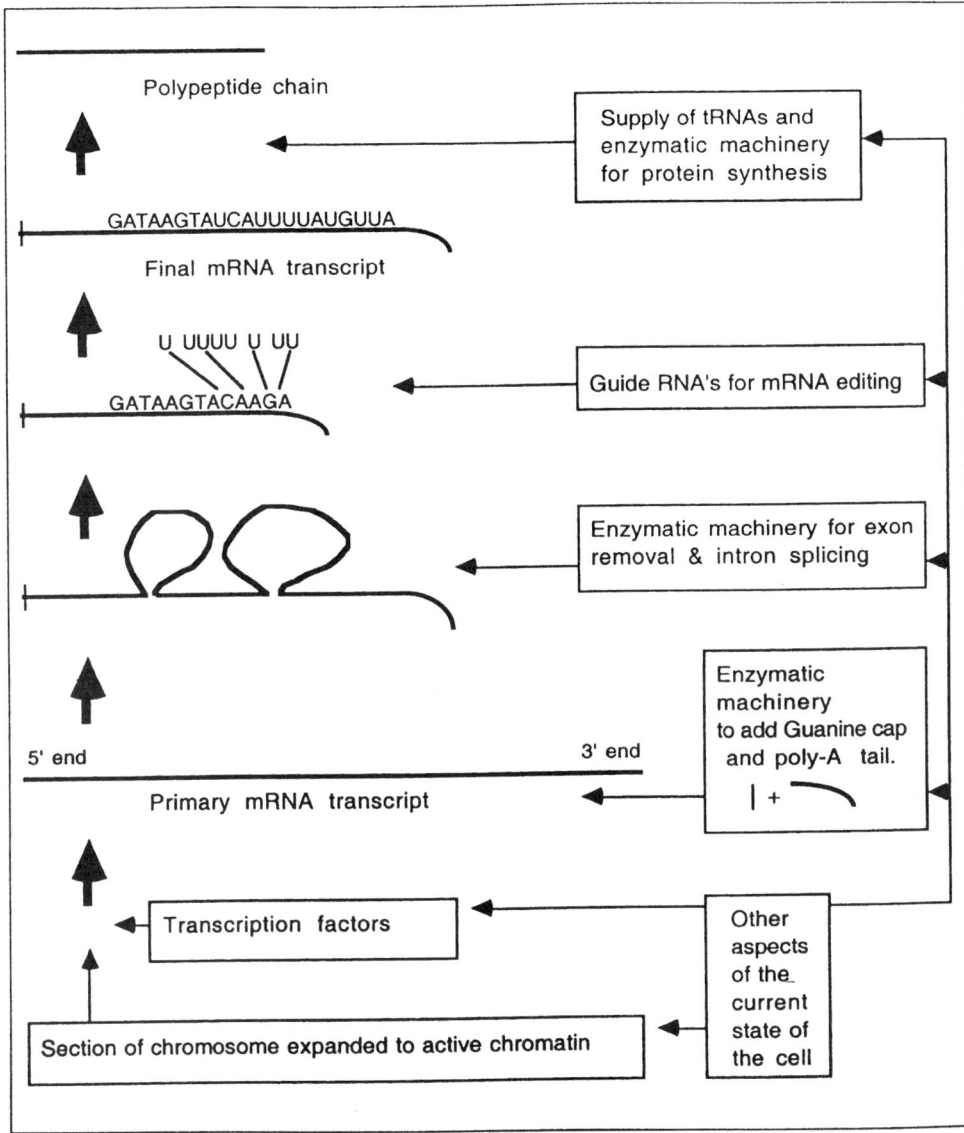

Figure 7.1
Schematic representation of some transcription, editing, and translation processes, highlighting the context-dependency of the expression of gene products. (This figure first appeared in Griffiths and Neumann-Held 1999; copyright © 1999 American Institute of Biological Sciences.)

zygote" (p. 127), and in his 1997 paper, Wolf discusses numerous examples where identical mutations do not exclude phenotypic variation. Similar viewpoints are also held by some other human geneticists. Scriver and Waters (1999: 267), for example, summarize: "It was a hope that delineation of genotypes [as in the human genome project] with new methods for the detection of mutations would enable the prediction of variant phenotypes; in the case of human genetic disease, this would have added value for prognosis and treatment."

But these hopes were an oversimplification, as Scriver and Waters show even for the case of the "classic 'monogenic' autosomal recessive disease" Phenylketonurie (PKU). Scriver and Waters (1999: 267)[5] remark that a mutation "at the human *PAH* locus was deemed sufficient to explain the impaired function of the enzyme phenylalanine hydroxylase (enzymic phenotype), the attendant hyperphenylalaninemia (metabolic phenotype) and the resultant mental retardation (cognitive phenotype)." However, they continue, "expectations for a consistently close correlation between the mutant genotype and variant phenotype has been somewhat disappointed."

What has caused these disappointments? Scriver and Waters (p. 268) write that "two important challenges to the hypothesis that a *PAH* genotype would consistently predict a 'monogenic' phenotype had emerged: (1) patients, even sibs, sharing identical mutant *PAH* genotypes could have greatly different cognitive and metabolic phenotypes; (2) there are many instances of discordance between the mutant *PAH* genotype, its predicted effect on enzyme function, and the associated metabolic phenotype."

Although the linear polypeptide chain that gives rise to the enzyme phenylalanine hydroxylase is effected due to a mutation of the corresponding DNA sequence, even the activity of the enzyme itself, not to mention the further effects, vary due to the environment in which this enzyme is located. To quote from Scriver and Waters

again: "The hope that the mutant genotype at the major locus would predict the variant phenotype originated in the new ability to study mutant allelic expression in reductive *in vitro* expression systems and in transgenic animals."

In other words, these experiments were performed under highly controlled conditions, in which all parameters but the particular gene (i.e. DNA sequences) could be held more or less constant. Scriver and Waters (p. 271ff) continue: "But genomes function *in vivo*, where much more than the major gene (here the *PAH* gene) is expressed and where the whole organismal phenotype is more than the sum of its parts."

These results of the "in vivo challenges" support my conviction that it is necessary to limit PMG to those processes that lead from DNA to linear polypeptide chains, and to distinguish these clearly from processes in which the resulting polypeptides participate in the construction of a phenotypic trait.

The Openness of PMG

After having discussed the limits of PMG, the next thing to do is to focus on the *openness* of this gene concept. The incorporation of DNA as well as of non-DNA located factors into PMG might raise questions whether PMG would allow scientific research. These kind of doubts present themselves often under the headline *holism*. Holism is not an unproblematic term, and at least from the standpoint of philosophy of science, one would have to ask for further clarification. However, that might not be necessary here. I will therefore substitute the term openness for holism and explain openness by paraphrasing the criticism as follows: In emphasizing *all* developmental causes for the construction of a polypeptide, PMG seems to require consideration of *everything* (in principle) or at least of a nonaccountable number of factors. However, since one cannot do empirical research on "everything," PMG cannot offer new insights or research strategies to biologists.

Therefore, DNA-centered gene concepts are superior, because here we find a concentration on one causal factor, which then can be tested in regard to its influence on developmental processes. DNA-centered gene concepts, but not PMG, offer biologists something they can transform into empirical questions and research in their laboratories.

In my view, these doubts are based on a misunderstanding, not only in regard to PMG, but also in regard to DNA-centered gene concepts. In particular, there seems to be a misunderstanding with regard to the consequences that PMG, but not its alternatives (!), would have for empirical research. Therefore, I want to begin my defense of PMG with a brief look on some methodological requirements of empirical science.

It seems trivial that an assumption, according to which some factor causally influences the production of an object of scientific inquiry, has to be tested (1) by varying the parameters under observation, holding all other parameters constant, and (2) by observing putative changes in the pattern of production of the object of interest. In addition, it might be required that some mechanism be given by which the factor under investigation can influence the observed outcome, in order to justify attributing causality rather than mere correlation.

Applying this to research on the effects of DNA sequences on polypeptide production, it is obvious—and reflects empirical methods and practice—that an experimental setup has to be developed, which allows for an answer to the question: What is the causal relationship between variations in DNA sequences on one hand and variations in polypeptide synthesis on the other hand? In such an experiment, then, at least in the ideal case, all other parameters except the DNA, are kept constant. The result of such an approach will clarify the role of DNA in the making of a polypeptide.

Why should PMG not allow for such an approach? Or to pose the question a different way: Why should DNA-centered gene concepts make such an empirical investigation easier than PMG?

One possible answer to this question is an assumption, according to which a result that "clarifies the role of DNA in the making of a polypeptide," could also be expressed by substituting "DNA" by "gene." But such a substitution could only be defended when arguing for a fuzzy gene concept, in which everything that refers to DNA segments may be called gene. The empirical methods of molecular biology, however, allow for finer distinctions, which then require finer distinctions in definitions. Depending on the experimental setup, it is possible to distinguish between DNA sequences that (in a particular context) function as promoters, for example, and those that function as enhancers. Both DNA segments will (mostly) influence the production of a polypeptide independently of one another. Because they can be distinguished from each other empirically, it makes sense to distinguish between them in terminology as well. However, both entities might very well causally influence the synthesis of the *same* polypeptide, which is the reason that in DNA-centered gene concepts they are summarized as being parts of *one gene*. So whereas empirical research will result in the identification of different DNA segments, these might very well belong to the same functional unit in regard to the production of a particular polypeptide. Here, the need to distinguish between DNA and gene becomes obvious even in the DNA-centered gene concept. Another point, however, is more important in the context of this section. In analogy to the doubts toward PMG one might also question DNA-centered gene concepts, and ask whether in the end it might not be the entire genome which has a causal influence on the synthesis of every single polypeptide. It seems to me to be a very appropriate answer to say: Yes, but who cares? The point is that of course the choice of which DNA segment is tested for its relevance for polypeptide synthesis, will be determined by pragmatic considerations, research interests, methods, and skills. To summarize those

DNA segments under a common heading of "gene" does not seem to pose a problem in the context of DNA-centered gene concepts.

PMG differs in this respect only in that it allows for inclusion of not only DNA, but also non-DNA located entities, thereby integrating into the gene concept those relevant entities that are necessary for the functional specification of the DNA sequences involved. Here again, of course, it will be pragmatic considerations, research interests, methods, and skills that decide which factor is tested for its relevance for polypeptide synthesis. Therefore, PMG's openness does not pose any more problems for the application of empirical methods, and the interpretation of the results, than any DNA-centered gene concept.

PMG in Context

The analysis of the preceding section reveals an important insight. I have argued that the goals of the researchers (and their funding sources), the technical, and the methodical tools play a decisive role, when considering whether DNA or any other parameters should be investigated in the search for an explanation of the synthesis of a polypeptide chain. These goals are not themselves part of the scientific method, but goals and tools set the stage for the actual scientific inquiry, and for the design of the experiment. Goals and tools allow for a *preselection* of possible scientific inquiries from possible alternative questions. Therefore, those preselections have to be justified on grounds other than scientific ones.[6] Here, we see now an important difference between PMG and DNA-centered gene concepts, in that DNA-centered gene concepts already reflect a preselection by focusing on the effect of DNA in the synthesis of polypeptides. Although I would not defend it, I might still be willing to concede that this might be unproblematic *as long as it is not assumed that the concentration on DNA factors alone can be justified by ontological claims.* By "ontological claims" I mean here the idea that

DNA sequences are somehow "more causal," "more real," or "more significant" for the synthesis of a polypeptide than other factors. Kelly Smith (2000) has called this interpretation of genetic results "gene-centrism": "the view that the genes are the most important explanatory factors in biology because of their unique causal powers—whether in controlling individual ontogeny (development), accounting for abnormal functioning in adults (disease), or explaining changes in populations over time (evolution)."

I have already provided evidence as to why, on the one hand, such an interpretation of molecular research is not defensible, and why, on the other hand, the same evidence allows a conceptualization of the gene in terms of PMG. In this section I briefly want to highlight some further benefits that PMG might have when gaining more attention among scientists and the general public.

It should be remembered that, after all, scientists apply, work, and discuss their (ontological) assumptions and interpretations not only in private, but publicly, in front of society as a whole. That alone is reason enough to ask how these assumptions can be justified, and particularly if reasons are presented to doubt them. Here, scientists are responsible for their research. Certainly since physicists became aware of the consequences of their "value-free" research, when their "formulas" exploded over Hiroshima and Nagasaki, (some) scientists started to accept that they cannot deny all responsibility for the consequences of their research.[7] But also philosophy of science, which is constituted by reflection on the different levels of the scientific discourse, has the potential for criticism of science and should use it (Janich et al. 1972–73; Mittelstrass 1973; Weingarten 1998). I would claim that such criticism need not be detrimental to the scientific enterprise itself.

With this in mind, we might now ask in which way alternative gene concepts and their underlying assumptions effect areas outside of the actual scientific process.

As has been discussed at length, the development of phenotypic traits is very complex, and a prediction of its outcome can only be a more or less appropriate estimation, if it is (only) based on an analysis of DNA sequences. It is empirical evidence that determines in the end the quality of predictions which can be made in any single case. This should have consequences for genetic counseling, as has been demanded by philosophers and ethicists like Rehmann-Sutter (1998) and Cor van der Weele (see chapter 24 of this volume), and also by human geneticists. Scriver and Waters (1999: 272) state clearly that in the case of PKU: "The biological view has implications for the investigation, counselling and treatment of patients with genetic disease, because each person, perhaps excepting monozygotic twin pairs, with a nominal monogenic disorder is likely to have a particular phenotypic form of it." They continue: "The ideas expressed here are not really new.... Lionel Penrose, in his inaugural address as Galton Professor in 1946 ... observed, among other things, that IQ was not predictable from genotype within PKU families, and opined that a multidisciplinary approach would benefit our understanding of this disease. His original paper has been reprinted to remind us that there were giants on whose shoulders we stand to see a little further."

The framework of this article does not provide room to discuss whether or not science really produces *progress*. Rather, I want to point out that more than fifty years after Penrose's talk his followers seem to have forgotten the earlier recognition of the complexity of traits.

Judging from the number of publications, their titles, and the discussion sections (for example, in *Science*, *Nature*, and certainly more specialized journals like the German *Medizinische Genetik*), it is obviously still far easier to obtain research money for investigations that claim to be able to locate *the* genetic causes of phenotypic traits and their variations than for alternative approaches. For scientific journals and communities, distributors of research money, popular science magazines, and the public in general, the "explanation" of traits such as aggressivity (Brunner et al. 1993a, 1993b; for a critical discussion, see

Neumann-Held 1997, 1999b) and even human physical performance (Montgomery et al. 1998), by reference to specific DNA sequences seems more convincing than, for example, a research project on the causal influences of the womb environment on health later in life (Barker 1998). One might view these preferences for "genetic research programs" as a temporary choice of society, which cannot financially support all more or less promising research programs. I think, however, that one should offer more convincing arguments for such a choice.

The defense of such choices becomes more critical when they are grounded in gene-centrism, on the assumption that these research strategies can supply a fundamentally deeper or truer knowledge of developmental pathways (and their aberrations) than other research approaches can do. The argument as to why to support genetic research is then not a scientific argument, nor an argument that points to some (assumed) pragmatic reasons (for example: easier access to manipulations), but is an ideological argument. Insofar as it is believed by scientists and by the general public that this argument is not ideological but logically derived from a scientific argument, the consequences can be that DNA *will become* the most important factor because that *will produce* the most dramatic effects, not with regard to developmental pathways, but with regard to the construction of human self-understanding. Such changes can already be observed, as has been demonstrated, for example, by Jackie Leach Scully.[8] Following this line of reflection, humans might not become "homo geneticus" (Rehmann-Sutter 1999) by genetic manipulations, but because humans would *believe* that they, their own lives, their children, the human race itself, could be styled genetically. Reproductive choices in, for example, preimplantation diagnosis, might not be taken based on scientifically backed knowledge, but rather on scientist-supported ideologies, which are rooted in particular background assumptions. If these assumptions, however, prove on examination to be somewhat outdated in the light of new empirical evidence, then it would seem sensible to replace them.

Here, PMG offers a promising alternative. Following PMG, one cannot claim "more causality" for DNA in processes of polypeptide synthesis, because it can be shown that the synthesis of a polypeptide requires in addition specific non-DNA factors. DNA sequences are "the cause" of the synthesis of a polypeptide only if the experimental setup does not allow the investigation of anything other than the effects of DNA sequences.

In the framework of PMG, it is obvious that such a result may not be interpreted as indicating the causally specific role of DNA but rather as pointing to the limitations of the scientific approach as such. This then might become reason to ask further questions. One might wonder, for example, whether claims of validity can be redeemed for those assumptions, according to which science is granted so much significant influence on human self-understanding. But this will have to be discussed in another place.

Acknowledgments

I like to express my thanks to Mathias Gutmann and to Christoph Rehmann-Sutter for many helpful discussions, and to Barbara Alexius for her help in translating this article. This work was partly supported by a grant of the foundation MGU, University of Basel.

Notes

1. For a detailed discussion of the positions of Fogle (1990) and Waters (1994), see Neumann-Held (1999a); of the position of Epp (1997), see Griffiths and Neumann-Held (1999). For extensive discussion of reductionist positions, see Sterelny and Griffiths (1999) and Sarkar (1998). For a criticism of reductionism in genetics from the standpoint of "methodical culturalism," see Gutmann (1998).

2. In addition to those mentioned here, there are many more molecular mechanisms known that influence the regulated synthesis of a polypeptide chain. Examples are complicated methylation processes that selectively inhibit expression from particular DNA segments. In *genetic imprinting*, for example, the methylation of DNA segments depends on their origin from the maternal or the paternal side. The consequence is that although the maternally and the paternally inherited DNA segments do not differ in the order of their nucleic acid sequences, they are nevertheless treated differently, and therefore have different effects. Heutink et al. (1992) report, for example, that paragangliomas, a specific kind of mostly benign tumors, will effect individuals only when the "disease gene" is inherited from the father, whereas "expression of the phenotype is not observed in the offspring of an affected female." For further examples of genetic imprinting see Strachan and Read (1996: 201ff).

3. The talk was given at the conference Developmental Systems, Competition, and Cooperation in Sociobiology and Economics, held at Marienrode Monastery, Hildesheim-Marienrode, Germany, on April 24–28, 1996.

4. Recently, Peter Beurton (1998) suggested a new gene concept which should unify evolutionary and developmental gene concepts. For a discussion of his approach, see Neumann-Held and Rehmann-Sutter (1999).

5. My special thanks go to Prof. Dr. Ulrich Wolf, Freiburg, who draw my attention to this article.

6. These justifications, which are asked for here, should not be confounded with a historical or sociological analysis of what is accepted as an explanation, as has been presented, for example, by Keller (1999).

7. For a case study on the German physicist and philosopher Carl Friedrich von Weizsaecker, see Drieschner (1992), and von Weizsaecker (1977: 101ff: "The political role of science in our culture").

8. I am quoting here from "Nothing like a Gene," a paper presented by Jackie Leach Scully at the Genes and Development Symposium at the University of Basel, March 19–20, 1999.

References

Barker, D. J. P. (1998). *Mothers, Babies and Health in Later Life.* Edinburgh: Churchill Livingstone.

Berger, P., and T. Luckmann. (1980). *Die gesellschaftliche Konstruktion der Wirklichkeit. Eine Theorie der Wissenssoziologie.* Frankfurt: Fischer.

Beurton, P. J. (1998). Was sind Gene heute? *Theory Bioscie.* 117: 90–99.

Beurton, P. J. (2000). A unified view of the gene, or how to overcome reductionism. In P. J. Beurton, H-J. Rheinberger, and R. Falk (Eds.), *The Concept of the Gene in Development and Evolution,* Cambridge: Cambridge University Press.

Brunner, H. G., M. R. Nelen, P. Iandvoort, et al. (1993a). X-linked borderline mental retardation with prominent behavioral disturbance: Phenotype, genetic localization, and evidence for disturbed monoamine metabolism. *American Journal of Human Genetics* 52: 1032–1039.

Brunner, H. G., M. R. Nelen, X. O. Breakefield, et al. (1993b). Abnormal behavior associated with a point mutation in the structural gene for monoamine oxidase A. *Science* 262: 578–580.

Burian, R. M. (1986). On conceptual change in biology: The case of the gene. In D. J. Depew and B. H. Weber (Eds.), *Evolution at a Crossroads,* pp. 21–42. Cambridge, MA: MIT Press.

Carlson, E. A. (1991). Defining the gene: An evolving concept. *American Journal of Human Genetics* 49: 475–487.

Dawkins, R. (1982). *The Extended Phenotype: The Gene as the Unit of Selection.* Oxford: Freeman and Co.

Dawkins, R. (1989). *The Selfish Gene* (2d ed.). Oxford: Oxford University Press.

Drieschner, M. (1992). *Carl Friedrich von Weizsäcker zur Einführung.* Hamburg: Junius.

Epp, C. D. (1997). Definition of a gene. *Nature* 389: 537.

Falk, R. (1984). The gene in search of an identity. *Human Genetics* 68: 195–204.

Fischer, E. P. (1995). What's in a gene? On the advantages of imprecision and the success of an open-ended idea. In P. Bernhard and C. Cookson (Eds.), *Genethics: Debating Issues and Ethics in Genetic Engineering,* pp. 6–9. Basel: H. P. Bernhard.

Fogle, T. (1990). Are genes units of inheritance? *Biology and Philosophy* 5: 349–371.

Gray, R. (1992). Death of the gene: Developmental systems strike back. In P. Griffiths (Ed.), *Trees of Life,* pp. 165–209. Dordrecht: Kluwer Academic.

Griffiths, P., and E. M. Neumann-Held. (1999). The many faces of the gene. *BioScience* 49: 656–662.

Gutmann, M. (1998). Information, Gene und Metaphern. In A. Daly (Ed.), *Was wissen Biologen schon vom Leben?* pp. 141–156. Loccum: Evangelische Akademie Loccum.

Heutink, P., A. G. L. van der Mey, L. A. Sandkuijl, et al. (1992). A gene subject to genomic imprinting and responsible for hereditary paragangliomas maps to chromosome 11q23-qter. *Human Molecular Genetics* 1: 7–10.

Janich, P., F. Kambartel, and J. Mittelstrass. (1972–73). Wissenschaftstheorie als Wissenschaftskritik. (Series of articles in nine contributions in *Aspekte V,* issues 9–12; *Aspekte IV,* issues 1–5).

Keller, E. F. (1999). Making sense of life: Explanation in developmental biology. In J. Maienschein and R. Creath (Eds.), *Biology and Epistemology,* pp. 244–260. Cambridge: Cambridge University Press.

Keller, E. F. (2000). Is there an organism in this text? In P. Sloan (Ed.), *Controlling Our Destinies,* pp. 273–290. South Bend, IN: Notre Dame University Press.

Kitcher, P. (1982). Genes. *British Journal of Philosophy of Science* 33: 337–359.

Kitcher, P. (1992). Gene: Current usages. In E. Fox Keller and E. A. Lloyd (Eds.), *Keywords in Evolutionary Biology,* pp. 128–131. Cambridge, MA: Harvard University Press.

Knippers, R. (1997). *Molekulare Genetik.* Stuttgart: Thieme.

Levin, B. (1985). *Genes.* New York: Wiley.

Lorenzen, P. (1987). *Lehrbuch der konstruktiven Wissenschaftstheorie.* Mannheim: Bibliographisches Institut.

Luhmann, N. (1987). *Soziale Systeme. Grundriss einer allgemeinen Theorie.* Frankfurt: Suhrkamp.

Maturana, H. R. (1985). *Erkennen: Die Organisation und Verkoerperung von Wirklichkeit.* (2d ed.) Braunschweig: Vieweg.

Maturana, H. R., and F. J. Varela. (1990). *Der Baum der Erkenntnis: Die biologischen Wurzeln des menschlichen Erkennens:* Goldmann.

Mittelstrass, J. (1973). Das praktische Fundament der Wissenschaft und die Aufgabe der Philosophie. In F. Kambartel and J. Mittelstrass (Eds.), *Zum normativen Fundament der Wissenschaft,* pp. 1–69. Frankfurt: Suhrkamp.

Montgomery, H. E., R. Marshall, H. Hemingway, et al. (1998). Human gene for physical performance. *Nature* 393: 221–222.

Neumann-Held, E. M. (1997). Gene können nicht alles erklären. Zur Interpretation genetischer Daten. *Universitas* 52: 469–479.

Neumann-Held, E. M. (1998). Jenseits des "genetischen Weltbildes." In E-M. Engels, T. Junker, and M. Weingarten (Eds.), *Ethik der Biowissenschaften,* pp. 261–280. Berlin: Verlag für Wissenschaft und Bildung.

Neumann-Held, E. M. (1999a). The gene is dead—long live the gene: Conceptualizing genes the constructionist way. In P. Koslowski (Ed.), *Sociobiology and Bioeconomics: The Theory of Evolution in Biological and Economic Thinking,* pp. 105–137. Berlin: Springer.

Neumann-Held, E. M. (1999b). Von Genen, Merkmalen und Kontexten. Eine philosophisch-wissenschaftstheoretische Analyse zum Begriff der "genetischen Verursachung." In T. Braun and M. Elstner (Eds.), *Gene und Gesellschaft,* pp. 85–94. Heidelberg: Deutsches Krebsforschungszentrum.

Neumann-Held, E. M., and C. Rehmann-Sutter. (1999). Individuation and reality of genes. A comment to Peter J. Beurton's article "Was sind Gene heute?" *Theory Biosci.* 118: 85–95.

Oyama, S. (1985). *The Ontogeny of Information: Developmental Systems and Evolution.* Cambridge: Cambridge University Press. (2d rev. ed., Durham, NC: Duke University Press, 2000.)

Oyama, S. (1988). Stasis, development and heredity. In M-W. Ho and S. Fox (Eds.), *Process and Metaphors in the New Evolutionary Paradigm,* pp. 255–274. London: Wiley.

Portin, P. (1993). The concept of the gene: Short history and present status. *The Quarterly Review of Biology* 68: 173–223.

Rehmann-Sutter, C. (1998). DNA-Horoskope. In M. Düwell and D. Mieth (Eds.), *Ethik in der Humangenetik. Die neueren Entwicklungen in der genetischen Frühdiagnostik aus ethischer Perspektive,* pp. 415–443. Tübingen: Francke Verlag.

Rehmann-Sutter, C. (1999). Das Jahrhundert der Genetik und der *homo geneticus. Novalis,* Nov., pp. 11–14.

Rheinberger, H-J. (2000). Gene concepts. Fragments from the perspective of molecular biology. In P. Beurton, R. Falk, and H-J. Rheinberger (Eds.), *The Concept of the Gene in Development and Evolution,* pp. 219–239. Cambridge: Cambridge University Press.

Sarkar, S. (1998). *Genetics and Reductionism.* Cambridge: Cambridge University Press.

Schaffner, K. (1993). *Discovery and Explanation in Biology and Medicine.* Chicago: University of Chicago Press.

Schmidt, S. J. (Ed.), (1987). *Der Diskurs des Radikalen Konstruktivismus.* Frankfurt: Suhrkamp.

Scriver, C. R., and P. J. Waters. (1999). Monogenetic traits are not simple: Lessons from phenylketonuria. *Trends in Genetics* 15: 267–272.

Singer, M., and P. Berg. (1991). *Genes and Genomes: A Changing Perspective.* Mill Valley, CA: University Science Books.

Smith, K. (2000). What is a genetic trait? In D. Magnus (Ed.), *Contemporary Genetic Technology: Scientific, Ethical, and Social Challenges.* Melbourne: Krieger.

Sterelny, K., and P. E. Griffiths. (1999). *Sex and Death: An Introduction to Philosophy of Biology.* Chicago: University of Chicago Press.

Strachan, T., and A. P. Read. (1996). *Molekulare Humangenetik.* Heidelberg: Spektrum Akademischer Verlag.

von Weizsäcker, C. F. (1992). *Der Garten des Menschlichen.* München: Hanser Verlag.

Waters, C. K. (1994). Genes made molecular. *Philosophy of Science* 61: 163–185.

Weingarten, M. (1998). *Wissenschaftstheorie als Wissenschaftskritik.* Bonn: Pahl-Rugenstein.

Wolf, Ulrich (1995). The genetic contribution to the phenotype. *Human Genetics* 95: 127–148.

Wolf, Ulrich (1997). Identical mutations and phenotypic variation. *Human Genetics* 100: 305–321.

8 Deconstructing the Gene and Reconstructing Molecular Developmental Systems

Lenny Moss

The articulation of a "developmental systems perspective" has entailed both deconstructive and reconstructive undertakings. The deconstructive side has consisted largely in critiques of nativism in ethology, and psychology (e.g., Lehrman, Gottlieb, Johnstone, Gray, and Oyama), critiques of genetic determinism in cell, molecular, developmental and population biology (e.g., Keller, Lewontin, Moss, Nijhout), and critiques of the incoherence of the gene-environment dichotomy itself (e.g., Lewontin, Oyama). Reconstructively, the conceptualization of organisms qua developmental systems has been given a certain structure by way of its characterization as systems of developmental resources (both inside and outside the skin of the organism) the interactions of which are stably reproduced in succeeding generations (Griffiths and Gray). Inside "the skin," the stock of relevant developmental resources has been shown to include epigenetic systems, such as chromosomal imprinting, cytoplasmic structure and steady-state dynamics, which, along with nucleic acid sequence, are heritable and susceptible of variation (Jablonka and Lamb). Also in the reconstructive vein, Weber and Depew have begun to bring the DST perspective into a productive encounter with work on the self-organizing properties of autocatalytic, thermodynamically "dissipative," non-equilibrium systems. The present chapter addresses both the deconstructive and the reconstructive dimension of DST. If successful, this chapter will highlight an intermediate level of biological organization that, by its very nature, will serve to recontextualize the understanding of genes, thus obviating even naive temptations toward gene/environment dichotomies, and even more important still, will open up a very rich area of empirical investigation to examination and conceptualization in developmental-systems terms.

This chapter will proceed in two parts. First, it will be necessary to introduce a basic distinction with respect to the concept of a gene so as to avoid confusing apples and oranges. This will constitute the deconstructive aspect of the chapter. "Genes," I will argue, are, and can be, *productively* conceived of in two different ways, *albeit with nothing good resulting from the conflation of the two*. And while these different senses of the gene bear an important historical relationship to the classical versus molecular distinction, they are not identical with it, and indeed may now both involve molecular-level referents. Next, the second of the two senses of the gene, *having become unencumbered by the first*, will be taken up and reembedded in the context of functionally conserved, contingently linked, multimolecular modules.

In a recent text entitled *Cell, Embryos and Evolution*, John Gerhart and Marc Kirshner (1997), two of our leading contemporary investigators in cell and developmental biology, have moved against the prevailing genomic tide to accomplish exactly what good scientists are meant to do: they have thought about the most difficult and important problems of their field and have offered new concepts and explanatory models based upon a comprehensive and insightful analysis of the available evidence. I will argue that Gerhart's and Kirschner's notions of conserved functional modules, which have become multiply contingent in their linkages and thereby central to exploratory and adaptive ontogenetic pathways, are exactly what the DST doctor requires for overcoming inner/outer dichotomies in favor of self-organizing, causally reciprocal systems of interactants.

Gene-P or Gene-D: That is the Question

Wilhelm Johannsen coined the term *gene* in 1907 and introduced the phenotype/genotype distinction by 1910 (Carlson 1966). His motivation for the latter was to step away from the preformationist assumptions of early Mendelism, which failed to distinguish between that which is phy- sically transmitted between generations and the characteristics that come to appear in the mature form of progeny. By 1923 Johannsen's skepticism and disenchantment with genetic dogmatism had reached a fever pitch and his inspired reflections provide a nice point of departure for making a key distinction. To begin with, Johannsen raised a question about the scope of Mendelian decomposability.

Certainly by far the most comprehensive and most deci- sive part of the whole genotype does not seem to be able to segregate in units; and as yet we are mostly operating with "characters," which are rather superficial in com- parison with the fundamental specific or generic nature of the organism. This holds good even in those frequent cases where the characters in question may have the greatest importance for the welfare or economic value of the individuals.

We are very far from the ideal of enthusiastic Mendelians, namely the possibility of dissolving geno- types into relatively small units, be they called genes, allelomorphs, factors, or something else. Personally I believe in a great central "something" as yet not divisi- ble into separate factors. The pomace-flies in Morgan's splendid experiments continue to be pomace flies even if they lose all "good" genes necessary for a normal fly- life, or if they be possessed with all the "bad" genes, detrimental to the welfare of this little friend of the geneticist. (Johannsen 1923: 137)

Depending on how one *hears* this passage, one may be tempted either to dismiss it as simply being out of date, or to respond sympathetically with the sense that, yes, there is something right about Johannsen's intuition. Upon inspection, one can see that what distinguishes these two "hearings" ultimately has much to do with the nature of the genotype/phenotype relationship. In one sense, we want to say that Johannsen is just wrong because we now know that all genom- ic DNA segregates into separate chromosomes. But to speak that way is to set aside any consid- eration of the direct relationship of a unit of genotype to a unit of phenotype. Johanssen went to pains to distinguish genotype from phenotype, but the gene that he has in mind still gets its

significance from having a definite relationship to some unit of phenotype. There are, of course, "genes" that satisfy this definition—that is, genes whose identity is specified by some "unit" of phenotype that is heritably expressed along Mendelian lines. The research program to which Johanssen was addressing himself assumed that all of the phenotype would come to be geneti- cally specified in this way. Johanssen hereby expresses his disbelief in this being the case. He thought that only a limited set of characteris- tics, none of which get to the core of a species identity, would separate out that way ... and *he was right*. But why would this be the case? Johanssen's reflections on the meaning of allelic variation provide further insight on this matter.

When we regard Mendelian "pairs," Aa, Bb and so on, it is in most cases a *normal* reaction (character) that is the "allele" to an *abnormal*. Yellow in ripe peas is nor- mal, the green is an expression for imperfect ripeness as can easily be proven experimentally, e.g., by ether- ization.... The rich material from the American *Drosophilia*—researches of Morgan's school has sup- plied many cases of multiple allelisms—most of all of them being different "abnormalities" compared with the characters of the normal wild fly.... To my mind the main question in regard to these units is this: Are experimentally demonstrated units anything more than expressions for local deviations from the original ("nor- mal") constitutional state in the chromosomes?

Is the whole of Mendelism perhaps nothing but an establishment of very many chromosomal irregulari- ties, disturbances or diseases of enormously practical and theoretical importance but without deeper value for an understanding of the "normal" constitution of natural biotypes? (Johannsen 1923: 138–140).

Eight decades years after Johanssen's paper, it is still surprisingly common for educated people to believe that alternative alleles at a single locus carry substantive instructions for qualitatively different traits. The canonical example for this misunderstanding would probably be eye color where one allele is "for brown eyes" and one allele is "for blue eyes." But, just as Johanssen suggests, what this locus actually entails is a nec- essary (but not sufficient) resource for making brown eye pigment. The "allele for blue eyes" is just any deviation from the sequence necessary for the brown pigment resulting in its absence in the organism. There is no "information" for blue pigments, let alone blue eyes, in "the gene for" blue eyes. Blue eyes are produced by the developing organism in the absence of this particular brown-eye genetic resource. The allele (or gene) for blue eyes is indeed a deviation from a norm, just in the sense that Johanssen suggests. Of course, this is not to say that a deviation from the norm cannot become important in a positive sense, or that what "counts" as normal cannot shift; it can. The norm is determined by the context. In an agricultural context where commodity value and not ecological versatility becomes the motor force of reproduction the loss of an ability to make a seed, for example, can become the norm. In the case of eye color, the inability to make brown presumably became a selective advantage within certain human breeding populations.

Johanssen suggested that the "whole of Mendelism" is nothing but a set of allelic deviations from a norm. It is that set of alleles associated with the Mendelian inheritance of specific phenotypes. It should not be surprising that the vast majority of these are cases in which the phenotypic difference is the result of the functional absence of some "normal" resource. When the absence of some specific sequence results in a predictable phenotype then one can speak of a "gene for this phenotype." But whether it is blue eyes, albinism, or cystic fibrosis, the resulting phenotype is the result of what organisms do in the absence of the normal sequence. Viewed this way it is hardly surprising that *what organisms do* will be context-dependent—varying with the presence and absence of all the other developmental resources, for example, genomic, gestational, nutritional, sunlight, and so on. This context dependence was assimilated in the gene-centrism of the Mendelian model by use of the terms *penetrance* and *expressivity* (with penetrance meant to denote the gene's propensity to express itself or

not and expressivity meant to denote the *extent* to which a gene expresses itself, i.e., in a phenotype). Context-dependent variations in phenotype were thereby treated as the result of nothing but the intrinsic properties of alleles.

The "whole of Mendelism" is circumscribed by those conditions that will allow something to "show up" as a Mendelian gene. If allelic variations do not result in predictably different phenotypes, either because all the variants at a locus suffice as "normal" resources or because the phenotypic consequences of alternative alleles is *too* context-sensitive, or if the absence of the normal resource is developmentally lethal, then such alleles will not "show up" as genes within the sphere of Mendelism. Johanssen's notion that Mendelian genes (no matter how economically and medically important) are limited to relatively superficial traits can readily be seen and appreciated through our contemporary eyes.

The basic notion of the gene that Johanssen described and delimited still has currency. It is still in use and still useful, although it is certainly no longer the only basis for our understanding of what it means to be a gene. This gene concept is what I will refer to as Gene-P. Gene-P is defined by its relationship to a phenotype, albeit with no requirements as regards specific molecular sequence nor with respect to the biology involved in producing the phenotype. Gene-P is the expression of a kind of instrumental preformationism (thus the "P").[1] When one speaks of a gene in the sense of Gene-P, one simply speaks *as if* it causes the phenotype. A gene for blue eyes is a Gene-P. What makes it count as a gene for blue eyes is not any definite molecular sequence (after all, it is the absence of a sequence-based resource that matters here), nor any knowledge of the developmental pathway that leads to blue eyes (to which the "gene for blue eyes" makes a negligible contribution at most), but only the ability to track the transmission of this gene as a predictor of blue eyes. Thus far Gene-P sounds purely classical, that is, Mendelian as opposed to molecular. But a molecular entity can be treated as a Gene-

P as well. BRCA1, the gene for breast cancer, is a Gene-P, as is the gene for cystic fibrosis, even though in both cases phenotypic probabilities based upon pedigrees have become supplanted by probabilities based upon molecular probes. What these molecular probes do is to verify that some normal DNA sequence is absent, by confirming the presence of one, out of many possible, deviations from that normal sequence that has been shown to be correlated (to a greater or lesser extent) with some phenotypic abnormality. To satisfy the conditions of being a gene for breast cancer, or a gene for cystic fibrosis, does not entail knowledge about the biology of healthy breasts nor of healthy pulmonary function, nor is it contingent upon an ability to track the causal pathway from the absence of the normal sequence resource to the complex phenomenology of these diseases. The explanatory "game" played by Gene-P is thus not confined to purely classical methods, which unfortunately has made it all the easier to conflate this meaning of the "gene" with the one I will refer to as Gene-D.

Quite unlike Gene-P, *Gene-D is defined by its molecular sequence*. A Gene-D is a developmental resource (hence the "D") which in itself is *indeterminate* with respect to phenotype. To be a Gene-D is to be a transcriptional unit (extending from start to stop codons) within which are contained molecular template resources. These templates typically serve as resources in the production of various "gene-products"—directly in the synthesis of RNA, and indirectly in the synthesis of a host of related polypeptides. To be a gene for N-CAM, the so-called "neural cell adhesion molecule," for example, is to contain the specific nucleic acid sequences from which any of a hundred potentially different isoforms of the N-CAM protein may ultimately be derived (Zorn and Krieg 1992). Studies have shown that N-CAM molecules are (despite the name) expressed in many tissues, at different developmental stages, and in many different forms. The phenotypes of which N-CAM molecules are co-constitutive are thus highly variable, contingent upon

the larger context, and not germane to the status of N-CAM as a Gene-D. The expression of an embryonic form (highly sialylated) in the mature organism is associated with neural plasticity in the adult brain (Walsh and Doherty 1997) but could well have pathological consequences if expressed in other tissues—yet it would not affect the identity of the N-CAM sequence as a Gene-D. So where a Gene-P is defined strictly on the basis of its instrumental utility in predicting a phenotypic outcome and is most often based upon the absence of some normal sequence, a Gene-D is a specific developmental resource, defined by its specific molecular sequence, and thereby functional template capacity, and yet it is indeterminate with respect to ultimate phenotypic outcomes.

A Gene-P allows one to speak predictively about phenotypes, but only (as Johanssen realized) in a limited number of cases and within some contextually circumscribed range of probabilities. In the absence of, for example, a full molecular-developmental understanding of the processes resulting in the pathophysiology of cystic fibrosis, it can be prognostically useful to speak of "the gene for cystic fibrosis." The "normal resource," that is, the Gene-D located at the cystic fibrosis locus for the great majority of individuals who do not have a family history of cystic fibrosis affliction, is not thereby a gene for *normal pulmonary function* (any more than the roughly thirty thousand other genes involved in normal pulmonary function); rather, it is a member of a "family" of transmembrane ion-channel templates. As a developmental resource, it is one among very many that play a direct role in pulmonary development and function (as well as many other things). To speak of, and direct one's attention to, this gene for a transmembrane-ion conductance regulator protein is to become involved in an entirely different kind of explanatory game, that is, that of a Gene-D. There is no preformationist story to be had at this level. To study the biological role and function of this gene for a chloride channel involves locating it within all of the contexts in which it is biologically active and attempting to elucidate the causal pathways in which it is an interactant (see Kerem and Kerem 1995; Jilling and Kirk 1997). And as with any developmental resource, its status with respect to cause and effect in any given interaction will be contextual and perspectival (i.e., its actions will be viewed as either the cause of something or as the result of something else, depending upon how a particular inquiry is framed). As a molecular-level developmental resource, Gene-D is ontologically on the same plane as any number of other biomolecules, that is, proteins, RNA, oligosaccharides, and so forth, which is only to say that it warrants no causal privileging *before the fact*. Gene-P and Gene-D are distinctly different concepts, with distinctly different conditions of satisfaction for what it means to be a gene. They play distinctly different explanatory roles. There is nothing that is simultaneously both a Gene-D and a Gene-P. That the search for one can lead to the discovery of another does not change this fact. Finding the Gene-P for cystic fibrosis led to the identification of a Gene-D for a chloride-ion conductance channel template sequence. But the latter is not a gene for an organismic phenotype. Its explanatory value is not realized (and cannot be realized) in the form of an "as if" preformationist tool for predicting phenotypes. Rather, the explanatory value of a Gene-D is realized in an analysis of developmental and physiological interactions in which the direction and priority of causal determinations are experimentally first revealed. Cystic fibrosis provides one example of the relationship between these two gene concepts, the BRCA1 gene provides another. In this case a Gene-P for use in identifying heritable risk for breast cancer was identified through studies on families in which such risk had been epidemiologically established. This resulted in the targeting of a locus and the construction of molecular probes for identifying some of the aberrant sequences the presence of which correlate with heightened incidence of breast cancer (Miki et al. 1994). It also resulted in

the identification of the "normal" sequence at the BRCA1 locus. This sequence provides the template from which a family of proteins are ultimately synthesized, and these proteins are (like it or not) referred to as BRCA1 proteins. BRCA1 proteins are expressed ubiquitously across tissue types and developmental stages (Welsch, Schubert, and King 1998). The BRCA1 proteins are expressed in different tissues at different times and appear to be associated with DNA repair, transcriptional transactivation of other Genes-D, and cell-cycle control (albeit all in the context of a complex and variable set of other factors) (Irminger-Finger, Siegel, and Leung 1999). Where the probes for BRCA1 as a Gene-P provide an instrument for predicting breast cancer risk, molecular studies on the BRCA1 sequence as a Gene-D have indicated that proteins derived from the "normal" sequence, in the cases of woman with "sporadic" breast cancer, fail to migrate to the cell nucleus and are rather found in the cytoplasm (Chen et al. 1995). Studies on BRCA1 as a developmental resource, that is, as a Gene-D, may well lead to important insights into the cellular dynamics of breast cancer on the basis of the elucidation of complex developmental processes that involve proteins derived from the "normal" BRCA1 sequence as well as many others. The potential medical significance of this is not easily *over*estimated inasmuch as 90 percent to 95 percent of breast cancers are *not* associated with any germline mutation, but rather are sporadic. Thus for the great majority of breast cancers there is no Gene-P to treat as a marker *before the fact*, as if it independently determined the phenotype. Studies on the dynamics of BRCA1 as a developmental resource, that is, as a Gene-D, with no phenotypic outcome preinscribed, have, not surprisingly, become widespread.

The common use of two qualitatively different explanatory strategies, that is, that of Gene-P and that of Gene-D, which yet share the name "gene" predictably lends itself to much easy confusion. The notion that there exists such a thing as a gene that is *simultaneously* a specific nucleic acid template *and* a preformationistic determinant of an organismic phenotype (i.e., the genetic *blueprint* we hear so much about) is based upon exactly this conflation. The Gene-P/Gene-D distinction is new to developmental systems theory literature. One of the benefits of this analysis is that it enables DST thinkers to acknowledge the fact of a valid role for Gene-P as tool, albeit coarse, yet useful under certain circumstances (such as clinical genetics counseling). As a concept, Gene-P, however, falls outside of the efforts of DST to formulate a perspective which does not presume the causal (or ontological) priority of any particular kind of entity and thereby maintains an explanatory openness on all empirical fronts. It is not Gene-P as such which should be condemned, but rather the conflation of the two gene concepts. Once distinguished from Gene-P and cut free of its baggage, Gene-D, the developmental resource gene, emerges as exactly the right gene concept for the purposes of DST.

In the absence of the Gene-P/Gene-D distinction, predictable consequences can be seen. Certain attempts to clarify "the" gene concept proceed by focusing on what amounts to one of the above while ignoring the existence of the other. Waters (1994) demonstrated how "genes became molecular" by focusing on how, in effect, Gene-P became molecularized while ignoring Gene-D. Godfrey-Smith (1999), conversely, has recently offered a treatment of the gene concept in terms of coding which amounts to identifying "the gene" with Gene-D through ignoring the history of Gene-P. Both of these studies have valuable things to offer on their respective sides of the divide—but attempts to seek clarity or coherence that fail to recognize the Gene-P/Gene-D distinction run the risk of bringing contaminating baggage across the border. Such is the case, I would suggest, for Neumann-Held's attempt (see chapter 7 of this volume) to provide DST with a unified gene concept. Her critical point of departure is the idea that the "classical molecular gene" cannot account for the making of a particular polypeptide. But if conceived as a Gene-D (or any other developmental resource) this would be fine. It is only a residual association with the Gene-P concept that leads one to expect more. Neumann-Held's solution is to pack everything into a revised gene concept that is causally necessary for the production of a particular polypeptide (which can only be known in retrospect) and call this the Process Molecular Gene (PMG).[2] The consequences of doing this are quite problematic. A typical eukaryotic Gene-D (as discussed below for the case of N-CAM) may provide the template resource for, let's say, ten different polypeptide forms,[3] based upon differential mRNA splicing. This would already increase the number of PMG names necessary by an order of magnitude. The PMG would have to include, for starters, both the complexes associated with transcription and the complexes associated with splicing, thus the PMG would jump up to a different level in the biological hierarchy, including as it would those intermediate levels of organization I refer to as multimolecular modules. But the composition of transcriptional complexes already are highly variable (with tissue type, developmental stage, physiological state of a cell, etc). And the regulatory impact of a transcriptional complex is not based solely on its composition but also upon the kinetics of its assembly (Miklos and Rubin 1996). Further, transcriptional complexes are also buffered so that variations in composition do not necessarily result in different transcriptional results (Gerhart and Kirshner 1997). So just at the level of attempting to include the most proximate factors involved in producing a polypeptide, there is an explosion of complexity and contingency that obviates the possibility of a manageable, let alone perspicuous, taxonomy.

A Gene-D denotes something, that is, a transcriptional unit of DNA, albeit a something that is only a resource from which different results may arise in different contexts. In trying to pack some causally complete set of reactants into a revised gene concept, Neumann-Held falls victim to a conflationary temptation. PMGs would not denote anything more durable than singularities in the life history of an organism and thus would be of negligible biological value. Even the attempt to limit an expanded gene concept to "just that" which pertains to the linear polypeptide backbone is unacceptably arbitrary, inasmuch as nascent polypeptides are enzymatically modified in the lumen of the endoplasmic reticulum simultaneous with their translational elaboration at the surface of the rough ER. In other words, biologically there is no temporal boundary between polypeptide backbone synthesis and polypeptide folding and modification. The PMG concept sacrifices both what the Gene-P and the Gene-D concepts provide, and offers nothing of value in their place. The solution to the Gene-P/Gene-D incompatibility problem is not to try to split the difference. These gene concepts do entirely different kinds of work in entirely different ways.

The Gene-P/Gene-D distinction manages to help address certain criticisms of DST. Critics of the developmental systems perspective, most emphatically Phillip Kitcher (in press), but also Sterelny, Smith, and Dickison (1996), have questioned whether DST proffers an extreme and unwieldy holism that, while possibly true to the multidirectional causal complexity of reality, obviates the possibility of actual scientific progress. But here we can show that Kitcher and like-minded scholars are guilty of mixing up apples and oranges. If DSTers insisted that *every* causal story, for whatever explanatory purpose, which happens to involve any *Gene-D* had to take into account every other developmental resource which might play a causal role at any time, or any place, in the life history of an organism, he/they would be justified in their criticisms. But the criteria of causal relevance for explanatory stories for DST are contingent upon the contexts of interest—indeed, that is much of the point. To separate out the instrumental *as-if* approach of Gene-P concepts, from the mechanistic basis of the DST approach is yet to say *nothing* about the *scope* of causal factors that must be invoked for

any particular explanatory story to be consistent with the tenets of the DST perspective.

But just how tractable are explanatory stories based solely upon the contingent causal relations of Genes-D and other contextually relevant developmental resources? What the sad endurance of that tired old dichotomy consisting of (conflated) genes and (ill-defined) environment has helped to obscure, are the many levels of biological ordering that mediate between individual molecules and whole developmental systems. To give up the preformationist umbilical cord is not to drop into an abyss of limitless complexity but rather to remain empirically open to discovering what level of biological ordering is most relevant for one's explanatory purposes. Those who have limited their explanatory sights to "genes and environment" have fallen behind the research tide. Functionally conserved, multimolecular modules have emerged as new units of development, morphology, innovation, and variation, at an intermediate level of biological ordering. It is time to turn our attention to the nature of these modules.

Modules in the Middle

Most conserved processes in metazoa operate under contingencies, that is, their activity depends on conditions, both external and internal to the organism. (Gerhart and Kirshner 1997)

The biological use of the term *module* is in itself rather ambiguous. Modules have been spoken of in many different places and with different explanatory intents, some of which would have little in common with present purposes. The sense that runs through *any* ascription of modularity is that of a unit that is a component part of a larger system and yet possessed of its own structural and/or functional identity. But modules have been posited from both "top-down" and "bottom-up" research directions, sometimes as *explanans* and sometimes as *explanandum*. When one begins with the phenomenal properties of a system as a whole, and then attempts to explain its working on the basis of a postulated modular organization, as cognitive scientists such as Minsky have done with the brain, then modules are arrived at from a top-down direction. Such modules have not been empirically observed but speculatively brought forth as *explanans* (not as *explanandum*). Modules of this sort may well express a preformationist viewpoint, although that does not necessarily need to be the case. One could, for example, begin "at the top" and postulate an ad hoc formation of modularized structure contingent upon circumstance. But in evolutionary psychology, where a new form of module-talk is in vogue, it is a clear spirit of preformationism that carries the new day. For example, evolutionary pyschologists have become prone to begin with evidence for some cognitive phenotype, (e.g., social cheat detection; depressive/socially submissive behavior, etc.) and then proceed to argue for its evolutionary origins as a proto-human Pleistocene adaptation. The cognitive capacity/phenotype (whether still of adaptive value or not) is then construed to be the expression of a developmentally invariant, preformationistically transmitted module that has been passed along from generation to generation ever since. The cautionary point is just that the concept of module itself does not specifiy its place along a preformationism-epigenesis axis. Having evidenced the slipperiness of the module idea and the fact that it is not just any old biological module that is being embraced, it is time to clarify that sense of module which is of interest.

The "genetic revolution" of the twentieth century did not result in a search for any form of subcellular modules, nor any expectation of finding such. Rather, the recognition of modularity came as a surprise. What was expected, by virtue of conflating Gene-P with Gene-D, was something like direct relationships, able to be mechanistically elucidated, between transcriptional units and organismic phenotypes. What has come to be found are modular patterns of organization both "above" and "below" the level of the full transcriptional unit. Of the latter, it

was famously found that "genes come in pieces" (Gilbert 1978), which is to say that the typical transcriptional unit extending from start to stop codons, turns out to be a sea of nonprotein template sequences ("introns") within which islands of template-sequence "exons" are embedded. What shores up the sense in which exons are indeed modules is the ample evidence that the protein fragments which exon templates "code-for" have structural and functional integrity (Gilbert and Glynias 1993). This is well illustrated by the example of N-CAM. The Gene-D for N-CAM, that is, that particular transcriptional unit, contains nineteen different exons which serve as modular resources from out of which many different N-CAM protein forms can ultimately be fashioned. Any N-CAM molecule, at the linear peptide level, is indebted to some specific *subset* of N-CAM Gene-D exon modules (but never all of them). NCAM proteins are classed into four main groups based on size (110KDa, 120KDa, 140KDa, 180KDa) and plasma membrane attachment (see figure 8.1). Where the 140KDa and 180KDa classes traverse through the plasma membrane the 120KDa class is linked to the membrane only superficially through an auxiliary connector. No N-CAM does both. And the basis for this difference is the either/or inclusion of one of two different exon modules into the final RNA "transcript." The 140KDa and 180KDa must be derived from a transcript which includes exon #16 in order to be able traverse the plasma membrane, but must lack exon #15, as where the 120KDa NCAM class is derived from a transcript that must include exon #15 in order to be able to associate by auxiliary connector, but must lack exons #16–19 (figure 8.1). The process by which the complete set of exons, at the level of the messenger RNA transcript, is reduced to a *functional* set through the excision of selected exons, is referred to as "splicing."

Modularity, at the level of individual genes (Gene-D), which is the rule not the exception for the eukaryotic cell and all metacellular organisms, provides for developmentally contingent

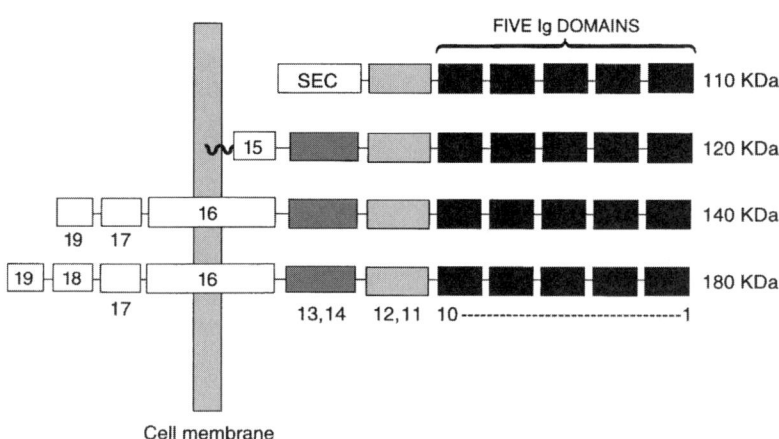

Figure 8.1
A schematic diagram of the four main classes of N-CAM protein. See text for details.

flexibility in the expression and realization of "gene-products" from out of the resource base which any Gene-D represents. N-CAM is just such a modularized Gene-D resource, but it is also just one member of a "superfamily" of modularized genetic resources whose kinship is defined by the possession of homologous modules. The "immunoglobulin superfamily" consists of many different genetic open reading frames (Genes-D), including those that provide the template resources for the various immunoglobulins, immune cell receptors, certain growth factor receptors, and a variety of cell-adhesion molecules in addition to N-CAM (Walsh and Doherty 1997). Modularity and homology have come together as complementary themes arising out of research into subcellular biochemistry and molecular biology. Much of the evolutionary novelty associated with increasing organismic complexity, it turns out, has been achieved through the reshuffling and mixing and matching of modular exon units to form families of homologous genetic (Gene-D) resources. This has been particularly pronounced with respect to those molecules associated with developmentally and functionally contingent associations between cells and other cells, and cells and extracellular matrices, where at least three such Gene-D homology groups have been identified: the integrins, the cadherins, and the immunoglobulin superfamily. A necessary molecular concomitant of organismal complexity appears to be that of great developmental versatility in the resources available for constructing cell-to-cell and cell-to-matrix linkages. The significance of this for developmental systems thinking will be addressed further below. It is also more than suggestive that during the course of 3.5 billion years, metacellular life has not evolved from prokaryotes whose modular structure at the DNA level appears to have been lost (Gilbert and Glynias 1993). Can it be that a necessary prerequisite for differentiated multicellular organization is a capacity for contingent cell-to-cell linkages that requires a proliferation of linker-molecule variants generated by modular reshuffling?

In their quasi-manifesto of 1996, developmental biologists Gilbert, Opitz, and Raff hailed the return of the three key concepts that had been shunted aside by geneticists. These concepts, which in a strong sense are interlinked, are that of macroevolution, homology, and morphogenetic fields. Homology, which had formerly provided a pivotal linkage between ontogeny and phylogeny, was cast aside by population genetics. If evolution was to be thought of strictly in terms of changes in gene frequencies, then there would be no place for relations of similarity in explaining evolutionary change. All could be accounted for on the basis of allelic difference, selection, and population dynamics. However, where nineteenth century biology discovered relations of homology on the macroscopic morphological level, later twentieth century biology has discovered relations of homology on the subcellular and molecular levels. Unlike nineteenth century homology, the new understanding is based upon modular units that appear to predate the divergence of any of the major body plans.

The strongest impetus for acknowledging the role of such ground-level homologies has come from further investigations into homeosis. Homeosis pertains to the progressive segmentation of body parts that had been genetically analyzed in *Drosophilia*. Homeotic genes were found to all contain a homologous sequence module called a "homeobox," to be arrayed along a single chromosome, and to be activated sequentially in sequential segments indicating a direct regulatory involvement in the processes of developmental segmentation. The big surprise came with the discovery that all of the above holds for vertebrates as well as invertebrates. The segmentation of invertebrates such as *Drosophilia* and that of vertebrates was thought only to be analogous, not homologous. The finding that genetic resources, their relative position on chromosomes and the activation cascade processes associated with developmental segmentation in both invertebrates and vertebrates are homologous, was startling. Similarly it was found that the Pax-6 gene homologs, which are functionally (experimentally) interchangeable across vertebrate and invertebrate taxa, appear to play a comparable role in eye morphogenesis, although vertebrate and invertebrate eyes are certainly not homologous in classical morphological terms.

Vertebrate and invertebrate bodily segmentation, and vertebrate and invertebrate eyes are fundamentally different. What can properly be gleaned from these findings is that there are biological processes that are more fundamental than even these differences. Not only genetic sequences, but also whole processes, including activation circuits are modular in nature and highly conserved. The vertebrate eye and the invertebrate eye, the vertebrate segmentation scheme and the invertebrate segmentation scheme, have homologous modules in common. And these process modules are likely to go back to the one-celled eukaryotic stage. Novel morphologies appear to be the result of novel linkages and novel timing, that is, novel developmental patterns, between modular components established within and across cellular boundaries. The developmental emergence of dynamic patterns of modular organization across cells results in morphogenetic fields. The recognition of highly conserved, dynamic, multimolecular modules at many intermediate levels of organization provides developmental systems theory, I would suggest, the materials for articulating its view.

Where Gilbert, Opitz, and Raff have been the harbingers of a new rapprochement with concepts of embryology past, Gerhart and Kirshner have contributed much toward an understanding of the dynamics of modularity and contingency. The core idea is just this. Evolution of complex multicellular organisms has proceeded through the pulling-apart of highly conserved processes whose linkages have become contingent. When linkages are contingent they can be "regulated" and made sensitive and responsive to surrounding conditions. Self-organizing sequential constructions of modular unions in regulatory linkage are truly cycles of contingency. Adaptive developmental flexibility is achieved through the capacity to generate variable alternative linkages that are subject to selective stabilization in context sensitive ways. A marvelous example of this can be seen in Kirshner's work on the microtubule system.

Microtubule structures are central to cellular morphogenesis: to the general formation of cell shape and internal compartmentalization, to the formation of axons and dendrites in neural tissue and of apical and basolateral domains in epithelia, to the directedness of cellular secretions, and so on. The microtubule system is also the source of the mitotic spindle fibers that pull homologous chromosomes apart during mitosis. Microtubules grow radially out from a microtubule organizing center (MTOC) and are found to be in a state of "dynamic instability." Microtubules emanating from the MTOC are simultaneously growing out toward the cell periphery and depolymerizing back to the center, albeit at different rates. Polymerization to the cell periphery from the cell center takes about ten minutes, whereas complete depolymerization takes only approximately two minutes (Gerhart and Kirschner 1997). The accelerated rate of depolymerization is dependent upon GTP hydrolysis. Why cells would expend significant amounts of stored energy (in the form of GTP) on *de*stabilizing microtubule structures had been a mystery. Cellular morphogenesis is a central part of organismic development, and microtubule filament assembly is a key feature of cellular morphogenesis. Kirshner's solution was to see processes of variation and selection as central to how an organism solves problems of self-organizing ontogeny. The expenditure of GTP energy buys directional variation. It prevents microtubules from lingering too long in the absence of contextual reinforcement (Mitchison and Kirschner 1984; Kirschner and Mitchison 1986). The microtubule system becomes a kind of context sensor. The ability to generate stable filamentous structures comes from within, but the orientational cues are contextual. Microtubule filaments quickly depolymerize unless they are selectively stabilized by contacts or signals at the cellular periphery (Gerhart and Kirshner 1997). Microtubule filament stabilization leads to cellular differentiation. Once a cell becomes spatially differentiated, it becomes itself an orientational force within a morphogenetic field. The microtubule system is in effect a subcellular module that participates in context formation and con-

text sensitivity—in the reciprocity of centripetal and centrifugal vectors of causal influence—emanating from and entering into individual cells.

The microtubule system causes cell-morphogenesis to be contingent upon the availability of cues in the larger morphogenetic field (which is itself the product of processes of interactive self-assembly). There are, however, even more general mechanisms for pulling core functional processes apart and rendering their connections contingent upon internal and external factors. The most general mechanism for doing this is that of phosphorylation. By adding a highly charged (negative) phosphate group to a protein, the enzymatic or other activity of that protein can generally be disrupted, and made contingent upon removal of the phosphate group. The enzymes that attach phosphate groups onto proteins are called "kinases." The enzymes that remove them are "phosphatases." There are an estimated two thousand genes (Gene-D) that are template resources for kinase synthesis and another one thousand for phosphatases (Hunter 1995). Thus 3 percent to 6 percent of the human genome is associated with this particular contingency-making process alone. Modularity is built of the ability to uncouple core functions and make the linkages to one another contingent upon context specifics. The proliferation of linkage possibilities is seen, among other places, in the elaboration, by way of superfamilies, of Gene-D resources for cell adhesion molecules as mentioned above. Many multiple cell-linkage devices, built of modular variation, make for sensitive context discrimination and the transmission of developmentally contingent vectors of influence in all directions. Any particular transmembrane cell-adhesion molecule will serve both selectively to stabilize bonds to other cells, or extracellular matrices, and to transmit signals to within the cell, which feed into common phosphorylation pathways that influence the state of the cell with respect to cell division, selective transcriptional activations or deactivations, secretory activity, and more. Multicellular assemblages of self-organized, mutually selectively stabilized cells constitute morphogenetic fields. It is little surprise that many of the Genes-P that have been implicated in tumor pathogenesis, as in the case of colorectal cancer, have led to the identification of Genes-D associated with cell-adhesion protein production. The loss of the stabilizing influence of a morphogenetic field could well result in a cellular trajectory that, while adaptive for the cell, becomes perilous for the organism (Moss in press).

Modularity has arisen as a theme in subcellular biology unexpectedly, first of all, as an *explanandum* not as an *explanans*. As investigators such as Gerhart and Kirshner and Gilbert, Opitz, and Raff (and many others) focus upon the meaning and significance of this, new and promising styles of explanation have begun to emerge. These module-centered explanatory approaches ideally complement the intuitions and explanatory strategies of DST and the Gene-P/Gene-D distinction that I have put forth. With Gene-P properly defined, demarcated and set aside to play its own role as an instrumentally simplified predictor device, Gene-D need no longer be burdened with responsibility for providing shortcuts to the phenotype but may be left to simply play the role of a resource designator. DST prescribes an explanatory outlook that is open to all fields and all directions of influence—choosing its level of gaze as befits circumstance. For many developmental accounts, it now appears, a story about the life history of context-sensitive modular associations may prove to be most perspicuous.

Notes

1. The idea that the Mendelian gene constitutes an expression of "instrumental preformationism" was introduced by Rafael Falk (1991, 1995).

2. Neumann-Held's position expressed in the current volume was adumbrated in Griffiths and Neumann-Held (1999). In this slightly earlier paper the limits of both the "classical molecular gene" and the

Dawkinsonian "evolutionary gene" are discussed and aptly criticized. However, in both cases the remedy that they propose is to save the "gene" from the inadequacy to which it would be condemned if understood as only some piece of DNA. While I am in sympathy with the diagnosis, I take issue with the proposed cure.

3. This is a conservative estimate. There are more than a hundred possible splicing isoforms (Zorn and Krieg 1992), and twenty-seven alternatively spliced forms are, for example, reported to be expressed during rat heart development alone (Reyes, Small, and Akeson 1991).

References

Carlson, E. A. (1966). *The Gene: A Critical History.* Philadelphia: Saunders.

Chen, Y., C. F. Chen, D. J. Riley, et al. (1995). Aberrant subcellular Localization of BRCA1 in breast cancer. *Science* 270: 789–791.

Falk, R. (1991). The dominance of traits in genetic analysis. *Journal of the History of Biology* 24: 457–484.

Falk, R. (1995). The struggle of genetics for independence. *Journal of the History of Biology* 28: 219–246.

Gerhart, J., and M. Kirschner. (1997). *Cells, Embryos, and Evolution: Toward a Cellular and Developmental Understanding of Phenotypic Variation and Evolutionary Adaptability.* Malden, MA: Blackwell Science.

Gilbert, W. (1978). Why genes in pieces? *Nature* 271: 501.

Gilbert, W., and M. Glynias. (1993). On the ancient nature of introns. *Gene* 135: 137–144.

Godfrey-Smith, P. (1999). Genes and codes: Lessons from the philosophy of mind? In V. Hardcastle (Ed.), *Biology Meets Psychology: Constraints, Conjectures, Connnections.* Cambridge, MA: MIT Press.

Griffiths, P., and E. Neumann-Held. (1999). The many faces of the gene. *Biosciences* 49: 656–662.

Hunter, T. (1995). Protein kinases and phosphatases: The yin and yang of protein phosphorylation and signalling. *Cell* 80: 225–236.

Irminger-Finger, I., B. Siegel, and W. Leung. (1999). The functions of breast cancer susceptibility gene 1 (BRCA1) product and its associated proteins. *Biological Chemistry* 380: 117–128.

Jilling, T., and K. Kirk. (1997). The biogenesis, traffic, and function of the cystic fibrosis transmembrane conductance regulator. *International Review of Cytology* 172: 193–241.

Johannsen, W. (1923). Some remarks about units in heredity. *Hereditas* 4: 133–141.

Kerem, E., and B. Kerem. (1995). The relationship between genotype and phenotype in cystic fibrosis. *Current Opinion in Pulmonary Medicine* 1: 450–456.

Kirschner, M., and T. Mitchison. (1986). Beyond self-assembly: From microtubules to morphogenesis. *Cell* 45: 329–342.

Kitcher, P. (in press). Battling the undead: How (and how not) to resist genetic determinism. In R. Singh, C. Krimbas, J. Beatty, and D. Paul (Eds.), *Thinking about Evolution: Historical, Philosophical, and Political Perspectives*. Cambridge: Cambridge University Press.

Miki, Y., J. Swenson, D. Shattuck-Eidens, et al. (1994). A strong candidate for the breast and ovarian cancer susceptibility gene BRCA1. *Science* 266: 66–71.

Miklos, G., and G. Rubin. (1996). The role of the genome project in determining gene function: Insights from model organisms. *Cell* 86: 521–529.

Mitchison, T., and M. Kirschner. (1984). Dynamic instability of microtubule growth. *Nature* 312: 237–242.

Moss, L. (in press). *What Genes Can't Do: Prolegomena to a Philosophy beyond the Modern Synthesis*. Cambridge, MA: MIT Press.

Reyes, A., S. Small, and R. Akeson. (1991). At least 27 alternatively spliced forms of the neural cell adhesion molecule mRNA are expressed during rat heart development. *Molecular Cell Biology* 11: 1654–1661.

Sterelny, K., K. C. Smith, and M. Dickison. (1996). The extended replicator. *Biology and Philosophy* 11: 377–403.

Walsh, F., and P. Doherty. (1997). Neural cell adhesion molecules of the immunoglobulin superfamily: Role in axon growth and guidance. *Annual Review of Cell and Developmental Biology* 13: 425–456.

Waters, C. K. (1994). Genes made molecular. *Philosophy of Science* 61: 163–185.

Welsch, P., E. Schubert, and M. King. (1998). Inherited breast cancer: An emerging picture. *Clinical Genetics* 54: 447–458.

Zorn, A., and P. Krieg. (1992). Developmental regulation of alternative splicing in the mRNA encoding *Xenopus laevis* neural cell adhesion molecule (NCAM). *Developmental Biology* 149: 197–205.

Eva Jablonka

During the last two decades, most thinking about inheritance and evolution has been deeply influenced by what has been learned about the molecular nature of the gene. The structural organization of the gene, the conditions for its transmission, the way in which it is transmitted, and the way it varies, have shaped the modern view of heredity and have been very influential in molding ideas about evolution. This influence went beyond the strictly biological realm and affected ideas about the evolution of culture. However, for a gene-like concept to be used in explanations of nongenetic evolution, a more general concept was necessary. Such a concept, the "replicator," was suggested by Dawkins (1976). The replicator was defined as "anything in the universe of which copies are made" (Dawkins 1982: 83). This definition seems, at first sight, broad enough to accommodate different types of heredity and reproduction, since "copying" can be understood to include many types of processes. However, as Dawkins, Hull, and many others made clear, the replicator entails a very special kind of copying, which presupposes that only instructions or representations (which is what replicators embody) rather than the implementations of representations, can be meaningfully "copied" or inherited. Following the distinction between genotype and phenotype, which was suggested by Johannsen at the turn of the twentieth century and molded the theory of the emerging discipline of genetics (Johannsen 1911), Dawkins suggested a distinction between replicators and vehicles. He defined the vehicle as "any unit, discrete enough to seem worth naming, which houses a collection of replicators and which works as a unit for the preservation and propagation of those replicators" (Dawkins 1982: 114). The vehicle was called "interactor" by Hull, to emphasize its active functional role as a propagator of replicators (Hull 1980). Vehicles or interactors are, of course, not only carriers of replicators, but they are also their products.

Development is something that happens to vehicles (and is controlled by replicators) to ensure the further propagation of replicators. While replicators are units of heritable variation, vehicles are targets of selection. The generation of new variant replicators is assumed to be independent of the selective environment (which acts on vehicles), and of the developmental process that vehicles undergo. The replicator is clearly very similar to the gene, the unit of Johannsen's genotype, and it carries much of the latter's baggage.

The view of inheritance embodied in the replicator concept affects the way in which evolution is understood, and leads to a view of evolution that reflects the modern neo-Darwinian version of Darwin's original selection theory. According to Darwin's theory, in a world in which there are interacting entities with the properties of multiplication, heredity, and heritable variation that affects the chances of multiplication, natural selection will necessarily occur, and in the long term, adaptive evolution will follow (Maynard Smith 1986). In this general form, Darwinian selection theory does not specify what the entities should be, how they multiply, how variations are inherited, or how they are generated. It also does not make a priori assumptions about the relationship between heredity and development. It is the generality of Darwin's selection theory that gives it its great explanatory power and its potential applicability to different domains of historical change.

For Darwinian selection theory to be fruitfully applied to a particular domain, its major concepts have to be specified for that domain. The replicator seems to fit particularly well the molecular, neo-Darwinian version of Darwinism (or genic neo-Darwinism). According to genic neo-Darwinism, nucleic acids are the sole units of heritable variation, the transmission of these units is independent of their expression, and the generation of genetic variations is not adaptively guided by the selective environment or the developmental history of the organism.

This replicator-centered, gene-derived view of heredity is, however, not only severely limited, but also severely misleading. There are multiple inheritance systems, with several modes of transmission for each system, that have different properties and that interact with each other. They include the genetic inheritance system (GIS), cellular or epigenetic inheritance systems (EISs), the systems underlying the transmission of behavior patterns in animal societies through social learning (BISs), and the communication system employing symbolical languages (SIS) (Jablonka, Lamb, and Avital 1998). These systems all carry information, which I shall define here as the *transmissible organization of an actual or potential state of a system.*

In addition to the intrinsic properties of the different inheritance systems, the feedback loops formed between the organism's activities and its ecological and social environment often create conditions for the reconstruction of ancestral phenotypes in descendant generations. Developmental and ecological legacies may be said to be passed on between generations. Inheritance systems with replicator-like properties are very unusual, and certainly do not represent or sum up the many ways in which heritable variations are transmitted across generations. I use "transmission" in a general way, to denote all the processes leading to the regeneration of the same type of organization-states across generations. This includes the direct transfer of resources, as well as the activities that lead to the reconstruction of ancestral phenotypes. In what follows I shall discuss different inheritance systems and compare them with respect to those properties that seem to me most pertinent to the understanding of inheritance: the type of variation transmitted; whether or not information is encoded; the type of mechanism leading to the regeneration of variations in the next generation; the relationship between development and the generation of new heritable variations (table 9.1).

I shall then discuss the transmission of parental and group legacies through niche construction, and argue that it is the whole developmental system, with all its different and interacting inheritance systems, that has to be considered when we think about the transmission of variations from one generation to the next (Oyama 1985; Griffiths and Gray 1994). This means that the replicator/vehicle dichotomy has to be discarded, and we must go back to a single (though complex) minimal unit—a unit that is simultaneously a unit of development, multiplication, and heritable variation—the reproducer (Griesemer 2000).

I start with the most fundamental and best understood inheritance system in living organisms, the genetic inheritance system, which is based on DNA replication.

The Genetic Inheritance System (GIS)

The information in the genetic inheritance system is organized in the sequence of nucleotides in nucleic acids, which in most extant organisms is DNA. The gene is a template made up of nucleotides whose sequential organization can be transformed through a complex process of decoding into functional RNA and proteins. Genetic information is thus *encoded*. Encoding means that one system of transmissible elements (signs) represents not just itself, but also another system of elements that combine to form the actual, functional, messages. In the GIS, nucleotide triplets in a structural gene are elements of the DNA system, and they represent amino acids in a protein, which is the functional "message." In natural language, utterances represent actual objects and events in the world, as well as other words and meaning-relations.

Information is also carried in DNA regions that can control the decoding of other DNA sequences. The noncoding but regulatory regions in DNA cannot be said to encode information in the same sense as the coding regions. However, particular sequences (of varying length) are spread throughout the genome and perform

Table 9.1
Types of information and modes of transmission for different systems of inheritance

Inheritance System	Variation transmitted		Information		Mode of transmission		Type of Heredity
	Unit	Origin	Alteration	Encoding	Type	Direction	
							Niche construction—variant interaction of organism and environment can be transmitted
							Unlimited at the cell level — Unlimited at the organism and lifestyle level
GIS (genetic)	DNA sequence (gene)	Blind & some patterned	Modular	Encoded	Modular	Mostly vertical	Unlimited
EIS (cellular)							
Steady-state	Activity state of metabolic cycle	Blind & patterned	Holistic	Nonencoded	Holistic	Mostly vertical	Limited
Structural	3D complex	Blind & patterned	Holistic	Nonencoded	Holistic	Mostly vertical	Limited
Chromatin marking	Pattern of chromosome marks	Blind & patterned	Holistic / Modular (methylation)	Nonencoded / Can be encoded	Holistic / Modular (methylation)	Mostly vertical	Limited / Limited?
Epigenetic (organismal) Inducing substance	Physiological state	Blind & patterned	Holistic	Nonencoded	Holistic	Mostly vertical	Limited
BIS (behavioral)							
Inducing substance	Pattern of behavior	Often induced (patterned)	Holistic	Nonencoded	Holistic	Mostly vertical	Limited
Non-imitative social learning	Pattern of behavior	Often learned (patterned)	Holistic	Nonencoded	Holistic	Vertical & horizontal	Limited
Imitation	Pattern of behavior	Often learned (patterned)	Holistic	Nonencoded	Holistic	Vertical & horizontal	Limited
SIS (symbolical)	Symbolic form and content	Learned (patterned)	Modular & holistic	Encoded	Modular & holistic	Mostly horizontal, some vertical	Unlimited

sequence-typical regulatory functions, so general types of functions can be inferred from sequence organization. Such regulatory sequences thus form a kind of higher order "code."

The organization of information in DNA is modular (or digital), that is, it is decomposable into separate discrete units drawn from a standard set, (the units in DNA are the nucleotides A, C, T, G), and the information is alterable digit by digit. Following Szathmáry (1995), a replication process that proceeds digit by digit will be called *modular replication*. The genetic system is the prime example for a system that is modularly replicated. The enzymatic machinery that replicates the DNA, or that edits and repairs it, is largely indifferent to its sequence organization. This means that a sequence that has beneficial effects when decoded will be replicated and repaired with the same fidelity as one with deleterious effects, or a sequence that is completely nonfunctional. Furthermore, the transmissibility of the template remains unaltered following its replication. Usually transmission is vertical, from parents to offspring, but occasionally it can be horizontal, so genetic information can be transmitted between nonrelated individuals, including individuals belonging to different species.

The modular nature of the replication and alteration of information allows for the inheritance of many combinations of modules—a DNA molecule with ten linearly linked nucleotides has more than a million possible variant sequences. This means that the evolutionary potential of a modularly alterable and transmitted unit, such as a gene that consists of hundreds of nucleotides, is very large. The number of possible sequences greatly exceeds the number of individuals in any realistic system. Such a system can be said to have *unlimited heredity* (Szathmáry and Maynard Smith 1993; Maynard Smith and Szathmáry 1995).

Until recently, the generation of variations in DNA has been assumed to be random with respect to the selecting environment. Variations were assumed to be exclusively the consequence of the meiotic reshuffling of genes (in sexually reproducing organisms), and of several classes of errors in DNA maintenance. Errors can be due to physico-chemical damage to the DNA, they can occur during DNA replication and repair, and they can result from the activity of genomic parasites: genetic elements that multiply excessively and move from site to site in the genome. Errors that are not removed or repaired accurately by the DNA maintenance machinery were assumed to be the ultimate raw material for evolution by natural selection. Although there is no doubt that a lot of variation in DNA is indeed random in this sense, the view that *all* variation is random has been challenged.

The challenge has come from several directions. It has been shown that different nucleotide sequences differ in the likelihood that they will be damaged, invaded by genomic parasites or replicated inaccurately. The rate and type of new variation may thus depend on how the nucleotides in the sequence are organized, and this organization may be adaptive. For example, Moxon and his colleagues have shown that in the pathogen *Haemophila influenza* the genes that influence its antigenicity are highly mutable because the short tandem repeats in them make them prone to mutation by recombination and strand slippage. The high mutation rate in these genes is advantageous, because it enables this pathogenic organism to evade the immune system of the host (Moxon et al. 1994). The sites in which mutations preferentially occur are the result of adaptive evolution. Moreover, mutation rate can increase selectively not only at sites but also in conditions in which a higher mutation rate is selectively beneficial. Wright (1997) has shown that amino acid starvation in *E. coli* increases the transcription of genes that help the cells survive longer, and concurrently increases the mutation rate in these genes. This condition-dependent increase in mutation rate is adaptive since such targeted mutation in the relevant genes may "rescue" the cell without greatly increasing the load of mutation. It seems that through natural selection the mechanisms that allow selective control of gene expression have been coupled with mechanisms that determine the fidelity of copying, so that the inducible system that turns genes on and off also turns the production of mutations on and off.

Such "targeted" mutations cannot be said to be random in the classical sense, since adaptively advantageous mutations are preferentially (though not exclusively) induced under the appropriate conditions and in the relevant domains. Randomness has not been eliminated, but it has been restricted and channeled. However, the mutations are not goal-directed in any teleological sense, and their targeted production is the consequence of natural selection that had acted on random variations. Variation has been targeted by selection to be preferentially generated in a subset of sites, under particular conditions. It is difficult to know how to define such variations. The term *patterned variation*, which has been suggested by the economist Ekkehart Schlicht with respect to cultural evolution, is the one I choose to use in this paper (Schlicht 1997). It is better than previously suggested terms such as directed, adaptive, induced, and guided variation because it does not carry the teleological connotation of premeditated design, yet does carry the connotation of some degree of preexisting structuring (by past natural selection). Once a system for generating patterned variation has evolved, it channels and guides evolution.

From an evolutionary point of view the existence of a cellular system for the production of patterned variations makes good sense. It would be remarkable if a cellular system for targeting the generation of variations had not evolved during the four billion years since life appeared on earth. It is quite easy to see how the enzymatic genetic engineering kit that all cells use to rearrange, amplify, and delete pieces of their DNA could have been modified by selection to allow the genome to respond to different reoccurring types of environmental stress (Shapiro 1997).

The ability to generate patterned variations forges direct links between heredity, develop-

ment, and evolution. The generation of patterned variation is part of the developmental process no less than changes in transcriptional activation of genes, although the effect of changes in DNA may often last longer than changes in transcriptional activity. The process of generating patterned variation is part of both development and evolution. Although there is a certain (short-term) degree of autonomy of heredity and development if mutations are random, if they are patterned, heredity loses this partial independence.

The Epigenetic Inheritance Systems (EISs)

Epigenetic inheritance systems are the systems underlying cellular heredity. It is well known that once cells become determined during development, they often maintain their functional and structural characteristics through many cell divisions, even though the stimuli that first induced their determined state early in development were transient, and are no longer present. Kidney cells and fibroblasts within the same organism have identical DNA base-sequences, yet each cell type "breeds true": Kidney cells transmit their functional state to daughter kidney cells, while skin fibroblast cells transmit their very different cellular phenotype to their descendants. The mechanisms that are responsible for this cellular inheritance have been termed epigenetic inheritance systems. The transmission of heritable epigenetic variations is possible not only within individuals, but also between generations of individuals, so EISs can have direct evolutionary importance.

Three types of epigenetic inheritance systems (EISs) have been described (Jablonka and Lamb 1995). The first type of EIS is the steady-state system, which is based on the activity of self-sustaining feedback loops. It was first described theoretically by Wright (1945), and has been found in many biological systems. In its simplest form, a gene produces a product as a result of

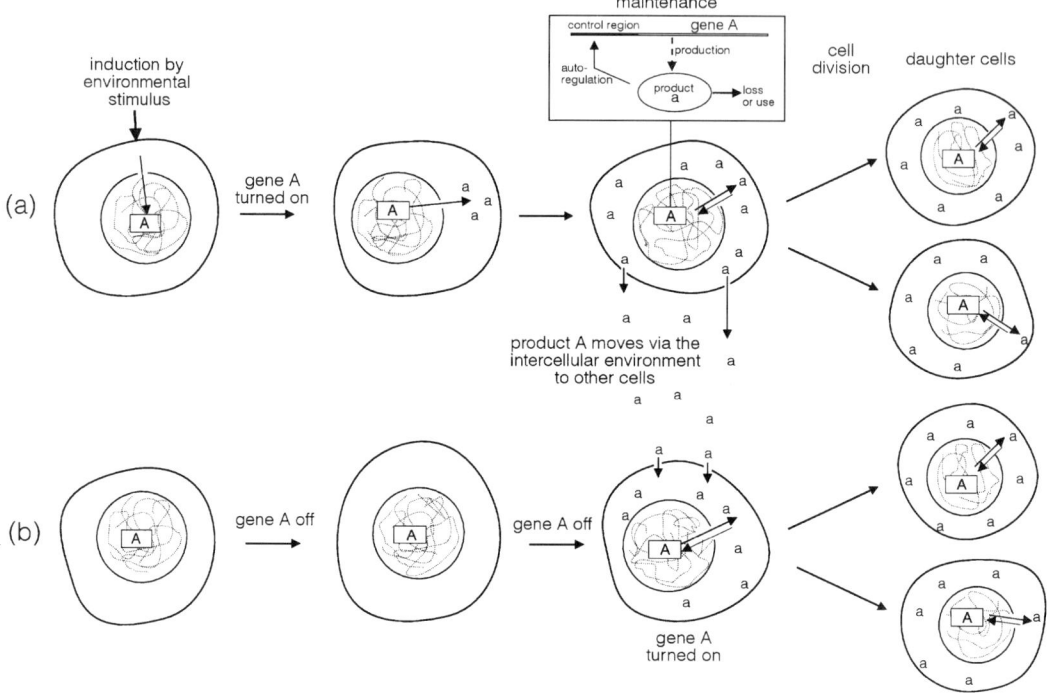

Figure 9.1
A steady-state system showing the perpetuation of an induced active state through cell division. (a) After induction, gene A is turned on and its product, *a*, positively regulates its own activity. The regulator *a* need not be a direct protein-product of gene A, but can be the metabolic product of the direct (protein) product, a small metabolite with regulatory function. The box shows the self-regulation of the genetic circuit. (b) The regulatory product *a* diffuses into the environment, enters into inactive cells, turns on gene A, and hence leads to the self-sustaining activity of the circuit in these cells.

induction by an external developmental or environmental stimulus, and this product then stimulates further activity of the gene (through positive self-regulation) even when the original external inducing stimulus has disappeared (figure 9.1). Once switched on, the cell lineage continues to produce the gene-product unless its concentration falls below some critical threshold value. Two genetically identical cells can therefore be in two alternative states ("on" and "off"), and both states can be self-perpetuating, even when the inducing environment changes. Thus two genetically identical cells in the very same environment may be heritably different because of the prior, different, developmental history of their ancestor cells. As long as the concentration of the products of the self-sustaining cycle does not fall below a critical threshold, the active, "on," state is maintained following cell division; once the concentration falls below the threshold, the cycle is in the "off" state, which is also maintained. The states of activity and inactivity are reproduced in daughter cells as an automatic consequence of cell division, and transmission is an integral part of growth and multiplication. The generation of the activity state is part of development, yet the developmental states can be faithfully transmitted within the cell lineage for many generations.

The information reproduced in this type of system is *nonmodular or holistic* (here I follow the distinctions, but not the terminology, of Szathmáry 1995). Although the cycle can be divided into discrete modules (modular gene A, modular product *a*, modular regulatory domain), the *functional* state cannot be transmitted module by module. It can only be transmitted when the processes of interactions among components are regenerated in the daughter cells. Changes in any one component usually prevent the transmission of the whole cycle. It is only the state of activity of the whole cycle that can vary. However, cellular states may also be transmitted horizontally, between lineages. If the positively regulating product is not only transferred to daughters cells as an inevitable part of cell division, but also diffuses to the cell's environment, it may "infect" neighboring cells from another lineage and induce its own activity state in them. Rather than inheriting the phenotype through descent, the nondescendant cells are interacting with the environment that the "infecting" cells have modified and become phenotypically identical to them through this interaction (figure 9.1b).

Often each individual self-sustaining cycle can have only two states ("on" or "off"), and the system can move only between two states, so nothing evolutionarily very interesting can occur at this level. The number of variant, functional, and heritable states that every single cycle can show is very small, much smaller than the number of individuals the population can include. The system therefore can be said to show *limited heredity*. However, within a cell there are often several independent cycles. More than a million variant cell states are possible if a cell has twenty different cycles! New developmental conditions can induce changes in the activity states of several cycles in cells, producing many variant states, which can then be subject to selection.

Thus, at the level of the cell, the inheritance of functional states may be practically unlimited, and cumulative evolutionary change may occur. In this case, of course, many of the variations are clearly induced by the environment (although random environmental fluctuations may also generate some variants). The environment both induces a set of different adaptive variant states and fine-tunes the adaptation by selecting the most appropriate ones. In this inheritance system both the reproduction of the activity states in daughter cells and the generation of variations are part of the cell's development, and it is the phenotype (a dynamic activity state, a process) that is reproduced.

The second EIS is that of structural inheritance, where existing cell structures are used to guide, or template, the formation of new similar structures. Variant complexes or architectures made up of the same components can be stably inherited. Inheritance is through some kind of three-dimensional templating, with existing structural patterns facilitating the construction of similar "daughter" patterns. For example, in ciliates, genetically identical cells can have different patterns of cilia on their cell surfaces, and these different patterns are inherited. Prions seem to be another example of such structural inheritance (Grimes and Aufderheide 1991, Tuite and Lindquist 1996). In this structural inheritance system there are clear modules (the modular components of the complex), but transmission is holistic: The complex is not transmitted module by module, nor are the modules alterable unit by unit. The structural information may be transmitted by the fragmentation of the original complex, followed by growth, as in a crystal, or by other means where the interacting units within the complex form the conditions for self-organization of free floating units. There is no general, autonomous system of transmission independent of the structural properties of the particular complex. The reliability of transmission will be specific to each structural complex and depend on its unique properties. Variations in the organization of the units into self-perpetuating complex-variants can be affected by environmental conditions, so variations are often patterned (figure 9.2). As with the steady-state EIS, structures are likely to be passed on vertically, by descent. However, horizontal transmission is also possible, as testify some prion diseases where the pathogenic prions are transmitted to nonrelatives and even to individuals of other species. The number of heritable states of each complex may be very limited, but in a cell with tens of complexes, there are practically unlimited heritable architectural states. "Copying" of complexes is part of development and multiplication; there is no specialized machinery that can copy different architectures. Variation, when patterned, is both developmental and evolutionary.

In the third EIS, the chromatin-marking EIS, states of chromatin that affect gene expression are clonally inherited. Genetically identical cells can have variant and heritable chromatin marks. Marks are protein or RNA complexes associated with DNA, or small chemical groups, such as methyl groups, that bind to certain nucleotides. The type, the density, and the pattern of marks on a chromosome region affect its potential transcriptional state, and changes in marks can be induced by the change in the environment. When the marks are protein complexes, their reproduction in daughter cells is probably similar to the reproduction of other three-dimensional complexes, although the DNA sequence to which protein marks bind may constrain variation and enhance the fidelity of reproduction. However, the best-understood chromatin marking EIS, the methylation-marking EIS, is somewhat unusual in its modular organization and mode of transmission. Nucleotides in many organisms can be in a methylated or nonmethylated state, and the alternative states can be clonally inherited. The most commonly methylated nucleotide is cytosine, and in most eukaryotes it is the cytosine in CpG doublets or CpNpG triplets that can be in an either methylated or nonmethylated state. The methylated state of the nucleotide has no effect on the coding properties of the triplet in which it participates, but can affect transcriptional activation in the chromosomal region in which it occurs. With this EIS there is a dedicated, function-independent, copying machinery (the enzyme methyl-transferase) that can copy patterns of methylation irrespective of their past or present function. Information is organized in a modular way (a nucleotide can be in two states—methylated or nonmethylated), so that methylation sites are alterable unit by unit, and transmission proceeds module by module (figure 9.3). However, the reproduction of methylation patterns is not always modular and does not always depend on the special enzymatic machinery. Methylation patterns can be transmitted sexually between organisms through sperms and eggs. As the germline becomes determined, there are widespread and sometimes radical changes in chromosome marks, including patterns of cytosine methylation on chromosomes. However, parental patterns of methylation can still be regenerated in the offspring because some traces of the past are retained, as partial (protein or methylation) marks, and these partial traces or "footprints" are reconstituted into full marks during the embryogenesis of the offspring. There seems to be a cycle of changes in chromosomal marks during germline formation and during early embryonic development that leads to the reconstitution of parental methylation marks (figure 9.4). Because changes in methylation marks, like changes in other types of heritable chromosome marks, can be induced by the environment and the variation can be inherited, some of the variation is patterned. Heredity in this system is unlimited when we consider the whole genome or several large chromosomal domains, but limited when a short DNA sequence is considered.

Unlike the GIS, with all EISs the generation of new variation is typically patterned (although it can also be completely accidental), and cannot be divorced from the physiological development of the cell as it interacts with the environment. In most cases, the transmission of information is

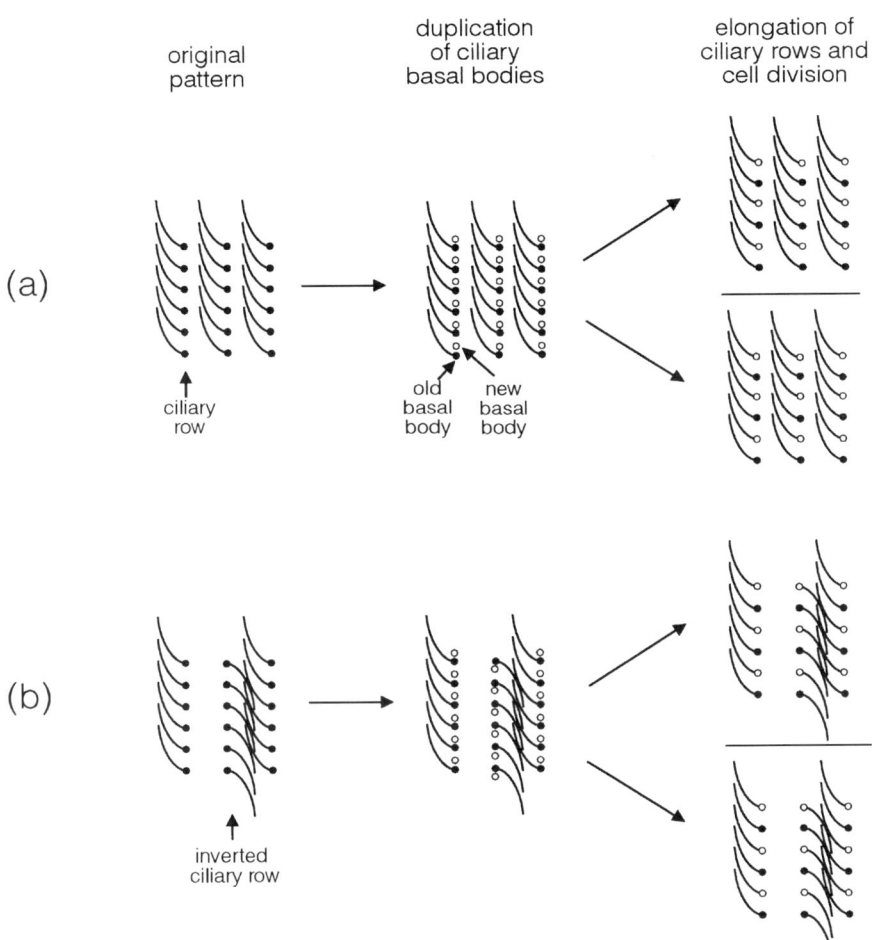

Figure 9.2
The perpetuation of two alternative organizations of ciliary structures in *Paramecium*. (a) The organization and perpetuation of normal ciliary rows through cell division (horizontal line). (b) The perpetuation of an experimentally inverted ciliary row.

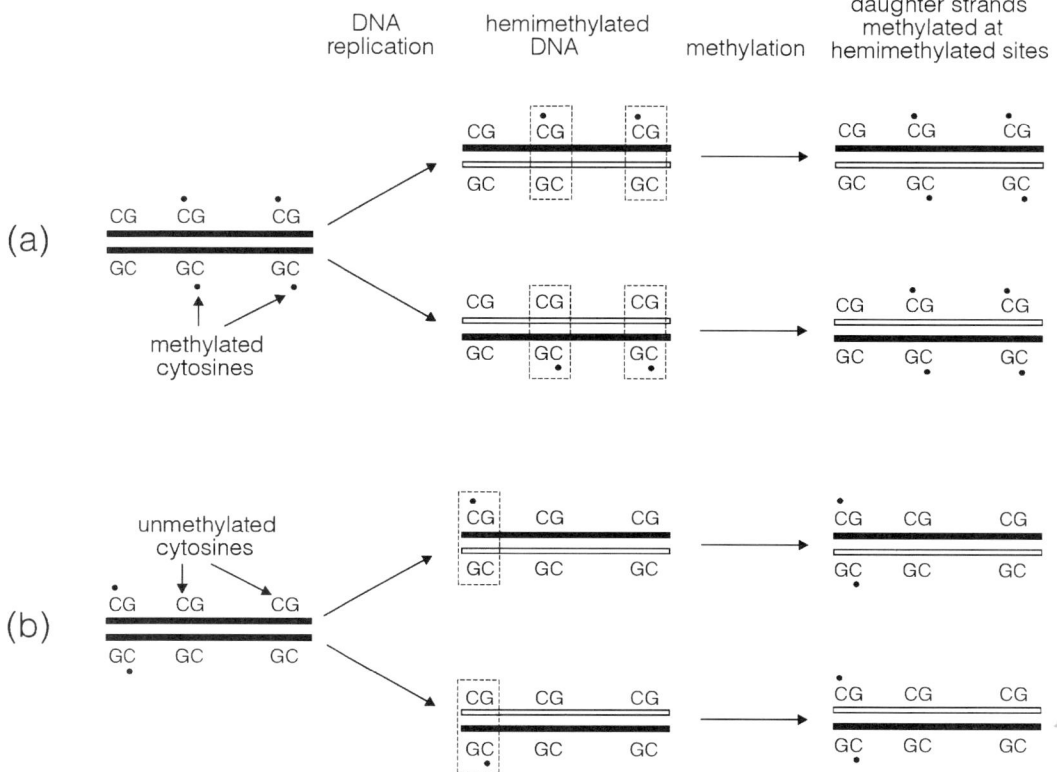

Figure 9.3
The inheritance of alternative patterns, (a) and (b), of DNA methylation. The black dots represent methyl groups. The dotted boxes show hemimethylated sites following the replication of DNA. These sites are preferential targets for a methylating enzyme, which methylates the opposite nonmethylated site in the DNA duplex. Different methylation patterns can therefore be perpetuated through cell division.

holistic. The processes that allow the faithful transmission of variant functional or structural states in the cell lineage do not utilize a dedicated, specialized, function-independent copying machinery (with the exception of the methylation EIS in somatic cells). Instead, these processes are by-products of general growth and multiplication processes. The fidelity of reproduction depends on the specifics of the cycle, or the three-dimensional structure of the complex. At the cellular level heredity is unlimited, although it may be very limited at the level of the functional,

transmitted unit itself. Of course, when we are looking at the functioning of the cell, the different inheritance systems interact and cannot be treated as autonomous: For example, products of a steady state EIS can affect heritable chromatin marks and 3D structures, and vice versa.

If we move from the level of the cell to the level of the multicellular organism, there is ample evidence showing that the cells that begin new organisms, the egg and the sperm, can carry epigenetic information, and that variations in epigenetic information are often inherited (Jablonka

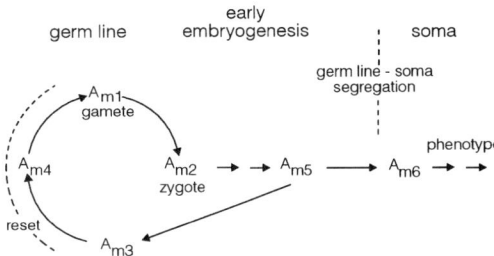

Figure 9.4
A normal cycle of changes in chromatin marks (e.g., methylation marks) during gametogenesis and early embryogenesis. As germ cells proceed through gametogenesis, the chromatin marks on DNA sequence A change (from m3 to m1). In the zygote, the mark on A is changed to m2, and during early embryogenesis to m5. When segregation of soma and germ line occurs, some cells with m5 marks become germline cells and again acquire mark m3. How an induced change in marks may alter the cycle is not shown here (for a figure and discussion of self-perpetuating cycles of changed marks, see Jablonka and Lamb 1995: 154–156).

and Lamb 1995, 1998). There is also another type of phenotypic information transfer between generations, which is more difficult to categorize because it does not occur at the cellular level, but at the level of the whole organism. The maternal environment in which the mammalian fetus develops sometimes has effects that can be carried over to later generations. For example, if female Mongolian gerbil embryos develop in a uterine environment in which most of the embryos are male, and they are therefore exposed to high level of testosterone, they mature late, are more territorial than other females and, in turn, produce litters with a greater proportion of males than the normal 1:1 sex ratio. The result is that their daughters, who usually also develop in a testosterone-rich uterine environment, also mature late, and produce mainly male offspring (Clark, Karpiuk, and Galef 1993; Clark and Galef 1995). The developmental legacy of the mother is transferred to her daughters, so there is a nongenetic transmission and repetition of this

distinctive reproductive pattern. Without any genetic differences, two maternal lineages may differ, consistently, over many generations, in the sex ratio of the offspring they produce.

Another example of phenotypic transmission at the organismal level is the transmission of microorganisms between generations through feces. Young of many species of mammals consistently eat their own and other individuals' feces, a habit known as coprophagy. Most of the mammals that practice coprophagy are herbivores, who consume cellulose-rich plant material and have a dense symbiotic bacterial and protozoan gut flora that helps them to break down and digest cellulose. The young of many herbivorous species eat their mother's feces, and in this way they directly inoculate their own guts with the maternal flora of useful microorganisms. Differences between the gut floras of different mothers, will be transferred to their offspring, and may be perpetuated for many generations. In many cases, these parental legacies affect behavior.

The Regeneration of Behavior: The Behavioral Inheritance Systems (BISs)

Behavior that can be transmitted has been categorized in many different ways. With social learning alone, more than thirty terms and distinctions have been suggested. For the purpose of this essay, which concentrates on the type and transmission of information, I will distinguish three general types of transfer of behavioral information.

The first is very similar to the whole-organism transgenerational reproduction of phenotypes discussed in the last section, but focuses on the reproduction of behavior. In this system the processes that lead to similarity between the behaviours depend on the transfer of behavior-affecting substances between interacting individuals. I therefore call this type of transfer the *inducing-substance* transfer. Unlike the other two BISs, transmission is not dependent on learning.

The transmission of food preferences via the transfer of substances through the placenta and the milk in mammals is a good example of this type of BIS. Mammal fetuses are able to smell semivolatile liquids transferred to them across the mother's placenta, and later show preference or aversion for food items carrying these smells (Smotheran 1982; Hepper 1988). Transmission of substances through milk has similar effects. The results of cross-fostering and other simple experiments with mice have shown that the food the mother prefers, and therefore frequently eats, biases the food preferences of the young so that those who feed on her milk tend to have the same preferences. Such results are typical for many mammals, including other rodents and ruminants (Galef and Sherry 1973; Provenza and Balf 1987).

There are other channels through which inducing-substances that bias behavioral preferences can be transferred (Avital and Jablonka 2000). Information transferred in inducing substances is not encoded, and its transmission is holistic. Usually (but certainly not always), transmission is vertical. The variation generated is commonly patterned (induced), and heredity is limited, although the number of variant preferences and the behaviors they influence may be quite large.

The second type of transfer of behavioral information occurs through nonimitative social learning. This has received a great deal of attention from experimental psychologists who differentiate between several different types of social learning that do not involve imitation and/or direct instruction (Zentall and Galef 1988; Heyes 1994). I call this type of social learning *nonimitative social learning*. In the cases covered by this category of social learning, the naive, observing individual (or "observer") learns about the environmental circumstances (including the objects, stimuli, and events) that elicit a particular behavior in the experienced individual. Two examples will help to illustrate the nature of such social mediation. When young monkeys become fearful of snakes after observing the panic-stricken reaction of adults to snakes, they too will avoid

Figure 9.5
A blue tit opening a milk bottle by tearing the foil cap (from Hinde 1982).

snakes. However, what they learn is not the motor flight behavior patterns of the experienced adults, but rather that snakes have to be avoided. The second example is the cultural spread of the blue tit's and the great tit's habit of opening milk bottle tops, a famous case of "cultural" transmission of behavior (Fisher and Hinde 1949). Tits learnt by observation the habit of removing the cap and getting at the cream at the top of the bottles (figure 9.5). This is probably another case of non-imitative social learning. The spread of the behavior from experienced to naive tits can be explained as the result of naive tits having their attention drawn to the milk bottle as a source of food, commonly through the behavior of an experienced individual (Sherry and Galef 1984). The method by which the top was removed was not imitated—each individual tit focused its attention on the milk bottle as a potential source of food and, after it own trial-and-error learning, finally learned how to remove the top in its own style. Such social mediation leads, in most cases, to similarity between the behaviors of the "observer" and the "model." The model guides or enhances the attention of the observer to the environmental stimulus (such as a milk bottle, or a dangerous predator), which elicits a similar emotional and behavioural response to its own.

In this type of behavioral inheritance, information is not encoded. Variation is generated by the inventor of the new behavior through asocial learning. It is therefore patterned rather than accidental. It is holistically transmitted through social learning, and can be transferred both vertically and horizontally. Heredity is rather limited, since the number of variants the behavior pattern can assume may be restricted. However, at the level of the overall lifestyle, heredity may be practically unlimited, since different variant patterns of behavior may combine to form many types of lifestyle.

The third type of BIS is learning by imitation and/or instruction. I consider it to be another type of BIS because of its modular way of transmission. As Heyes (1993) has argued, there is no compelling evidence to suggest that imitation is inherently more cognitively demanding than several other types of social learning. However, the modular way of transmitting and altering behavior during imitation or instruction—the parsing of behavioral acts—sets it apart from other types of social learning. During imitation, the naive individual reproduces not only similar responses to the environment, but also the model's actual acts. Vocal imitation is very common among many species of bird, whereas motor imitation has been validated beyond reasonable doubt in only a few species. Humans, chimpanzees, dolphins, budgerigars, rats, and a few other birds and mammals have been shown to be able to imitate motor acts. However, because relatively few experiments have been designed to distinguish between imitative and nonimitative learning of motor acts, the extent of motor imitation may be underestimated. Intentional instruction seems to be very rare in the animal world, but again, this issue has not yet been systematically studied.

The information acquired during imitation and nonsymbolical instruction is, as with other types of social learning, patterned and nonencoded, and is transmitted both vertically and horizontally. Heredity is often limited although, in theory, if a behavioral act is made of many individually alterable and transmissible modules (for example, if the song of a songbird is made of many types of phrases), heredity may be unlimited. However, a huge number of combinations leads to a huge amount of nonsense-messages—to functionally useless or even positively harmful information. Only if there is some patterning or ordering of the combinatorial process can the search for functional sequences in the infinite space of possibilities yield functionally meaningful results (Schlicht 1998). It is only when there is a reasonable probability that variant behavioral modules combine to form different yet *functional* sequences of behavior that the modular transmission of sequences opens up truly wide evolutionary possibilities. Rule-bound organization and transfer of information is clearly necessary. We see this kind of rule-bound organization in systems of symbolic communication.

Symbolical Systems of Inheritance (SIS)

As T. W. Deacon stresses in his 1997 book on the evolution of language, symbols are not simple. The American philosopher Charles Peirce distinguished between three ways in which a sign (defined as information communicated between sender and receiver) can refer to something. First, a sign can refer to an object by resembling it. This type of sign is called an icon, and an example is a picture of a house, which refers to an actual house, or the pattern on a mimetic butterfly's wings, which resembles (and can be said to refer to) the pattern on a model's wings. Second, a sign can be an index and refer to an object by association, through being linked to the object in space or time. For example, the size and brightness of a male peacock's tail is an index of its health and vigor. Finally, a sign can refer to an object by convention, or according to a reference-rule that enables it to refer to other signs in the system.

Such signs are symbols. Symbols must represent objects, operations, and relations among signs (as in natural language and, in the purest way, in mathematical notations). The category to which a sign belongs depends on the interpretive system of which it is part, rather than on the isolated sign; a portrait, for example, though iconic, is also a part of a symbolical system, and should therefore be interpreted as a symbol. Natural human language is another example of a symbolical system. In the sentence I am writing just now, most words refer to other words rather than to objects in the world.

From the point of view adopted in this chapter, symbolical systems are transmitted by social learning, which often involves imitation and a greater or lesser degree of intentional instruction. Symbols are transmitted both modularly and holistically. For example, in the case of natural language, the narrative, the sentence, the word, the phoneme are all transmitted, but it is quite clear that a spoken narrative is (unless a story is learned by heart) more holistically transmitted than a single new word. Interpretation depends on the rules of the system (for example, grammatical rules), so symbolical systems are organized by those rules. Sometimes, as in natural language or mathematics, the organization is easily formalized (rules of language-specific grammar, mathematical axioms), but it can be more fuzzy (as in dancing, music, and the visual and motor arts). Information is (by definition) encoded and is almost invariably transmitted horizontally. Vertical transmission is common in some systems, however. For example, early language learning usually involves vertical parent/offspring interactions. In other cases, such as the transmission of painting skills, it is almost always nonvertical from master to student. Symbolical systems have unlimited heredity and huge evolutionary potential. The rules of symbolical systems organize the systems and order them, so variation is inherently constrained and patterned by these internal rules. New variations arise as a result of insight, trial-and-error-learning and accident.

Table 9.1 summarizes the different properties of the four types of inheritance systems and allows a comparison among them. What is clear is that a system based on encoded information, modular transmission, and modular alteration of the composing modules is a very special type of inheritance system. The two inheritance systems that have these properties and are closest to each other in this respect are the GIS and the SIS. However, the SIS is evolutionarily derived from the BISs, and it shares important characteristics with them. It is nevertheless significant that both the GIS and the SIS have unlimited heredity at the level of the transmitted units themselves, and not, as with other inheritance systems, only at a higher level of organization. Because of the ability to encode information, both the GIS and SIS transmit a lot of unexpressed information. Nonfunctional genes are transmitted, as also are nonimplemented ideas. This provides a huge reservoir of variation, which may become useful in new conditions. I believe that this ever-present potential gives these systems a particularly important role in long-term evolution. However, no inheritance system acts in isolation: inheritance systems interact both directly and indirectly. For example, the social animal, with its BISs, determines the selective regime in which genes are ultimately selected.

Another point suggested by the table is that by considering a higher level of organization, limited inheritance systems may become unlimited. Hence we see that EISs are limited inheritance systems at the level of the unit of transmitted information (cycle of activity, 3D complex, local pattern of marks), but may be unlimited at the level of the cell phenotype. A practically unlimited number of cell phenotypes can be generated. The same is true of BISs—at the level of a single behavior pattern there may be few variants, but the lifestyle as a whole can display many more variations. Although biological information at the lower level is holistically organized, at the higher level each state is treated as a module that can combine with others and produce practically unlimited variation.

It seems that as a system becomes more functionally cohesive during evolution, evolving repair and compensatory mechanisms, its heredity becomes increasingly more limited. There is less selectable variation, and the result may be evolutionary stasis. There are two situations in which escape from such stasis is possible. One occurs when selection acts at higher level of biological organization (at the level of many combining units), that is, when a higher level of individuality emerges (Jablonka 1994; Jablonka and Lamb 1995). The second occurs when a system of encoding the information evolves. Both situations have occurred during evolutionary history.

The Transmission of Organism/Environment Variations: Niche Construction and Niche Regeneration

The right-hand side of table 9.1 shows that organisms often transfer variations in their epigenetic characteristics or their behavior patterns in an indirect way. By providing their descendants with the initial conditions that allow the repetition of their own developmental processes, similarity between generations is enhanced. Both Waddington (1959) and Lewontin (1983) stressed that living organisms are not passive entities, but ones that actively choose and construct their environment, and hence also the selective regime in which they live and in which they breed. The most obvious examples are the nests of birds and the dams of beavers. Such artifacts are often also passed on to the next generation.

Odling-Smee developed these ideas further, stressing the multigenerational transfer of many types of variations in niches. He argued that because through their activity and behavior organisms construct the ecological and social niche that they occupy, this "niche construction" may often ensure that the environmental condi-

tions in which they have lived will be regenerated and reexperienced by their descendants (Odling-Smee 1988, 1995; Odling-Smee, Laland, and Feldman 1996). For example, males of some species of bowerbirds build small huts to attract females, bringing fruits, seeds, and fungi to decorate them. These decorations are often able to grow, so by their behavior bowerbirds also ensure the long term supply of the materials which they, and their descendants, will choose as decorations (Diamond 1986, 1987, 1988). Caching seeds is another example of a habit that may be reinforced through the effect it has on the local environment. By caching seeds, animals provide themselves with a source of food for harsh winters, but because some of the cached seeds germinate, caching also provides new plants that will form seeds and create future caching opportunities (Källander and Smith 1990; Smith and Reichman 1984).

Even more obvious examples of niche construction are the propagation of dialects in bird or whale groups, where the dialect of the previous generation is the condition for the acquisition of this dialect by the younger generation. Similarly, learning to speak by human children is guaranteed by the child's developing in a preexisting linguistic community. Such ecological or social niche construction ensures that the ecological and social milieu is transmitted. The conditions eliciting the ancestral behavior are reconstructed, and selection for the maintenance of the behavior pattern that fits the constructed niche occurs.

The regeneration of ancestral niches and selective regimes can occur at different levels of biological organization. At the cellular level, we saw that when the regulatory product of a steady-state cycle can diffuse into the environment it changes it, thereby creating the conditions that induce a cycle of self-perpetuating activity in neighboring cells. This is a simple form of niche construction. All types of niche construction depend on the formation of self-sustaining feedback loops between the developing organism and its niche.

A Different Kind of Darwinism

The diversity of inheritance systems that are able to transmit variation at different levels of biological and social organization should surely prevent developmental and evolutionary biologists from interpreting development and evolution in terms of genetic variation alone. Yet, not only are other sources of heritable variation neglected in gene-centered accounts, but also the whole dynamics of inheritance, which is an aspect of the developmental process, is ignored. This leads to a very faulty account and understanding of development and of evolution, and completely misses the complexity, possibilities, and limitations of developmental and evolutionary processes.

Moving from the gene to the more abstract replicator, and assuming that the replicator is the unit of variation and evolution, is also not satisfactory. The replicator/vehicle dichotomy, which is fundamental to the concept of a replicator, is meaningless in all cases in which the transmission of information or the generation of new heritable information depends on development. Yet, as table 9.1 illustrates, this is the usual case. The replicator-vehicle distinction cannot therefore be used to analyze heredity, development, or evolution. However profitable the distinction between replicator and vehicle may be for some evolutionary theorizing, this distinction simply does not apply to real organisms.

At the beginning of this chapter I suggested that in order to have a unifying concept of heredity that encompasses all the types of inheritance system, we need a theoretical framework that is broader than that used by genic neo-Darwinism. The developmental system approach suggested by Oyama (1985) and Griffiths and Gray (1994) provides such a framework, as it focuses on the developing and interacting individual, with the multiplicity of its inheritance systems and self-perpetuating feedback loops. The reproducer concept suggested by James Griesemer (2000) provides the unit of analysis for such an approach, for the reproducer is simultaneously a unit of development, of multiplication, and of heritable variation, as well as a target of selection.

The focus on units of reproduction introduces back into evolution the developing individual as an active evolutionary agent. This leads to the consideration of the different types of developmental processes that lead to the regeneration and reproduction of variant characters. It inevitably leads to concurrent attention to selection at different levels of organization—the gene level, the cell level, the organism level, and so on, and to different types of heritable variation—the genetic, the epigenetic, the behavioral, and the symbolical. It is this richer version of Darwinian theory that needs to be adopted.

References

Avital, E., and E. Jablonka. (2000). *Animal Traditions: Behavioural Inheritance in Evolution.* Cambridge: Cambridge University Press.

Clark, M. M., and B. G. Galef. (1995). Parental influence on reproductive life history strategies. *Trends in Ecology and Evolution* 10: 151–153.

Clark, M. M., P. Karpiuk, and B. G. Galef. (1993). Hormonally mediated inheritance of acquired characteristics in Mongolian gerbils. *Nature* 364: 712–716.

Dawkins, R. (1976). *The Selfish Gene.* Oxford: Oxford University Press.

Dawkins, R. (1982). *The Extended Phenotype.* Oxford: Freeman.

Deacon, T. W. (1997). *The Symbolic Species.* New York: W. W. Norton.

Diamond, J. (1986). Biology of birds of paradise and bowerbirds. *Annual Review of Ecology and Systematics* 17: 17–37.

Diamond, J. (1987). Bower building and decoration by the bowerbird *Amblyornis inornatus. Ethology* 74: 177–204.

Diamond, J. (1988). Experimental study of bowerbird decoration by the bowerbird *Amblyornis inornatus* using colored poker chips. *American Naturalist* 131: 631–653.

Fisher, J., and R. A. Hinde. (1949). The opening of milk bottles by birds. *British Birds* 42: 347–359.

Galef, B. G., and D. F. Sherry. (1973). Mother's milk: A medium for transmission of cues reflecting the flavour of mother's diet. *Journal of Comparative Physiological Psychology* 83: 374–378.

Griesemer, J. (2000) Reproduction and the reduction of genetics. In P. Beurton, R. Falk, and H-J. Rheinberger (Eds.), *The Concept of the Gene in Development and Evolution*. Cambridge: Cambridge University Press.

Griffiths, P., and R. D. Gray. (1994). Developmental systems and evolutionary explanations. *Journal of Philosophy* 91: 277–304.

Grimes, G. W., and K. J. Aufderheide. (1991). *Cellular Aspects of Pattern Formation: The Problem of Assembly*. Basel: Krager.

Hepper, P. G. (1988). Adaptive fetal learning: Prenatal exposure to garlic affects postnatal preferences. *Animal Behaviour* 36: 935–936.

Heyes, C. M. (1993). Imitation, culture and cognition. *Animal Behaviour* 46: 999–1010.

Heyes, C. M. (1994). Social learning in animals: Categories and mechanisms. *Biological Review* 69: 207–231.

Hinde, R. A. (1982). *Ethology: Its Nature and Relations with Other Sciences*. New York: Oxford University Press.

Hull, D. L. (1980). Individuality and selection. *Annual Review of Ecology and Systematics* 11: 311–332.

Jablonka, E. (1994). Inheritance systems and the evolution of new levels of individuality. *Journal of Theoretical Biology* 170: 301–309.

Jablonka, E., and M. J. Lamb. (1995). *Epigenetic Inheritance and Evolution: The Lamarckian Dimension*. Oxford: Oxford University Press.

Jablonka, E., and M. J. Lamb. (1998). Epigenetic inheritance in evolution. *Journal of Evolutionary Biology* 11: 159–183.

Jablonka, E., M. J. Lamb, and E. Avital. (1998). "Lamarckian" mechanisms in Darwinian evolution. *Trends in Ecology and Evolution* 13: 206–210.

Johannsen, W. (1911). The genotype conception of heredity. *American Naturalist* 45: 129–159.

Källander, H., and H. G. Smith. (1990). Food storing in birds: An evolutionary perspective. In D. M. Power (Ed.), *Current Ornithology* vol. 7, pp. 147–207. New York: Plenum Press.

Lewontin, R. (1978). Adaptation. *Scientific American* 239(3): 156–169.

Maynard Smith, J. (1986). *The Problems of Biology*. Oxford: Oxford University Press.

Maynard Smith, J., and E. Szathmáry. (1995). *The Major Transitions in Evolution*. Oxford: Freeman.

Moxon, E. R., P. B. Rainey, M. A. Nowak, and R. E. Lenski. (1994). Adaptive evolution of highly mutable loci in pathogenic bacteria. *Current Biology* 4: 24–33.

Odling-Smee, F. J. (1988). Niche constructing phenotypes. In H. C. Plotkin (Ed.), *The Role of Behavior in Evolution*, pp. 73–132. Cambridge, MA: MIT Press.

Odling Smee, J. (1995). Biological evolution and cultural change. In E. Jones and V. Reynolds (Eds.), *Survival and Religion: Biological Evolution and Cultural Change*, pp. 1–43. New York: John Wiley & Sons.

Odling-Smee, F. J., K. N. Laland, and M. W. Feldman. (1996). Niche construction. *American Naturalist* 147: 641–648.

Oyama, S. (1985). *The Ontogeny of Information: Developmental Systems and Evolution*. Cambridge: Cambridge University Press. (2nd rev. ed., Durham, NC: Duke University Press, 2000.)

Provenza, F. D., and D. F. Balf. (1987). Diet learning by domestic ruminants: Theory, evidence and practical implications. *Applied Animal Behavioural Science* 18: 211–232.

Schlicht, E. (1997). "Patterned variation": The role of psychological dispositions in social and economic evolution. *Journal of Institutional and Theoretical Economics* 153(4): 722–736.

Schlicht, E. (1998). *On Custom in the Economy*. New York: Oxford University Press.

Shapiro, J. A. (1997). Genome organization, natural genetic engineering and adaptive mutation. *Trends in Genetics* 13: 98–104.

Sherry, D. F., and B. G. Galef. (1984). Cultural transmission without imitation: Milk bottle opening by birds. *Animal Behaviour* 32: 937–938.

Smith, C. C., and O. J. Reichman. (1984). The evolution of food caching by birds and mammals. *Annual Review of Ecology and Systematics* 15: 329–335.

Smotheran, W. P. (1982). Odor aversion learning by the rat fetus. *Physiology of Behavior* 29: 769–771.

Szathmáry, E. (1995). A classification of replicator and lambda-calculus models of biological organization.

Proceedings of the Royal Society of London, Series B
260: 279–286.

Szathmáry, E., and J. Maynard Smith. (1993). The origin of genetic systems. *Abstracta Botanica (Budapest)*
17: 197–206.

Tuite, M. F., and S. L. Lindquist. (1996). Maintenance and inheritance of yeast prions. *Trends in Genetics* 12:
467–471.

Waddington, C. H. (1959). Evolutionary systems:
Animal and human. *Nature* 183: 1634–1638.

Wright, B. E. (1997). Does selective gene activation direct evolution? *FEBS Letters* 402: 4–8.

Wright, S. (1945). Genes as physiological agents:
General considerations. *American Naturalist* 74: 109–
124.

Zentall, T. R., and B. G. Galef. (Eds.), (1988). *Social
Learning: Psychological and Biological Perspectives.*
Mahwah, NJ: Lawrence Earlbaum.

10 Niche Construction, Ecological Inheritance, and Cycles of Contingency in Evolution

Kevin N. Laland, F. John Odling-Smee, and Marcus W. Feldman

A recurrent theme of this book is the rejection of dichotomous thinking characterized by emphasis on processes that are regarded as either internal or external to living organisms. As Lewontin (1983) has pointed out, the tendency to think dichotomously is not confined to developmental biologists. Evolutionary biologists can also slip into a dichotomous mode of reasoning. One of the principal dichotomies in evolutionary theory, to which Lewontin draws attention, stems from the separation of the causes of ontogenetic variation, seen as coming from internal factors, especially Mendelian genetics, and the causes of phylogenetic variation, seen as something that is imposed by natural selection pressures arising from autonomous external environments (Lewontin 1983). In this chapter we shall focus primarily on the second half of this dichotomy. We want to reconsider the extent to which Darwinian natural selection pressures in external environments really are autonomous, that is, they really are independent of the organisms they select, and we shall suggest that often they are not. We shall also propose evolutionary processes that can cause naturally selected organisms to modify their own natural selection, as well as the selection of other organisms.

Classically, adaptation has been conceived of as a process by which natural selection, stemming from an external and independent environment, gradually molds organisms to fit an established environmental "template." The environment is seen as posing problems, and those organisms best equipped to deal with the problems leave the most offspring (Lewontin 1982, 1983). Although the environmental template may be dynamic, in the sense that processes independent of the organism may change the world to which the population adapts (Van Valen 1973), the changes that organisms themselves bring about are rarely considered in evolutionary analyses. Yet to varying degrees, organisms choose their own habitats, choose and consume resources, generate detritus, construct important components of their own environments (such as nests, holes, burrows, paths, webs, pupal cases, dams, and chemical environments), and destroy other components (Lewontin 1983; Odling-Smee 1988). In addition, many organisms choose, protect, and provision "nursery" environments for their offspring. On the basis of this kind of evidence, Lewontin (1982, 1983) has argued that the "metaphor of adaptation" should be replaced by a "metaphor of construction" (see also Gray 1988). We have sought to build on Lewontin's writings by exploring the consequences of these constructive processes, which we have collectively termed *niche construction* (Odling-Smee 1988; Odling-Smee, Laland, and Feldman 1996). We argue that through niche construction organisms not only shape the nature of their world, but also in part determine the selection pressures to which they and their descendants are exposed. Other authors in this volume have pursued similar themes (Gray 1988; Griffiths and Gray 1994).

Niche construction is not the exclusive prerogative of large populations, keystone species or clever animals; it is a fact of life. All living organisms take in materials for growth and maintenance, and excrete waste products. It follows that, merely by existing, organisms must change their local environments to some degree. In spite of its universality, niche construction is virtually never incorporated into evolutionary accounts. Niche construction is too obvious and ubiquitous for biologists to be unaware of it existence. If evolutionary biologists currently neglect niche construction, it is unlikely to be because they dispute that it occurs. We suspect that most evolutionary biologists feel that they can afford to neglect the effects of niche construction, because these effects are regarded as either trivial, inconsequential, not liable to change the nature of the evolutionary process, or unlikely to do so sufficiently often to warrant consideration. It is convenient, and also simpler, to regard niche construction as merely the product of natural selection, and not a process shaping selection pressures, because then the organism/environment dichotomy is preserved unsullied, and each evolutionary event can be considered a response on the part of organisms to a fresh change in the external environment.

We regard niche construction as an important, neglected process in evolution. To us, the idea that niche construction can be dismissed because it is the product of natural selection makes no more sense than the counterproposal that natural selection can be disregarded because it is a product of niche construction. From the beginning of life all organisms have, in part, modified their selective environments, and their ability to do so is, in part, a consequence of their naturally selected genes. Niche construction and natural selection are two processes, operating in parallel, but also interacting. The pertinent question, then, is whether what is gained in understanding by a focus on niche construction justifies the added complexity that is required by its incorporation into evolutionary accounts. In this chapter we describe why we believe there is utility to complicating evolutionary accounts by explicitly incorporating niche construction into empirical and theoretical analyses of evolutionary phenomena. We illustrate, using examples from natural history, how the effects of niche construction are frequently nontrivial, directional, accumulatory, pervasive, and liable to change the nature of the evolutionary process. We will also describe the findings of theoretical models, which suggest that niche construction can make a substantial difference to the evolutionary process and can generate unusual evolutionary dynamics. Our emphasis on niche construction leads us to the position that evolution proceeds in reciprocal and simultaneous cycles of selection and niche construction. Evolution is characterized by these cycles of contingency.

Niche Construction and Ecological Inheritance

The constructive and self-organizing nature of developmental systems extends far beyond the boundaries of the individual organism. Much of an organism's world is constructed, organized, regulated, and destroyed by the organism itself, by its ancestors, and by its fellow members of an ecological community. How should we characterize this niche construction? Let us begin with a definition of the niche. A niche refers to the "occupation" of an organism, for example, to the ways in which an organism obtains its resources or defends itself in its environment, in contrast to its location or "address" in its environment, or habitat (Ehrlich and Roughgarden 1987). Our view is consistent with Hutchinson's (1957) concept of the niche as a multidimensional hypervolume, provided it is realized that Hutchinson's concept is fundamentally relativistic, and that a tolerance space cannot be defined except in relation to an organism. In theory, organisms can be decomposed into arrays of features (traits or characters), while environments can be decomposed into arrays of factors (Bock 1980). A feature of an organism is only an adaptation if and when it is matched to a specific selection pressure arising from an environmental factor at a particular location, it is the product of national selection, and that it increases the fitness of the organism at that address and moment, for example, if it permits more efficient acquisition of a food resource (Bock 1980). We interpret Bock (1980) as treating adaptation as a dynamic and historical process: current utility, that is, synergy between a feature and a factor, is not sufficient to label the feature an adaptive trait (Odling-Smee 1988). Niche construction occurs when an organism modifies the functional relationship between itself and its environment by actively changing one or more of the factors in its environment, either by physically perturbing these factors at its current address or by relocating to a different address, thereby exposing itself to different factors.

The niche construction of both past and present generations may influence a population's selective environment. Spiderwebs, insect pupal cocoons, and caddis fly larvae houses modify the selective environments of the constructors themselves. In contrast, bird's nests, female insect's oviposition site choices, and the soil transformed by earthworms also modify the environment of the constructor's descendants. The latter cases are all examples of ecological inheritance. We define as ecological inheritance any case in which an organism experiences a modified functional relationship between itself and its environment as a consequence of the niche-constructing activities of either its genetic or ecological ancestors (Odling-Smee 1988). Ecological inheritance is built into our theoretical models (Laland, Odling-Smee, and Feldman 1996, 1999), and, as we will describe, the addition of this second kind of inheritance system can make a considerable difference to the evolutionary process.

At first sight it may be tempting to conclude that the impact that most organisms have on their environments is trivial, a mere drop in the ocean compared with the action of major geophysical, chemical, or meteorological processes. A closer inspection reveals that countless organisms across the breadth of all known taxonomic groups significantly modify their local environments (Lewontin 1983; Hansell 1984; Odling-Smee 1988; Jones, Lawton, and Shachak 1994, 1997; Odling-Smee, Laland, and Feldman 1996; Laland, Odling-Smee, and Feldman 1996). Following Lewontin (1983), and in agreement with Jones, Lawton, and Shachak's (1994, 1997) concept of "ecosystem engineering," we argue that organisms not only adapt to environments, but in part also construct them. They may also do so across a huge range of temporal and spatial scales stretching, for example, from a hole bored in a tree by an insect, to the contribution of cynobacteria to the earth's 21 percent oxygen atmosphere, as a consequence of millions of years of photosynthesis (Odum 1989). Niche construction starts to take on a new significance when it is acknowledged that, by changing their world, organisms modify many of the selection pressures to which they and their descendants are exposed, and that this may change the nature of the evolutionary process.

Dawkins (1982) argues that genes can express themselves as "extended phenotypes" outside the bodies of the organisms that carry them. For instance, the beaver's dam is an extended phenotypic effect of beaver genes, while the houses of caddis fly larvae are equally expressions of caddis fly genes. Extended phenotypes play an evolutionary role by influencing the chances that the genes responsible for the extended phenotypic trait will be passed on to the next generation. However, this is just one aspect of the evolutionary feedback from niche construction. To go back to the beaver, its dam sets up a host of selection pressures that feed back to act not only on the genes that underlie dam building, but also on other genes that may influence the expression of other traits in beavers such as their teeth, tail, feeding behavior, their susceptibility to predation or disease, their social system, and many other aspects of their phenotypes, including, probably, the expression of another extended phenotype, the building of lodges by beavers in the lakes they create by their dams.

Dam construction may also affect many future generations of beavers that may "inherit" the dam, its lodge, and the altered river, as well as many other species of organisms that now have to live in a world with a lake in it. For example, beavers can create wetlands that can persist for centuries. They can modify the structure and dynamics of riparian zones, and they can influence the composition and diversity of both plant and animal communities. Some of these consequences of niche construction may be just ecological, but some are likely to have evolutionary consequences as well (Naiman, Johnston, and Kelley 1988; Jones, Lawton, and Shachak 1994, 1997). Our general point here is that niche construction generates forms of feedback in evolution that are not yet fully appreciated by contemporary evolutionary theory (Lewontin 1983; Odling-Smee 1988; Odling-Smee, Laland, and Feldman 1996; Laland, Odling-Smee, and Feldman 1996).

There are numerous examples of organisms choosing or changing their habitats, or of constructing artifacts, leading to an evolutionary response (Odling-Smee 1988; Jones, Lawton, and Shachak 1994; Laland, Odling-Smee, and Feldman 1996). For instance, orb-web spiders construct webs, which have led to the subsequent evolution of camouflage, defense, and communication behavior on the web (Preston-Mafham and Preston-Mafham 1996). Similarly, ants, bees, wasps, and termites, construct nests that are themselves the source of selection for many nest regulatory, maintenance, and defense behavior patterns (Hansell 1984; Holldobler and Wilson 1994). The construction of artifacts is equally common among vertebrates. Many mammals (including badgers, gophers, ground squirrels, hedgehogs, marmots, moles, mole rats, opossums, prairie dogs, rabbits, and rats) construct burrow systems, some with underground passages, interconnected chambers, and multiple entrances (Nowak 1991). Here, too, there is evidence that burrow defense, maintenance, and regulation behaviors have evolved in response to selection pressures that were initiated by the construction of the burrow (Hansell 1984; Nowak 1991). The same applies to reptiles and amphibians.

Of course, this will be no surprise to the biologically minded, yet the breadth and scale of niche construction will surprise many. Few people realise that there are more than 34,000 species of spider that construct silken egg sacs, burrows, or webs (Preston-Mafham and Preston-Mafham 1996). There are more than 9,000 species of birds, the vast majority of which construct nests (Forshaw 1998), and probably as many fish that do the same (Paxton and Eschmeyer 1998). There are 9,500 known species of ants, and 2,000 known species of termites, all living in social colonies, and almost all building some kind of nest (Holldobler and Wilson 1994; Gullan and Cranston 1994). Niche construction is all-pervasive.

Most cases of niche construction, however, do not involve the building of artifacts, but merely the selection or modification of habitats. For

example, as a result of the accumulated effects of past generations of earthworm niche construction, present generations of earthworms inhabit radically altered environments where they are exposed to modified selection pressures (Darwin 1881; Lee 1985). This is a good example of an ecological inheritance. Perhaps one of the most frequently documented cases of ecological inheritance is provided by female insects. Females of the vast majority of the millions of species of insect are oviparous, and usually the eggs are deposited on or near the food required by the offspring upon hatching (Gullan and Cranston 1994). The offspring of virtually all insects inherit from their mother a legacy of a readily available, nutritious larval food source.

Figure 10.1 shows how niche construction and ecological inheritance interact with natural selection and genetic inheritance. Figure 10.1a represents the standard evolutionary perspective: Populations of organisms transmit genes from one generation to the next, under the direction of natural selection. Figure 10.1b extends this perspective to acknowledge that phenotypes modify their local environments through niche construction. Genes are transmitted by ancestral organisms to their descendants, exactly as the standard theory describes, but in addition, phenotypically selected habitats, phenotypically modified habitats, and artifacts, persist, or are actively or effectively "transmitted," by these same organisms to their descendants via their local environments. The environments encountered by descendent organisms are not just "templates" to which organisms adapt. Environments are partly determined by independent environmental events (for instance, climatic, geological, or chemical events), but also partly by ancestral niche construction.

The Consequences of Niche Construction

The evolutionary significance of niche construction hangs primarily on the feedback that it generates. Several topics in population biology, such as habitat selection, frequency- and density-

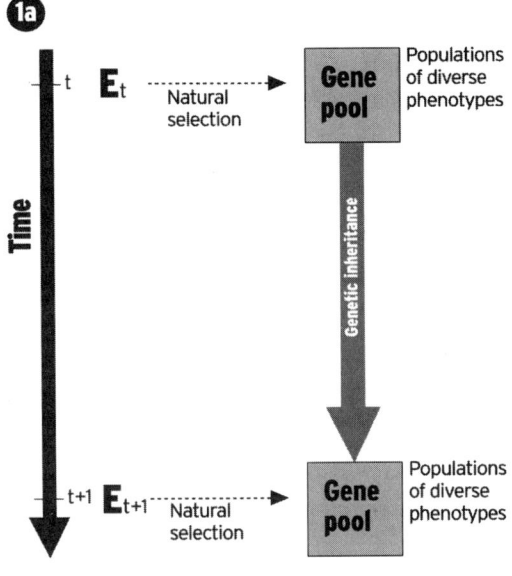

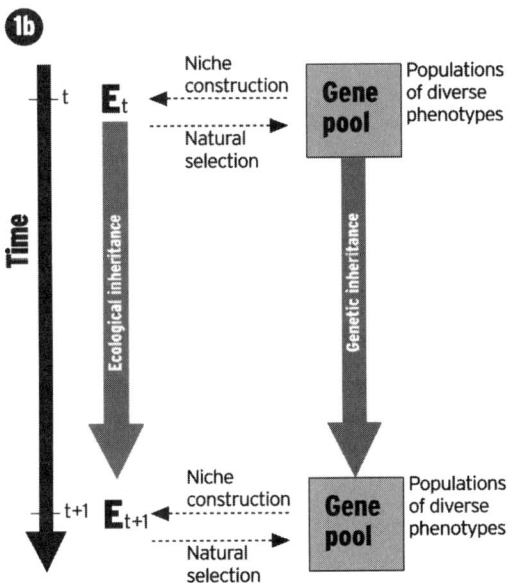

dependent selection, and coevolution, are concerned with the evolutionary consequences of feedback from changes that organisms bring about in their own and in other populations' selective environments (Maynard Smith 1989). However, most analyses of these subjects have focused only on those loci that influence the production of the niche-constructing phenotype itself. What is missing is an exploration of the feedback effects on other loci.

We have begun the development of a body of theory that sets out to explore the evolutionary consequences of niche construction in a systematic manner (Laland, Odling-Smee, and Feldman 1996, 1999). Our theoretical analyses, which employ two-locus, population-genetic models, have uncovered a number of interesting evolutionary consequences of the feedback from niche construction. The first consequence is that traits whose fitness depends on sources of selection that are alterable by niche construction (recipient traits) coevolve with traits that alter sources of selection (niche-constructing traits). This results in very different evolutionary dynamics for both traits from what would occur if each had evolved in isolation. Selection resulting from niche construction may drive populations along alternative evolutionary trajectories, may initiate new evolutionary episodes in an unchanging external environment, and may influence the amount of genetic variation in a population, by affecting the stability of polymorphic equilibria.

Moreover, because of the multigenerational properties of ecological inheritance, niche construction can generate unusual evolutionary

Figure 10.1

(a) The standard evolutionary perspective: Populations of organisms transmit genes from one generation to the next, under the direction of natural selection. (b) With niche construction: Phenotypes modify their local environments (E) through niche construction. Each generation inherits both genes and a legacy of modified selection pressures (ecological inheritance) from ancestral organisms.

dynamics. This is because when ecological inheritance is involved, the evolution of the recipient trait depends on the frequency of the niche-constructing trait over several generations. For instance, timelags were found between the onset of a new niche-constructing behavior, and the response of a population to a selection pressure modified by this niche construction (Laland, Odling-Smee, and Feldman 1996). These timelags generated an evolutionary inertia, where unusually strong selection is required to move a population away from an equilibrium, and a momentum, such that populations continue to evolve in a particular direction even if selection pressures change or reverse. Although these findings are novel, they are consistent with those of related theoretical analyses (Feldman and Cavalli-Sforza 1976; Kirkpatrick and Lande 1989; Robertson 1991). Theoretical population genetic analyses have established that processes that carry over from past generations can also change the evolutionary dynamic to generate opposite responses to selection, and sudden catastrophic responses to selection (Feldman and Cavalli-Sforza 1976; Kirkpatrick and Lande 1989; Robertson 1991). This growing body of theory supports our view that, in the presence of niche construction, adaptation ceases to be a one-way process, exclusively a response to environmentally imposed problems, and instead becomes a two-way process, with populations of organisms setting as well as solving problems (Lewontin 1983; Odling-Smee, Laland, and Feldman 1996). Our findings support Lewontin's (1983) original intuition that niche construction may be a major source of adaptation in evolution, and that its evolutionary effects deserve serious consideration.

A second difference that incorporating niche construction makes in evolution is that it allows acquired characteristics to play a role in the evolutionary process, in a non-Lamarkian fashion, by their influence on selective environments through niche construction. When phenotypes niche construct, they become more than simply "vehicles for their genes" (Dawkins 1989) since

they may now also be responsible for modifying some of the sources of natural selection in their environments that subsequently feed back to select their own genes. However, relative to this second role of phenotypes in evolution, there is no requirement for the niche-constructing activities of phenotypes to result directly from naturally selected genes before they can influence the selection of genes in populations. Animal niche construction may depend on learning and other experiential factors, and in humans it may depend on culture.

The Galápagos woodpecker finch provides a specific example (Alcock 1972). These birds create a woodpecker-like niche by learning to use a cactus spine or similar implement to peck for insects under bark. While true woodpeckers' (*Picidae*) bills are adaptive traits fashioned by natural selection for grubbing, the finch's capacity to use spines to grub for insects is not an adaptation. Rather, the finch, like countless other species, exploits a more general and flexible adaptation, namely, the capacity to learn, to develop the skills necessary to grub, in environments reliably containing cactus spines and similar implements. We emphasize that we are not suggesting that the grubbing of woodpeckers is innate while that of woodpecker finches is learned. Rather, we suggest that the grubbing of woodpeckers and not woodpecker finches has been fashioned by natural selection. The finch's capacity to use spines is not guaranteed by the presence of relevant genes, yet develops reliably as a consequence of its ability to interact with the environment in a manner that allows it to benefit from its own experience. Moreover, the finch's learning certainly opens up resources in the bird's environment that would be unavailable otherwise and is therefore an example of niche construction. This behavior probably created a stable selection pressure favoring a bill able to manipulate tools rather than the sharp, pointed bill and long tongue characteristic of woodpeckers. Because tool manipulation can itself in part depend on learning, there is a further twist to this example. Niche-constructing skills influenced by learning could modify natural selection in favor of an enhanced learning ability. This point is obviously speculative, however it is worth making because it leads us to reconsider another well-known example of dichotomous thinking in biology, that between nature and nurture, which, in this instance, is exemplified by genes versus learning.

Let us try to be more precise about the nature of the information that guides niche construction. As a consequence of the differential survival and reproduction of individuals with distinct genotypes, genetic evolution results in the acquisition, inheritance and transmission of genetically encoded "knowledge" by individuals in populations, relative to particular environments. This information, expressed throughout development, underpins much niche construction, and is common to all species. Although we describe this information as genetically encoded, we recognize that it is expressed as phenotypic traits that also depend on a whole range of developmental factors. In addition, many species have also evolved a set of complicated ontogenetic processes, that allow individual organisms to acquire additional information that is not acquired through population genetic processes, and is not encoded in genes. These processes are themselves adaptive traits and products of genetic evolution, and their functioning is dependent on genetically encoded information, but they are nevertheless distinct from genetic evolution. They comprise "facultative" or "open" developmental processes that are based on specialized information-acquiring subsystems in individual organisms, such as brain-based learning in animals or the immune system in vertebrates. We regard these subsystems as particularly interesting forms of phenotypic plasticity because they are capable of additional, individually based information acquisition, again relative to particular environments. Unlike other developmental influences on the phenotype, these systems are adaptive traits selected precisely because of their information-gathering quality. This allows learned knowledge to guide niche construction in many animal species. However,

if learned information, stored in neural tissue, is expressed in niche construction in a manner that subsequently modifies natural selection pressures and results in the selection of genes that change the learner's brains, then a simple distinction between population genetics and learning becomes untenable. An example, where something like this seems to have happened is in food-storing birds, such as the marsh tit, where the niche-constructing act of storing and hiding food items, seems to have selected for a bigger hippocampus and better memory (Sherry et al. 1989). To clarify, the distinction we have made is not between innate and learned traits, but between information acquired through population genetic processes (expressed under the influence of countless developmental factors in particular environments), and information acquired through specialised information-acquiring subsystems in individual organisms (expressed under the influence of genetic information and other developmental factors in particular environments). Each class of information guides niche construction.

Going beyond individual learning, a few species, including many vertebrates, have also evolved a capacity to learn from other individuals and to transmit some of their own learned knowledge to others by social learning. The resulting "proto-cultural" information may also underlie niche construction. An informative example is the spread of milk bottle top opening in British tits (Hinde and Fisher 1951). These birds learned to peck open the foil cap on milk bottles, and to drink the cream, and this behavior spread right throughout Britain and into continental Europe. Hinde and Fisher found that this behavior probably spreads by local enhancement, where the tits' attention is drawn to the milk bottles by a feeding conspecific, and after this initial tipoff, they subsequently learn on their own how to open the tops. However, further analysis by Sherry and Galef (1984) revealed that, in addition to social learning by local enhancement, milk bottle-top opening could be acquired by another means. They found that this behavior could also spread if the birds were merely exposed to opened milk bottles, even if there were no other birds present to watch performing the opening behavior. In this example, the birds' niche-constructing behavior is propagated by local enhancement. However, by creating opened milk bottles, this niche construction biases the probability that other birds will learn to open bottles. Moreover, the selection pressures acting on genetic variation at loci affected by milk bottle opening are modified in essentially the same manner as if genes underpinned opening. For example, this self-induced selection might influence selection acting on the birds' learning capacities, foraging behavior, or digestive enzymes.

This social transmission reaches its zenith as human culture, where this ability to learn from others is facilitated by a further set of processes (e.g. language, complex cognition). We treat cultural processes as a third category of information acquisition, distinct from, but also partly dependent on, population-genetic processes, and specialized individual-based information-acquiring subsystems, such as learning. This third type of information can also guide niche construction. Consider the example of Kwa-speaking yam cultivators in West Africa, who increased the frequency of a gene for sickle-cell anemia in their own population as a result of the indirect effects of yam cultivation. These people traditionally cut clearings in the rainforest, creating more standing water and increasing the breeding grounds for malaria carrying mosquitoes. This, in turn, intensifies selection for the sickle-cell allele, because of the protection offered by this allele against malaria in the heterozygotic condition (Durham 1991). Here human cultural niche construction has resulted in a well-established and quantifiable evolutionary response.

The consequences of environment modification by organisms, however, are not restricted to evolution, and organisms can affect both their own and each other's ecology by modifying sources of natural selection in their environments (Jones, Lawton, and Shachak 1997). In ecosystem ecology this process is known as "ecosys-

engineering" and it is currently receiving increasing recognition by ecologists (Jones, Lawton, and Shachak 1997). Thus a third difference that incorporating niche construction makes to understanding biological systems is that it modulates and may partly control the flow of energy and matter through ecosystems. Such modifications can have profound affects on the distribution and abundance of organisms, the influence of keystone species, the control of energy and material flows, residence and return times, ecosystem resilience, and specific trophic relationships (Jones, Lawton, and Shachak 1997).

Jones, Lawton, and Shachak (1997) point out that a major ecological consequence of the niche construction of organisms is that it establishes "engineering webs," or control webs, in both communities and ecosystems. Engineering webs do not conform to the same principles of mass flow, stoichiometry (the proportions of chemical elements in organisms), and the conservation of energy that govern the more familiar energy and material flows and trophic relations among organisms. This makes it difficult to understand how engineering webs achieve their control, or to predict which organisms are likely to have the biggest effect on an ecosystem. However, ecosystem engineering often depends on the adaptations of organisms, which may drive evolution when they modify natural selection pressures by niche construction, and when they generate legacies of modified natural selection pressures for subsequent generations. Thus, regardless of whether organisms are themselves members of a specified trophic web, their niche-constructing adaptations may qualify them as members of an associated engineering web, which may allow them to exert a degree of evolutionary as well as ecological control over ecosystems.

Organism-Environment Coevolution

It is common to portray evolution as a process characterized by descent with modification of entities in a lineage. Traditionally, organisms are regarded as the entities that evolve, while their environments are described as changing but not as evolving. Environments are only regarded as evolving to the extent that they are composed of other populations of organisms that clearly do evolve. It is also generally assumed that the distribution of characters of organisms in an evolving population reflects the frequency of particular naturally selected genes currently found in that population. A consequence of this approach is that evolutionary biologists are often content to use only a genetic currency when they describe evolution.

The emphasis on evolving developmental systems by the authors of this volume is one welcome departure from the traditional perspective. We would like to propose another one by suggesting that when organisms niche construct, it is not just the organisms that evolve, because they are also likely to cause a more general coevolution in organism-environment systems by their niche construction. (Gray 1988 and Griffiths and Gray 1994 present a similar perspective). In arguing thus, we do not advocate the mere redescription of environmental change as evolution, which would constitute a purely semantic substitution. Instead we maintain that niche-constructed components of the environment are both products of the prior evolution of organisms and, in the form of modified natural selection pressures, causes of the subsequent evolution of organisms, and that as both products and causes of evolution, these environmental components need to be incorporated in evolutionary theory more fully than they are at present. It is in this sense that we see organisms and their environments as comprising coevolving systems. Neither are we merely advocating a broadening of the concept of the organism so that extended phenotypes are regarded as part of the evolving system (Dawkins 1982), with the rest of the environment treated as nonevolving. Such a position might be justified were there always a direct correspondence between the frequency of particular genes in a population and the incidence of a particular niche-constructed environmental component. Under such circumstances, the evolutionary process could once again be described using a genetic currency, and it would also satisfactorily include those evolving aspects of the environment that are this type of extended phenotype.

In reality, there is a more difficult problem to solve. The problem is that ancestral organisms not only bequeath genes to their descendants, relative to their selective environments in the standard way (figure 10.1a) but also bequeath a legacy of modified natural selection pressures, as an ecological inheritance, relative to those same genes (figure 10.1b). The dichotomy between evolving organisms and nonevolving environments has clearly broken down in the evolutionary scheme illustrated in figure 10.1b. The separation of organism and environment in evolution does not sit well with the evolutionary process once the implications of niche construction are spelled out.

Acknowledgements

This work is supported by a Royal Society University Research Fellowship to Kevin N. Laland and by National Institutes of Health Grant GM28016.

References

Alcock, J. (1972). The evolution of the use of tools by feeding animals. *Evolution* 26: 464–473.

Bock, W. J. (1980). The definition and recognition of biological adaptation. *American Zoologist* 20: 217–227.

Darwin, C. (1881). *The Formation of Vegetable Mold through the Action of Worms, with Observations on Their Habits.* London: Murray.

Dawkins, R. (1982). *The Extended Phenotype.* Oxford: Freeman.

Dawkins, R. (1989). *The Selfish Gene.* (2d ed.) Oxford University Press.

Durham, W. H. (1991). *Coevolution: Genes, Culture and Human Diversity.* Palo Alto: Stanford University Press.

Ehrlich, P. R., and J. Roughgarden. (1987). *The Science of Ecology.* New York: Macmillan.

Feldman, M. W., and L. L. Cavalli-Sforza. (1976). Cultural and biological evolutionary processes, selection for a trait under complex transmission. *Theoretical Population Biology* 9(2): 238–259.

Forshaw, J. (1998). *Encylopedia of Birds.* (2d ed.) San Diego: Academic Press.

Gray, R. D. (1988). Metaphors and methods: Behavioral ecology, panbiogeography, and the evolving synthesis. In M-W. Ho and S. W. Fox (Eds.), *Evolutionary Processes and Metaphors,* pp. 209–242. Chichester: Wiley.

Griffiths, P. E., and R. D. Gray. (1994). Developmental systems and evolutionary explanation. *Journal of Philosophy* 91: 277–304.

Gullan, P. J., and P. S. Cranston. (1994). *The Insects: An Outline of Entomology.* London: Chapman & Hall.

Hansell, M. H. (1984). *Animal Architecture and Building Behavior.* New York: Longman.

Hinde, R. A., and J. Fisher. (1951). Further observations on the opening of milk bottles by birds. *British Birds* 44: 393–396.

Holldöbler, B., and E. O. Wilson. (1995). *Journey to the Ants: A Story of Scientific Exploration.* Cambridge, MA: Belknap.

Hutchinson, G. E. (1957). Concluding remarks. *Cold Spring Harbor Symposia on Quantitative Biology* 22: 415–427.

Jones, C. G., J. H. Lawton, and M. Shachak. (1994). Organisms as ecosystem engineers. *Oikos* 69: 373–386.

Jones, C. G., J. H. Lawton, and M. Shachak. (1997). Positive and negative effects of organisms as physical ecosystem engineers. *Ecology* 78: 1946–1957.

Kirkpatrick, M., and R. Lande. (1989). The evolution of maternal characters. *Evolution* 43: 485–503.

Laland, K. N., F. J. Odling-Smee, and M. W. Feldman. (1996). On the evolutionary consequences of niche construction. *Journal of Evolutionary Biology* 9: 293–316.

Laland, K. N., F. J. Odling-Smee, and M. W. Feldman. (1999). The evolutionary consequences of niche construction and their implications for ecology. *Proceedings of the National Academy of Sciences, USA* 96(18): 10242–10247.

Lee, K. E. (1985). *Earthworms: Their Ecology and Relation with Soil and Land Use.* London: Academic Press.

Lewontin, R. C. (1982). Organism and environment. In H. C. Plotkin (Ed.), *Learning, Development and Culture,* pp. 151–172. New York: Wiley.

Lewontin, R. C. (1983). Gene, organism and environment. In D. S. Bendall (Ed.), *Evolution from Molecules to Men,* pp. 273–285. Cambridge: Cambridge University Press.

Maynard Smith, J. (1989). *Evolutionary Genetics.* Oxford: Oxford University Press.

Naiman, R. J., C. A. Johnston, and J. C. Kelley. (1988). Alteration of North American streams by beaver. *Bioscience* 38: 753–762.

Nowak, R. M. (1991). *Walker's Mammals of the World.* (5th ed.) Baltimore: Johns Hopkins University Press.

Odling-Smee, F. J. (1988). Niche constructing phenotypes. In H. C. Plotkin (Ed.), *The Role of Behavior in Evolution,* pp. 73–132. Cambridge, MA: MIT Press.

Odling-Smee, F. J., K. N. Laland, and M. W. Feldman. (1996). Niche construction. *American Naturalist* 147: 641–648.

Odum, P. E. (1989). *Ecology and Our Endangered Life-Support Systems.* Sunderland, MA: Sinauer.

Paxton, J. R., and W. N. Eschmeyer. (1998). *Encyclopedia of Fishes.* San Diego: Academic Press.

Preston-Mafham, R., and K. Preston-Mafham. (1996). *The Natural History of Insects.* Malborough, England: Crowood Press.

Robertson, D. S. (1991). Feedback theory and Darwinian evolution. *Journal of Theoretical Biology* 152: 469–484.

Sherry, D. F., and B. G. Galef. (1984). Cultural transmission without imitation—milk bottle opening by birds. *Animal Behaviour* 32: 937–938.

Sherry, D. F., A. L. Vaccarino, K. Buckenham, and R. S. Herz. (1989). The hippocampal complex of food-storing birds. *Brain, Behavior and Evolution* 34: 308–317.

Van Valen, L. (1973). A new evolutionary law. *Evolutionary Theory* 1: 1–30.

III THE DEVELOPMENT OF PHENOTYPES AND BEHAVIOR

11 The Ontogeny of Phenotypes

H. Frederik Nijhout

A view of phenotypes from a developmental systems perspective has to be concerned with the *processes* that give rise to the phenotype. Variation in the rate, timing, and spatial distribution of these processes gives rise to variation in the phenotype. A distinction is usually drawn between genotype and phenotype, in which the phenotype is derivative and the genotype (or a gene) is assumed to somehow "control" the characteristics of phenotypes. Yet for complex traits there is usually a weak (and often a variable) correlation between genotype and phenotype. Good examples of such variable association are found in many human genetic diseases where a genetic defect is associated with the disease in only a small fraction of a population, or where the association exists reliably only in certain families or ethnic groups (Vogelstein and Kinzler 1997; Haines and Pericak-Vance 1998). A developmental systems perspective helps to explain, and in some cases to predict, the degree of association (or lack thereof) between genetic variation and phenotypic variation. The approach I will take in this chapter is to examine the consequences of interactions *within* developmental systems for the properties of phenotypes and for the correlation between genotype and phenotype.

Genotypes

Insofar as a discussion of phenotypes makes sense only in relation to genotypes (for otherwise we would simply use the term *character* or *trait*), it is useful to define genotype first. The usual textbook definition of the genotype as "the genetic constitution of an organism" (King and Stansfield 1985) is not particularly useful because terms like *genetic constitution* and *genetic makeup* are not well defined. A more useful definition of genotype is one that explicitly makes use of what we know about the structure of genes. A genotype is typically considered in the context of a phenotype, and thus refers to those genes that are in some way related to a phenotype. There are actually two sets of genes that need to be considered in the definition of a genotype. We can consider either all the genes that affect the *variation* in a character, or all the genes that contribute to the *ontogeny* of a character. How one finds *all* the genes, in either case, is problematic. At one extreme, it could be all the genes that are expressed in all the cells that make up the trait, in a lineage going back to the zygote. This is a large number of genes and could conceivably approach the size of the whole genome. At the other extreme, it could be only those genes that have a major effect on the variation in the trait we are interested in (that is, eliminating from consideration those genes that have small quantitative effects). This is a much smaller number, and, in practice, this is what most people mean when they refer to the "genotype of an organism." The limiting case of this would be a system in which all genes but one have only a single allele (and are thus invariant). In such a system only the one variable gene would constitute the genotype; the lack of variation in characters other than those affected by this one gene makes their genetics moot.

From a developmental systems perspective it is actually something nearer the first extreme that is of interest, because we are interested in the ontogeny and evolution of complex traits (traits whose origin and variation depends on many factors). In practice, it is probably not particularly useful to consider *all* relevant genes back to the zygote, because that gets us into an infinite regression: The zygote's development is influenced largely by maternal genes, whose regulation depends on yet additional maternal genes, and so forth. It might seem that genes expressed early in the ontogeny of a trait are less relevant to variation of the mature trait than genes expressed later, and could therefore be ignored, but it is easy to conceive of exceptions to this apparently sensible viewpoint (for instance, coiling direction of adult snail shells [*Limnaea*] is determined by a maternal gene expressed during early embryonic cleavage [Sturtevant 1923; Freeman & Lundelius 1982]). So how are we to determine what portion of the (variable) genes of an organism or species constitute the relevant genotype for a given trait? It turns out, as we will see, that for complex traits the genes whose variation have the greatest effect on the phenotype differ from one genetic background to another. In other words, the genes that are perceived to affect a trait depends on the context in which that trait is observed.

Phenotypes

What a phenotype is depends on one's perspective and on the level at which one chooses to do the analysis. Let's say that we have a gene for an enzyme. The simplest level of phenotype is the protein coded by a gene. At the next level it is the activity of the enzyme, measured as velocity of substrate depletion or product accumulation. At yet a higher level, the phenotype is the *effect* of the reaction product (or the effect of the absence of the substrate). For instance, if the enzyme is anthocyanidin synthase (which catalyzes the reaction from colorless leucocyanidin to purple cyanidin in morning glories [Holton and Cornish 1995]), then we can call the presence of anthocyanidin synthase the phenotype, or we can call purple flower color the phenotype. In all these cases we are looking at different manifestations of the same genetic factor: protein molecule, enzymatic activity, pigment molecule, optical color.

Now let's say we have a diploid organism that is not homozygous for the two alleles for a gene. We now have a new phenotype, namely that produced by the joint contribution of the two forms of the enzyme encoded by the two alleles. In the chain of manifestations outlined earlier, this new phenotype is expressed above the level of the protein molecule. If the phenotype is expressed as

"activity," it is sometimes simply the average of the two independent activities of the enzymes produced by each allele, but more often the joint activity is closer to one of them, in which case we say that one allelic form is dominant over the other to some degree. This is simple monohybrid Mendelian genetics, and explaining what happens at the level of the phenotype in terms of the characteristics of the genotype is relatively unproblematic.

A more complex and more typical case is sickle-cell anemia, a disease "caused" by a known deficiency in a gene for hemoglobin. Two genes are required to make the components of the adult human hemoglobin molecule: an alpha-globin gene and a beta-globin gene. Hemoglobin has a quaternary structure composed of two alpha- and two beta-globins. A point mutation in the code for the sixth amino acid of the beta-globin chain is the only genetic defect associated with sickle-cell anemia. As far as we know, this defect has no effect at all on the structure or function of the hemoglobin molecule, as long as the molecule is in the presence of relatively high oxygen tensions, and is fully oxygenated. When the molecule enters an environment with low oxygen tension (such as might exist within tissues that have a high metabolic rate) the defective molecule may crystalize instead of remaining in a gel-like state (so this phenotype is contingent on an environmental factor). Whether or not the hemoglobin crystalizes depends on the actual tension of oxygen, and on whether one or both of the beta chains in the hemoglobin are defective. Suppose now that we have a person who is homozygous for the genetic defect. Such a person's hemoglobin will crystalize upon deoxygenation. The crystaline form of hemoglobin is a phenotype, as is the deformation of the red blood cell caused by this crystalization (this is, of course, the phenotype that gave rise to the name of the disease). But there are more phenotypes yet. The sickled cells may block capillaries and cause tissue necrosis (a phenotype), and if this tissue is in the brain, it may cause various neurological or behavioral symptoms (another phenotype). We could go on and on listing the damage and its symptoms in different tissues as separable phenotypes, because individuals differ in exactly how they manifest the disease. In fact, manifestation of any form of disease is the ultimate phenotype, because whereas some homozygous individual suffer greatly and die in childhood, others live asymptomatically to old age. Thus not only can a phenotype (or rather, the phenotypic consequences of genetic variation) be viewed at many different levels, but the actual expression of these phenotypes at any one level is also context-dependent, and is not just a direct consequence of the genetic variation.

A Simple Complex Trait

Context-dependency can actually be observed in much simpler systems as well. Consider a short linear metabolic pathway in which the product of one biochemical reaction serves as the substrate for the next reaction, and so forth. In such a system the flux through the system (measured for instance, as the rate of accumulation of the final product in the reaction chain) depends in some way on the activities of all the enzymes in the pathway. Metabolic control theory can be used to analyze and predict this dependency (Kacser and Burns 1973, 1981; Kacser and Porteous 1987; Keightley and Kacser 1987). Interestingly, if there is genetic variation in only *one* of these enzymes, then the degree to which the flux changes depends not only on that enzyme but also on how many additional enzymes there are in the pathway. The dependence of flux on enzyme activity for a linear chain of n enzymatic reactions is given by Kacser and Burns (1981) as

$$\text{Flux} = \frac{(X_1 - X_n)K_{1n}}{M_1/V_1 + M_2 K_{12}/V_2 + M_3 K_{13}/V_3 + \cdots + M_n K_{1n}/V_n} \quad (11.1)$$

where X_1 and X_n are the concentrations of the initial substrate and final product, respectively, the

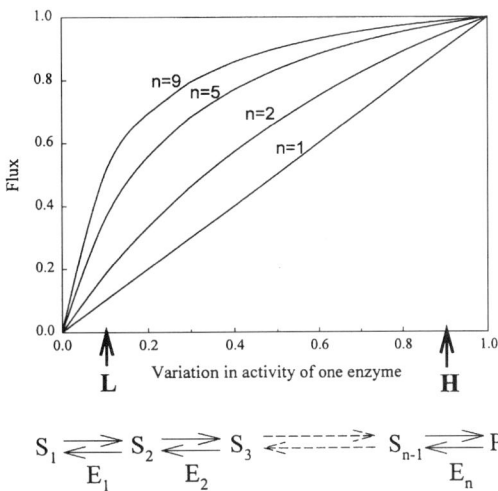

Figure 11.1
The effect of variation in the "activity" of one enzyme on the flux through a metabolic pathway. The relationship becomes progressively more nonlinear as the pathway gets longer. Enzymatic variation can come about through mutations, so the horizontal axis can be read as the effect of allelic variation on the genes that code for an enzyme on the activity of that enzyme. Arrows indicate the values of two extreme "alleles" encoding high (H) and low (L) enzymatic activities, used to construct figure 11.2.

K is an equilibrium constant, V is a maximal velocity (V_{max}), M is a Michaelis constant (K_m), and the subscripts refer to the number of the enzyme in the chain. Figure 11.1 illustrates the relationship between the activity of any one enzyme in such a chain and the flux through the chain. If there is only one enzyme the relation is linear, but the relationship becomes increasingly hyperbolic as the chain becomes longer. This means that the sensitivity of flux (a phenotype) to variation in the activity of an enzyme (a lower level phenotype) is a systemic property, determined by *all* enzymes in the system. The contribution of each enzyme to the total flux is described by its sensitivity coefficient, which is a measure of how much the flux will change given a particular amount of change in the activity of that one enzyme. This coefficient can be calculated by taking the partial derivative of the flux with respect to the enzyme of interest, and is thus proportional to the tangent at any one point on the curves in figure 11.1 (Kacser and Burns 1981). The sensitivity coefficient of a given enzyme thus not only varies with the activity of that enzyme but also varies with the total number of additional enzymes in the pathway. It is therefore a joint property of *all* the enzymes in the pathway.

One of the most important results of this type of metabolic control theory is the *summation property*: the magnitudes of the sensitivity coefficients of all enzymes in a system must sum to one (Kacser and Burns 1973, 1981). This implies that the more enzymes there are, the smaller their respective coefficients will be (assuming all have approximately the same relative effect on flux). It also implies that if not all enzymes have the same sensitivity coefficient, only a few can have a large coefficient. This is because if one or a few enzymes have a large coefficient, then the coefficients of the rest must be small since they all have to sum to unity. The summation property therefore indicates that if a system is controlled by many enzymes, and if all enzymes have an equal effect on flux, the sensitivity of the system to variation in any one of them will be very small (if there are n enzymes in a system and their sensitivity coefficients are equal and add to 1, each will have a sensitivity coefficient of $1/n$). But if the enzymes do not have an equal effect on flux, then only one or a few of them can have large coefficients and the rest must have proportionally small ones, so that all still add up to one. Hence in a system where enzymes do not have equal effects, one (or a few) will appear to be in control of the flux, and the effects of the remainder will be small and perhaps even undetectable. What is most important to recognize is that the sensitivity coefficient of an enzyme is not a property of the enzyme itself, but is a systemic property (its value is determined largely by the denominator in equation 11.1, which contains terms for *all* the

enzymes in the pathway). Therefore, it is the ensemble that determines which of the enzymes will be the one with the largest coefficient.

One way to examine how the flux depends on variation in a single enzyme is by means of a sensitivity analysis: examining the effects of variation in one parameter or variable while keeping all others constant. The parameters in this system are the various constants that define the kinetics properties of each enzyme: V_{max}, K_m, and K_{eq} (see equation 11.1). Each of these constants is determined by the tertiary structure of the enzyme, which, in turn, is determined by the base sequence of the gene that encodes it. In a real population of organisms the genes will be polymorphic (that is, there will be many alleles), so that at the population level these constants actually vary among individuals. This means that for a sensitivity analysis it is necessary to make a decision about the values at which the various kinetic constants will be held. This could be done in a variety of ways, but a simple and heuristically instructive method is to assume that there are only two alleles for each enzyme-encoding gene in the population, so that in a diploid organism we need deal only with three genotypes for each gene. We can then hold all genes except for one either homozygous (*AA* or *aa*) or heterozygous (*Aa*), and examine the effect of variation of that one gene on flux through the system. Sensitivity analyses examining the effect of variation in V_{max} and K_m of one of the enzymes in a linear pathway of nine enzymatic steps are illustrated in figure 11.2. The three graphs in each panel describe the effects of variation in three different genetic backgrounds: *LL*, a background in which the remaining eight genes are homozygous for the allele with the larger value; *SS*, a background homozygous for alleles of the smaller value; *LS*, a background heterozygous for the two alleles. These graphs show that the genetic background is a major determinant of the degree to which variation in one gene affects the flux through the system. For instance, the mean slope of the curves in figure 11.2a shows that variation in the V_{max} of an en-

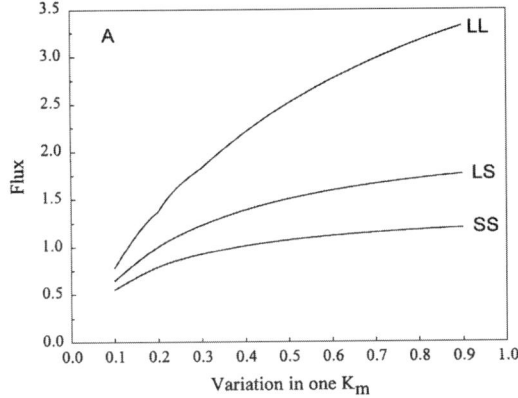

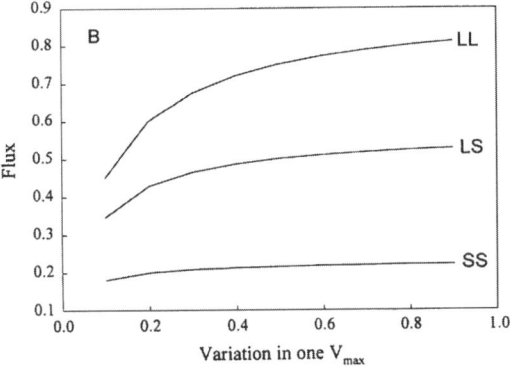

Figure 11.2
(A) Effect of variation in the value of the Michaelis constant (K_m) of one enzyme on the flux through a linear biochemical pathway of nine enzymes. (B) The effect, of variation in the V_{max} of one enzyme in a nine-enzyme pathway. The three curves in each panel represent three genetic backgrounds in which the other eight enzymes in the pathway are held constant at genotypes that are either homozygous (SS and LL) or heterozygous (LS) for enzymatic values indicated by the arrows in figure 11.1. The magnitude of the effect of enzymatic variation depends strongly on the genetic background (i.e., on the allelic values of all other genes that participate in the ontogeny of the phenotype [flux]).

zyme has only a small effect on flux, and the separation between the curves shows that variation in the genetic background has a much larger effect.

Kacser and Burns (1981) showed that dominance is an emergent property of the interaction among enzymes linked in a biochemical pathway. This is because the interaction among enzymes causes the relationship between enzymatic variation and flux to become nonlinear (figure 11.1). Thus, although the two alleles of a gene act additively at the physiological level (*additivity* means that the physiological effect of each allele in a diploid organism is independent of the effect of the other allele, so that their joint effect is simply the sum of their independent effects; a heterozygote would then have a genetic value exactly halfway between that of the two homozygotes), the interaction among genes causes nonlinear interactions at the level of the phenotype, which is manifested as partial dominance of one allele over the other (figure 11.3).

Because *all* genes acquire a nonlinear relationship with flux, this means that the genetic background in which any one gene acts also becomes nonlinear. The nonlinearity of the genetic background is revealed by the uneven spacing of curves in figure 11.2a. The sensitivity curve for the heterozygous genetic background does not lie halfway between the homozygous genetic background curves, as one would expect if the gene in question simply acted additively with the background genes. In other words, the epistatic interaction among genes is also nonlinear, at least for variation in K_m (but apparently not for variation in V_{max}; figure 11.2a).

A Developmental System as the Determinant of Phenotype

The simplest and most widespread developmental mechanism for pattern formation is a diffusion-gradient threshold mechanism. Diffusion gradients are used throughout embryonic and postembryonic development to specify the location and spatial pattern of developmental events.

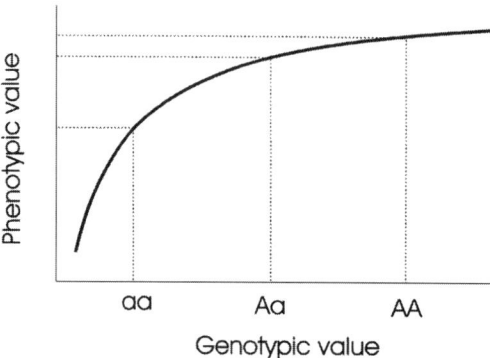

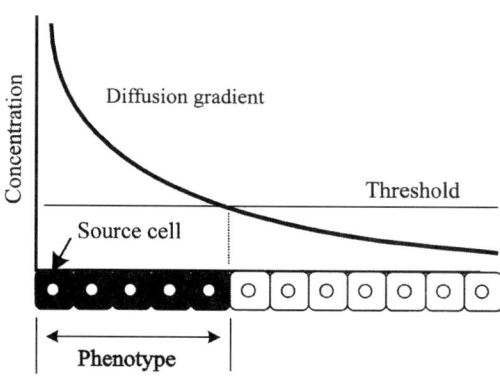

Parameters: Synthesis rate at source
Diffusion coefficient
Decay rate
Threshold
Time at which threshold is read
Background synthesis

Figure 11.3
Dominance is an emergent property of nonlinearity in the relationship between genetic variation and phenotypic variation. The curve in this figure represents the function that relates variation in the alleles of one gene (horizontal axis) to the phenotype (vertical axis). The shape of this curve is determined by all the factors that interact to produce the phenotype (cf. figures 11.1, 11.5, and 11.6). If two alleles (*A* and *a*) are codominant at the genetic level, then the value of the physiological "activity" of the heterozygote will be exactly halfway between that of the two homozygotes. The final phenotype of the heterozygote (vertical axis) is, however, closer to one homozygote than the other, so that at the phenotypic level one of the alleles is partially dominant.

Figure 11.4
A diffusion-gradient threshold system. A cell produces a signaling molecule that diffuses from cell to cell and is metabolized over time. Hence progressively farther cells experience a progressively lower concentration of the signal. At a certain point in time the cells "read" the concentration of the signal. If the signal is above a threshold value the cells express one kind of gene (dark cells), and if it is below threshold they express a different gene (white cells). The phenotype is the size of the population of cells that is dark. Most processes of pattern formation during embryonic and postembryonic development are regulated by such a gradient threshold mechanism. The exact value of this phenotype is determined by the six parameters shown.

A diffusion-gradient threshold mechanism and the phenotype it produces thus constitute a simple but realistic developmental system. (Nijhout and Paulsen 1997; Klingenberg and Nijhout 1999). The study of genetic variation and evolution in such a simple system reveals much about properties of developmental systems in general. We assume that the values of the parameters that control a diffusion gradient threshold system are influenced by genes (for simplicity we assume that alleles at a single gene are responsible for the entire genetic variation encountered in each parameter). Figure 11.4 shows how six parameters can control the value of the phenotype in such a developmental system. The genetics of the phenotype produced can be studied by introducing genetic variation in the parameter values. A sen-

sitivity analysis of each parameter is shown in figure 11.5. As in the case of the metabolic pathway outlined earlier, the relationship between genotypic value of a particular gene and the phenotypic value depends almost entirely on the values of the other genes: the genetic background. Dominance and epistasis can be seen, respectively, from the curvature of the relation between genetic and phenotypic values and from the spacing of the curves in different genetic backgrounds.

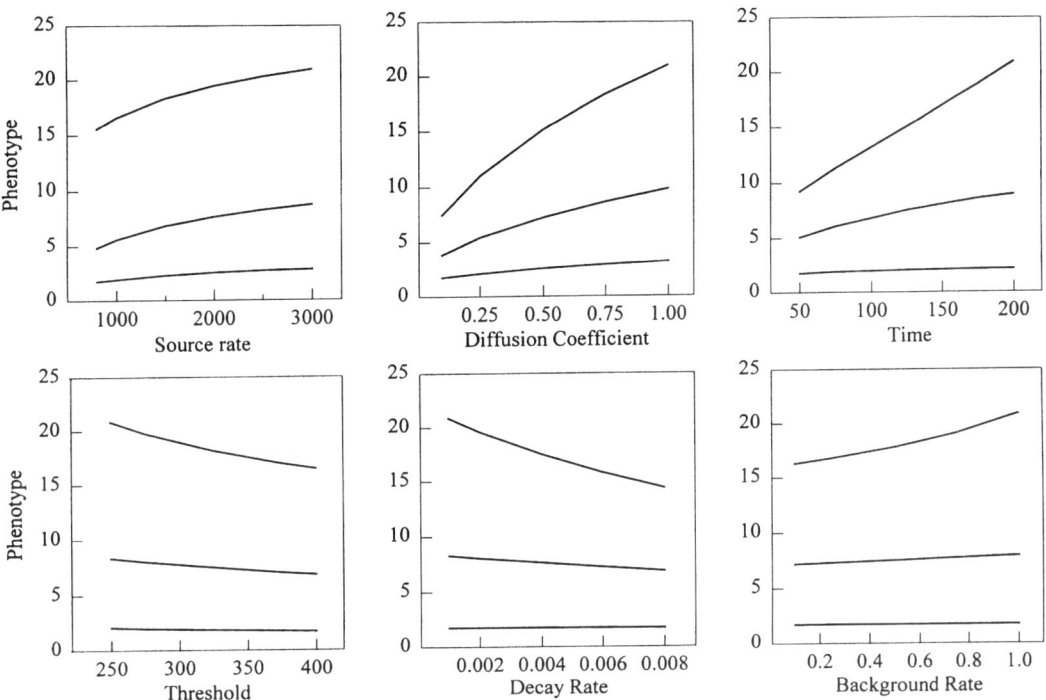

Figure 11.5
Effect of genetic variation in the six parameters of a diffusion-gradient threshold mechanisms on the value of the phenotype. As in figure 11.2, the curves in each panel represent the sensitivity of the phenotype to variation in one parameter when the others are kept constant at a homozygous high, homozygous low, or heterozygous level. Because of the interactive nature of this system, a given genetic value at any one gene can correspond to a broad range of phenotypic values, and a given phenotypic value can be associated with a broad range of genetic values, depending entirely on the values of all other genes in the system. Thus there is no one-to-one correspondence between genetic value and phenotypic value for any of the genes that participate in this system. (After Nijhout and Paulsen 1997.)

Evolution in such a system can be studied by setting up a population of diffusion-gradient threshold models by computer simulation. Individuals are assumed to be diploid, and each parameter is represented by two alleles, one tending to produce a high phenotypic value the other a low phenotypic value. The frequencies of the two alleles in a simulated population can be defined (e.g., 10% high alleles and 90% low alleles) and a population of individuals can be produced by random combination of the alleles. Six unlinked genes, each with two alleles, can produce 729 pos-

sible genotypes, the frequency of each genotype being determined by the relative frequencies of each of the alleles. Random combination of alleles yields a population with a particular mean phenotype and Hardy-Weinberg frequencies of genotypes. We can now study the evolution of this developmental system in a population genetic context by asking how the allele frequencies change under artificial selection on the phenotype.

The efficacy of selection is a function of the correlation between genotype and phenotype. By inspection of figure 11.5 it might seem on first sight that this correlation is very poor: a given genetic value at any one gene can be associated with a broad range of phenotypes, and any phenotype can be associated with an equally broad range of genetic values of every gene. Yet selection on the phenotype leads to an orderly and gradual change in the mean phenotype of the simulated population (figure 11.6a). In each generation, only individuals with phenotypes larger than a particular critical value were selected and then allowed to mate randomly among themselves to produce the next generation. Although the phenotypic response to selection was orderly, the genetic response was rather more complex (figure 11.6b). Initially only two genes responded to selection and not until their allele frequencies had changes considerably did additional genes begin to respond. At any one time during the selection process, most of the genetic change was focused on only one or two genes. The cause of this pattern of genetic change is the fact that not all genes are equally correlated with the phenotype, and that any gene's correlation with the phenotype depends on its frequency in the population and well as on the frequency of all the other genes. The correlations are shown in figure 11.6c. The six genes in this system are not all equally correlated with the phenotype, and the correlation of any one gene with the phenotype changes as selection progresses and allele frequencies change. One of the most interesting features of this correlation pattern is that each gene appears to take a turn, as it were, being the most highly correlated with the phenotype as selection progresses. Another interesting observation is that at most times, the majority of genes are very weakly correlated with the phenotype. It appears that at any one time only one or a few genes are highly correlated with the phenotype while most genes are essentially neutral and not seen by selection.

An outside observer, chancing upon such a population of organisms at some arbitrary time during selection, would identify only one or two genes as primarily "controlling" the phenotype. Another observer, at a different time, would identify a different set of genes as the ones whose variation was most highly correlated with phenotypic variation and which would thus appear to be primarily in control. For instance, in generation 6 most of the phenotypic variation is due to variation in the gene that affects timing, whereas in generation 11 most of the phenotypic variation is due to variation in the gene that affects the breakdown rate of the diffusing substance. At any one time most genes are poorly correlated with the phenotype and thus are neutral to selection on the phenotype. These genes are nevertheless critical contributors to the phenotype, and in a different genetic background one or more of them could become most highly correlated with the phenotype while previously highly correlated genes become nearly neutral. This phenomenon is called *pseudoneutrality* or *contingent neutrality*; it is not a property of the gene but of the background in which it is expressed.

The Effect of Environmental Variation on the Phenotype

We have just looked at the effects of genetic variation on the phenotype and modeled this effect as if environmental variation did not exist. In both systems we have examined, the metabolic pathway and the diffusion-gradient threshold mechanism, it is actually easy to see how environmental variation could alter the phenotype. The activity of an enzyme, for instance, depends on the temperature, pH, and ionic strength of its environment as well as on the presence of chemical inhibitors and synergists. A change in temperature can therefore alter the activity of an enzyme in the same way that a mutation would. Both the equilibrium constants and the reaction constants of enzymatic reactions vary with temperature, as given by the van't Hoff and the Arrhenius equation, respectively. In a multienzyme metabolic pathway, a change in temperature of course affects all enzymes simultaneously, but because enzymes differ in their temperature sensitivity,

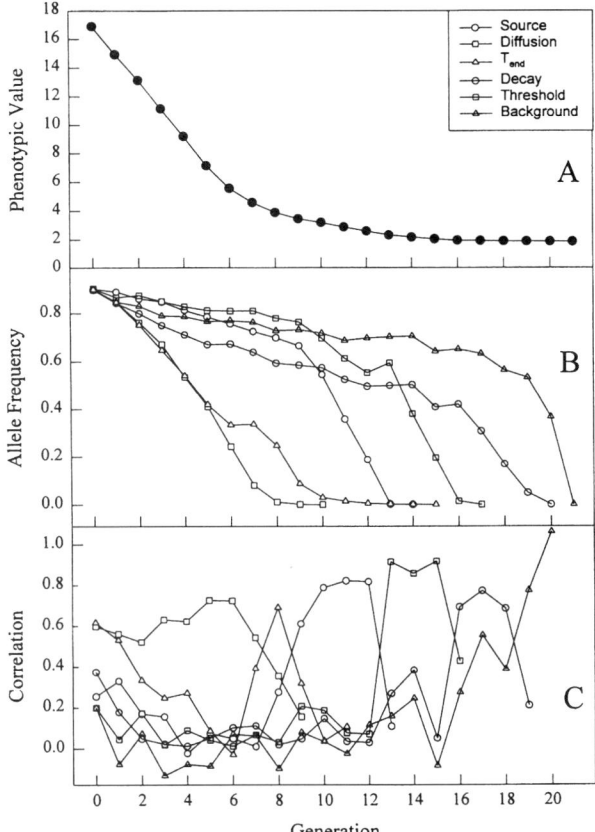

Figure 11.6
Response of a diffusion-gradient threshold mechanism to selection on the phenotype. Each of the six parameters was represented by two alleles (a *high* allele, producing a large phenotypic value, and a *low* allele, producing a low phenotypic value), whose frequency was initially set at 90 percent high and 10 percent low. An initial population of ten thousand "models" was generated by random recombination of diploids from this gene pool, and their respective phenotypes were calculated. All individuals with phenotypes above the mean were eliminated and the next generation produced by random "mating" of the remaining individuals, and this process was repeated for eighteen generations. (A) The response of the mean phenotype to selection. (B) The response of the allele frequencies for the six parameters to selection. As can be seen, not all parameters responded synchronously. The pattern shown here was identical in several independent runs of the selection, and was not due to stochastic events associated with random mating. (C) The correlation between variation at each parameter and variation of the phenotype. The parameters that respond most rapidly to selection are those that are most highly correlated with the phenotype. Only one or two parameters are highly correlated with the phenotype at any one time. As selection progresses, each parameter in turn becomes "most highly" correlated. Thus each of the parameters could be considered a major gene, a modifier gene, or a nearly neutral gene at different times, depending entirely on the genetic background defined by the other parameters. (After Nijhout and Paulsen 1997.)

different enzymes in the pathway will be affected to different degrees. The overall effect of temperature will therefore be integrated across all enzymes in a pathway.

Variation in enzyme activity with temperature results in variation in the flux through a metabolic pathway. Although this flux is a phenotype in its own right, it also affects other phenotypes for whose development or properties the metabolic pathway is necessary. Variation of the phenotype with variation in an environmental factor is called a *reaction norm* (Schlichting and Pigliucci 1998). All phenotypes express reaction norms in response to one or more environmental variables. As a rule, a particular correspondence between a genotype and a phenotype is defined only under a restricted set of environmental (and genetic background) conditions.

In the more complex diffusion-gradient threshold system it is likewise easy to see how various environmental factors could affect the phenotype. Temperature not only affects the kinetics of synthesis and breakdown of diffusing substances, but also affects their diffusion coefficients. Because temperature affects equilibrium constants, it is also likely to affect the value of thresholds. The reaction norm to temperature would be defined by the joint effects of temperature on all the processes that define the diffusion and threshold system. Other environmental variables such as pH and ionic composition can affect a gradient threshold system by the same mechanisms they affect enzyme kinetics.

In general, genetic (mutational) and environmental variation have similar effects on the components of a developmental system. Both types of variation alter the rates and equilibrium values of dynamical processes and therefore both can have the same effect in the final outcome of the processes that produce the phenotype. The main difference between genetic and environmental variation is that the former can alter the value of only a single component of the system (by altering the properties of a single protein), whereas the latter generally alters the properties of many

components simultaneously. For instance, a mutation will directly affect the activity of only a single enzyme, whereas a change in temperature will directly alter the activities of many enzymes simultaneously. Of course, due to the highly integrated nature of developmental systems, both mutation and environmental variation can have a host of indirect effects on many parts of the system. The equivalency of genetic and environmental variation is demonstrated by the relative ease with which it is possible to produce accurate phenocopies of known mutations by simple environmental perturbations (Santamaria 1979; Oyama 1981; Mitchell and Petersen 1982).

In order to produce a predictable phenotype in a variable (varying in time) and diverse (varying in space) environment, the developmental system needs to be made robust to this variation. Resistance to a variable environment is accomplished by homeostatic physiology, which attempts to maintain a fairly constant internal environment in the face of external variation. A diverse environment can be accommodated by selection for mutations and combinations of alleles that adjust the dynamics of the developmental system to produce the target phenotype in each particular environment.

Can the Findings of Simple Systems Be Generalized to More Complex Phenotypes?

As a trait becomes more complex during ontogeny (or evolution) it becomes dependent on, and therefore sensitive to, an ever greater number and diversity of factors, both genetic and nongenetic. Moreover, as our insight becomes more sophisticated and we take more factors into account in our analysis of a trait, the relationship between any one factor (or determinant) and the phenotype becomes increasingly indirect and contingent.

Most developmental systems are substantially more complex than the two exemplars discussed above. It is therefore useful to consider whether the general features of these simpler systems, such as nonlinearity, emergent dominance of alleles, contingent neutrality, and poor correlation between variation at any one gene and the phenotype (in a typical genetic system this would be called *variable penetrance*), also occur in more complex systems. In addition, we would like to know whether more complex systems have additional properties that emerge from higher-level interactions that are not exhibited by these simpler systems.

The emergent properties of the two simple systems are due to two characteristics: (1) nonlinearity of the relationship between genetic variation (or environmental variation) and phenotypic variation, and (2) unequal effects of each of the determinants on the phenotype. The interaction of the nonlinearity with the inequality of effects makes the strength of the association between a given determinant and the phenotype context dependent, and it changes the absolute as well as the *relative* magnitude of the effect of a given determinant in different genetic backgrounds and in different environments. All systems that are nonlinear and that have components that interact nonadditively and with different quantitative effects on the phenotype will exhibit the emergent properties outlined above (dominance, contingent neutrality, and incomplete and variable penetrance).

More highly complex systems can deviate from the general characteristics of the simple systems we have studied in the following ways. More complex systems can have compensatory or buffering mechanisms (such as feedback loops, duplicate pathways, and alternative pathways) that make the phenotype insensitive to variation in one or more of its determinants. In such cases, variation in those particular determinants would not be expressed as phenotypic variation and would be essentially neutral to selection. More complex systems can also be more nonlinear, with U- or S- or N-shaped curves that relate genetic (or environmental) variation to phenotypic variation. In such cases variable penetrance and contingent neutrality still apply, but a new

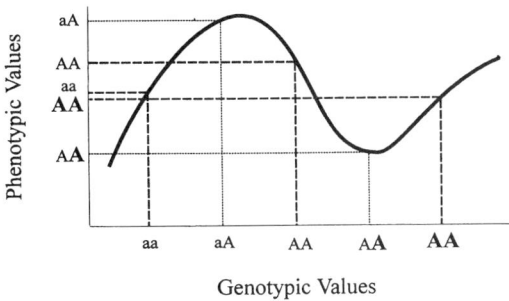

Figure 11.7
Complex relationship between genotype and phenotype can arise from nonlinear multifactorial interactions among the parameters of the developmental system that give rise to the phenotype. The curve in this graph illustrates such a (hypothetical) relationship. Three alleles and five genotypes of increasing genetic value are shown on the horizontal axis. Due to the nonlinearity of the system, the order of phenotypic values does not correspond to the order of genetic values, and the dominance among alleles depends on their absolute, not their relative, values. As in figure 11.3, the shape of the relationship is determined by all interacting components of the system. Accordingly, mutations at other genes can change the shape of the curve, and this alters the dominance relationships of the focal gene.

level of contingency emerges, namely in the dominance of alleles. Dominance among any two alleles in such systems depends on exactly where along the curve they occur (figure 11.7). Dominance relations can change sign in different regions of the genetic value axis, or switch from overdominance to underdominance. Finally, more complex developmental systems are typically composite, built up of a large number of parallel and sequential modules that are themselves relatively simple (for instance, early embryonic development in *Drosophila* is controlled by a coupled sequence of diffusion-gradient threshold events in which one thresholded gradient defines the site of expression of a new gene whose product diffuses and activates a subsequent gene wherever it is above or below some threshold, and so on). If the interactions among the modules have nonlinear components, then new context dependencies will arise.

Conclusion

Phenotypes are properties of developmental systems. The rate constants and equilibrium constants of the dynamical processes that define a developmental system are themselves defined by, and can be modified by, both genetic and environmental factors. In this view of developmental systems, the primary role of genes is to supply enzymes that catalyze biochemical reactions and that control the rates of various processes. Allelic variation leads to variation in the dynamics of developmental processes, and these lead to variation in phenotypes. Nonlinearity in the association between genetic variation and phenotypic variation leads to two properties of interacting systems. At the level of the alleles of a single gene, nonlinearity produces dominance. This means that the effect of the two alleles in a diploid is not simply additive, but that the phenotype associated with the heterozygote is closer to that of one of the homozygotes than to the other. At the level of the interacting genes within the developmental system, nonlinearity leads to context-dependent variation in the absolute effect of a gene on the phenotype. This in turn leads to contingent neutrality of a significant fraction of the genes whose products play a role in the developmental system. Due to contingent neutrality it is possible to detect only a small fraction of the genes that affect a phenotype, and the composition of this fraction is different in different genetic and environmental backgrounds. A gene that has a major effect on a phenotype in one genetic context could appear as a modifier gene of small effect in a different context, and could not be detected at all in yet another context (even though its characteristics and amount of variation do not change). The degree to which genetic variation is associated with phenotypic variation is, therefore, a property not of the genes but of the developmental system in which the genes play a role.

References

Freeman, G., and J. W. Lundelius. (1982). The developmental genetics of dextrality and sinistrality in the gastropod *Limnaea peregra*. *Wilhelm Roux Archives* 191: 69–83.

Haines, J. L., and M. A. Pericak-Vance. (1998). *Approaches to Gene Mapping in Complex Human Diseases*. New York: Wiley-Liss.

Holton, T. A., and E. C. Cornish. (1995). Genetics and biochemistry of anthocyanin biosynthesis. *Plant Cell* 7: 1071–1083.

Kacser, H., and J. A. Burns. (1973). The control of flux. *Symposia of the Society for Experimental Biology* 27: 65–104.

Kacser, H., and J. A. Burns. (1981). The molecular basis of dominance. *Genetics* 97: 639–666.

Kacser, H., and J. W. Porteous. (1987). Control of metabolism: What do we have to measure? *Trends in Biochemical Sciences* 12: 5–14.

Keightley, P. D., and H. Kacser. (1987). Dominance, pleiotropy and metabolic structure. *Genetics* 117: 319–329.

King, R. C., and W. D. Stansfield. (1985). *A Dictionary of Genetics*. Oxford: Oxford University Press.

Klingenberg, C. P., and H. F. Nijhout. (1999). Genetics of fluctuating asymmetry: A developmental model of developmental instability. *Evolution* 53: 358–375.

Mitchell, H. K., and N. S. Petersen. (1982). Developmental abnormalities in *Drosophila* induced by heat shock. *Developmental Genetics* 3: 91–102.

Nijhout, H. F., and S. M. Paulsen. (1997). Developmental models and polygenic characters. *American Naturalist* 149: 394–405.

Oyama, S. (1981). What does the phenocopy copy? *Psychological Reports* 48: 571–581.

Santamaria, P. (1979). Heat shock induced phenocopies of dominant mutants of the bithorax complex in *D. melanogaster*. *Molecular and General Genetics* 172: 161–163.

Schlichting, C. D., and M. Pigliucci. (1998). *Phenotypic Evolution: A Reaction Norm Perspective*. Sunderland, MA: Sinauer.

Sturtevant, M. H. (1923). Inheritance of direction of coiling in *Limnaea*. *Science* 58: 269–270.

Vogelstein, B., and K. W. Kinzler. (1997). *The Genetic Basis of Human Cancer*. New York: McGraw-Hill.

12 The Development of Ant Colony Behavior

Deborah M. Gordon

Developmental systems theory seeks alternatives to the idea that an organism's development is the expression of information already in place before development began. It is obvious that such alternatives are needed to understand the development of an ant colony. A colony's behavior arises from the relations of ants to each other and to the rest of their environment (Gordon 1999). The colony's behavior can not be the product of pre-packaged instructions, because there is nowhere to locate the package. As the colony grows older, the ants die, to be replaced by their younger sisters. The queen lays the eggs but does not direct the behavior of the ants. Each ant's behavior is based on local information. The aggregate of all ants' behavior produces the development of the colony.

An ant colony consists of many sterile female workers and one or more reproductive females or queens. The queen produces new workers. Ant colonies produce offspring colonies, so a colony can be considered as an individual organism: A colony is born, grows older, reproduces and, when the queen dies and there is no one to produce more workers, the colony dies. The behavior of the colony develops, year after year, as new ants are born into the colony and die. This chapter is about the development of colony behavior in the red harvester ant (*Pogonomyrmex barbatus*), a desert species that eats the seeds, mostly of grasses, that it collects from the ground. In a mature harvester ant colony, each summer the queen produces reproductives, winged queens and males who fly off to mate with the reproductives of other colonies. After mating, the males die and the newly mated queens found new colonies. A colony may live for fifteen to twenty years (Gordon 1991), beginning with a single queen and reaching a stable size of about ten thousand workers when the queen, and the colony, are five years old. Individual ants live only a year.

I study a population of about three hundred harvester ant colonies in the desert of southeastern Arizona. Each year I census all of the colonies, which are individually labeled. This census, now in its fifteenth year, tells me how old each colony is, and that has made it possible to track the development of colony behavior, as well as the demography of the population.

A colony's behavior changes as it grows older and larger. Colonies of any age must perform certain tasks, such as building a nest, taking care of the juvenile stages (eggs, larvae, and pupae), and collecting food. Task allocation is the process that adjusts the numbers of workers engaged in each task, in response to changes in the environment and the needs of the colony (Gordon 1996). Task allocation operates differently in young, small colonies and old, large ones. Old colonies respond in a more stable way to environmental perturbations than do younger ones. In addition, young and old colonies differ in their relations with neighboring colonies of the same species.

Ant colony behavior develops as a result of the ants' responses to the changing contexts they experience as the colony grows older. The most obvious change over time in the colony is its size (figure 12.1; Gordon 1992). Colony size thus seems to be the first place to look for explanations for age-dependent changes in colony behavior. In a large colony with many ants, each ant experiences a different environment from the one that surrounded an ant in the smaller, younger colony. Colony size affects the pattern of an ant's interactions with other ants: more ants can lead to a higher rate of contact among ants. Ants modify their environment by creating and modifying a nest, adjusting the amounts of vegetation, refuse and chemical traces on the surface of the nest mound, and by altering through food collection the distribution of food in the foraging area around the nest. More ants modify their environment more, and in different ways.

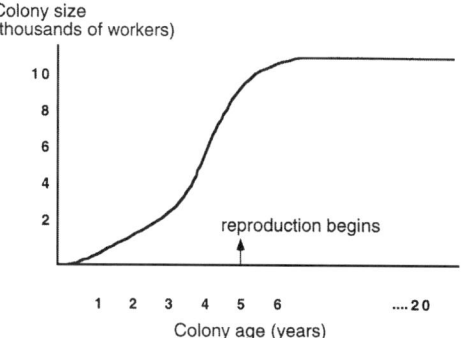

Figure 12.1
How colony size, in numbers of workers, depends on colony age.

A population of colonies thus provides an example of niche construction (*sensu* Lewontin 1983). In harvester ants, this process of niche construction unfolds over the fifteen- to twenty-year lifespan of the colony. To find proximate explanations for the ways colony behavior changes as the colony grows older, we must examine how the colony modifies its environment as it grows larger.

Founding a Colony

The beginning of a colony's development is the stage I know least about. Six to eight weeks after the mating flight, the first workers emerge. These ants are tiny; the queen fed them as larvae from her own fat reserves and they are much smaller than later cohorts of workers. These first ants go out to forage, with only four to six weeks of foraging time remaining before the colony's first winter sends them back inside the nest. By the next summer, the colony will have a small nest with many tunnels and chambers, a small mound above the ground, and five hundred to a thousand ants, and the foragers will begin to explore the neighborhood.

An ant in the first batch of workers emerges from its pupal case into a single tunnel and the company of a few other workers, perhaps some eggs and larvae. Somehow some ants get out and collect some food, and this food is given to the larvae. Some ants must begin to dig. A year later, by the time the colony has reached its first summer, an ant emerging from its pupal case finds a very different environment. An ant in the one-year-old colony is part of a functioning colony. Several tasks are carried out simultaneously. The processes that shuttle ants from one task to another are in place. There is a nest with a system of tunnels and chambers. In a one-year-old colony, ants come barging out of the nest entrance apparently intent on some activity, perhaps heading straight for a foraging trail or ambling over to the midden pile to move bits of refuse from one place to another. But ants in the first batch of workers seem more uncertain (though I have been able to observe them only in the laboratory).

It might be tempting to believe the first forager to walk out of the nest is "programmed" to collect food, bring it back to the nest and feed it to the queen who then produces more and more of these "hard-wired" automata. But there is no reason to believe the first ant to come out has some prior knowledge of what to do, any more than the rest of the ants do. The ant is born in a chamber at the bottom of a straight tunnel. It can go only around the chamber or up the tunnel. Eventually it goes up. It gets outside. Being an ant, the events it encounters, mostly olfactory, elicit responses that get it eventually to food, and eventually it picks up some food, and eventually it finds its way back inside.

What is contained in the notion that "being an ant," there are some things that the ant is likely to do? It has an ant's body, and that offers a set of possibilities for what it perceives, and how it moves, and what it can pick up. Maybe the joint actions of ants' bodies are likely to produce chambers with a certain curvature of the walls, because of the ways they move their heads as they scoop up sand. Being an ant also somehow establishes a set of possibilities for how ants respond to each other. We know that in many species of ants, individuals differ in their behavior, especially in how active or mobile they are. For some tasks, in some species, it has been shown that experience contributes to an ant's task performance.

The queen has a minimal role in organizing colony behavior. She does not direct the behavior of the ants in the colony. She has no authority. In fact, there is no evidence that any ant ever tells another what to do. So it would be absurd to imagine that somehow the queen carries inside her all of the instructions that later produce the behavior of the mature colony of ten thousand ants. A queen's life, after her brief, early excursion outside, consists of eating and laying eggs. In the laboratory, the queen of a large colony usually has some workers nearby. She rarely moves around much, and often seems to be standing doing nothing. Sometimes workers feed her, or groom her, or pile up eggs as she lays them. Winged reproductives are produced three to six weeks before the day of the mating flight, so a queen will have spent only a tiny portion of her life in her parent colony and will have little prior experience of life in a mature colony. She is outside of a nest for only a day or two to mate and get to the place where she will dig a new nest. Most newly mated queens die before they ever dig a nest; many are eaten by birds and lizards. If a newly mated queen does manage to start a nest, she digs furiously for the first two or three days, creating a single tunnel not much wider than her own body, about eighteen inches deep. She never comes out of the nest again, except if the colony moves to a new nest.

Apparently many new colonies do not survive, because there are many more nests of founding queens on the site each year (five hundred to one thousand) than there are one-year-old colonies the following year (twenty to fifty) (Gordon and Kulig 1996). I do not know if the queens die before they can produce any workers, or if the first batch of workers does not provide enough food to support the queen through the winter. Perhaps the small initial group of workers has a low probability of survival because there are not enough ants to compensate for each other's incompetence or to ensure that ants perform the necessary tasks. Maybe it often happens that the first few foragers out never find their way back; the probability of getting back to the nest may be low. In larger colonies, ants can use their foraging nestmates as a cue to the direction back to the nest, perhaps by following a gradient in the density of nestmates. But in a very small colony, a forager has a small chance of encountering a nestmate on the way home.

The accumulation of mistakes similar to those of the first foragers may prevent small colonies from successfully performing tasks inside the nest. In large colonies, some ants dig new chambers and others carry out the soil. The ants that scrape soil from the wall of the chamber might just put it down on the floor. Other ants walk by, and eventually some ant picks up a bit of soil and takes it somewhere else, maybe toward the entrance, maybe not. Eventually soil builds up near the entrance where ants pick it up and carry it out. In a very small colony, there may not be enough ants to ensure that on average, the soil will get out. Perhaps in some tiny new colonies, new chambers do not get built when they are needed. Small colonies may die because the chances of getting things done are low. But as a colony grows larger, eventually there comes a point when it has sufficient numbers of ants working in a sufficiently modified environment that on average, the colony accomplishes its tasks often enough to survive.

Colony Growth

Colonies grow quickly for the first four or five years (figure 12.1). If a colony manages to survive to be two years old, it will probably last out the next fifteen years (Gordon and Kulig 1998). A three- to four-year old colony, with four thousand to six thousand ants, is in the steepest portion of the colony's growth curve (Gordon 1992).

By five years of age, a colony can reach its mature size of ten thousand ants and begin to reproduce. This is a stable size; the colony ceases to grow, although, because ants live a year, all of its ants must be replaced each year. Colony growth may be limited by the queen's capacity to lay eggs; to maintain a size of ten thousand ants she must lay ten thousand eggs a year. A mature colony has an established nest mound and a set of three to eight habitual foraging directions, of which it chooses about two to five each day. It begins to reproduce, sending winged reproductives to the annual mating flight (Gordon 1992). It may remain in this state for another fifteen years.

Colonies vary in size. Some never seem to reach the size of the largest colonies (ten thousand to twelve thousand ants), and some never reproduce. Many of these apparently stunted colonies are in very crowded neighborhoods where competition for food may be especially intense. I suspect that such colonies do not reproduce because they are not large enough, perhaps because they do not have enough workers to collect sufficient reserves of food. However, there are also colonies that appear to be large but do not reproduce. We counted reproductives as they leave for the mating flight in two years and in both years, about a third of the colonies on the site that were of reproductive age did not send out any reproductives at all (Gordon and Wagner 1997; Wagner and Gordon 1999). If the colony reaches two years of age it is likely to survive for many years, but its development from ages two to five may determine whether it will reproduce.

Development of the Nest

The nest work of a small colony is different from that of a larger one, because the one-year-old colony is still making its nest, while the older colony is maintaining an existing nest. To construct a nest the ants have to kill any bushes growing where the nest is to be. They accomplish this by destroying the roots underground, and aboveground by climbing into the bush and clipping off the vegetation bit by bit. During the weeks of rainy weather that come each summer, the young colonies work on creating a nest entrance that will be raised above the level of the ground, which prevents water from pouring into the nest entrance. Using tiny twigs they make a thatched collar around the nest entrance, and then cover it with dirt.

When a colony is three years old it begins its period of most rapid growth, in numbers of ants. By this time most colonies have created a small nest mound. They have begun to cover the mound with tiny pebbles, which they bring in from the surrounding area. Any bushes remaining on the mound are dead, bare branches which eventually break off. There is usually a discrete pile or two of refuse, called the midden, with the husks of the seeds the ants are currently milling. Ants at work on the nest mound carry out sand from construction inside the nest, or sort and pile the midden.

When we dig up colonies, we find that there is a cone-shaped mass of chambers, with the top of the cone corresponding to the nest mound on the surface, and the length of the cone about the same as the diameter of the mound. A three-year-old colony's mound might be about half a meter wide, with a correspondingly deep cone of chambers; the chambers of a mature colony's nest may extend down one meter. Somewhere off the bottom of the cone is a single tunnel leading down another meter or even two, to a chamber where we find the queen and brood when we excavate a colony. These chambers can be as deep in a three-year-old colony as an older one. Probably the queen and brood do not usually stay so deep underground. It is clear the brood is not usually this deep, because on warm days it is easily found in chambers just underneath the top surface of the mound. In the laboratory, ants bring the brood to heated boxes, and the queen is usually near the smaller brood, eggs and small larvae. Perhaps in the field, the queen moves along with the brood as it is carried along a temperature gradient by the workers.

In excavated nests we find some chambers packed with stored seeds, and some with refuse. Colonies seem to use chambers and then build more as they are needed.

On the surface of the mound, the nest entrance has a smooth runway leading into it because the larger grains of dirt have been moved aside, leaving only the smallest ones. I can see a few centimeters into the nest with a fiber optics microscope. Inside the nest entrance of a three-year-old colony, the walls of the tunnels appear to be rough. In a colony of any age, there is a chamber just inside the nest entrance. In a three-year-old colony, that chamber usually divides into just two tunnels. Inside the nest of a colony five years or older, the walls of the chambers and tunnels seem smooth. The chamber nearest the nest entrance may branch into three or more main tunnels.

In a nest of any age, most chambers seem to have at least two openings, but some have more. I have never been able to trace the complicated topology of an entire nest. This topology is important, though, because the structure of the nest probably influences the flow of ants, and thus the rate at which ants interact as they come in and out of the nest.

Task Allocation

Task allocation of a mature colony, five years or older, differs from that of a small, young one in several ways (Gordon 1987, 1989a).

In a large colony, only about 25 percent of the colony works outside the nest, and large numbers remain inactive inside. These inactive ants may function as reserves that would emerge from the nest if needed in some situation. Such a situation has not occurred in the seventeen summers I have observed these ants, but though seventeen years is a long time for a person it is utterly insignificant on the evolutionary scale for an organism with a five-year generation time. In a young colony, a larger proportion of the colony, perhaps 50 percent, works outside the nest. Though the total numbers in the nest change with colony age as shown in figure 12.1, so that a six-year-old colony may be ten times as large as a one-year-old one, the numbers active outside the nest in a large colony are rarely more than two times those in a small one.

In the larger colony, ants in undisturbed colonies are unlikely to switch tasks from one day to the next. This suggests that an older colony has a larger number of ants available to perform exterior tasks than does a smaller one. In a small colony, an ant performing nest maintenance one day may be shunted into foraging the next. Perhaps in a large colony, an ant tends to be surrounded by more ants doing its own task, so there is a small probability it will encounter an opportunity to do a different task, or the stimulus provided by ants engaged in another task.

When the colony's environment changes, so do the numbers of ants engaged in the relevant task. If food appears, more ants forage. If the nest is disturbed, more ants take up nest maintenance and repair. These shifts lead to shifts in other, unrelated activities. I found in perturbation experiments that foraging, nest maintenance work, and patrolling are all related in this way: a shift in the numbers engaged in one task, caused by a change in the environment relevant to that task, leads to a change in the numbers engaged in the other tasks as well. These shifts arise from two ways that ants of one task respond to changes in the numbers performing another task: first, ants switch tasks, and second, ants may decide to become active, or remain inactive.

These adjustments in the distribution of workers performing colony tasks, differ in young and old colonies. The behavior of an older colony is more homeostatic, and more stable, than that of a younger one. The more a young colony is disturbed, the more it changes. However, the more an older colony is disturbed, the more its behavior seems to resemble that of an undisturbed colony. I do not know what process accounts for this result. Perhaps in a larger colony, each ant's

behavior is somehow more buffered by the behavior of other ants.

The finding that small and large colonies differ
in behavior suggests that the rate of interaction
an ant experiences may influence its task. In an
older, larger colony, an ant has more opportunities to interact than in a smaller one. If an ant's
behavior is influenced by interaction rate, then
even if ants act according to the same rules in
small and large colonies, the outcome will differ.
If an ant is likely to perform task X when it meets
workers of task X at a certain rate, then large
colonies may have more ants performing task X
simply because each ant has more task X ants to
meet. Such rules, of course, might involve positive or negative feedback. An ant's behavior is
clearly subject to environmental influences as well
as the influence of its interactions with others, but
interactions seem to play a role in task allocation.
For example, in laboratory studies we found that
the probability an ant does midden work is correlated with the rate at which it encounters midden workers (Gordon and Mehdiabadi 1999). We
also find that ants of different task groups differ
in cuticular hydrocarbons, so that in the course
of a brief antennal contact an ant can determine
the task of the ant it meets. We are currently investigating how interaction rate contributes to
task allocation, and we hope that this work will
elucidate why old colonies appear to be more stable in behavior than younger ones.

Relations with Neighbors

Very small colonies differ from larger ones in
their relations with their neighbors. I did some
experiments in which I enclosed some colonies
and observed the reactions of their neighbors
(Gordon 1992). When a very young colony had
its older, larger neighbor enclosed, its foraging
trails would shift toward the foraging area of the
absent neighbor. Neighbors of all ages reacted to
the absence of an enclosed colony in this way.
When the enclosed colony was released, the very
young colonies left the foraging area of the newly

released colony, and surprisingly, they retreated
even more. A very young colony's foraging area
after the neighbor was released, was even smaller
than it had been before the experiment began. It
was as though the sudden return of an absent
neighbor was a stronger deterrent than that neighbor had been before it disappeared.

Colonies that are three to four years old, at the
steep part of their growth curve, are more persistent in encounters with neighbors than younger
or older colonies. When a colony were enclosed,
its three- to four-year-old neighbors, like neighbors of other ages, began to use the foraging area
of the enclosed colony. But when the enclosed
colonies were released, the three- to four-year-
olds, unlike smaller or larger colonies, did not
leave their newly acquired foraging area. They
returned to the site day after day, although this
often involved fighting with the newly released
colonies.

Colonies of five years or more seem to be
much more staid in their relations with neighbors
than the quickly growing three- to four-year-olds.
When colonies were enclosed, older neighbors
entered their foraging areas. When the enclosed
colonies were released, the older neighbors simply retreated, foraging elsewhere much as they
had before the enclosed colony disappeared.
Older colonies adjust their foraging trails so as
to avoid those of neighbors. Over the fifteen to
twenty years of its life, a colony settles into relations with its neighbors. Ants distinguish nestmates from all other ants by a colony-specific
odor. At least in harvester ants, workers also distinguish ants of neighboring colonies from ants
of more distant, stranger colonies, presumably
using the same colony-specific odor (Gordon
1989b).

Neighbor relations may change as colonies
grow older because growth rates change. A three-
year-old colony has four thousand workers to
collect and process the food to make six thousand
ants for the following year. By contrast, a six-
year-old colony has ten thousand workers to collect and process the food to make a colony of the

same size the following year. The larvae consume most of the food of the colony; adult ants eat much less. So in a young, growing colony, there are more hungry larvae per forager than in an older one that is larger but of stable size. In laboratory colonies there seems to be an increase in numbers foraging in the days following the appearance of larvae from eggs. This is the only direct evidence I have that the presence of larvae affects the behavior of foragers, but it seems clear this relation must be important. When there are no larvae, none are fed; when there are larvae, they are fed; somehow the behavior of foragers must be linked to that of the ants that feed larvae, and somehow that must be linked to the numbers of larvae present. Parts of this chain are understood for some social insect species, though we know little about the details in this species of harvester ant. Neighbor relations in the field are consistent with the speculation that a forager in a growing three- to four-year-old colony is under more pressure to obtain food than one in an older colony of stable size, because the ratio of brood to foragers is higher in the growing colony.

The relation between numbers of larvae to feed and the behavior of foragers is probably mediated in some way by the state of the colony's supply of stored food. Amount of stored food probably depends on colony size. I have excavated colonies of known age to count the ants, but have not been able to measure food stores accurately. My impression, though, is that older, larger colonies have larger food stores. In one colony of twelve thousand ants we found more than a liter of stored seeds.

Conclusion

Little is known about the development of social insect colonies, because there are few species in which individually labeled colonies have been followed over time. The work described here shows how much we have left to learn about the development of the behavior of a harvester ant colony.

It seems plausible that many of the changes in behavior that we have observed arise from changes in colony size, which in turn lead to changes in the environment of the colony. Age-dependent changes in colony behavior seem to be a consequence of the relation between a colony's growth and its modification of its environment. This relation is complicated and not yet well understood. We do not know what determines the rate of colony growth. The availability of food, how well the colony collects and processes the food, the physiological condition of the queen, competitive pressure from neighbors, all are likely to influence colony growth. Colony size, in turn, determines how ants interact with each other, how many ants are available to collect and process food, and how a colony fares in competition with its neighbors. We know that all of these factors are related to colony size, though we do not yet know exactly how.

One set of important questions about colony development concerns the first few months of a colony's life. What do the first workers do? Why is colony mortality so high at this stage? Careful observation of very young colonies, both in the laboratory and in the field, would be very informative. A second set of questions concerns the relation between the physical structure of the nest, the flow of ants in and out of the nest, and the rate at which ants interact. Here both theoretical and empirical work is needed. Mathematical models of the complicated three-dimensional movement of ants through tunnels would help to guide empirical research. A third set of questions concerns the differences in task allocation in young and old colonies. Why is the behavior of older, larger colonies more stable and homeostatic than that of smaller, younger ones? Again, both theoretical and empirical work is needed. Models are needed that propose simple, plausible rules at the individual level, which might lead to the outcomes we observe in real colonies. Such rules determine how an ant's interactions, with its environment and with other ants, affects its decision about which task to perform. Differences between theoretical and empirical results show the gaps in our understanding, and point the way to further empirical work. A final set of questions concerns the relation between food and colony growth. We need to understand why some colonies grow faster than others. A colony's neighborhood influences its growth because neighbors compete for food; thus we need to know how food influences colony growth to understand how neighbors affect colony development.

The development of colony behavior is the process that relates changes in colony size to changes in the colony's environment. Each ant has the capacity to participate in this process, but that capacity is not separate from the process. An ant operates in the context of the colony. An ant alone could not function as an elemental unit of colony behavior, because there would be no colony behavior within which to act. As the colony grows, the ants encounter new circumstances. To our knowledge, an ant in a young colony responds to a particular situation in much the same way as an ant in an old one. But an ant in a young colony finds itself in different situations from an ant in an old one. An older colony is larger than a young one, and its environment has been modified by past cohorts of ants. To explain colony development we need to know how the behavior of mature colonies emerges from the changing relations of individual ants and their environment.

References

Gordon, D. M. (1987). Group-level dynamics in harvester ants: Young colonies and the role of patrolling. *Animal Behaviour* 35: 833–843.

Gordon, D. M. (1989a). Dynamics of task switching in harvester ants. *Animal Behaviour* 38: 194–204.

Gordon, D. M. (1989b). Ants distinguish neighbours from strangers. *Oecologia* 81: 198–200.

Gordon, D. M. (1991). Behavioral flexibility and the foraging ecology of seed-eating ants. *American Naturalist* 138: 379–411.

Gordon, D. M. (1992). How colony growth affects forager intrusion in neighboring harvester ant colonies. *Behavioral Ecology and Sociobiology* 31: 417–427.

Gordon, D. M. (1996). The organization of work in social insect colonies. *Nature* 380: 121–124.

Gordon, D. M. (1999). *Ants at Work: How an Insect Society Is Organized.* New York: Free Press.

Gordon, D. M., and A. W. Kulig. (1996). Founding, foraging and fighting: Colony size and the spatial distribution of harvester ant nests. *Ecology* 77: 2393–2409.

Gordon, D. M., and A. Kulig. (1998). The effect of neighboring colonies on mortality in harvester ants. *Journal of Animal Ecology* 67: 141–148.

Gordon, D. M., and N. Mehdiabadi. (1999). Encounter rate and task allocation in harvester ants. *Behavioral Ecology and Sociobiology* 45: 370–377.

Gordon, D. M., and D. Wagner. (1997). Neighborhood density and reproductive potential in harvester ants. *Oecologia* 109: 556–560.

Lewontin, R. C. (1983). Gene, organism and environment. In D. S. Bendall (Ed.), *Evolution from Molecules to Men,* pp. 273–285. Cambridge: Cambridge University Press.

Wagner, D., and D. M. Gordon. (1999). Colony age, neighborhood density and reproductive potential in harvester ants. *Oecologia* 119: 175–182.

Behavioral Development and Darwinian Evolution

Patrick Bateson

The One True Cause

The effectiveness of education, the role of parents in shaping the characters of their children, the causes of violence and crime, and the roots of personal unhappiness are self-evidently matters of huge importance. And, like so many other fundamental issues about human existence, they all relate to behavioral development. The catalog continues. Do bad experiences in early life have a lasting effect? Is intelligence in the genes? Can adults change their attitudes and behavior? When faced with such questions, many people want simple answers. They want to know what really makes the difference. Explanations in terms of combinations of conditions is perceived as wooly, obscurantist, and running counter to the successful analytical programs of science.

The search for simple environmental origins, which had wide appeal in the mid-twentieth century, has been partly superseded by an equally skewed belief in the overriding importance of genes. If pressed, scientists may concede that their talk of genes "for" shyness, maternal behavior, promiscuity, verbal ability, criminality, or whatever, is merely a shorthand. They may nonetheless try to legitimize the language of genes "for" behavior, by pointing to seemingly straightforward examples like the genes for eye color. Nonetheless, the notion of genes "for" behavior undoubtedly corrupts understanding.

A single developmental ingredient, such as a gene or a particular form of experience, might produce an effect on behavior, but this certainly does not mean that it is the only thing that matters. Even in the case of eye color, the notion that the relevant gene is the cause is misconceived, because all the other genetic and environmental ingredients that are just as necessary for the development of eye color remain the same for all individuals. A more honest translation of the "gene for" terminology would be something like:

"We have found a particular behavioral difference between individuals which is associated with a particular genetic difference, all other things being equal." The media and the public might start to get the message if plain language like this were used routinely.

Fortunately, it no longer seems obscure to many others to refer to developmental processes as systems. Susan Oyama's witty and eloquent writing (Oyama 1985) has inspired many others (see the many excellent contributions in Lerner [1998]). I like to think that this change in thinking may reflect commonplace experience. From an early age many people are exposed to computer games, in which outcomes depend on a combination of conditions. Children playing such games meet, for instance, in the dungeon of the dark castle a dragon that can only be killed with a special sword that had to picked up on the top of the crystal mountain; in doing so, they have begun to accustom themselves to the contextual conditional character of the real world. The linear thinking of a previous generation, with every event having a single cause, is slowly being replaced by an understanding of coordinated process. That, at least, is my optimistic view. In this chapter I shall consider why systems approaches are essential to an understanding of behavioral development. I shall also argue that they can be successfully married to a Darwinian approach to evolution and to current utility, often viewed as antagonistic to the developmental systems thinking (Ingold 1986; Kauffman 1993).

How Much Nature, How Much Nurture?

The importance of both genes and environment to the development of all animals, including humans, is obvious enough. This is true even for apparently simple physical characteristics, let alone complex psychological variables. Take shortsightedness, for example. Myopia runs in families but is segregated from other characteristics, suggesting that it is genetically inherited. But it is also affected by the individual's experience. Both a parental history of myopia and, to a lesser extent, the experience of spending prolonged periods studying close-up objects will predispose a child to become shortsighted (Zadnik 1997).

Different styles of doing science are brought into play when dealing with a case like the development of myopia, in which different types of factor affect the outcome. Those who have an aversion to systems thinking sometimes like to present the purely statistical approach as the only scientific solution to the age-old problem of the relative contributions of the different factors. The question becomes: "How much of the variation between individuals in a given character is due to differences in their genes, and how much is due to differences in their environments?" The suggested answer, satisfying old-style linear thinking, was provided by a measure called "heritability." The meaning of heritability is best illustrated with an uncontroversial characteristic such as height, which is clearly influenced by both the individual's family background (supposedly genetic influences) and nutrition (environmental influences). The variation between individuals in height that is attributable to variation in their genes may be expressed as a proportion of the total variation within the population sampled. This index is known as the heritability ratio. The higher the figure, which can vary between 0 and 1.0, the greater the contribution of genetic variation to individual variation in that characteristic. So, if people differed in height solely because they differed in their genes, the heritability of height would be 1.0; if, on the other hand, variation in height arose entirely from individual differences in environmental factors such as nutrition, then the heritability would be 0.

Calculating a single number to describe the relative contribution of genes and environment has obvious attractions. Estimates of heritability are of undoubted value to animal breeders, for example. Given a standard set of environmental conditions, the genetic strain to which a pig belongs will predict its adult body size better than other variables such as the number of piglets in a sow's litter. If the animal in question is a cow and the breeder is interested in maximizing its milk yield, then knowing that milk yield is highly heritable in a particular strain of cows under standard rearing conditions is important.

But behind the deceptively plausible ratios lurk some fundamental problems. For a start, the heritability of any given characteristic is not a fixed and absolute quantity—tempted though many scientists have been to believe otherwise. Its value depends on a number of variable factors, such as the particular population of individuals that has been sampled. For instance, if heights are measured only among people from affluent backgrounds, then the total variation in height will be much smaller than if the sample also includes people who are small because they have been undernourished. The heritability of height will consequently be larger in a population of exclusively well-nourished people than it would be among people drawn from a wider range of environments. Conversely, if the heritability of height is based on a population with relatively similar genes—say, native Icelanders—then the figure will be lower than if the population is genetically more heterogeneous; for example, if it includes both Icelanders and African pygmies. Thus, attempts to measure the relative contributions of genes and environment to a particular characteristic are highly dependent on who is measured and in what conditions.

Another problem with heritability is that it says nothing about the ways in which genes and environment contribute to the biological and psychological cooking processes of development (Bateson and Martin 1999). This point becomes obvious when considering the heritability of a characteristic such as "walking on two legs." Humans walk on fewer than two legs only as a result of environmental influences such as war wounds, car accidents, disease, or exposure to teratogenic toxins before birth. In other words, all the variation within the human population results from environmental influences, and consequently the heritability of "walking on two legs" is zero. And yet walking on two legs is clearly a fundamental property of being human, and is one of the more obvious differences between humans and other great apes such as chimpanzees or gorillas. It obviously depends heavily on genes, despite having a heritability of zero. Low heritability clearly does not mean that development is unaffected by genes.

If a population of individuals is sampled and the results show that one behavior pattern has a higher heritability than another, this merely indicates that the two behavior patterns have developed in different ways. It does not mean that genes play a more important role in the development of behavior with the higher heritability. Important environmental influences might have been relatively constant at the stage in development when the more heritable behavior pattern would have been most strongly affected by experience.

Yet another serious weakness with heritability estimates is that they rest on the extraordinary assumption that genetic and environmental influences are independent of one another and do not interact. The calculation of heritability assumes that the genetic and environmental contributions can simply be added together to obtain the total variation. In many cases this assumption is clearly wrong. For example, in one study of rats the animals' genetic background and their rearing conditions were both varied; rats from two genetically inbred strains were each reared in one of three environments, differing in their richness and complexity (Cooper and Zubek 1958). The rats' ability to find their way through a maze was measured later in their lives. Rats from both genetic strains performed equally badly in the maze if they had been reared in a poor environment (a bare cage) and equally well if they had been reared in a rich environment filled with toys and objects. Taken by themselves, these results implied that the environmental factor (rearing

conditions) was the only one that mattered. But it was not that simple. In the third type of environment, where the rearing conditions were intermediate in complexity, rats from the two strains differed markedly in their ability to navigate the maze. These genetic differences only manifested themselves behaviorally in this sort of environment. Varying both the genetic background and the environment revealed a statistical interaction between the two influences.[1]

An overall estimate of heritability has no meaning in a case such as this, because the effects of the genes and the environment do not simply add together to produce the combined result. The effects of a particular set of genes depend critically on the environment in which they are expressed, while the effects of a particular sort of environment depend on the individual's genes. Even in animal breeding programs which use heritability estimates to practical advantage, care is still needed. If breeders wish to export a particular genetic strain of cows that yield a lot of milk, they would be wise to check that the strain will continue to give high milk yields under the different environmental conditions of another country. Many cases are known where a strain that performs well on a particular measure in one environment does poorly in another, while a different strain performs better in the second environment than in the first.

A further point about the mutual actions of individual on the environment and environment on the individual is brought out nicely by the example of myopia already briefly mentioned. Nesse and Williams (1994) point out that the growth of the cornea is affected by the sharpness of the image on the retina. Objects close to the eye, such as books, cause the cornea to grow so that the image of print is less fuzzy. Individual differences in the feedback mechanism mean that some people in a modern environment respond more to the experience of reading than others and consequently become more shortsighted. In an environment empty of books such people would not be shortsighted.

Alternative Lives

Striking examples of interaction are found throughout the animal kingdom when genetically identical individuals develop in totally different ways, depending on environmental cues they received when they were young. After a fire on the high grassland plains of East Africa, for example, the young grasshoppers are black instead of being the normal pale yellowish-green. Something has switched the course of their development onto a different track. The grasshopper's color makes a big difference to the risk that it will be spotted and eaten by a bird, and the scorched grassland may remain black for many months after a fire. So matching its body color to the blackened background is important for its survival. The developmental mechanism for making this switch in body color is automatic and depends on the amount of light reflected from the ground (Rowell 1971). If the young grasshoppers are placed on black paper they are black when they molt to the next stage. But if they are placed on pale paper the molting grasshoppers are the normal green color. The grasshoppers actively select habitats with colors that match their own. If the color of the background changes they can also change their color at the next molt to match the background, but they are committed to a color once they reach adulthood.

Turtles, crocodiles, and some other reptiles commit themselves early in life to developing along one of two different developmental tracks and like grasshoppers, they do so in response to a feature of their environment (Janzen and Paukstis 1991). Each individual starts life with the capacity to become either a male or a female. The outcome depends on environmental temperature during the middle third of embryonic development (Yntema and Mrosovsky 1982). If the eggs from which they hatch are buried in sand below 30 °C, the young turtles become males. If, however, the eggs are incubated at above 30 °C they become females. Temperatures below 30 °C activate genes responsible for the production of male sex hormones and male sex hormone re-

ceptors. If the incubation temperature is above 30 °C, a different set of genes is activated, producing female hormones and receptors instead (Crews 1996). It so happens that in alligators the sex determination works the other way around, such that eggs incubated at higher temperatures produce males. (In humans and other mammals, by contrast, the sex of each individual is generally determined genetically at conception; if it inherits only one X sex chromosome, it becomes male.)

Each grasshopper and turtle starts life with the capacity to play two distinctly different developmental tunes—green or black, male or female. A particular feature of the environment then selects which of those tunes the individual will play during its life. And once committed, the individual cannot switch to the other tune. Once black as an adult, the grasshopper cannot subsequently change its color to green, just as a male turtle cannot transform itself into a female.

The broad pattern of an individual's social and sexual behavior may also be determined early in life, with the individual developing along one of two or more qualitatively different tracks. Many examples are found in the animal kingdom. The caste of a female social insect is determined by her nutrition early in life. The main egg producer of an ant colony, the queen, is part of a teeming nest in which some of her sisters care for her offspring, others forage, yet others clean or mend the nest, and finally other sisters specialize in guarding it (Wilson 1971). Locusts may or may not become migratory, depending on crowding; when the numbers living in a given area build up, their offspring develop musculature and behavior suitable for long flights and then the whole swarm moves off (Pener and Yerushalmi 1998). Vole pups born in the autumn have much thicker coats than those born in spring; the cue to produce a thicker coat is provided by the mother before birth. The value of preparing in this way for colder weather is obvious (Lee and Zucker 1988).

The sexual behavior of some primates can also develop along two or more distinctly different tracks. An adult male gelada baboon, for example, will typically defend and breed with a harem of females. After a relatively brief but active reproductive life, he is displaced by another male and never breeds again. To position himself so that he can acquire and defend a harem, the male must grow rapidly at puberty. He develops the distinctive golden mane of a male in his prime and becomes almost twice the size of the females. However, when many such males are present in the social group, an adolescent male may adopt a distinctly different style of reproductive behavior. He does not develop a mane or undergo a growth spurt. Instead, he remains similar in appearance and size to the females. These small males hang around the big males' harems, sneakily mating with a female when the harem-holder is not paying attention. Because the small, sneaky male never has to fight for females, he is likely to have a longer, if less intense, reproductive life. If he lasts long enough he may even do better in terms of siring offspring than a male who pursues the alternative route of growing large and holding a harem. These two different modes of breeding behavior represent two distinctly different developmental routes, and each male baboon must commit himself to one or other of them before puberty (Dunbar 1984).

All of these examples illustrate a surprising aspect of development that has intriguing implications for humans. In each case, the individual animal starts its life with the capacity to develop in a number of distinctly different ways. Like a jukebox,[2] the individual has the potential to play a number of different developmental tunes. But during the course of its life it plays only one tune. The particular developmental tune it does play is selected by a feature of the environment in which the individual is growing up—whether it be the color of the ground, the temperature of the sand, the type of food, or the presence of other males. Furthermore, the particular tune that is selected from the developmental jukebox is adapted to the conditions in which it is played.

Is it the case that people, like grasshoppers or baboons, are conceived with the capacity to play a number of qualitatively different developmental tunes—in other words, to live alternative lives?

Each of us started life with the capacity to live many different lives, but each of us lives only one. In one sense individual humans are obviously bathed in the values of their own particular culture and become committed by their early experience to behaving in one of many possible ways. Differences in early linguistic experience, for example, have obvious and long-lasting effects. By the end of a typical high school education, a young American will probably know about fifty thousand different words (Pinker 1994). The words are different from those used by a Russian of the same age. In general, individual humans imbibe the particular characteristics of their culture by learning (often unwittingly) from older people. When environmental conditions select a particular developmental route in animals, the mechanisms involved are likely to be different; learning may not enter into the picture at all. Even so, is it possible that some aspects of human development are triggered by the environment, as though the individual were a jukebox? Was each of us conceived with the capacity to develop along a number of different tracks each of which is preadapted to the circumstances in which the individual finds itself? And does the environment select the particular developmental track that each of us follows?

A series of studies led by the epidemiologist David Barker, which assessed people across their entire lifespan from birth to death, has lent strength to the suggestion that human development may also involve environmental cues that prepare the individual for a particular sort of environment (Barker 1998). The work was based in part on a large sample of men born in the English county of Hertfordshire between 1911 and 1930. Those men who had had the lowest body weights at birth and at one year of age were most likely to die from cardiovascular disease later in life. The heaviest babies faced a subsequent risk of dying from cardiovascular disease that was only half the average for the group as a whole, whereas the risk for the smallest babies was 50 percent above average (in other words, three times greater than that for the largest babies). Individuals who had been small babies were also more likely to suffer from diseases such as diabetes and stroke in adulthood.

How could a link have arisen between an individual's birth weight and his physical health decades later? The evidence pointed to a connection with the mother's nutritional state: women with poor diets during pregnancy had smaller placentas, and forty years later their offspring had higher blood pressure (a risk factor for cardiovascular disease and stroke). But the links with maternal nutrition went much further back than pregnancy. Measurements of the mothers' pelvises revealed that those who had a flat, bony pelvis tended to give birth to small babies. These small babies, after they had grown up, were much more likely as adults to die from stroke. The implication was that poor nutrition during their mother's childhood affected the growth of her pelvis which, in turn, curtailed the growth of her offspring during pregnancy, which, in turn, increased her offspring's risk of stroke and cardiovascular disease in adulthood.

Poor maternal physique and health are associated with reduced fetal growth, with consequences for the offspring's later health. The question arises, then, as to whether these connections make sense in adaptive terms. Could it be that, in bad conditions, the pregnant woman unwittingly signals to her unborn baby that the environment which her child is about enter is likely to be harsh? (Remember that we are thinking here about what might have been happening tens of thousands of years ago as these mechanisms were evolving in ancestral humans.) And perhaps this weather forecast from the mother's body results in her baby being born with adaptations, such as a small body and a modified metabolism, that help it to cope with a shortage of food. This hypothetical set of adaptations has been called the "thrifty phenotype" (Hales and Barker 1992; Hales, Desai, and Ozanne 1997). And perhaps these individuals with a thrifty phenotype, having small bodies and specialized metabolisms adapted to cope with meager diets, run

into problems if instead they find themselves growing up in an affluent industrialized society to which they are poorly adapted. That, at least, is the hypothesis.

People who grow up in impoverished conditions tend to have a smaller body size, a lower metabolic rate, and a reduced level of behavioral activity (Waterlow 1990). These responses to early deprivation are generally regarded as pathological—just three of the many damaging consequences of poverty. But they could also be viewed as part of a package of characteristics that are appropriate to the conditions in which the individual grows up—in other words, adaptations to an environment that is chronically short on food, rather than merely the pathological byproducts of a bad diet. Having a lower metabolic rate, reduced activity, and a smaller body all help to reduce energy expenditure, which can be crucial when food is usually in short supply.

Now this conjecture might well be regarded as offensive. It could be seen as encouraging the rich to look complacently at their impoverished fellow human beings, by arguing that all is for the best in this best of all possible worlds (as Voltaire's Doctor Pangloss would have had it). Merely to assert that every human develops the body size, physiology, biochemistry, and behavior that is best suited to their station in life would indeed be banal. The point, however, is not that the rich and the poor have the same quality of life. Rather, it is that, if environmental conditions are bad and likely to remain bad, individuals exhibit adaptive developmental responses to those conditions. To put it simply, they are designed to make the best of a bad job.

Of course, many of the long-term effects on health of a low birth weight may simply be byproducts of the social and economic conditions that stunted growth in the first place. Ignorance and shortage of money make the prevention and treatment of disease more difficult; overcrowding, bad working conditions, and poverty produce psychological stress and increase the risk of infection. People with little money have poorer diets, and adverse social or physical factors that foster depression and hopelessness increase the risks of disease (Wilkinson 1996). In industrialized nations the poor and unemployed have more illnesses and die sooner than the affluent. But social and economic conditions do not account for everything, because the connections between low birth weight and subsequent health are still found among babies born in affluent homes (Barker 1998).

Whether or not the thrifty phenotype hypothesis proves to be correct, everybody agrees that environmental conditions early in development have a significant impact on many other aspects of human biology, including size. People are getting bigger (Eveleth and Tanner 1990). For decades now, the average height of men and women in industrialized countries has been steadily increasing. Although some of the height differences between people are due to genetic differences, the general trend for average height to increase is almost certainly due primarily to improvements in nutrition and, to a lesser extent, health. Hence, successive generations of the same family have grown taller despite having a similar genetic makeup.

The environmental improvements that have led to this general increase in height have affected males and females somewhat differently. Men have been growing taller faster than women. For example, the average height of men in Britain has been increasing at a rate of just over one centimeter every ten years, whereas the average height of women has been increasing at about one third of that rate. In consequence, men are now relatively bigger than women than they were a century ago (Huh, Power, and Rodgers 1991).

While the gap between the sexes has been widening, the average difference in height between social classes has remained roughly constant, with men from affluent professional homes being nearly 2 cm taller than men from manual backgrounds, and women from professional homes being 1.6 cm taller. In other countries the trends have been somewhat different. In Russia, where the improvements in nutrition have occurred more recently, the rate of increase in height lagged behind Britain but has been at almost three times the rates found in Britain in the last few decades. In the United States the trend toward ever taller offspring in successive generations, which started earlier than in Britain, has leveled off in recent years. These findings suggest an upper bound on the effect of nutrition on human height.

The Demise of Heritability

The examples of condition-dependent development do not pose any problems for evolutionary biologists, even though they should give pause to those who search for universals. From an adaptationist standpoint, the development of a phenotype appropriate to the circumstances in which the animal finds itself makes a great deal of sense. Nevertheless, the striking ways in which environmental factors can trigger one of a set of alternative responses pose serious difficulties for those behavior geneticists who seek to partition variation into genetic and environmental components. And worse is to come. The conventional analytical method that partitions behavioral variation into genetic and environmental components may be misleading in a different way. The two major contributors to variation may not combine even in nonlinear fashion to produce their overall effect. For example, the performances of adopted children in tests of cognitive ability are related to those of their adopting parents and their biological parents. Commonly in such studies, both types of parents have independent effects on the children (Mackintosh 1998). The effects of the genes (provided by the biological parents) and the effects of the environment (provided by the adopting parents) seem to add together. In the case of IQ scores, for example, each factor accounts for about 10 percent of the variation in the children's scores.

A common-sense view of what happens is that initially the cognitive abilities of the child are most strongly affected by its biological parents, but that later in development they are increasingly affected by the experiences the child has had with its adopting parents. However, the quality of the exchanges between the adopting parents and the child will depend on the match between their characteristics. A potentially able child who is adopted by dull people might be less stimulated and more frustrated than if he or she had been adopted by lively, intelligent people. Conversely, adopting parents who are disappointed by the less able child might provide a less supportive environment than those whose expectations are satisfied by the responsiveness of an able child. Here again, the difference between the child and its adopting parents probably matters, but this time in the reverse direction. One study, for example, found that the bigger the absolute difference in IQ between the biological and adopting parents, the more the child was adversely affected (Bateson 1987a). The difference between the parents and child accounted for as much of the variation in the children as the direct influences of the biological and adopting parents. The consequences of the relationships between adopting parents and the children were not revealed by a simplistic analysis that assumes that what went in is directly related to what comes out. The appropriate analysis was not carried out until a plausible question was asked about the nature of the developmental process. I do not know why this should have happened but suspect the beauty of the statistical procedure called Analysis of Variance seemed to offer a sufficient explanation. The very language of "accounting for variation," common enough in statistics books, seemed to preempt thought about what might be happening during development. I will return to this point when I consider the types of model that may be used fruitfully in order to understand developmental process.

Attempts to partition phenotypic variation in behavioral characteristics may yield answers of a kind, but little sense of what happens as each individual grows up. The language of nature ver-

sus nurture, or genes versus environment, gives only a feeble insight into the processes. The best that can be said of the nature/nurture split is that it provides a framework for uncovering a few of the genetic and environmental ingredients that generate differences between people. At worst, it satisfies a demand for simplicity in ways that are fundamentally misleading.

Any scientific investigation of the origins of human behavioral differences eventually arrives at a conclusion that most nonscientists would probably have reached after only a few seconds' thought. Genes and the environment both matter. The more subtle question about how much each of them matters defies an easy answer; no simple formula can solve that conundrum. This then raises the need for a systems approach.

Metaphors and Models

The common image of a genetic blueprint for behavior fails because it is too static, too suggestive that adult organisms are merely expanded versions of the fertilized egg. In reality, developing organisms are dynamic systems that play an active role in their own development. Even when a particular gene or a particular experience is known to have a powerful effect on the development of behavior, biology has an uncanny way of finding alternative routes. If the normal developmental pathway to a particular form of adult behavior is impassable, another way may often be found. The individual may be able, through its behavior, to match its environment to suit its own characteristics. At the same time, playful activity increases the range of available choices and, at its most creative, enables the individual to control the environment in ways that would otherwise not be possible.

A low-tech cooking metaphor serves to shift the focus onto the multicausal and conditional nature of development (Bateson and Martin 1999). Using butter instead of margarine may make a cake taste different when all the other ingredients and cooking methods remain unchanged. But if other combinations of ingredients or other cooking methods are used, the distinctive difference between a cake made with butter and a cake made with margarine may vanish. Similarly, a baked cake cannot readily be disaggregated into its original raw ingredients and the various cooking processes, any more than a behavior pattern or a psychological characteristic can be disaggregated into its genetic and environmental influences and the developmental processes that gave rise to it.

To use a different metaphor, development is not like a fixed musical score that specifies exactly how the performance starts, proceeds, and ends. It is more like a form of jazz in which musicians improvise and elaborate their musical ideas, building on what the others have just done. As new themes emerge, the performance acquires a life of its own, and may end up in a place none could have anticipated at the outset. Yet it emerges from within a fixed set of rules and the constraints imposed by the musical instruments.

It is clear, then, that because of the system in which they are embedded, no simple correspondence is found between individual genes and particular behavior patterns or psychological characteristics. Genes store information coding for the amino acid sequences of proteins; that is all.[3] They do not code for parts of the nervous system and they certainly do not code for particular behavior patterns. Any one aspect of behavior is influenced by many genes, each of which may have a big or a small effect. Conversely, any one of many genes can have a major disruptive effect on a particular aspect of behavior. A disconnected wire can cause a car to break down, but this does not mean that the wire by itself is responsible for making the car move. Without a strong set of binding ideas, it is not easy to think about all aspects of the various strands of evidence, which often seem to point in opposite directions. Some theorists have argued that the seemingly simple and orderly characteristics of development (such as they are) are generated by dynamic processes of great complexity (Kauff-

man 1993). Many mathematical techniques, such as catastrophe theory and "chaos," have been developed to deal analytically with the complexities of dynamical systems. For all that, it is questionable whether the descriptive use of mathematics brings with it any explanatory power. Much more promising are those approaches that bind evidence across different levels of analysis.

Richard Dawkins wrote: "If a computer is doing something clever and life-like, say playing chess, and we ask how it is doing it, we do not want to hear about transistors, we accept them. . . . We need *software explanations* of behavior. I do not mean that animals necessarily work like computers. They may be very different. But just as the lowest level of explanation is not always the most appropriate for a computer, no more is it for an animal. Animals and computers are both so complex that something on the level of software explanation must be appropriate for both of them" (Dawkins 1976a). Rules like "store a representation of the input and then compare output against it" have been very important in obtaining understanding of song-learning (Marler 1976). Feedback mechanisms with moving setpoints can explain the dynamic but ordered way in which developmental processes can home in on a particular adult endpoint (Bateson 1976; Chalmers 1987). The conditional IF . . . THEN rules are useful in understanding alternative modes of development (Caro and Bateson 1986). The competitive exclusion rule explains the dynamics of sensitive periods (Bateson 1987b). In their different ways, all these ideas when applied to development at the right level may provide some of the useful ways to reduce the complexity. While they may show us how the job might be done, are we wholly satisfied?

Careful work on developmental processes such as imprinting and song-learning in birds have led to models that take account of both behavior and the underlying neural mechanisms. In the area in which I have worked extensively, the results of imprinting experiments depended greatly on the conditions that were used (Bateson 1966).

When many factors affect the outcome of a developmental process, what should be done about it? The chances are that all the different influences will not add together, and, if they do not, small changes in certain factors may sometimes make big differences to the outcome and large changes in others will have no effect whatsoever. For a long while it was thought that, in the case of filial imprinting, movement was critical to the young bird. A breakthrough came when a reanalysis of old data led to the conclusion that features in the jungle fowl, the ancestral form of the domestic fowl, were particularly attractive to chicks (Horn and McCabe 1984). The preference for aspects of naturalistic stimuli had been missed because, under laboratory conditions, the detectors take longer to develop in chicks than do the ones driven by an ambulance light. Even though the predisposition is less specific than at first it seemed (Johnson and Horn 1988), it looks as though such a head-neck feature detector, along with others responding to movement, color and contrast, feed into those bits of the brain that establish representations of the object to which the bird has been socialized (Horn 1991).

Imprinting involves another type of plasticity in the nervous system, namely connecting up the representation of the familiar object exclusively to the system controlling filial behavior and, much later in development, to the one controlling sexual behavior (Bateson 1981). One compelling strand of evidence is the result of taming. When a bird is well-imprinted and then exposed to another object, at first the bird withdraws showing every sign of great alarm. By degrees this alarm habituates and the bird becomes tame. However, tame birds do not express any social behavior toward the object, which is by now very familiar. They evidently recognize it, but that is all. It seems, therefore, that at least two stages are involved in imprinting. One involves recognition and one involves the control of social behavior by the representation of the familiar object. From this consideration emerged a three-stage model involving analysis, recognition, and execution.

The first step involves detection of features in a stimulus presented to a young bird. Aspects of the stimulus which the bird is predisposed to find attractive are picked out by particular detectors at this level of processing. The second step involves comparison between what has already been experienced and the current input. Of course, before imprinting has taken place, no comparison is involved. Once it has occurred, recognition of what is familiar and what is novel is crucial. Finally, the third stage involves control of the various motor patterns involved in executing filial behavior. In the case of the tame bird, the assumption is that a representation is formed, but this representation has no way of gaining access to the executive system.

The model was formalized (Bateson 1991). Parameters described as "features of the organism" referred to properties of the real animal's nervous system that also have to develop. These features may or may not be affected by relatively small changes in the conditions of development or by variation in genotype. If and when they can be estimated, they should not be treated as though they were similar to gravitational constants. When external conditions affect the parameters that are regarded as features of the organism, the developmental effect would be relatively nonspecific, an issue that has been frequently considered in the literature on behavioral development (Bateson 1976; Lehrman 1970).

With Gabriel Horn, I took the formal model of imprinting and developed a neural net (Bateson and Horn 1994). We tried to ensure that most of the subprocesses are as plausible in neural terms as current knowledge allows and that the whole system has the behavioral structure of an intact animal. The model exhibits a well-known feature of behavioral development seen in animals, tending to settle into familiar habits, while also able to build with increasing elaboration on the basis of previous perceptual experience. The preemptive effect of experience means that other stimuli are less able to produce change unless exposure to those novel stimuli is forced and prolonged. This aspect of the model simulates the closure of the sensitive period in development, which is known to be dependent on experience, at least in part. The greater the activating value of the features in the experienced stimuli, the more quickly the sensitive period comes to an end.

The model also provides a ready explanation for evidence that is difficult to explain on the basis of a purely additive approach to development (Bolhuis and Honey 1994, 1998; Honey and Bolhuis 1997). When the maternal call of the domestic hen accompanies the presentation of a visual stimulus, the domestic chick is more responsive and develops a stronger preference for the visual stimulus. However, if the auditory stimulus is played in the absence of the visual stimulus before presentation of the compound stimulus, the preference for the visual stimulus is weaker. From the standpoint of animal learning, an even more striking result is that if the auditory stimulus is played on its own after the compound, the preference for the visual stimulus is also weaker than when the post-compound exposure was omitted. In terms of the Bateson and Horn model, the playing of the auditory stimulus on its own weakens the link between the analysis modules processing the features of the visual system and the recognition system. This is because the downstream modules are active when the upstream modules are inactive.

The case for some degree of formalization is that an explicit working model brings with it mental discipline and may expose weaknesses in a verbal argument that are all too easily missed. It can also serve several other valuable functions. It can show how the dynamics of development do not mean that the processes are so complicated that they are beyond comprehension. When critics say, "You make the whole process of development sound so complicated," it is possible to reply, "No, the explanatory devices are really rather simple." Importantly, they enable combinations of conditions to be explored systematically, which would take a long time with live animals. Furthermore, in the process of such ex-

plorations, surprises are frequently encountered, which then propose experiments with real animals or reexamination of data not previously analyzed. Therefore, from the point of future empirical research, the modeling approach can suggest profitable new lines of inquiry. The predictions may be false, but they are worth testing just because the assumptions are rooted in psychological and biological reality. Above all, such a systems approach shows how the old either/or oppositions applied to behavior simply evaporate when knowledge starts to advance. It bears directly on the general arguments about the interplay between features of the organism that developed prior to the stage of interest and features of the environment. Finally, it illustrates the interplay that is needed between theory and practice. If the theories of process are anchored in the real world, they reduce complexity and offer a genuine sense of understanding of how real systems generate the seemingly elaborate things that we observe. However, they do more than that. They make functional sense.

Darwinian Approaches to Development

I have long argued for the benefits of using a Darwinian standpoint when approaching the problems of behavioral development (Bateson 1982, 1987a, 1988, 1995, 1984, 1986). Many biologists want to be told *why* the job needs to be done. In other words they want to be given a functional account or think about a characteristic in terms of design. Robert Hinde noted how much ethology has been helped by considering the adaptiveness of a behavior pattern, even when the primary concern is with mechanism (Hinde 1982). Why is it advantageous to bring in the functional argument? Animals live in a complex world and most biologists would argue that the degree of match between their behavior and the conditions in which they live makes a big difference to their survival and reproductive success.

Those who do not like Darwinian arguments about apparent natural design being brought into discussions of development[4] see no value in them. If grasshoppers are green when the grass is green and black when the savanna has been blackened by fire, the obvious benefit to the individual of minimizing risk from predation is irrelevant to the question of how they come to match the color of their background. The appropriate developmental questions are about how the process works. They are right, but they miss the point. The design point frames the mechanistic argument. It leads to studies that might otherwise not have been done. To give one example from development, the functional approach provides a way of thinking about the difference in timing between sexual imprinting and filial imprinting. The suggestion is that an animal should not tune its reference point for mating preferences too early in development lest it obtains information about the juvenile appearance of its siblings that could not be used effectively when the time comes to choose a mate. On the other side it must not tune its mating preferences too late in development after the family group has broken up and it is likely to be exposed to non-kin. Indeed, the evidence suggests that birds delay sexual imprinting until their siblings have molted into adult plumage (Bateson 1979). In this way the bird is able use its experience of close kin so that it chooses a mate that is genetically a bit different from itself, but not too different. If the two types of imprinting are treated as part of the same general process, the difference seems to be of no importance and is quickly forgotten. However, with attention focused on the problem, we can attempt to analyze the mechanisms responsible for the difference in timing. The point is, then, that the functional approach can stimulate research on the processes of development.

The heuristic point about the mentally enabling role of functional explanations also relates also to the seeming conflict between selfish gene language and the system theories arising from studies of development—a discussion that runs as a theme throughout this book and other recent treatments of biological philosophy (Sterelny and Griffiths 1999). It is possible, I believe, to resolve the conflict between the developmental systems approach, which takes into account many influences, and the evolutionary selfish genes approach, which seems to suggest only one cause. However, we need to be very clear about the difficulties of translating one language in to the other and the attendant confusions that arise when translation occurs carelessly.

Dawkins suggested that the way to understand evolution is not in terms of the needs of species, groups or individuals, but in terms of the needs of genes (Dawkins 1976b). Genes recombine in each generation to form temporary federations. The alliance forms an individual organism. By reproducing, individuals serve to perpetuate the genes which in the next generation recombine in some other kind of alliance. Genes are selfishly intent on replicating themselves by the best possible means. Dawkins was clearly and deliberately using a heuristic device when he attributed motives to genes. He obviously did not think that gene have intentions. It is easier for most of us to get our minds round a problem when we can think of a complex system in terms of the way they strive to reach a specific end state. This is not only true in biology. A great nineteenth-century physicist, William Rowan Hamilton, formulated a general and widely accepted teleological principle for use in mechanics. It is a powerful way of thinking about systems, the behavior of which is determined by many factors.

The language of intentions can, of course, be played many ways. When the ambient environmental temperature during development is crucial for the expression of a particular phenotype, changes in temperature by a few degrees may lead to a startling evolutionary change. It may lead to extinction—such as is predicted for turtles whose sex depends on temperature early in development and after global warming would all end up as females. Would not such cases give as much status to a necessary temperature value as to a necessary gene? The temperature value is also required for the expression of a particular set of phenotypes—

a balanced sex ratio in turtles before global warming. It is also stable (within limits) from one generation to the next. It may even be transmitted from one generation to the next if the survival machine makes a nest for its offspring. The bird is the nest's way of making another nest (Bateson 1978).

Dawkins's riposte to my tease was that nest material doesn't have the permanence of DNA (Dawkins 1978). Later he developed the point (Dawkins 1982). Nests do not have the causal significance of genes. "There is a causal arrow going from gene to bird, but none in the reverse direction. A changed gene may perpetuate itself better than its unmutated allele. A changed nest will do no such thing unless, of course, the change is due to a changed gene, in which case it is the gene that is perpetuated, not the nest" (p. 98). In a recent essay, R. D. Gray (in press) has elegantly questioned this claim. He points how variation in the mode of nest construction could plausibly be passed on by extragenetic as well as genetic means.

In concluding his discussion of alternative teleologies and my concern that he was giving too much status to the gene as programmer, Dawkins (1982) wrote amusingly: "As is so often the case, an apparent disagreement turns out to be due to mutual misunderstanding. I thought Bateson was denying proper respect to the Immortal Replicator. Bateson thought that I was denying proper respect to the Great Nexus of complex causal factors interacting in development" (p. 99). Reflecting on this debate many years later, I think that more was at issue than different emphases and interests. Dawkins response showed how easily we get snarled up in the language. The details of the evolutionary mechanism involving small changes in DNA had been mixed up with the intentions that were rhetorically attributed to genes. While Dawkins was quite justified, in my view, in writing about genes' intentions, he was wrong when he treated this language as readily translatable into the language of what genes actually do. The language of selfish genes does not easily translate in the causal language of population biologists even though the users are ostensibly talking about the same issue.

For population geneticists, a genetic difference is identified by means of a biochemical, physiological, structural, or behavioral difference between organisms (after other potential sources of difference have been excluded by appropriate procedures). Dawkins suggested that his move backward and forward between the language of gene intentions and the more orthodox language of genetic differences was acceptable because they are simply alternative ways of describing the same thing. To make his point, he described perception of the Necker cube. The front edges of the line drawing of the cube suddenly flip to the back as we look at them. Each perceived image of the cube is as real as the other. Dawkins suggests that, in similar ways, the teleological and mechanistic images of evolution translate backward and forward into the other. However, it doesn't make any sense to attribute motives to a comparison, as may be illustrated by a parable.

Let us consider the spread of a new brand of biscuit in supermarkets from the perspective of the recipe. While shoppers may compare biscuits and buy one brand, it is the recipe used for making desirable biscuits that survives and spreads in the long run. Therefore, the word in the recipe that makes the biscuit successful is selfish, because it serves to perpetuate itself. So far so good. But my story loses all coherence if I conclude by adding that the difference in the wording of recipes used for making successful and less successful is selfish.

It is worth developing the biscuits in a supermarket parable just a little further. What shoppers *really* do in supermarkets is to select a word in a faithfully reproduced recipe. Really? Words in recipes as the units of the shoppers' selection? It is an odd idea, since shoppers respond among other things to the outcome of the cooking process which gave rise to the biscuits. For very similar reasons many biologists, myself included, disliked the idea of genes being treated as the units of selection in Darwinian terms. After all, Darwin had used his metaphor of "natural selection" because he had been impressed by the ways in which plant and animal breeders selected the characteristics they sought to perpetuate. Of course, the gene selection idea grew up because it made sense of those cases in which the consequences of an act favored the survival of genetically related individuals rather than the actor. Dawkins (1982) helpfully advanced the argument when he drew a distinction between "replicators" and "vehicles."[5]

To separate further the many different strands of thought that have been run together in the units of selection debate, it is worth making a three-way distinction that mirrors in part the mechanism Darwin introduced to explain the evolution of adaptations. Darwin suggested that three components of the mechanism are essential. First, competing biological entities must differ in their characteristics. (The entities have usually been individual organisms in conventional neo-Darwinian thinking, but they may be parts of organisms or assemblies of organisms.) Second, the entities must survive and/or reproduce themselves with differing degrees of success. Third, the entities must be more likely to share characteristics with their descendants or relatives than with unrelated entities.

The developmental processes creating the phenotypes of the entities have been the major concern of DST and, indeed, of this chapter. Darwin had nothing to say about how phenotypic variation is generated, and yet its existence is central to his evolutionary mechanism. The modern emphasis on developmental process has served to put empirical flesh on notions of what generates the variation in the raw material for differential survival and reproductive success. Once the phenotypes are created, DST has not much to say about differential survival and differential reproductive success. However, the necessary conditions for re-creating the characteristics of the successful entity in the next generation or for generating similar characteristics in kin such as siblings or cousins is once again the province of DST. The conditions shared by related entities will commonly consist of genes but will include all those nongenetic factors that DST thinking has brought back on stage. The value of making the three-way distinction is that it separates out the first and third aspects of the Darwinian mechanism, which had been run together into the single concept of "replicator."

A confusion of teleological and causal explanation led critics and supporters alike to a non sequitur. Proponents argue that if genes are usefully regarded as selfish, it follows that they uniquely bring into being the phenotypic characters of the whole animal. Opponents argue that if it is valuable to treat development as a process involving feedback and many mutual actions between different agents, then the selfish gene language is inappropriate. Both arguments are mistaken, in my view. The developmental systems approach is not in conflict with the selfish gene approach to evolution.

The evolutionary process does not require a simple correspondence between genes and adaptive behavior. Darwinian evolution operates on individuals that have developed within a particular set of conditions. If those conditions are stable for many generations, then the changes that matter will arise from segregation of factors that give rise to individual differences. Individuals vary; some survive and reproduce more successfully than others because they possess a crucial characteristic; and close relatives are more likely to share that characteristic than unrelated individuals. Apparent design is produced, even when it is at the end of the long and complicated process of development. But the environment does not cease to be important for evolution just because it remains constant. Change the environment and the outcome of an individual's - development may be utterly different. Indeed, if an individual does not inherit its parents' environment along with their genes, it may not be well adapted to the conditions in which it now finds itself.

Conclusion

Old styles of thought die hard. Even the much-used distinction between sex and gender has a strong whiff of the old genes/environment opposition. "Sex" is biology; but "gender" is part of culture, the acquired behavior deemed appropriate to the social role of that sex. Any fair-minded person listening to setpiece debates about behavioral development is likely to be left irritated by the claims and counterclaims. And well-meaning attempts to break out of the nature/nurture straitjacket have often resulted in an obscure and bewildering portrayal of development as a process of impenetrable complexity (what Salman Rushdie once described in another context as a P2C2E—a Process Too Complicated To Explain). Indeed, development seemed so unfathomably complex to eighteenth-century biologists that they believed that it must depend on supernatural guidance.

The processes involved in behavioral development do indeed look forbiddingly complicated on the surface. Some would argue that it is worse underneath and that such order as is found is generated by dynamical systems of great complexity (Kauffman 1993). One approach to development has been to suggest that everything interacts with everything else. A challenge to such thinking is the evidence for the segregation of characteristics in closely related individuals. How is it that characteristics such as a big nose or a retiring disposition skip a generation? How is it that siblings are so different from each other at birth? How is rapid artificial selection for behavioral characteristics such as tameness possible? None of the evidence that leads to such questions implies any simple correspondence between an inherited factor and the development of a phenotypic characteristic. But the evidence does imply that fractionation and independent inheritance of some of the factors necessary for development happens all the time.

In contrast to an agenda that easily renders development a process too complicate to explain, I prefer to argue that simplicity and regularity may be found in the developmental processes that give rise to unique individuals. Confidence in that conclusion comes not from general principles but from the careful analysis of particular cases. The essence of development—change coupled with continuity—starts to make sense. It becomes possible to understand how the individual is so responsive to events at one stage and so unaffected by them at another.

Order underlies even those learning processes that make individuals different from each other. Knowing something of the underlying regularities in development does bring an understanding of what happens to the child as it grows up. The ways in which learning is structured, for instance, affect how the child makes use of environmental contingencies and how the child classifies perceptual experience. Yet predicting precisely how an individual child will develop in the future from knowledge of the developmental rules for learning is no easier than predicting the course of a chess game. The rules influence the course of a life, but they do not determine it. Like chess players, children are active agents. They influence their environment and are in turn affected by what they have done. Furthermore, children's responses to new conditions will, like chess players' responses, be refined or embellished as they gather experience. Sometimes normal development of a particular ability requires input from the environment at a particular time; what happens next depends on the character of that input. The upshot is that, despite their underlying regularities, developmental processes seldom proceed in straight lines. Big changes in the environment may have no effect whatsoever, whereas some small changes have big effects. The only way to unravel this is to understand the developmental processes.

Acknowledgments

I am very grateful to Susan Oyama and Russell Gray for the trouble they took over earlier versions of this chapter. They are good friends as well as good editors.

Notes

1. Susan Oyama considers the various meanings of "interaction" in her chapter. In its purely statistical sense, "interaction" implies that two or more factors have an effect that would not be deduced from knowledge of what each does when the other is kept constant. For example, when Factor A is present at its highest level with Factor B at its lowest level, an outcome measure might have a high value, but when Factor B is also at its highest level, the outcome measure might have a low value. Confusion begins when interaction is used for the process that might give rise to such a result.

2. The editors did not like my jukebox metaphor because a record is put into the machine ready-made, whereas an environmentally triggered alternative mode of development involves all those mutual actions between the developing individual and its environment which is the theme of so much of this book. I agree with them and understand their discomfort. Nevertheless, I hope they will forgive me for persisting in using the metaphor despite its misleading aspects and disturbingly preformationist character. I persist because the image does embody the idea that the various alternative ways in which an individual might develop are coherent and the phenotype produced is appropriate to the circumstances in which it develops (except when the forecast is incorrect!).

3. Some would argue that even at the level of base pairs and amino acids, the coding metaphor is inappropriate. Neumann-Held discusses the matter in her chapter. See also Sarkar (1996).

4. Tim Ingold would probably include himself in this group.

5. Griffiths and Gray (1997) point out that the replicators are not necessarily genetic, and Gray (in press) feels that "vehicle" is too passive a concept to represent the organism's active engagement with its environment.

References

Barker, D. J. P. (1998). *Mothers, Babies and Health in Later Life*. (2d ed.). Edinburgh: Churchill Livingstone.

Bateson, P. P. G. (1966). The characteristics and context of imprinting. *Biological Reviews* 4l: 177–220.

Bateson, P. P. G. (1976a). Specificity and the origins of behavior. *Advances in the Study of Behavior* 6: 1–20.

Bateson, P. P. G. (1976b). Rules and reciprocity in behavioural development. In P. P. G. Bateson and R. A. Hinde (Eds.), *Growing Points in Ethology*, pp. 401–421. Cambridge: Cambridge University Press.

Bateson, P. (1978). Review of the Selfish Gene by Richard Dawkins. *Animal Behaviour* 26: 316–318.

Bateson, P. (1979). How do sensitive periods arise and what are they for? *Animal Behaviour* 27: 470–486.

Bateson, P. (1981). Control of sensitivity to the environment during development. In K. Immelmann, G. W. Barlow, L. Petrinovich, and M. Main (Eds.), *Behavioral Development*, pp. 432–453. Cambridge: Cambridge University Press.

Bateson, P. (1982). Behavioural development and evolutionary processes. In King's College Sociobiology Group (Ed.), *Current Problems in Sociobiology*, pp. l33–l5l. Cambridge: Cambridge University Press.

Bateson, P. P. G. (1984). Genes, evolution and learning. In P. Marler and H. S. Terrace (Eds.), *The Biology of Learning*, pp. 75–88. Berlin: Springer-Verlag.

Bateson, P. P. G. (1986). Functional approaches to behavioural development. In J. G. Else and P. C. Lee (Eds.), *Primate Ontogeny, Cognition and Social Behaviour*, pp. 183–192. Cambridge: Cambridge University Press.

Bateson, P. (1987a). Biological approaches to the study of behavioural development. *International Journal of Behavioral Development* 10: 1–22.

Bateson, P. (1987b). Imprinting as a process of competitive exclusion. In J. P. Rauschecker and P. Marler (Eds.), *Imprinting and Cortical Plasticity*, pp. 151–168. New York: Wiley.

Bateson, P. (1988). The active role of behaviour in evolution. In M.-W. Ho and S. Fox (Eds.), *Process and Metaphors in Evolution*, pp. 191–207. Chichester: Wiley.

Bateson, P. (1991). Are there principles of behavioural development? In P. Bateson (Ed.), *The Development and Integration of Behaviour*, pp. 19–39. Cambridge: Cambridge University Press.

Bateson, P. (1996). Design for a life. In D. Magnusson (Ed.), *Individual Development over the Lifespan: Biolog-*

ical and Psychosocial Perspectives, pp. 1–20. Cambridge: Cambridge University Press.

Bateson, P., and G. Horn. (1994). Imprinting and recognition memory: a neural net model. *Animal Behaviour* 48: 695–715.

Bateson, P., and P. Martin. (1999). *Design for a Life: How Behaviour Develops.* London: Cape.

Bolhuis, J. J., and R. C. Honey. (1994). Within-event learning during filial imprinting. *Journal of Experimental Psychology: Animal Behavior Processes* 20: 240–248.

Bolhuis, J. J., and R. C. Honey. (1998). Imprinting, learning and development: From behaviour to brain and back. *Trends in Neurosciences* 21: 306–311.

Caro, T. M., and P. Bateson. (1986). Organisation and ontogeny of alternative tactics. *Animal Behaviour* 34: 1483–1499.

Chalmers, N. R. (1987). Developmental pathways in behaviour. *Animal Behaviour* 35: 659–674.

Cooper, R. M., and J. P. Zubek. (1958). Effects of enriched and restricted early environments on the learning ability of bright and dull rats. *Canadian Journal of Psychology* 12: 159–164.

Crews, D. (1996). Temperature-dependent sex determination: The interplay of steroid hormones and temperature. *Zoological Science* 13: 1–13.

Dawkins, R. (1976a). Hierarchical organisation: A candidate principle for ethology. In P. P. G. Bateson and R. A. Hinde (Eds.), *Growing Points in Ethology,* pp. 7–54. Cambridge: Cambridge University Press.

Dawkins, R. (1976b). *The Selfish Gene.* Oxford: Oxford University Press.

Dawkins, R. (1978). Replicator selection and the extended phenotype. *Zeitschrift für Tierpsychologie* 47: 61–76.

Dawkins, R. (1982). *The Extended Phenotype.* Oxford: Freeman.

Dunbar, R. I. M. (1984). *Reproductive Decisions.* Princeton, NJ: Princeton University Press.

Eveleth, P. B., and J. M. Tanner. (1990). *Worldwide Variation in Human Growth.* (2d ed.). Cambridge: Cambridge University Press.

Gray, R. D. (in press). Selfish genes or developmental systems? Evolution without interactors and replicators? In R. Singh, C. Krimbas, J. Beatty, and D. Paul, (Eds.), *Thinking about Evolution: Historical, Philosophical, and Political Perspectives.* Cambridge: Cambridge University Press.

Griffiths, P. E., and R. D. Gray. (1997). Replicator II—judgement day. *Biology and Philosophy* 12: 471–492.

Hales, C. N., and D. J. P. Barker. (1992). Type 2 (non-insulin-dependent) diabetes mellitus: The thrifty phenotype hypothesis. *Diabetologia* 35: 595–601.

Hales, C. N., M. Desai, and S. E. Ozanne. (1997). The thrifty phenotype hypothesis: How does it look after 5 years? *Diabetic Medicine* 14: 189–195.

Hinde, R. A. (1982). *Ethology.* Oxford: Oxford University Press.

Honey, R. C., and J. J. Bolhuis. (1997). Imprinting, conditioning, and within-event learning. *Quarterly Journal of Experimental Psychology. Section B Comparative and Physiological Psychology* 50: 97–110.

Horn, G. (1991). Cerebral function and behaviour investigated through a study of filial imprinting. In P. Bateson (Ed.), *The Development and Integration of Behaviour,* pp. 121–148. Cambridge: Cambridge University Press.

Horn, G., and B. J. McCabe. (1984). Predispositions and preferences. Effects on imprinting of lesions to the chick brain. *Animal Behaviour* 32: 288–292.

Huh, D. L., C. Power, and B. Rodgers. (1991). Secular trends in social class and sex differences in adult height. *International Journal of Epidemiology* 20: 1001–1009.

Ingold, T. (1986). *Evolution and Social Life.* Cambridge: Cambridge University Press.

Janzen, F. J., and G. L. Paukstis. (1991). Environmental sex determination in reptiles: Ecology, evolution, and experimental design. *Quarterly Review of Biology* 66: 149–179.

Kauffman, S. A. (1993). *The Origins of Order.* New York: Oxford University Press.

Lee, T. M., and I. Zucker. (1988). Vole infant development is influenced perinatally by maternal photoperiodic history. *American Journal of Physiology* 255: R831–R838.

Lehrman, D. S. (1970). Semantic and conceptual issues in the nature-nurture problem. In L. R. Aronson, E. Tobach, D. S. Lehrman, and J. S. Rosenblatt (Eds.), *Development and Evolution of Behavior,* pp. 17–52. San Francisco: Freeman.

Lerner, R. M. (1998). *Handbook of Child Psychology* vol 1. (5th ed.). New York: Wiley.

Mackintosh, N. J. (1998). *IQ and Human Intelligence.* Oxford: Oxford University Press.

Marler, P. (1976). Sensory templates in species-specific behavior. In J. C. Fentress (Ed.), *Simpler Networks and Behavior,* pp. 314–329. Sunderland, MA: Sinauer.

Nesse, R. M., and G. C. Williams. (1994). *Evolution and Healing.* London: Weidenfeld & Nicolson.

Oyama, S. (1985). *The Ontogeny of Information: Developmental Systems and Evolution.* Cambridge: Cambridge University Press. (2d rev. ed., Durham, NC: Duke University Press, 2000.)

Pener, M. P., and Y. Yerushalmi. (1998). The physiology of locust phase polymorphism: An update. *Journal of Insect Physiology* 44: 365–377.

Pinker, S. (1994). *The Language Instinct.* London: Penguin.

Rowell, C. H. F. (1971). The variable coloration of the Acridoid grasshoppers. *Advances in Insect Physiology* 8: 145–198.

Sarkar, S. (1996). Decoding "coding": Information and DNA. *Bioscience* 46: 857–864.

Sterelny, K., and P. E. Griffiths. (1999). *Sex and Death.* Chicago: University of Chicago Press.

Waterlow, J. C. (1990). Mechanisms of adaptation to low energy intakes. In G. A. Harrison and J. C. Waterlow (Eds.), *Diet and Disease in Traditional and Developing Countries,* pp. 5–23. Cambridge: Cambridge University Press.

Wilkinson, R. G. (1996). *Unhealthy Societies: The Afflictions of Inequality.* London: Routledge.

Wilson, E. O. (1971). *The Insect Societies.* Cambridge, MA: Harvard University Press.

Yntema, C. L., and N. Mrosovsky. (1982). Critical periods and pivotal temperatures for sexual differentiation in loggerhead sea turtles. *Canadian Journal of Zoology* 60: 1012–1016.

Zadnik, K. (1997). Myopia development in childhood. *Optometry and Vision Science* 74: 603–608.

14 Parental Care and Development

Peter H. Klopfer

As a naive young student, I was once shocked when a revered elder statesman prefaced his remarks on the neural control of locomotion with the observation that as a dedicated Marxist he "obviously" had to believe ... I had been raised in the belief that preconceptions were taboo for those of us dedicated to experimental science, and I truly thought this to be true. In the half century since, quite aside from personal revelations, the work of deconstructionists has made it clear that few of us can truly divorce our preconceptions from the questions we ask or the answers we favor. This is particularly evident when one considers the enormous literature on parental care and the question of how indelibly the actions of the parent are impressed upon the young. This is the issue we will address here.

Patterns of parental care vary enormously both between and within species (Gubernick and Klopfer 1981). Biologists tend to assume these differences have some adaptive significance. Thus, McCabe and Blanchard (1950) detailed the nest-building and parental activities of three species of the mouse, *Peromyscus* (*maniculatus*, *truei*, and *californicus*), which show, respectively, a minimum of parental oversight; some degree of oversight; and oversight that includes the construction of an elaborate, waterproof home looked after by both parents. This last species has also fur that is particularly soft and easily wetted by the elements. The differences in their parental activities are also reflected in the manner in which they utilize other resources in their environment, and these latter differences keep the three species ecologically distinct—that is, they prevent their competitively excluding one another (cf. Hutchinson 1951). "Adaptiveness" has a temporal dimension, too, and biologists have also taken note of the fact that the ease with which whatever influence exercised by parental care upon subsequent behavior can be overcome also varies. The effects of deermice being reared in a pasture nest influence subsequent habitat

choices irreversibly, but this is not true for rearing in a nest in woodland (Wecker 1963). One can only speculate as to why this should be so, but it does indicate that some rearing experiences are more constraining than others. Similarly, Klopfer (1965, 1971) demonstrated that ducklings imprinted onto models of different colors and patterns would not be equally constrained in their later preferences: Some models allowed for a greater range of preferences than others, and others merely served to activate a preference for a different model.

Among humans, Whiting and Childs (1953) have chronicled cultural differences as great or greater than those that distinguish the aforementioned three species of *Peromyscus*, but questions of adaptivity have seemed of less interest or relevance here. Instead, discussion has centered upon the long-term effects of particular parental rearing styles, that is, upon what in animals was referred to as the temporal dimension of adaptivity. In general three views have prevailed, with a fourth recently gaining ascendancy.

1. Early fixation. This view emphasizes the effects of primacy, and is often modeled upon the ethological concept of imprinting. A metaphor that describes this process is that of softened sealing wax, which, for a short time only, can receive an imprint, whereupon, after hardening, the imprint becomes a permanent feature. The exposure of young precocial birds (viz., those that are motorically able to leave the nest upon hatching, as opposed to altricial) to a moving object shortly after hatching, which object then becomes the preferred companion or even subject for later sexual attachments, was popularized by Konrad Lorenz (1937), though rather more recently Gottlieb (1971) corrected many of the misconceptions that Lorenz's descriptions engendered. Gottlieb pointed out the many qualifications and specifications entailed in imprinting, as well as other attributes that make it very different from the simple and permanent process the imprinting

metaphor suggests. Furthermore, long-term effects of primary experiences, which may suggest imprinting, can often be the result of complex but subtle feedback processes. Gunther (1961) describes how a newborn human infant may sometimes have its upper lip pushed against its nostrils when it first nurses, the consequence of a mismatch between the shape of the maternal breast and the infant's mouth. When this occurs, the baby quickly stiffens and seems to thrust itself away from its mother. After a few repetitions of this behavior, many women, seemingly rejected by their newborn, develop defensive reactions, and, in the end, abandon efforts to nurse, feeling themselves rejected and useless. This temporally limited scenario with long-term consequences may resemble imprinting, though here the effect is to create an aversion rather than an attachment. However, the similarity is clear when one notes that in domestic goats, lambs, and calves, an animal that has first suckled for a few minutes after its birth can be taught to drink from a pail only with difficulty (and usually only by allowing it first to suckle on a finger or other object that is then immersed in the pail). In short, the suckling response is initially attached in a fashion suggesting classical imprinting, though Gunther's example reveals the important role of sensory feedback in maintaining (or terminating) the attachment.

2. Gradual shaping. The appropriate metaphor here is a windswept tree on a barren coast, which, in time, displays in the angles of its branches the effects of continual shear. Or, as the adage runs, as the twig is bent, so grows the limb. There is much observational support for this view, though it ignores occasional, and major, discontinuities due to the sudden onset of hormone production, which may be seasonally controlled (as in many songbirds) or developmentally triggered (as in many mammals).

3. Metamorphosis. The model here is the holometabolous insect that spends one part of its life as a crawling larva or caterpillar, then pupates only to emerge in an altogether different, winged form, with little to suggest in its new life the styles and habits of the old. Kagan, in a study some decades ago (1984) described the withdrawn, shy, almost secretive behavior of young, evidently largely neglected children in the highlands of Guatemala. Some even displayed many of the hallmarks of incipient autism. Returning some years later, Kagan found these same children to have become social, friendly, and self-reliant youths, who displayed no signs of early neglect. Somehow they had become transformed. The sudden onset of adult micturition patterns in male puppies, once androgenic hormones reach a certain level, is another example, as are the dramatic changes in gender seen in some Dominican children, who appear to be and are raised as girls and who, at puberty, suddenly develop androgen sensitivity and become functional young men (Imperato-McGinley et al. 1974).

4. Regulation by interaction: in this example, the focus is not on the developing infant, but rather on the interactive unit, parent/child, which is examined within the context of both biological and social/environmental constraints. It can be viewed as similar (though I emphasize the similarity is a superficial one!) to a goal-seeking - automaton that can adjust its reactions to the external environment so as to remain on course. Ashby (1958) describes a mechanical "turtle" that will leave its well-lit hutch (wherein its batteries are recharged) to wander about a room whose obstacles it "learns" to avoid. As it "tires" (viz., its battery runs down), it becomes increasingly photopositive and thus returns to its hutch for a "rest," only to reemerge and continue its explorations when recharged. A biological example is provided by Chisholm's study of Navajo infancy, which is described later.

In nonhuman species, where comparative studies are feasible, the different models of development are at least sometimes susceptible to being examined from a functional, or adaptationist, point of view. Thus, it has been argued (Klopfer 1959) that imprinted preferences are to be expected in motorically precocial species and not in their altricial near-kin. If a young duckling, it is argued, does not very quickly learn to remain near its sibs and its parents, it will not likely survive. A young songbird, on the other hand, physically unable to disperse from its nest, can take time to learn the hallmarks of its parents. Similarly, surface-nesting waterfowl can rely upon visual cues when they learn the characteristics of their kind, while hole-nesting species might be expected to rely more upon auditory features. Some of these speculations are both plausible and true, though as Gottlieb (1971) has shown, imprinting is not the simple phenomenon that Lorenz originally described (cf., Hess 1958). In studies of humans, the issue of adaptation at this level is rarely raised, and for good reason: Few investigators of cultural differences in development and child-rearing would deny that there is more than one way to skin a cat. A culture's rearing myths and practices are necessarily of a piece with other features of that culture, and the evolutionary aspects of adaptation are rarely if ever relevant (but see Chisholm 1983, for an example of adaptational thinking). It is rather a question of how and why early experience associated with different parenting styles shapes later behavior.

With respect to studies that seek to determine the applicability of any particular model to any given sample, it is embarrassingly evident that the investigator's conceptual frame largely determines the outcome. Even so astute an investigator as Kagan has at different stages of his life interpreted the same body of data as indicating genetic determination of certain traits (such as shyness), or as indicating the salience of early influences, or as suggesting remarkable resilience and abilities for compensatory responses in the young child (Kagan 1984, 1998; Kagan, Reznick, and Snidman 1988). Whorf's warning that the language of our questions shapes our perceptions is particularly pertinent when one examines the literature on human development and the role of parental influences (Whorf 1950).

Increasingly, however, one particular model is emerging that seems to allow assimilation of much of the accumulated knowledge about de-velopment and that also provides a useful heuristic. This is the notion that what must be observed and described is the interaction between parent and offspring, and that it is the quality of the interaction that plays a determining role in the outcome, not any single learning mechanism. In what follows, I describe some earlier work of my own on the compensatory responses of goat mothers to their impaired young, work that was undertaken with an altogether different objective in mind, but which, I think, does illustrate the kinds of interactions that are important. Then, I review a study of Navajo infancy by Chisholm (1983), which provides an especially persuasive human example of the role of interaction between parent and offspring.

Some animals are more equal than others: Even in groups of highly inbred creatures, at least in species with some degree of social organization, certain individuals emerge as dominant or as leaders, initiating movements, taking the initiative in territorial defense, or displacing their conspecifics. Small herds of goats display this phenomenon clearly: Certain animals have priority of access to localized sources of food, yet others are at the head of the herd when it moves to new terrains. Individual differences that reflect the potential for being, in one or another sense, a "leader" are evident early in the life of the kid: The most vigorous of a group of month-old kids almost invariably becomes the dominant animal. Thus, we asked, how come leaders to their posts, focusing our attention on a group of about three dozen toggenburg goats (Klopfer and Klopfer 1974).

Manning (1965) and Benzer (1973), among many others, had previously documented differences in temperament among animals reared identically. Benzer's work on the genetics of behavior in *Drosophila*, in fact, was allegedly the result of his noting the dramatic differences in the temperaments of his two daughters (Weiner 1999). The differences he observed in the fruitflies were heritable, and presumably linked to genetic differences, but epigenetic influences were not ruled out. What might these developmental influences have been?

Among goats, twins are the commonest litter size, at least among multiparous females, and triplets occur frequently, too. However, birth weights are highly variable within a litter, even in those cases where the twins are evidently monozygotic. We noted that the heavier kid of any litter displayed a greater vigor than its lighter siblings, vigor being measured in terms of the time taken to first stand unsupported and to commence udder-seeking behavior. The mother attends most to the more vigorous of her kids, irrespective of birth order. This has the paradoxical effect of slowing this kids' progress to the teat. The mother's attention entails licking and nuzzling the kid, often with sufficient force that the kid is knocked off its feet, must rest for a time, and then once again must struggle up and find its balance. Thus, it often happens that the initially weaker sibling, lacking interference in its progress to the teats, is able to initiate nursing first. The result is that initial differences in strength and weight are muted, and the weaker animals are not necessarily relegated to the status of subordinate.

These results do not answer the question of how a dominant or leadership position develops, but they suggest the absence of an invariant relationship between prenatal (including, perhaps, hereditary) influences and subsequent status. However, the mother's moderating influence was seemingly mechanical: she merely licked the more active kid more, and the deleterious effect on its progress to the source of nutrients was incidental. Suppose that initial differences in vigor were sufficiently great as to suggest that the weaker kid had a reduced chance of survival. Or, suppose that resources to sustain the mother were in short supply, so that rearing two kids rather than just one was unlikely. Would the mother modify her responses in an adaptive fashion?

To examine the role of the mother in modulating the effectiveness of her kids' nursing attempts, we first used drugs to enhance the difference in the vigor of twin kids (Klopfer and Klopfer

1977). As expected, the weaker kids, up to a point, benefited from the reduced maternal attention, getting to the udder and nursing sooner than their more vigorous mate who was being more intently licked (and knocked down) by the mother. However, where the degree of a kid's weakness exceeded a threshold, which could be specified in the number of minutes required to resume standing, the feedback failed and the more vigorous kid did nurse first and come to dominate the milk supply.

If the mother's ability to monitor her kids' behavior was impaired, which we achieved by applying blindfolds and by pharmacologically reducing olfactory acuity, the feedback also broke down. When impaired, mothers stood passively while their kids sought the udder, and the more vigorous kids invariably reached the udder first and were thus in a position to monopolize the milk supply. As we concluded at the time, mother love is not blind: The degree to which offspring share in the maternal bounty is very much the result of an interaction between mother and young. The ultimate status of the neonatal kid is not specified at birth, nor is there any reason to doubt that subsequent interactions would be equally effective in modifying behavior.

Navajo infants, it has long been claimed, are quieter and less irritable than infants of Anglo parents (see Chisholm 1983, from whom all of the following is taken). Depending upon which of the developmental metaphors one finds more appealing, a variety of putative explanations for this difference can be found. For example, Navajo infants are traditionally placed in a cradleboard, which severely restricts their movements. The duration of confinement does vary considerably, from only a few to twenty-three hours in a day, and may terminate soon after the first year of life or continue for as long as two years. If one assumes this confinement to constitute a perturbation in the normal course of development, equivalent, in one sense, to the administration of tranquilizing drugs to the kids described above, one might expect certain effects with long-term consequences for the developmental pathway. Chisholm's study did, in fact, identify some immediate consequences of the confinement. The cradleboarded children, and this was true for Anglo babies placed in cradleboards as well, were less active and less easily aroused. Mothers and cradleboarded infants were also less responsive to one another. He concluded (p. 216) that "the immediate effects of the cradleboard on infant behavior and mother-infant interaction are clear and a substantial corpus of theory suggests that these immediate effects should last. The question now becomes why don't these clear immediate effects of the cradleboard last?" His exhaustive series of tests produces no sign of a cradleboard effect at later ages, just months after cradleboarding ends.

Chisholm suggests three responses to the cradleboard, which may serve as the regulating factors that assure a stable developmental outcome. First, the child itself is allowed to determine when it is to be removed. Despite the reduced activity and arousal state of the cradleboarded infant, it can still fret and cry, and Navajo mothers are unanimous in saying that such behavior is the signal for removing the child from its confinement. Furthermore, if it resists being replaced, the resistance is generally honored. Thus, despite having its movements restricted, the infant, especially as it grows older, is able to develop a sense of control over its condition.

A second potential corrective lies in the fact that directly upon the infant's removal from the cradleboard the mother and infant engage in a particularly intense period of affectionate interaction, which Chisholm likens to the catching up in growth experienced by children who have been ill for a time.

Finally, the cradleboard assures a degree of conflict-free contact that a cribbed or crawling infant might not experience. The cradleboard is generally placed vertically at a height that allows direct eye-to-eye contact between the mother and the child, whether the mother is standing at a loom or seated on the ground shelling nuts. A crawling infant would likely be subject to frequent disciplining for scattering the nuts, while a cribbed infant would merely be staring into the sky, both situations less socially rewarding than the continuous face to face communication the cradleboard allows.

Chisholm concludes that it is not the individual features of particular developmental practices of a culture that are important for development—the cradleboard has advantages, but so does its absence, and both practices have short-term effects on the behavior and temperament of the child. What is significant is the entire parenting process, particularly those regularities in the interaction of infant and parent which assure predictable outcomes: the infant struggles, its confinement ceases. The Navajo child reared in conventional Western fashion will, by the time the two are walking, not be distinguishable.

The patterns that assure constancy in the outcomes may differ from one culture to the other. But, it is not the culture, or the genetic substrate, except in a very general way, nor the rearing practices that define the child, but the manner of their interaction. This both limits the possible variance and provides a buffer against perturbations that might have serious or long-term consequences. Kagan, attacking the allure of infant determinism, states it thus:

[It] does not mean that the events of the first two years are without any force. It only means that a fearful, quiet, tense two-year old who has had an uncertain environment remains malleable should benevolent changes occur, and a laughing, securely attached, smart two-year-old is not protected from angst should her life turn harsh. Both science and autobiography affirm that a capacity for change is as essential to human development as it is to the evolution of new species. The events of the opening years do start an infant down a particular path, but it is a path with an extraordinarily large number of intersections. (Kagan 1998: 149–150)

The conclusion is somewhat paradoxical. A predictable pattern of parenting is necessary for normal development of the infant, but different patterns may produce similar outcomes, and none of these seem to have effects that are irreversible in later life. Another conclusion must be that the relationship between early rearing and later behavior is unique to each species, for it is evident that for other animals different rearing patterns may not have similar outcomes. This limits greatly our ability to extrapolate from greylag geese, to chimpanzees, to humankind (Klopfer 1996). Not all species are subject to such a variety of social and physical environments as are *Homo sapiens*, so neither should selective forces promote or sustain the degree of postinfant malleability as humans display. Lumsden and Wilson (1981) propose that selection acts to favor particular epigenetic rules. These are regularities in developmental processes that assure a particular phenotype will always appear in a particular environment. Those rules, however, derive their power not merely from the fact that the products of certain genes are predictable, but equally from the fact that the genes are expressed in an ordered and predictable environment, intra- and extracellular, physical, and social. In the case of the Dominican children alluded to earlier, the genes that are involved in the development of male morphology are stymied by a cellular environment (presumably influenced by other genes) that renders tissues insensitive to androgens. When this sensitivity abates, development switches to the male track. One can readily surmise that in most U.S. communities, a child undergoing such a metamorphosis, female until age eleven or twelve, then transforming into a boy, would suffer severe psychosocial trauma, leading to a variety of behavioral pathologies. In the Dominican communities accustomed to such children, their condition is part and parcel of an expected scheme and evidently occasions no untoward consequences. Again, it is the existence of the ordered social system that allows for a stable, predictable outcome, even with the variations in the developmental sequence.

In humans, the dimensions of the social environment are multitudinous. Thus, it should come as no surprise that many different patterns of development can still lead to similar outcomes, just as very minor differences can also produce major disparities (viz., the Gunther studies of suckling, alluded to above). As Schrödinger (1951) wrote, long before the structure of DNA had been described, the determinacy in the development of living organisms requires their feeding upon the order in their environment. Where that environment becomes randomly variable, determinacy must decline, but where it is ordered and predictable, outcomes can be constant even while development trajectories vary.

References

Ashby, W. (1958). *An Introduction to Cybernetics.* London: Chapman and Hall.

Benzer, S. (1973). Genetic dissection of behavior. *Scientific American* 229: 24–37.

Chisholm, J. S. (1983). *Navaho Infancy: An Ethological Study of Child Development.* New York: Aldine.

Gottlieb, G. (1971). *Development of Species Identity in Birds: An Inquiry into the Prenatal Determinants of Perception.* Chicago: University of Chicago Press.

Gubernick, D. J., and P. H. Klopfer. (1981). *Parental Care in Mammals.* New York: Plenum.

Gunther, M. (1961). Infant behavior at the breast. In B. M. Foss (Ed.), *Determinants of Infant Behavior,* pp. 37–44. London: Methuen.

Hess, E. (1958). "Imprinting" in animals. *Scientific American* 198: 81–90.

Hutchinson, G. E. (1951). Copepodology for the ornithologist. *Ecology* 32: 571–577.

Imperato-McGinley, J. L., L. Guerrero, T. Gautier, and R. E. Peterson. (1974). Steroid 5-alpha reductase deficiency in man: An inherited form of male pseudohermaphroditism. *Science* 186: 1213–1215.

Kagan, J. (1984). *The Nature of the Child.* New York: Basic Books.

Kagan, J. (1998). *Three Seductive Ideas.* Cambridge, MA: Harvard University Press.

Kagan, J., J. S. Reznick, and N. Snidman. (1988). Biological bases of childhood shyness. *Science* 240: 167–171.

Klopfer, P. H. (1959). An analysis of learning in young *anatidae. Ecology* 40: 90–102.

Klopfer, P. H. (1965). Imprinting: A reassessment. *Science* 147: 302–303.

Klopfer, P. H. (1971). Imprinting: Determining its perceptual basis in ducklings. *Journal of Comparative and Physiological Psychology* 75: 378–385.

Klopfer, P. H. (1996). Mother-young attachments: On the use of animal models. *American Scientist* 84: 319–321.

Klopfer, P. H., and M. S. Klopfer. (1974). How come leaders to their posts? *American Scientist* 61: 560–564.

Klopfer, P. H., and M. S. Klopfer. (1977). Compensatory responses of goat mothers to their impaired young. *Animal Behavior* 25: 286–291.

Lorenz, K. (1937). The companion in the bird's world. *The Auk* 54: 245–273.

Lumsden, C. J., and E. O. Wilson. (1981). *Genes, Mind and Culture.* Cambridge, MA: Harvard University Press.

Manning, A. (1965). *Drosophila* and the evolution of behavior. In J. D. Carthy and C. L. Duddington (Eds.), *Viewpoints in Biology 4,* pp. 125–169. Reading, MA: Butterworths.

McCabe, T. T., and B. D. Blanchard. (1950). *Three Species of Peromyscus.* Santa Barbara, CA: Rood Associates.

Schrödinger, E. (1951). *What Is Life?* New York: Cambridge University Press.

Wecker, S. C. (1963). The role of early experience in habitat selection by the prairie deermouse, *Peromyscus maniculatus bairdi. Ecological Monographs* 33: 307–325.

Weiner, J. (1999). Lord of the flies. *New Yorker, April 5,* pp. 44–51.

Whiting, J. M., and J. L. Childs. (1953). *Child Training and Personality.* New Haven: Yale University Press.

Whorf, B. L. (1950). *Language, Thought, and Reality.* Cambridge, MA: MIT Press.

IV RETHINKING DEVELOPMENT AND EVOLUTION

Terms in Tension: What Do You Do When All the Good Words Are Taken?

Susan Oyama

Anyone attempting to articulate a developing position is likely to feel immobilized from time to time by the very conceptual and historical baggage that makes "good words" good—by the rich connectivity that gives them their power. One needs to recognize important differences and commonalities with other traditions without compromising the project at hand. Never easy, such articulation becomes more risky still if one seeks to make contact with other disciplines, for each has its own theoretical wrangles. It often seems that the good words have already been taken, not to mention retaken and mistaken. One of this book's aims is to present some of the work that clusters around the notion of the developmental system, but another is precisely this kind of exploratory contact with allied approaches.

In this chapter I investigate the concepts of *interaction* and *system* in the context of a view of development and evolution as *constructive* processes. These concepts are unified by an enlarged understanding of inheritance (Oyama 1982, 1985), and, like notes in a chord, they must be understood in relation to each other. They must also be viewed in relation to the problems that the notion of the developmental system is meant to resolve. Words like this are untidy to work with, but, as I implied earlier, much of their resonance derives from their historical complexity. If we are careful, we can draw on these resources, heightening some harmonics and damping down the rest. The alternative is neologism. This last should not be rejected in principle, but in the end we cannot make anything pristinely new. Nor is it clear why we would want to. Whether we use good words or shun them, we must confront their existing senses.

In sampling the possibilities and difficulties of words like *interaction*, *system*, and *construction*, I do not offer an etymology or systematic history, much less an all-purpose method for dealing with semantic dilemmas, but something more homely: a partial account of my own reasons for adopting and sometimes revising (perversely, some may think) these terms. Misapprehensions occur no matter how precise one tries to be, and by suggesting the conceptual locus from which they are made, mistakings can be informative in themselves. Though what follows is partly about rectifying or preventing misreadings, then, I hope it will also give readers a sense of the ongoing disagreements and alliances out of which the developmental systems perspective has been, and is now being, interactively constructed. In addition, I would like to impart a sense of the potential scope of these interlocking ideas: how they provide a conceptual framework within which phenomena can be brought together in ways that are both consonant and novel.

Interaction

Philip Kitcher (forthcoming) begins his contribution to a festschrift for Richard Lewontin with a quotation from my *Ontogeny of Information* (Oyama 1985: 26–27):

"But wait," the exasperated reader cries, "everyone nowadays knows that development is a matter of interaction. You're beating a dead horse."

I reply, "I would like nothing better than to stop beating him, but every time I think I am free of him he kicks me and does rude things to the intellectual and political environment. He seems to be a phantom horse with a thousand incarnations, and he gets more and more subtle each time around.... What we need here, to switch metaphors in midstream, is the stake-in-the-heart move, and the heart is the notion that some influences are more equal than others...."

In my book, the passage in question follows a discussion of the ways in which DNA is given a special role in accounts of ontogenesis. The causal privileging in such accounts is not a matter of saying that only genes are needed for development (no one says this). It is the attribution of special directive, formative, or informative power to genes—in short, the treating of some causes as more equal than others. This is often done despite the fact that the analyses themselves show a variety of elements playing analogous roles in regulation, differentiation, and variation. Equal recognition, I argued, was not being rendered for equal work, however the analyst seemed to be measuring "work" at the time: The rules kept changing. To reject a special executive role for the DNA is not to deny that all sorts of distinctions can be made among factors and among the ways they impinge on development. Nor is it to deny the possibility of prediction or of making various kinds of qualitative and quantitative statements about particular sets of data. It is to say that many of the classical ways of describing biological processes are ill conceived, including many of those that claim to be paying appropriate attention to a variety of factors. As the allusion to Orwell's (1946) *Animal Farm* suggests, claims of equality in biological causation can be given the lie by actual practice, just as they can in everyday political life.

Kitcher (forthcoming) believes that the standard "interactionist *Credo*" signals a kind of causal egalitarianism. He defends that interactionism against a number of attackers, including, it turns out, me. His charge that dialectical biology is hostile to interactionism is perhaps understandable; Levins and Lewontin (1985) explicitly reject the term, as do Lewontin, Rose, and Kamin (1984). The reason for including developmental systems theorists in his treatment is less immediately obvious; still, even though I have always considered myself an interactionist of a certain type, I have also been critical of what I tend to call "standard" or "conventional" or "traditional" interactionism (in order to distinguish it from what I now call *constructivist interactionism;* see also B. H. Smith's [1997: xxi] related "constructivist-interactionist account of belief"). The question, then, is how standard interactionism and its alternatives are to be con-

ceived. Although the dialectical biologists and I have tended to focus on different authors and issues,[1] Kitcher is not wrong to see us as having similar theoretical preferences, even if they reject the term and I do not.

Lewontin, Rose, and Kamin (1984) say that interactionism is "the beginning of wisdom," but add that one way it errs is in treating organism and environment as separate and independent rather than as interpenetrating and interdependent (p. 268). It is an *ism* they object to, however, and many of the ways in which they use *interaction* as a "civilian" word rather than as a special term are very much in accord with the ways I do. They say of the constituent parts of a whole, for instance, that "the interaction of these units in the construction of the wholes generates complexities," or "[O]rganisms alter the external world as they interact with it" (p. 11). There is in these writings an appreciation of mutually defining and mutually influencing organisms and environments: of interactive emergence. Outcomes tend to be attributed to a distributed set of participants, whose actions and effects are interdependent, rather than to actors neatly separated from that on which they operate.

This systemic view of causality also characterizes the developmental systems perspective (or DST, for theory, broadly construed), which in addition stresses the dependence of both invariant and variant features on the same kinds of complex sequences of interactions, rather than depicting some as principally organized from the inside and others from the outside. Both stability and change, similarity as well as variety, must then be explained in the same general terms. As a way of illuminating the meaning of interactionism in DST, I am going to comment on several aspects of Kitcher's critique. These remarks will not include a broad defense of the other writers he discusses, though I will refer to them from time to time. (See Gray, forthcoming, and Griffiths 1999, for overlapping responses.) The dialectical biologists and I appear to have attended to different aspects of current usage, and the situation is too disordered to allow pronouncements on what "interactionists" as a group *really* say and do (for there is no "group" as such, and Kitcher does not name names). Nevertheless, I will permit myself some observations.

Readers of *Vaulting Ambition* (1985) will recognize the stance Kitcher assumes in the chapter in question, as a reasonable, politically sensitive person who readily admits that people sometimes go beyond the evidence and who seeks to rectify such mishaps without offending good sense. He defends standard interactionism against its well-meaning but misguided critics, viewing some of our concerns with sympathy but fearing that we work against ourselves by yearning for a better response to genetic determinism, in the form of an undefined and obscure "trans-interactionist" view (apparently his term). According to him, any problems in the literature flow from careless use of basically sound methods and concepts, not from more fundamental difficulties: Genetic determinism can be adequately dealt with using the instruments at hand.

Briefly described, DST's constructivist interactionism can resemble a number of other views, real or imagined. It is not, however, to be confused with the bloodless interactionism that boils down to "the environment is important too" (e.g., Buss et al. 1999; for contrast see Caporael's [1997] concept of repeated assembly). Versions like Kitcher's, which explicitly allow genetic and environmental variables to be analyzed in similar ways, do not automatically qualify either; recall that much privileging takes place *despite* such analyses. We do not claim that statistical interactions will be found in every data set.[2] Nor is constructivist interactionism opposed in principle to the possibility of developmental stability or regularity. Although it is frequently most easily demonstrated by altering conditions to alter outcomes, that is, it should not be mistaken for some sort of environmental determinism or an insistence that developmental plasticity is the rule. How could it be? Part of the point is to distinguish between issues of regularity of *outcomes*

and the nature of the causal *processes* producing those outcomes. (In the second part of this chapter, in fact, we encounter the opposite worry, that the developmental systems view posits more regularity than is reasonable. Such attributions reflect just the well-entrenched assumptions about contrasting developmental explanations that are being rejected; see also Wimsatt [1986].) One of my own complaints about the various conventional interactionisms is that their formulations are often presented as resolutions of nature/nurture problems, when in fact they accept the basic dichotomies from which those problems grow (see Johnston 1987; Oyama 1985, 2000b). Someone might say, for instance, that the nature/nurture opposition is nonsensical because some things don't just mature, but require interaction. Here it is evident that *interaction*, far from challenging the concept of internally driven maturation, assumes it. The fatal weakness of oppositional stances that stress either genetic or environmental effects, in fact, is that no matter how useful their examples may be in disputing particular deterministic claims, they miss the larger chance by accepting too much.

In light of all this, I contend that Kitcher misrepresents at least a significant segment of standard interactionism, even as he misrepresents the positions of its critics. Though his own views are not always laid out fully enough in the paper under discussion to allow a judgment, moreover, it is possible to argue that he himself is not a standard interactionist at all. I will look first at his discussion of norms of reaction. One of his more disconcerting claims is that DST and dialectical biology consider them to be incoherent. Then I turn to the separation of causal factors and causal democracy.

Norms of Reaction

A norm of reaction shows the set of phenotypes resulting when a given genotype develops in various (or more sweepingly, all possible or feasible) environments. At the same time that Kitcher claims that DST and dialectical biology consider norms of reaction nonsensical, he acknowledges Lewontin's influential writings on the utility of such graphs. Gray (1992, forthcoming) and van der Weele (1999; see chapter 24) have also found them useful. None of us, to my knowledge, has said the idea of norm of reaction is itself ill formed. We might well argue with some of the ways it is used, however (see, for instance, Oyama 2000a, where I discuss privileging in statistical representations, including norms of reaction).

Kitcher says that one way to understand genetic determinism, "the idea that genetic causes take priority" (forthcoming), is as a claim of a flat norm of reaction. I doubt that this *ism* can be so neatly pinned to particular predictions, but it allows him to start with a convenient representational device that makes genetic determinism a matter of degree. A flat norm of reaction is then one extreme on a continuum. Notice that this choice bypasses some questions that immediately come up in a DST framework: What does the shape of a norm of reaction have to do with the idea that "genetic causes take priority"? As I argue later, priority can be given in a variety of ways. When Kitcher speaks of norms of reaction revealing causal contributions, I would ask, contributions to what? What is being quantified here? If the genotype in question is being placed in every environment consistent with survival, say, then many, many other factors are invariant as well. Despite the fact that these are by definition needed for development, Kitcher attributes causal efficacy only to the genes. Under what conception of developmental processes are genes more important in traits that emerge reliably in many environments than in those that vary across them? Is the degree of *environmental* "determination" proportionately greater if a genetic mutation brings about no phenotypic change?[3]

Imagine a morphological feature—having just one of a particular body part, perhaps—for which there is no known genetically correlated variation consistent with viability in some fixed developmental environment. (The point may be sharper with body parts than behavior, because people tend to think of the body as genetically caused; the analytic logic, however, is the same for both.) Because Kitcher's analysis of genetic determination may tacitly rely on the existence of an alternative genotype with a different norm of reaction, furthermore, we can add to my counterexample an alternative developmental environment, perhaps one that yields some genetically correlated phenotypic variation. (It could be that in this second environment, organisms lacking the organ in question can survive.) Would we now want to say that the morphological feature in the first case is completely environmentally determined because it appears with all of the genotypes (all surviving organisms have just one of the parts in question)? And in the second case, does the environment have less causal efficacy, and the genes proportionately more?

Like Lewontin, Kitcher evinces considerable mistrust of heritability studies, especially the drawing of conclusions about causation from them. Yet, given Kitcher's affection for the "model of causal analysis that looks at the effects of a single genotype across varying environments," I am intrigued by his dismissal, for this sounds like much of what goes on in investigations of heritability. (The "single genotype" would really be some set of organisms judged to be genetically similar, possibly only at a single locus, by a criterion that may involve no investigation of the genetic material itself.) Heritable (genotype-correlated) variance can be quantified by comparing the developmental products of various pairings of genotypes and environments in an analysis of variance (ANOVA). It is possible to view the set of phenotypic outcomes for any particular genotype as a mini-norm of reaction: the phenotypes that result when organisms with the "same" genotype are raised in varying developmental environments. It is partial, of course, but every *actual* norm of reaction shows only a small sample of possible conditions, and we are typically warned against extrapolating beyond them. The entire ANOVA matrix can then be seen as an array of such norms of reaction (successive rows, say), each one showing the average outcomes for individuals of the same genotype in one of the environments studied.

Faced with a set of results that shows genetic main effects, one can associate differences in outcome with differences in genotype. Some account for *similarities* in outcome by shared genes, but in my view this is suspect, though it is the explanatory strategy Kitcher seems to prefer. He attributes the "genetically determined," uniform developmental outcomes in a flat norm of reaction to the genes that do not vary (are held constant, shared, identical, and so forth, albeit across a wider range of conditions than can be encompassed by an ANOVA—or any other analytic technique I am aware of). He appears to do with norms of reaction, that is, what behavior geneticists often do with heritability—explain similarities by shared genes.[4]

The problem lies less with the analytic technique per se than with the preanalytical assumptions that guide its use, and thus the meanings that are attached to the results. Kitcher may be taking objections to the latter as outright rejections of the norm of reaction as a tool. In DST, a single factor might under certain circumstances be "determinative" in the sense of being a good predictor across some range of cases, but a good deal of mischief comes from confusing this with the more common causal meanings; they come together when *genetically determined* is used for organisms or characteristics rather than for patterns of outcomes. The question of developmental fixity, one sense of which is a flat norm of reaction, is often a significant ingredient of biological or genetic determinism, but it is not the only one. The "priority" of genetic causes can also be understood in terms of an underlying meaning, truth, or *essence*, perhaps pressing for "expression." Certain subtle and often more consequential ways of prioritizing genes are therefore not touched by discussions that concentrate only on what I call issues of *incidence*—frequency, predictability, inevitability.

My concern is thus not to endorse some methods and condemn others, but to ask what conclusions are being drawn when a method is used.

The distinction between accounting for a difference among phenomena and accounting for the phenomena themselves is significant. It is this distinction (and more generally, the difference between population and developmental analysis) that lies behind the warning that is so familiar to those who followed the race and IQ controversies of the 1980s: not to confuse accounting for variation among organisms with accounting for the ontogeny and nature of the organisms themselves. It is one thing to speak of contributions to patterns of similarities and differences among organisms under particular circumstances. ANOVA and norms of reaction are among the methods that allow this, and if part of Kitcher's point is that we should attend to the range of conditions sampled, it is a point often made and well taken. It is another matter, however, to say that some aspects of us are made more by (more programmed by, more completely inscribed in, more informed by) the genes than others. The understandings that support what we might call *genetic essentialism* is one of my concerns here, and they are not addressed by Kitcher's critique. This is one reason I consider the widespread usage of *genetically determined* to indicate a phenotype with a flat norm of reaction to be so terribly unfortunate. A related difficulty is that the phrase thus used marks a *lack* of variation, rather than the systematic variation that I take to be the proper statistical use of *determine*. The latter, stricter, usage is in fact frequently invoked by behavior geneticists and others who bemoan the mishandling of the language of determination. All this makes it very hard to hold on to the distinction between differences and things.[5] It does, however, fit the idea that there are two theoretically separable kinds of explanations for organisms and their features, an idea DST disputes.

Separation of Causal Factors

According to Kitcher, dialectical biology and DST reject interactionists' claim that the genes can be separated out as causal factors. There may be something to this, but our objection is not what Kitcher thinks.[6] He attributes to Lewontin a belief that I have never seen expressed by that author or by any of the other people under discussion—indeed, cannot imagine any of us holding: that something cannot be a "causal factor" in the development of a phenotype if that thing depends on the phenotype. Recall the causal language of the quotation from Levins and Lewontin (1984: 100), about organisms altering the world as they are altered by it. Unless one simply stipulates that causes must be independent of the things they affect,[7] a chemical would be a "causal factor" affecting an organism's health even if it only becomes toxic when the organism's digestive juices act on it. I believe that such reciprocal dependency is one of the points of dialectical biology's organism-environment interpenetration. It is certainly one of the points of my concept of constructivist interaction: that this is the way these causal relations work, interdependently and interactively—not that they are not causal relations at all. So the claim is hardly that genetic effects on organisms cannot be identified, but that the genes have their effects by being affected by other factors—by their cellular environments, if you will—and these often include the very processes they influence. The impact of gene products, furthermore, tends to vary with other conditions. Starting an account with genetic transcription, and treating the DNA as an "independent variable" that "initiates" an interesting cascade of events, leads only too easily to obliterating from the causal landscape the events and conditions that brought that transcription about.

Kitcher then interprets *interdependence* as causal relevance rather than actual physical impact, as when only certain aspects of a bird's surroundings are relevant to it, or, in the example just given, when a chemical is toxic to one organism but not another (see also Godfrey-Smith 1996). In this sense, what counts as the bird's environment depends on what/how it is and what it does, a matter to which we will return. Kitcher (in

press) rightly says that such organismic "determination" of the relevant environment is consistent with "the attempt to draw norms of reaction that identify the causal contributions." I know no one who has suggested otherwise, although the meaning of "causal contributions" may be a point of disagreement. I would say, and I suspect Kitcher would concur, that the genes cannot be separated out as prime movers: autonomous imparters of form and function, carrying in their very molecular structure representations of their potential or proper developmental outcomes. Like any interactant, they become causes only with respect to other participants in an interaction, and their effects are similarly system-dependent, the ubiquitous language of genetic privilege notwithstanding. The prevalence of such privileging was one of the early motivators for my own work; these notions are not easily addressed with the analytic tools we have been discussing, but they do yield to a different kind of approach.

In DST constructivist interaction, interdependence, and reciprocal contingency counteract one-way causal stories, reminding us that although we may use our methods of randomization and control to "isolate" the effects of a factor, we do so precisely because our factors are abstracted from a dense causal complex. Statistical control, furthermore, hardly eliminates the causal roles of stabilized factors. It simply excludes them from the *analysis*. It is the neglect of mutual influence, and the unprincipled separating out of genes as controlling, instructive agents, that DST resists, not conventional analyses of contributing factors or the possibility of causal influence, whatever views one may have of causality itself. This brings us to the issue of parity.

Causal Democracy and Parity

DST makes extensive use of parity of reasoning. Descriptions and explanations of development are often asymmetric: The logic that is used (when there is a logic at all) to characterize certain factors as informing, coding, controlling, and so forth, could be, but typically is not, applied to other factors that play demonstrably comparable roles. In contrast, DST includes as full-fledged interactants many factors that are generally left in the background (for examples and discussion of the pragmatics of research, see van der Weele 1999). When Kitcher blandly preaches "causal democracy," then, I am fascinated. Not only is most standard interactionism shot through with asymmetries, but the notions of causal *symmetry*, or *parity*, which do have a democratic ring, inform the very concept of a developmental system. Thus my pleasure that Kitcher seems to like parity was somewhat, shall we say, *muted*. (Remember that he presents his democracy as a tenet of the reasonable interactionism that needs to be defended against well-intentioned but fuzzy-headed folks like me.[8])

Confusion is diminished when one realizes that DST's parity is quite different from Kitcher's democracy. In fact, he says that DST cannot adopt his democracy principle, because the principle is based on separating (but not prioritizing) causal factors (personal communication, May 1999). The question is what it means to separate factors. It appears that Kitcher has in mind more or less what behavior geneticists do when they insist on the legitimacy of separating genetic from environmental causes (even though they believe this is incompatible with "interactionism"; recall note 6), and the framework is a population genetic one.

In DST, in any case, systematic even-handedness is not a matter of outlawing particular methods but of detecting and avoiding unprincipled privileging. If this is what you mean by "control" or "inheritance," does it characterize this other case, too, even though you use another term for it? Notice that parity is not a matter of pronouncing all factors to be "equal" or "the same" in the sense that they cannot be distinguished. Because there is an infinite number of ways one can compare things, this would be absurd. Parity is not identity; it is consistency with respect to a criterion. DST applies certain criteria and definitions to a wider range of entities and factors than is usual, to see whether conventional practice has prematurely or arbitrarily restricted its categories. The payoff is a broad reconceptualization of development, heredity, and evolution.

Referring to an exchange with DSTers Griffiths and Gray over the "genes for" locution (Griffiths and Gray 1994, 1997; Sterelny and Kitcher 1988; Sterelny, Smith, and Dickison 1996), Kitcher concedes that his favored strategy, of fixing genotype and varying environment to see if a phenotype is genetically determined, can be reversed, so the environment is fixed while genotypes are varied. Once one accepts the logic, he admits, this technique can reveal an environmental factor "for" a phenotype, just as the reverse operation can be said to show genes "for" phenotypes. The exercise is useful because it explicitly plays out the "gene for" logic, and so demonstrates its limitations. I would be very surprised if his standard interactionists, whoever they are, embraced this little bit of parity—let alone its wider developmental and evolutionary implications, which Kitcher does not discuss. Those phantoms of the allegedly dead horse of the nature/nurture dichotomy haunt the literature, dragging behind them clanking chains of invidious distinctions among classes of causal "equals"—between those that define potential and those that just select outcome, say—distinctions that both require the dichotomy and perpetuate it. Hence my wish for just the kind of far-reaching reworking of perspective that DST is about, and that Kitcher thinks is unnecessary.

Being several weary miles further down the road than I was when I wrote the passage that serves as Kitcher's epigraph, I am now less likely to talk about swift and lethal stakes in the heart. As I suggested, a great deal turns on conceptions of causality. Many scientists avoid looking too closely at such matters, and with good reason: The philosophical literature in this area is baroque. That a sophisticated and in some ways sympathetic reader should misconstrue DST as Kitcher does highlights the importance of attending to the conceptual frame that surrounds nar-rower analytical matters. Having remarked in *Vaulting Ambition* on the vagueness of the "ge-netic bases" that loomed so large in the socio-biology debates, he states that the controversy was really about *evidence* (1985: 7–8), and in the paper under discussion he counsels us to turn from "quibbling about the proper definition of genetic determinism" to investigating norms of reaction. I, on the other hand, doubt that such evidence will do the trick without closer attention to how "genetic bases" and "determination" are used, and what conclusions they are thought to justify. Such attention might well shed light on what he calls the "fragility of our representations of the non-genetic causal factors." (In *Vaulting Ambition*, in fact, he does some useful quibbling of his own.)

I share Kitcher's conviction that entirely too many pronouncements are made about the rela-tive importance of genes and environments in this or that trait (as opposed to careful investi-gation of particulars). Cavalier extrapolation is no substitute for case-by-case research, and I am hardly against such work. I do think that re-search alone will not dispel the problems under discussion. After all, the standard interactionist travesties have flourished in a rich soil of findings on various degrees and kinds of susceptibility to various classes of developmental contingencies. Often they are attempts to accommodate de-velopmental variation within quite retrograde frameworks. Kitcher misses the myriad explana-tory asymmetries maintained in standard interac-tionism, and he does not follow the evolutionary implications of the more globally symmetrical approach implied by the "environmental feature for" analysis. It would be interesting to know whether in his democracy some causes would still end up being more equal than others.

Kitcher began *Vaulting Ambition* with an anec-dote about a cousin who failed a major sorting exam in the English educational system and was given a bicycle as consolation. In saying, "[A] bicycle is not enough" (1985: 11), he was remark-

ing on the political importance of the matters he tackled in that book, including some of the deeply held assumptions about biological nature that dialectical biology and DST challenge. Pre-sumably he was also commenting on the lame-ness of formulaic egalitarian talk in the face of strictly stratified school systems. He says (forth-coming), "[N]o interactionist denies that many causal factors are involved in development (that, after all, is the point of interactionism)." I suggest that this kind of thinly "liberal,"[9] many-factors-are-involved interactionism is not enough, either. If, like many of the standard interactionists I have written about over the years (Oyama 1982, 1985, 2000b), he believes that biology and culture are "transmitted" via different "channels," and that they "interact" in ways that allow their rel-ative contributions to the constitution of or-ganisms to be quantified, then we have serious differences that have not been captured in his cri-tique. If, on other hand, his vision is really of interdependent, symmetrically treated causal in-fluences, any one(s) of which may serve to dis-tinguish organisms in some study but none of which is a mastermind bearing representations of traits or organisms, or is even more responsible for making the organism, and all of which help make up an organism's inheritance, then he is as atypical an "interactionist" as I am.

System

The emphasis on distributed causality, the emer-gence of form and function in the interaction of heterogeneous internal and external causal influences on various scales, the causal inter-dependencies and lack of absolute distinctions between causes and effects—such ideas hardly originate with DST (or dialectical biology), but are associated with more general notions of sys-tems. The constructivist interactionism I have been limning is in fact a quite "systemsy" one, which brings me to my second good word.

This section is organized by several questions, in contrast to the first one, much of which was directed at single paper (albeit one that raised several pertinent questions). Must *system* imply tightly predictable self-regulation, or can it serve broader purposes? Is it consistent with DST's stress on interaction? How does it fit with the idea of construction? Effective exchange within and across disciplines requires attention to such questions. Serious terminological and conceptual mismatches with ecology, for instance, could present difficulties, for treating organisms and their (developmentally and evolutionarily effective) environments together suggests the fundamental significance of ecological concepts (see chapters 9, 10, 19, and 22).

Systems: Regular by Definition?

Ecology's pivotal term, *ecosystem*, has the word in question built right into it. At the beginning of the twentieth century, Frederic Clements treated plant communities as individuals with life cycles. He took experimental physiology as his model: Ecological succession was a community's ontogeny (Hagen 1988). In Joel Hagen's (1992) history of the discipline, the fortunes of the ecosystem concept and the organism metaphor are closely entwined. Organismic development, as usual, was considered autonomous, regular, goal-directed, internally driven. Arthur Tansley, the plant ecologist who originated the term *ecosystem* in the 1930s, complained that Clements took a useful analogy too literally, seeing succession as maturation to a single climax state. Most ecologists later chose ecosystems over organismic analogies, but Hagen (1992: 48) argues that the community-as-cybernetic-system idea was derived from the organism metaphor and incorporated aspects of it.

Organism and machine metaphors coexisted in these early ecologists' work (Hagen 1992). Both Hagen and Peter Taylor (1988), however, report a shift toward more formal cybernetic models after World War II, when many scientists were drawn to information theory. Inspired by wartime engineering successes, they envisioned a world governed by "self-regulating feedback systems"; diagrams of energy flow through ecosystems showed clearly defined circuits, eventually manipulable from the outside (Taylor and Blum 1991: 277). Communities became steady-state devices.

This accords with popular visions of nature's balanced harmony. Ecological anthropologist Andrew Vayda (1996: 9–15) warns against systems approaches in ecology, charging that notions of self-regulation divert attention from the need to establish particular causal linkages. He laments analysts' habit of treating social processes and ecosystems as predefined, with established boundaries, rather than allowing them to "grow through being studied," as their various connections are explored (see also Taylor and García-Barrios 1995).

Vayda is thus skeptical about DST's use of terms like *system* and *process* (personal communication, June 1997). Yet my own misgivings about the internalism of traditional treatments of *organismic* development have fixed on the same sorts of problems. For him, systems thinking is the culprit, whereas for me it has been part of the remedy.

Earlier, with regard to dialectical biology, I observed that terminological differences may partially obscure considerable accord about overall approach, and perhaps this is the case here. But there is more. The privileging of wholes over parts that Vayda and I mistrust is an aspect of the reifying holism that Lewontin, Rose, and Kamin (1984, chap. 10) regard as an inadequate response to reductionism: uniform parts subservient to wholes with minds of their own. Hagen (1992: 98) describes the tendency of particular species and populations to "evaporate" once ecosystems are seen in terms of energy or information flow. When higher levels are privileged over lower ones, a possible consequence is exaggerated confidence that the expected outcomes will appear. Another is inattention to the actual interactions themselves, and to differences among the interactants.

Systems: Emblematic of Unpredictability?

Must *system* be constitutively contaminated with this brand of holism?[10] Environmentalists are as likely to emphasize the fragility of ecosystems as their stability, to the point that Donella Meadows (1988: 16) speaks of the dismayed paralysis that can come from contemplating "Awful Interconnections." Systemic linkages can produce cascades of catastrophes. Some might see such cascades as *intrusions* into an otherwise predictable natural order, but Hagen (1992: 27) reports that even the staunchest believers in predictable succession allowed for destabilizing events. He thinks scholars have missed the heterogeneity and flexibility of early ecologists' ideas, maintaining that a constant tension between stability, integration, and balance on the one hand, and change and unpredictability on the other, runs through Darwin and the entire history of ecology (1992, chap. 4).

Robert Jervis (1997) uses systemic contingencies, nonlinearities, and multiple and indirect effects to show how often, and how startlingly, predictions of social relations can be wrong. He comments on the dangers of ceteris paribus clauses (p. 76), criticizing those who treat systems as entities with wills, capable of subordinating their constituent parts. Change and variation on the domestic level, he asserts, can influence national and international events: The variable characteristics of the parts can be important (pp. 103–104). Commanding a richly varied store of examples, Jervis covers many other significant aspects of complex systems, such as historicity and path dependence (pp. 155–156), as well as research and intervention strategies (pp. 73–74, 282–294). The surprises and uncertainties he describes are not pathological irruptions; they follow from the multileveled complexity of the systems themselves.

Open Systems, Open Questions

The point is not that systems "really are" unpredictable.[11] The alternatives of regularity and un-predictability echo both the tensions Hagen writes about and the nature/nurture debates from which my own investigations grew. Ecological succession, international politics, and organismic life cycles give varying impressions of orderliness, depending partly on the levels at which they are observed and the indicators that are used, so to some extent the diverse emphases of these theorists are easily explained. The point, which rests in part on such variety, is that the concept of system has more possibilities than are typically exploited in any one literature, and should not be unduly circumscribed.

The notion of a developmental system can do useful work *only if it is construed broadly enough.* It cannot prejudge the issue (issues, really) of regularity, but requires, by the stark expedient of blocking the exits, that developmental questions be answered by looking at developmental processes and products, not at what labels they bear. To keep the questions of regulation and variation appropriately open, a quite lean sense of *system* is called for: a collection of interdependencies that can, under certain conditions, be so regular as to appear to be imbued with goal-seeking agency, but that is not defined by such regularity. This minimal sense can capture the varied interactions and mutual influences of such networks without assuming their degree of integration and self-regulation. It is minimal, that is, not because it allows only a few factors or only simple interactions—on the contrary—but because it does not, as a matter of definition, involve specific kinds of relations among constituents or among levels, or specific kinds of outcomes (but see chapter 17). This kind of mobile complexity is not easily visible in the cogwheels and electrical circuits that Taylor and Blum's (1991) postwar ecologists devised. It is more easily seen in the looser sense of system I advocate, though, as well as in Taylor's own work (chapter 22), in which regularity of various kinds is possible but is not bought at the cost of heterogeneity and historicity.

Early systems theorist Ludwig von Bertalanffy (1967: 66–69) at one point describes a system simply as "a complex of components in interaction."

The mechanistically stimulus-response cybernetic system is one subtype; but the attributes of what Bertalanffy calls *general* (or *open*) *systems* are easier to connect with the developmental systems approach. They include "*dynamic interaction* between many variables" rather than the simpler "circular causality" of feedback regulation; an emphasis on exchanges of matter rather than information; and the ability to self-organize, a kind of self-making of which a cybernetic system is incapable, but which is indispensable if we are to talk about developmental construction. It is this last, the possibility of growth and differentiation, that has attracted those studying the ontogenies of actual organisms.

As noted earlier, Lewontin, Rose, and Kamin (1984: 279) believe that early systems theorists were occasionally guilty of wooly holism. This may be the case. I am not attempting to recover an ur-systems theory, but to show that those writings may be *roomier* than is sometimes thought. Developmental biologists have tended to be impressed by regularity, but there is more than one way to explain it. What does it mean for ecological succession to be (like) development? Organismic maturation is the model, but this can be conceived in noninternalist terms (Oyama 1982), and ecological succession has been invoked precisely to *challenge* the notion of programmed development. Yrjö Haila (1999) argues that ecological succession and seasonal changes should be seen as developmental, but confirms (personal communication, December 1998) that this must not be misconstrued. Development itself must be understood in a more open and contingent manner than usual. To put it crudely, while a social theorist or ecologist might have asked whether societies or sand dunes were "like organisms," and meaning by this, were they self-contained, self-guiding, and self-maintaining, certain developmentalists with a systems bent eventually asked whether *organisms* were "like organisms" in that sense.

One of the more precise questions nested in this last one is whether an adequate explanatory complex is to be found within the organism's skin. Tansley's critique was mentioned earlier (Hagen 1992: 78–86). He did not question internalist understandings of ontogeny, only their immoderate application in ecology. Before settling on *ecosystem,* he even considered *quasi-organism.* From the present point of view two things are noteworthy about his account: (1) ecosystems were identified and somewhat artificially isolated by the scientist confronting a nature of overlapping parts and indistinct boundaries (compare with Vayda on predefined systems, cited earlier), and (2) they included abiotic components. Raymond Lindeman, a young colleague of G. Evelyn Hutchinson, developed this idea with his mentor a decade later. How, Lindeman asked, could one separate the living from the nonliving, given the fact that the former disintegrated into molecules, whereas abiotic materials were taken up and moved through food chains? Biogeochemistry blurred the boundaries.

Ecosystem modeling relied, however, on single currencies of energy or information. This is finally not as congenial to the present project as the boundary-breaching itself, which sits nicely with the inclusion in a developmental system of not only abiotic resources but also other organisms. As Johnston (chapter 2) shows, certain psychologists and biologists saw that an account of species-typical development must be extended beyond the confines of the body. Bertalanffy's (1967: 69) mention of the heterogeneity of interacting components, along with his considerably earlier declaration that "determination factors" could be found both inside and outside the egg (1962: 68–70), also suggest historical precedent for a more ample notion of system.[12]

There is irony here. The concept that some find useful in undoing the internalism of traditional notions of development has, for others, its own internalist aura (just as the interactionism discussed in my first section seems externalist to some). It is instructive to consider the contrast to *system* in each case. In ecology it was something rather more "individualistic"—if not an untheorized collection of creatures and features, then at least a focus on particular populations and species (Hagen 1992, chaps. 2, 5). For developmental systems thinkers the contrast is with an opposition between internal control and external input. The kind of system that has served us involves heterogeneous, interdependent causal factors both inside and outside the skin; the possibility of more or less (sometimes *much* more or *much* less) orderly processes without a preformed plan; and the emergence of structure and function from specific causal interactions among very specific conditions—in short, a very "interactiony" sort of system. The collisions and cooperations of varied and distributed causal factors should invite, even demand scrutiny, rather than being easily black-boxed (with the justification that all the "information" is inside the package). The role of the investigator in defining the system is crucial, however; one doesn't glance out the window and see one of these organism-environment complexes, shrink-wrapped and ready to go.

Interactive Construction in Systems

The notion of construction is inseparable from the readings of *interaction* and *system* being presented here. There is a recurrent tendency to associate predictability and constancy mainly with insides, and (certain kinds of) change and variation with outsides. And yet Lehrman (chapter 3), Gottlieb (1997; see also chapter 4), and other developmentalists describe exchanges with the external environment that are crucial to the ontogeny of characters usually considered "innate"—not as acquired additions to a "biological base," but as aspects of the growth of the organism itself. And as Lewontin (chapter 6) and other evolutionists have argued, developmental, behavioral, and other "internal" factors (that is, characteristics of species, and thus in many treatments internal with respect to the evolutionary environment) contribute to the natural selection usually said to shape populations from the outside. In neither case does the mere (standard interactionist?) inclusion of both internal

and external factors capture the interdependence of these phenomena.

Development has conventionally been explained by internalist models and evolution, largely by externalist (especially selectionist) ones. But developmental constancy is no less a product of (systemic) interaction than is variation. In like manner, lability, unpredictability, and variability is no less (interactively) systemic than is constancy. I have argued for a parallel synthesis of *evolutionary* insides and outsides that I take to be in line with the idea of organism-environment interpenetration (Levins and Lewontin 1985). Both developmental and evolutionary construction are historical comings into being, by concrete events and activity on a variety of scales: changes in the dimensions, material constituents and modes of functioning of an organism and its worlds. The emphasis is on temporality and physicality, partly to counteract the disembodied, formal quality of the language of programs, algorithms and gene pools (Hendriks-Jansen 1996; Varela, Thompson, and Rosch 1991).

It should go without saying that there is no constructor, but I have learned I must say it: The gene does not build organisms in some special centrally controlled way that other interactants do not. But then, the organism does not make (most of) its own environment, either, though it does select and alter its surroundings. Nor does "the environment" make organisms or adaptations over ontogenetic or phylogenetic time. The conjoining of construction to interaction and systems is meant to work against this persistent desire to meet the maker.

Also to be headed off is any reading of *construction* deriving from the very dichotomies it denies. This is not *construction* as opposed to *maturation*. It is not *socially constructed* as opposed to *biological* or *real* or *natural* or *essential* or *universal*. Rather, it is a reminder that every aspect of an organism, rare or boringly predictable, must develop, from some always incompletely characterized and changing (over both developmental and evolutionary time) complex of interactants.

It is said in DST that the developmentally relevant environment depends on the organism. This means that whether, and how, any aspect of the surround is involved in producing an organism is a function of that organism's characteristics and its activity. But the operative features of the surroundings are not just "preregistered," as Horst Hendriks-Jansen (1996) puts it in a discussion of situated robotics, *there* in some straightforward way, but are identified by reference to the organism and may emerge only through its activity. He describes a device that follows walls without being instructed specifically to do so. It navigates by "landmarks" that are not actual features of the surroundings but are based on correlations between what their sensors pick up and their own movements (p. 135). Likening the landmarks to Gibsonian invariants, which are produced as an animal moves, Hendriks-Jansen points out that they exist neither in the head nor in the surroundings. They are constructed during an interactional history (p. 9), and "the emergent phenenomena open up possibilities for behavior that did not exist prior to their emergence" (p. 30).[13] This is what interactive construction looks like in a living system, too. It is this kind of mutual definition of influences that lies behind DST's and dialectical biology's talk of the ultimate impossibility of separating causes, and it is not captured by Kitcher's analytic "democracy."

Terms in Tension

One could say that in DST *interaction* and *system* are linked in a kind of dynamic tension, a tension that is at once a source of their utility and their trickiness. I am, with some seriousness, adopting a constructivist-interactionist-systems attitude toward these terms, for words, too, are always in a context of language and practice, each with a history. Any of my good words alone threatens to slip toward precisely the implications one wants to keep at bay (recall Hagen's account of the tension between unpredictability and stability). *Interaction,* at least the reciprocally constraining, mutually constructive interactionism of a developmental system, can keep a complex from melting into an undifferentiated whole or closing in on itself and becoming an inscrutable black box. It is a reminder of concrete events, of organisms in real worlds, cells in real tissues, groups in real societies. *System,* at least the open, loosely defined (or better, always to-be-specified) systems I have described, can help us see that the interactions are connected, though not always in neat spatial bundles. Hence my interactiony systems, my systemsy interactions. Together with *construction,* they offer a different way of dealing with developmental and evolutionary stability and change.

It is not a bad thing that these words remain richly complicated and variously read. They can no more be made clear *once and for all* than a developmental system can be pinned down to a fixed location or frozen in time, and its elements enumerated *once and for all.* We can, and inevitably will, keep talking about them, but more to the point, *using* them. What can be accomplished with distinctions, comparisons, and lists of disclaimers is in the end limited, though I have done my best. In the long run what will be persuasive, or not, is the systematic deployment of this vocabulary, so familiar in some ways (composed as it is of rich, powerful, messily good words) and so subversively odd in others, to pose questions, analyze, interpret, distinguish, and tie together. A sampling of such efforts is collected in this volume. Their appeal will depend on their ability to produce that combination of novelty, coherence, and utility that recommends a conceptual framework to the interested scholar.

Acknowledgments

I thank Yrjö Haila, Philip Kitcher, Barbara Smith, Peter Taylor, and Rasmus Winther for comments on earlier drafts of this chapter. A skeletal version of this material was presented at the July 1997 meetings of the International Society for History, Philosophy and Social Studies of Biology, in Seattle, Washington. The session was entitled "Conceptualizing Developmental Processes (DST 3)."

Notes

1. Lewontin, Rose, and Kamin (1984: 272) mention Popper, Lorenz, Campbell, and Piaget; their remarks on, and dissatisfaction with, interactionism should be read in light of this. My own sources included the tradition of animal behavior studies discussed by Gottlieb and by Johnston (see chapters 2 and 4), as well as certain trends in developmental biology and developmental psychology (see discussions in Oyama 1999, 2000b). I mentioned Piaget in my book because of his emphasis on constructive activity, and because of his willingness to take a developmental approach to certain species-common cognitive features, not just those that vary among individuals. When it comes to his tendency to describe the environment as setting evolutionary and cognitive problems for organisms (and, perhaps, his insistence on the *necessity* of logicomathematical knowledge), I suspect the dialectical biologists and I have similar objections.

2. Because statistical interactions involve nonadditive variance, I find the phrase "nonadditive interactions" unhelpful, even when used to make very much the sorts of points I make here (as in Gray, forthcoming, and Sterelny and Griffiths 1999). Pat Bateson, in a number of personal communications over the years, has objected to *interaction* itself on similar grounds: that it implies the statistical meaning. I would have been inclined to adopt his alternative, *interplay,* if only it lent itself to adjectival and adverbial constructions: I would be hard put to talk about an *interplayive* relationship!

3. The spelling out of the connection between causal priority and such representations will require some attention to quite basic conceptions of causality, with special attention to the relations between individual and population analyses. It may well be, as Godfrey-Smith suggests (chapter 20), that Kitcher's particular views on causality explain some of these striking incongruencies.

4. The reason this is suspect is related to the point made earlier: Many aspects of the developmental environment are also shared, even as others vary. This

is easy to forget because of the habit, encouraged by nature/nurture oppositions, of treating "the environment" as unitary and thus capable of being manipulated or fixed as a whole.

One might argue that, given the selectivity of factors and levels in an ANOVA, Kitcher would reserve the language of determination for outcomes in all possible (or "almost all," etc.) environments. But then it could virtually never be used, and we would be left with the question of how to treat results of analyses that could actually be done, as opposed to ones that could not. Practically feasible ones produce the partitionings of variance that are typically taken to show degrees of genetic and environmental determination, but different studies will produce different partitionings, and I know of no way of placing them on the overarching continuum Kitcher seems to have in mind, precisely because this continuum presupposes knowledge of all the pertinent environments.

5. The causal issues are daunting, so it is difficult to bring diverse treatments into proper confrontation. Behavior geneticists frequently insist on a strict language of populations and differences, not individuals and similarities. At other times, however, they are quite content to talk of inherent gene-based natures or of similarities among individuals being due to shared genes (Plomin 1986: 7–9; Rowe 1994: 3; see Oyama 1988 for comments).

Sober (1988: 317–318) argues for more distinctions: between something's being a cause and its making a difference, and between its "relative contribution" and the difference it makes. The genes, he says, can have "zero magnitude" in the case of a person who would have had the same height if she had had different genes, so that "genes can be a cause of height, even if they are judged to have zero magnitude." This is a kind of converse of the flat norm of reaction, which for Kitcher shows genetic determination (100% causal magnitude?). Perhaps some such set of distinctions informs Kitcher's presentation.

6. It is amusing to juxtapose his allegedly interactionist goal of separating the causal roles of genes and environments with behavior geneticist Robert Plomin's testy reference to the "mistaken interactionist notion that the separate effects of heredity and environment cannot be analyzed" (1986: 7).

7. Kitcher does not do this explicitly, and my impression is that his (1989) more formal treatment of causality does not require it.

8. There is a gap here. What he actually says is that standard interactionists *should be* democratic. His discussion, however, implies quite strongly that they are. In the restricted sense of being willing to attribute some role to both genetic and environmental factors in certain analyses, they sometimes are, but the question is how such roles are interpreted.

9. I use this language in response to Kitcher's appropriation of *democracy*. I do not, however, endorse the political terminology or wish to adopt it more generally. The rhetorical gestures it licenses are likely to sow complication without conceptual gain.

10. Hagen (1992: 137) suggests that holism aided in the formation of a new specialty and stimulated research; such things can help at one time and become a hindrance at another.

11. Oddly enough, Schaffner (1998) associates DST's "developmental emergentism" with unpredictability. See Griffiths and Knight (1998).

12. In this last passage he seems to be resisting a privileging of internal factors over external ones. The term *determination factor* is from Roux, who located these formative influences inside the egg, distinguishing them from mere "factors of realization," which could be external and which supplied matter and energy. These very familiar kinds of arguments are grist for parity analysis.

13. The robot was designed by Maya Mataric. In situated robotics the emphasis is on bottom-up emergence of behavior from a few simple low-level rules, involving actions like advancing and turning. Though these are behavioral examples, developmental interactions are similar. In Bateson's mismatches between children and adoptive parents (chapter 13), the significant variable is not a characteristic of parents, children, or "environment," but rather a relationship that emerges as they interact with, and react to, one another. Peter Klopfer (chapter 14) gives other examples.

References

Bateson, P. P. G. (1988). The active role of behaviour in evolution. In M. W. Ho and S. W. Fox (Eds.), *Evolutionary Processes and Metaphors*, pp. 191–207. London: Wiley.

Bertalanffy, L. von. (1962). *Modern Theories of Development: An Introduction to Theoretical Biology*. New York: Harper & Row.

Bertalanffy, L. von. (1967). *Robots, Men and Minds: Psychology in the Modern World*. New York: George Braziller.

Buss, D. M., M. G. Haselton, T. K. Shackelford, A. L. Bleske, and J. C. Wakefield. (1999). Interactionism, flexibility, and inferences about the past. *American Psychologist* 54: 443–445.

Caporael, L. R. (1997). The evolution of truly social cognition. *Personality and Social Psychology Review*, 1: 276–298.

Godfrey-Smith, P. (1996). *Complexity and the Function of Mind in Nature*. Cambridge: Cambridge University Press.

Gottlieb, G. (1997). *Synthesizing Nature-Nurture: Prenatal Roots of Instinctive Behavior*. Mahwah, NJ: Lawrence Erlbaum Associates.

Gray, R. D. (1988). Metaphors and methods: Behavioural ecology, panbiogeography and the evolving synthesis. In M. W. Ho and S. W. Fox (Eds.), *Evolutionary Processes and Metaphors*, pp. 209–242. London: Wiley.

Gray, R. D. (1992). Death of the gene: Developmental systems strike back. In P. E. Griffiths (Ed.), *Trees of Life: Essays in Philosophy of Biology*, pp. 165–209. Dordrecht: Kluwer Academic.

Gray, R. D. (forthcoming). Selfish genes or developmental systems? Evolution without replicators and vehicles. In R. Singh, C. Krimbas, J. Beatty, and D. Paul (Eds.), *Thinking about Evolution: Historical, Philosophical, and Political Perspectives*. Cambridge: Cambridge University Press.

Griffiths, P. E. (1999). The fearless vampire conservator. In C. Rehmann-Sutter and E. M. Neumann-Held (Eds.), *Genes in Development. Rereading the Molecular Paradigm*. Manuscript submitted for publication.

Griffiths, P. E., and R. D. Gray. (1994). Developmental systems and evolutionary explanation. *Journal of Philosophy*, 91: 277–304.

Griffiths, P. E., and R. D. Gray. (1997). Replicator II: Judgement day. *Biology and Philosophy* 12(4): 471–492.

Griffiths, P. E., and R. D. Knight. (1998). What is the developmentalist challenge? *Philosophy of Science* 65: 253–258.

Hagen, J. B. (1988). Organism and environment: Frederic Clements's vision of a unified physiological ecology. In R. Rainger, K. R. Benson, and J. Maienschein (Eds.), *The American Development of*

Biology, pp. 257–280. Philadelphia: University of Pennsylvania Press.

Hagen, J. B. (1992). *An Entangled Bank: The Origins of Ecosystem Ecology.* New Brunswick, NJ: Rutgers University Press.

Haila, Y. (1999). "Biodiversity" and the nature/culture divide: Conflicting tendencies. *Biodiversity and Conservation* 8: 165–181.

Hendriks-Jansen, H. (1996). *Catching Ourselves in the Act.* Cambridge, MA: MIT Press.

Jervis, R. (1997). *System Effects: Complexity in Political and Social Life.* Princeton, NJ: Princeton University Press.

Johnston, T. D. (1987). The persistence of dichotomies in the study of behavioral development. *Developmental Review* 7: 149–182.

Kitcher, P. (1985). *Vaulting Ambition: Sociobiology and the Quest for Human Nature.* Cambridge, MA: MIT Press.

Kitcher, P. (1989). Explanatory unification and the causal structure of the world. *Minnesota Studies in the Philosophy of Science* 13: 410–505.

Kitcher, P. (forthcoming). Battling the undead: How (and how not) to resist genetic determinism. In R. Singh, C. Krimbas, J. Beatty, and D. Paul (Eds.), *Thinking about Evolution: Historical, Philosophical and Political Perspectives.* Cambridge: Cambridge University Press.

Levins, R., and R. Lewontin. (1985). *The Dialectical Biologist.* Cambridge, MA: Harvard University Press.

Lewontin, R. C., S. Rose, and L. J. Kamin. (1984). *Not in Our Genes.* New York: Pantheon.

Meadows, D. (1988). World interconnectedness also works in our favor. *Annals of Earth* 6(1): 16.

Orwell, G. (1946). *Animal Farm.* New York: Harcourt Brace.

Oyama, S. (1982). A reformulation of the concept of maturation. In P. P. G. Bateson and P. H. Klopfer (Eds.), *Perspectives in Ethology* vol. 5, pp. 101–131. New York: Plenum.

Oyama, S. (1985). *The Ontogeny of Information: Developmental Systems and Evolution.* Cambridge: Cambridge University Press. (2d rev. ed., Durham, NC: Duke University Press, 2000.)

Oyama, S. (1988). Reply to Robert Plomin's review of *The Ontogeny of Information,* and Populations and phenotypes: A review of *Development, Genetics, and Psychology.* [Reciprocal book reviews and author's replies.] *Developmental Psychobiology* 21: 97–100, 101–105.

Oyama, S. (1992). Pensare d'evoluzione. L'integrazione del contesto nell'ontogenesi, nella filogenesi, nella cognizione (Thinking about evolution: Integrating the context in ontogeny, phylogeny and cognition). In M. Ceruti (Ed.), *Evoluzione e Cognizione. L'Epistemologia Genetica di Jean Piaget e le Prospettive del Costruttivismo,* pp. 47–60. Bergamo, Italy: Lubrina Editore.

Oyama, S. (1999). Locating development, locating developmental systems. In E. K. Scholnick, K. Nelson, S. A. Gelman, and P. H. Miller (Eds.), *Conceptual Development: Piaget's Legacy,* pp. 185–208. Hillsdale, NJ: Lawrence Erlbaum.

Oyama, S. (2000a). Causal democracy and causal contributions in DST. *Philosophy of Science* 67 (proceedings): 332–347.

Oyama, S. (2000b). *Evolution's Eye: A Systems View of the Biology-Culture Divide.* Durham, NC: Duke University Press.

Plomin, R. (1986). *Development, Genetics, and Psychology.* Hillsdale, NJ: Lawrence Erlbaum Associates.

Rowe, D. C. (1994). *The Limits of Family Influence: Genes, Experience, and Behavior.* New York: Guilford Press.

Schaffner, K. F. (1998). Genes, behavior, and developmental emergentism: One process, indivisible? *Philosophy of Science* 65: 209–252.

Smith, B. H. (1997). *Belief and Resistance: Dynamics of Contemporary Intellectual Controversy.* Cambridge, MA: Harvard University Press.

Sober, E. (1988). Apportioning causal responsibility. *Journal of Philosophy* 85: 303–318.

Sterelny, K., and P. Griffiths. (1999). *Sex and Death: An Introduction to Philosophy of Biology.* Chicago: University of Chicago Press.

Sterelny, K., and P. S. Kitcher. (1988). The return of the gene. *Journal of Philosophy* 85: 339–361.

Sterelny, K., K. C. Smith, and M. Dickison. (1996). The extended replicator. *Biology and Philosophy* 11: 377–403.

Taylor, P. J. (1988). Technocratic optimism, H. T. Odum, and the partial transformation of ecological metaphor after World War II. *Journal of the History of Biology* 21: 213–244.

Taylor, P. J. (1995). Building on construction: An exploration of heterogeneous constructionism, using an analogy from psychology and a sketch from socio-economic modeling. *Perspectives on Science* 3(1): 66–98.

Taylor, P. J., and A. S. Blum. (1991). Ecosystems as circuits: Diagrams and the limits of physical analogies. *Biology & Philosophy* 6: 275–294.

Taylor, P. J., and R. García-Barrios. (1995). The social analysis of ecological change: From systems to intersecting processes. *Social Science Information* 34(1): 5–30.

van der Weele, C. (1999). *Images of Development: Environmental Causes in Ontogeny.* Albany: State University of New York Press.

Varela, F. J., E. Thompson, and E. Rosch. (1991). *The Embodied Mind.* Cambridge, MA: MIT Press.

Vayda, A. P. (1996). *Methods and Explanations in the Study of Human Actions and Their Environmental Effects.* Jakarta: CIFOR/WWF.

Wimsatt, W. C. (1986). Developmental constraints, generative entrenchment, and the innate-acquired distinction. In W. Bechtel (Ed.), *Integrating Scientific Disciplines,* pp. 185–208. Dordrecht: Martinus-Nijhoff.

16 Darwinism and Developmental Systems

Paul E. Griffiths and Russell D. Gray

Open almost any biology textbook and you will find the following definition: Evolution is change in gene frequency.[1] This definition reflects the conventional view of natural selection and the conventional view of heredity. Natural selection occurs because individuals vary, some of these variations are linked to differences in fitness, and some of those variants are heritable (Lewontin 1970). Because variants that are not heritable cannot play a role in natural selection, and because the mechanism of inheritance is presumed to be genetic, evolution is defined as change in gene frequencies.[2]

Developmental systems theory (DST) is a radical challenge to dichotomous accounts of development—accounts centered on a dichotomy between genes on the one hand and every other causal factor on the other. Proponents of DST argue that the empirical differences between the role of DNA and that of other developmental resources do not justify the metaphysical distinctions currently built upon them. In particular, any sense in which genes code for phenotypic traits or program development or contain developmental information can be equally well applied to other factors required for development (Gray 1992, 2001; Griffiths and Gray 1994a; Johnston 1987). In this paper we explore the implications of this "parity thesis" (Griffiths and Knight 1998) for the orthodox picture of evolution by natural selection. It is generally accepted that the neo-Darwinian synthesis marginalized developmental biology (Depew and Weber 1995; Gilbert, Opitz, and Raff 1996). Attempts to reintroduce developmental considerations have often been framed as attacks on (neo-) Darwinism (Goodwin 1984). We think this is a mistake. Here we explore how the traditional Darwinian concepts of inheritance, selection, adaptation, and lineage can be reworked from a developmental systems perspective. Rather than diminish the power of natural selection, in the spirit of Darwin's original insight this approach expands the range of phenomena that can be given adaptive-historical explanations.

Inheritance

What does an organism inherit? Certainly more than the nuclear DNA. A viable egg cell must contain a variety of membranes, both for its own viability as a cell and to act as templates for the assembly of proteins synthesized from the DNA into new membrane. A eukaryote cell must contain a number of organelles, such as mitochondria, with their own distinctive DNA. But the full variety of the contents of the cell is only now being uncovered. For normal gene transcription to occur, DNA must be accompanied by the elements of the chromatin marking system. For normal differentiation of the embryo, initial cytoplasmic chemical gradients must be set up within the cell. The essential role of still further parts of the package, such as microtubule organizing centers, is becoming apparent. But unpacking the inherited resources in the cell is not the end of unpacking inheritance. In multicellular organisms the parental generation typically contributes extracellular resources. An ant in a brood cell is exposed to a variety of chemical influences that lead it to develop as a worker, a queen or a soldier. A termite inherits a population of gut endosymbionts by coprophagy. In viviparous organisms the environment of the womb provides not only nutrition but also a range of stimulation essential for the normal development of the nervous system (for examples, see chapter 4 this volume). This stimulation continues after birth. The effects of severe deprivation of conspecific stimulation in infants has been documented in many tragic "experiments" (Harlow and Harlow 1962; Money 1992). Nor are these effects confined to animals. Many eucalypt species have seeds that cannot germinate until they have been scorched by a bushfire. To increase the frequency of bushfires to the point where this system works reliably, local populations of eucalyptus trees must create forests scattered with resinous litter and hung with bark ribbons. These are carried aloft by the updraft as blazing torches and spread the fire to new areas. Even after the resources created by the population as a whole are added in, a range of other factors must be present before the sum of the available resources adds up to a viable package. Development frequently requires gravity or sunlight or, for a hermit crab, a supply of discarded shells from other species. These factors are unaffected by the activities of past generations of the species that rely on them. Nevertheless, the organism must position itself so that these factors interact with it and play their usual role in development. While the evolving lineage cannot make these resources, it can still make them part of its developmental system.

It is uncontroversial to describe all these resources as playing a role in development. But it is highly controversial to say that these same resources are "inherited." With the exception of genes, and more recently the chromatin marking system, their roles are not supposed to extend to the intergenerational processes of evolution. Nongenetic factors, it is generally supposed, do not have the capacity for replication through many generations, and lack the potential to produce the kind of variation upon which natural selection can act: "The special status of genetic factors is deserved for one reason only: genetic factors replicate themselves, blemishes and all, but non-genetic factors do not" (Dawkins 1982: 99). Or, more bluntly: "Differences due to nature are likely to be inherited whereas those due to nurture are not; evolutionary changes are changes in nature, not nurture" (Maynard Smith 2000).

The continued popularity of this argument is puzzling. Many nongenetic resources are reliably passed on across the generations. Variations in these resources can be passed on, causing changes in the life cycle of the next generation. It is still more puzzling to find many of these very phe-

nomena discussed, and their evolutionary significance recognized, in John Maynard Smith's own work (Maynard Smith and Szathmáry 1995). The concept of inheritance is used to explain the stability of biological form from one generation to the next. In line with this theoretical role, developmental systems theory applies the concept of inheritance to any resource that is reliably present in successive generations, and is part of the explanation of why each generation resembles the last. This seems to us a *principled* definition of inheritance. It allows us to assess the evolutionary potential of various forms of inheritance, rather than immediately excluding everything but genes and a few fashionable extras.

One way to conceptualize the role of extragenetic inheritance is as a number of separate ("parallel") channels for the transmission of developmental information. The most traditional multiple channel model has two systems of heredity: genes and culture. In recent years more biologically sophisticated models, with their roots in actual empirical work on inheritance, have emerged (see chapters 9, 10, and 23 of this volume). Multiple channel models are an effective way to draw attention to the phenomena overlooked by a purely genocentric account of heredity. However, we have strong reservations about multiple inheritance or "extended replicator" (Sterelny, Dickison, and Smith 1996) models. We believe that it is both more biologically realistic and, in the long run, more productive to think of the life cycle being reconstructed by a system of resources. Let's start with issues of biological realism. So-called channels are not generally independent of one another. Many "channels" are so strongly intertwined that they cannot affect development unless other channels develop in a way that is automatically "epistatic" with the first channel. The chromatin marking system, often described as a "parallel" inheritance mechanism, modifies the pattern of gene expression. It can be useful to treat DNA sequences and chromatin marks separately in some modeling exercises, but this is an idealization akin to leaving out linkage in a genetic model. It should not be built into the basic way we conceive the system. Furthermore, the developmental system as we conceive it includes not only the "channels" of the other formulations, but also developmental resources which are not easily represented as "channels" or "replicators." It is hard to think of germination in eucalypts as a character transmitted via the bushfire channel, but it would be necessary to think this way to make a multiple channel model complete. The extended replicator theory handles this case by treating it as the replication of selfish bushfires which use eucalypt trees to achieve their goals. Whether or not that is adequate, the extended replicator theory also has to model the standing features of the physical world which form part of most developmental systems. Sunlight, gravity, mineral concentrations in the local soil, and many other factors must be present if "channels" are to convey and "replicators" to replicate. There are a number of ways in which evolving lineages can ensure the inheritance of these factors. These range from highly active methods, such as habitat and host imprinting, to entirely passive methods such as the biogeographic association between a lineage and a region. It is clear that evolving lineages can do better or worse than one another because of fitness differences caused by these developmental factors. As we have argued elsewhere (Griffiths and Gray 1997), the idea that developmental systems can be reduced to a collection of independent replicators is either inadequate or has to recognize relationships to persistent features of the environment as an addition to the cast of replicators. The "selfish standing-biogeographic-association-with-a-low-rainfall-region" is unlikely to appeal.

We would also argue that a developmental systems conceptualization is more heuristically valuable than a multiple channel or multiple replicator model. Holding most of the developmental system constant in order to tease out the roles of a single factor is a valuable technique, and much research in the developmental systems tra-

dition makes use of it (see, for example, chapter 4 of this volume). However, describing this work in the way we have just done keeps the context-dependence of causation in development in the center of the stage. One of the main motivations of developmental systems theory is to draw attention to fact that developmental causes do not have their effects in isolation, but as part of a wider system of causes. Causation in development is thus intrinsically likely to be context-dependent. The very idea of "developmental information" runs some risk of disguising this fact. Susan Oyama (1985) points out that once information is localized in, for example, a sequence of DNA, it is all too easy to forget that the developmental effect of this sequence is a function of context. The same DNA sequence in a different time or place might convey quite different information. "Information" here is being used in the statistical sense, that of correlation between developmental input and developmental outcome.[3] But the associations of the vernacular concept of information are often present when the statistical concept is applied. In the vernacular sense, information is "intentional": it is the meaning conferred on a symbol or a message by its creator. This meaning can be misinterpreted by the recipient of a message, but the meaning of the message is not thereby changed. Hence thinking of developmental causation as the expression of information carries the association that the significance of a cause is independent of the context in which it acts. The idea of dual (or multiple) inheritance systems runs a similar risk of pushing context dependency into the background. Consider, for example, the methylation inheritance system (Jablonka and Szathmáry 1995). The developmental significance of a methylation pattern depends on the gene whose transcription it modifies, and on much else. It is, of course, possible to identify predictive relationships between patterns of methylation and developmental outcomes. However, the idea that these developmental outcomes are transmitted down the methylation inheritance channel obscures the way in which the relationship between methylation pattern and outcome depend on what is happening in numerous "other" channels. Eva Jablonka recognizes this difficulty when she says that the different inheritance systems cannot be treated as autonomous. However, the very idea of separate systems suggests autonomy and it would be desirable to find a formulation that avoids this.

Natural Selection

Armed with a thoroughly epigenetic view of development and an expanded view of inheritance, let us now turn to the concept of natural selection. In principle, there seems no reason why this concept should not be decoupled from gene-centered theories of development and evolution. After all, Darwin developed the theory of natural selection prior to the mechanisms of inheritance being discovered. The three requirements for natural selection (variation, fitness differences, heritability) are agnostic about the details of inheritance. In Daniel Lehrman's classic phrase, "Nature selects for outcomes" (Lehrman 1970: 28). The developmental routes by which fitness differences are produced do not matter as long as they reliably reoccur.

Consider the following two cases: Newcomb et al. (1997) found that a single nucleotide change in blowflies can change the amino acid at an active site of an enzyme (carboxylesterase). This change produced a qualitatively different enzyme (organophosphorous hydrolase), which conferred resistance against certain insecticides. This case fulfills the three requirements for natural selection. There are phenotypic differences in insecticide resistance, these differences are likely to produce differences in fitness, and these differences are heritable. Moran and Baumann (1994) discuss a similar, fascinating example of evolution in action. Certain aphid species reliably pass on their endosymbiotic *Buchnera* bacteria from the maternal symbiont mass to either the eggs or developing embryo. The bacteria enable their aphid hosts to utilize what would otherwise be nutritionally unsuitable host plants. Aphids that have been treated with antibiotics to eliminate the bacteria are stunted in growth, reproductively sterile, and die prematurely. A lineage that inherits bacteria is clearly at an advantage over one that does not. Once again there is variation (lineages with either different *Buchnera* bacteria or without *Buchnera*), these differences confer differences in fitness, and they are heritable. All biologists would recognize the first case as an example of natural selection in action, but they would probably balk at categorizing the aphid/bacteria system in the same way. Yet why should these cases be treated differently when both meet the three criteria for natural selection?

An obvious response would be to claim that if there is selection in this case, then it can be reduced to selection of genetic differences. Aphids with genes for passing on their endosymbionts have evolved by outcompeting aphids with genes for not passing on endosymbionts. However, it is possible to have differential reproduction of the aphid/bacteria system without any genetic difference between the two lineages involved. An aphid lineage that loses its bacteria will produce offspring without bacteria. These offspring remains genetically identical to the lineages with which they compete, but have a lower expected reproductive output. A naturally occurring instance of this sort of selectively relevant non-genetic variation is found in the North American fire ant *Solenopsis invicta* (Keller and Ross 1993). Colonies containing large, monogynous queens and colonies containing small, polygynous queens were shown to have no significant genetic differences. Differences between queens are induced by the type of colony in which they have been raised, as shown by cross-fostering experiments. Exposure of eggs from either type of colony to the pheromonal "culture" of a polygynous colony produces small queens who found polygynous colonies, leading to more small queens, and so forth. Exposure of eggs from either type of colony to the pheremonal "culture" of a monogynous colony produces large queens who found monogynous colonies, leading to more large queens, and so forth. What appears to happen here is that a "mutation" in a nongenetic element of the developmental matrix can induce a new self-replicating variant of the system which may differ in fitness from the original.

The moral that proponents of developmental systems theory draw from the comparison of these cases is that the power of selective explanations need not be limited to genetic changes. The range of phenomena that can be given selective explanation should be expanded to include differences dependent upon chromatin marking systems (chapter 9), prions (Lindquist 1997; Lansbury 1997), dietary cues in maternal milk, cultural traditions and ecological inheritance (Gray 1992; chapter 10). Selection for differences in one of these heritable developmental resources is likely to have consequences for other aspects of the developmental system. Whitehead (1998) has recently argued that cultural selection has led to genetic changes in this way. He observed that in species of whales with matrilineal social systems mitochondrial DNA diversity is ten times lower than in those with nonmatrilineal social systems. He suggested that differences in maternally transmitted cultural traits, such as vocalizations and feeding methods, have conferred a sufficient advantage to lead to the spread of some maternal lineages, and thus their mtDNA. The mtDNA that exists today remains because it hitchhiked along with the cultural traits that were selected for.

At this point orthodox gene-centered biologists might concede that natural selection can indeed be generalized to cover cases of expanded inheritance. Having made this concession they might then attempt to minimize its significance. Genic selectionists, for example, might be tempted to reduce cases of expanded inheritance to a dual inheritance model—genes and their cultural equivalent (memes). But many of the examples of extragenetic inheritance discussed in this book do not fit this dichotomy (e.g., chromatin marking systems, chemical traces from the maternal diet passed on via fetal olfactory conditioning or in maternal milk, the inheritance of gut symbionts, and the inheritance of fire ant colony type). A division of these cases into those that are "sort of genetic" and those that are "sort of cultural" will be largely arbitrary. A somewhat more insightful response to the challenge of expanded inheritance has been outlined by Kim Sterelny (chapter 23). Following Richard Dawkins's well-worn track up Mt. Improbable (Dawkins 1996), Sterelny argues that only cumulative selection can produce complex adaptive structures. Sterelny then outlines some of the requirements an inheritance system would need to make cumulative selection possible (e.g., a large range of possible phenotypes, longevity, high fidelity replication, vertical transmission, and developmental modularity). According to Sterelny, extended forms of inheritance like cultural traditions and ecological inheritance are unlikely to satisfy these requirements; they are inheritance systems, but not highly evolvable inheritance systems. In contrast, however, he argues that symbiont transmission might score quite highly on his criteria for evolvability.

While there is much that we agree with in Sterelny's analysis, there are two important conceptual differences between our views. First, Sterelny adopts a particularly Dawkinsian view of what is important in evolution (i.e., cumulative selection leading to adaptation). While we do not deny that cumulative selection is an important part of evolution, there is a lot more to be explained than just this. Our Darwinian mission is, after all, to explain the diversity of life—the myriad fascinating changes in shape, size, physiology, behavior, and ecology. Extended forms of inheritance can play important roles in evolution without providing the heritable basis for cumulative selection. For example, Maynard Smith and his collaborator Eörs Szathmáry (1995) argue that epigenetic inheritance has played a crucial role in some of the major evolutionary transitions. Sterelny is sympathetic to this view. He notes that symbiosis might be of considerable significance in the generation of evolutionary novelty. After all, the eukaryotic cell is probably an example of frozen symbiosis. Not only might expanded forms of inheritance play an important role in the generation of evolutionary novelty they could also significantly alter the dynamics of evolutionary change. Pal and Miklos (1999) recently modeled the impact of epigenetic inheritance, such as chromatin marking, on the evolutionary trajectory of a population through an adaptive landscape. Their results suggest that this expanded inheritance can facilitate transitions from suboptimal to higher peaks, thus creating more effective evolutionary dynamics than would be possible under strict genes-only conceptions. Expanded forms of inheritance may also be the cause of reproductive isolation and hence of speciation. Parasitic *Wolbachia* bacteria infect up to 20 percent of insect species, and it has been suggested that this cytoplasmically inherited microorganism may be a major cause of speciation in insects. In many species, individuals infected with one strain of the bacteria cannot successfully fertilize individuals infected with another strain. Some clearly separate species also become interfertile when "cured" of their *Wolbachia* infections (Vines 1999). In all these ways, epigenetic inheritance can be a major factor in evolution.

Our second point of departure from Sterelny's position is that we do not divide expanded inheritance into separate inheritance systems or multiple replicators. As we argued above, these factors are physically and functionally linked. The effects of differences in these factors are likely to be context sensitive, and should be seen as part of a system of causes rather than as separate information channels or replicators. Putting these conceptual differences aside, there is still a great deal that can be gained by examining Sterelny's criteria for an evolvable inheritance system (i.e., the range of

possible phenotypes, longevity, fidelity, vertical transmission, and developmental modularity). We will now address each of these criteria in turn.

Natural Selection and the Range of Possible Phenotypes

Maynard Smith and Szathmáry have introduced a distinction between limited and unlimited systems of heredity (Maynard Smith and Szathmáry 1995; Szathmáry and Maynard Smith 1997). They argue that most nongenetic inheritance systems can only mutate between a limited number of states. In contrast, they note that the genome and language both have recursive, hierarchical structures, and hence an indefinite number of possible heritable states. This unlimited range of combinatorial possibilities enables microevolutionary change and cumulative selection to take place. These points are all perfectly legitimate, but from a developmental systems perspective the significance of unlimited inheritance should not be oversold for three reasons.

First, the unlimited nature of an inheritance system is a property of the developmental system as a whole, not only of the resource in which we find the recursive structure. The vast coding potential of genes, language and perhaps pheromones is created by the way in which combinations of these factors "mean something" to the rest of the developmental system. Asking if a system is limited or unlimited holds the current developmental system fixed, and asks what can be achieved by ringing the changes on one of the existing developmental resources. But the lesson of the evolutionary transitions—the introduction of whole new levels of biological order, such as multicellularity—is that evolution can change developmental systems so as to massively expand the possible significance of existing developmental resources. A base-pair substitution in a multicellular organism has potentials that it lacked in a unicellular ancestor. If the substitution occurs in a regulatory gene it could mean a new body plan. The role of systems of "limited heredity" in these evolutionary transitions is considerable, as Maynard Smith and Szathmáry make clear.

Second, from a selectionist viewpoint the combinatorial richness of an inheritance system must be measured in terms of the number of different phenotypic effects, not just the number of combinations of components (chapter 23). If the rest of the developmental system were such that the indefinitely many base-pair combinations of DNA collapsed into only a few developmental outcomes, then for all its combinatorial structure DNA would not be an unlimited heredity system. It not hard to imagine cellular machinery with this result because the existing genetic code is substantially redundant in just this way. Several codons produce the same amino acid.

A third and final reason not to place too much emphasis on the limited/unlimited distinction is that it treats genetic and extragenetic inheritance separately. From a developmental systems perspective these sources of heritable variation should viewed as acting together. Adding one form of inheritance to another causes a multiplication of evolutionary possibilities, not just an addition to them. The greater the range of possibilities the more scope there is for cumulative selection and microevolution.

Natural Selection, Longevity, and Fidelity

Another strategy that could be used to minimize the evolutionary significance of expanded forms of inheritance is to argue that they lack sufficient longevity and high fidelity replication to be the targets of cumulative selection. Sterelny (chapter 23) claims that this is likely to be the case with cultural inheritance. Following Tomasello, Kruger, and Ratner (1993) he argues that high fidelity cultural inheritance requires genuine imitative learning, and this is rare outside humans. Other forms of social learning, like local enhancement and emulation,[4] are unlikely to result in faithful reproduction of the same motor patterns. Thus, in the words of Boyd and Richerson (1996), culture is common but cultural evolution is rare.

Although this is certainly a possibility, evolutionary biologists know remarkably little about the longevity of cultural traditions in animals and the degree to which they have involved cumulative changes. Research on these questions, while exciting, is in its infancy. For example, Whiten et al. (1999) have documented substantial cultural variation in tool use, grooming, and courtship behaviors between populations of wild chimpanzees (*Pan troglodytes*). However, there is little evidence that these variations have persisted for long periods of time, nor that they are a product of cumulative cultural change. It is possible that sensitive periods and social and ecological scaffolding may facilitate reliable cultural inheritance. One example is the reliability of human linguistic inheritance. If the controversial claim that human language relationships are congruent with our evolutionary history is correct, then linguistic inheritance has tracked genetic divergence for as long as 200,000 years (Cavalli-Sforza et al. 1988; Penny, Watson, and Steel 1993). In nonhumans we know much less. However, a study on chaffinches on Atlantic islands by Lynch and Baker (1986) found substantial congruence between a tree based on morphology and a tree constructed from song syllables, indicating a common evolutionary history going back one to two million years. Whitehead's (1998) study on the possible effect of cultural selection on mtDNA diversity in whales suggests that cultural inheritance has exhibited considerable longevity and fidelity in the species with matrilineal social systems. The results of his computer simulations indicate that, assuming a 10 percent reproductive advantage is culturally transmitted down maternal lineages, it would take more than two hundred generations to produce the observed tenfold reduction in mtDNA diversity.

While the cases both for and against cultural inheritance are limited by a lack of evidence, the same cannot be said for some other forms of extended inheritance. Perhaps the most impressive involves the aphid/bacteria symbiosis analyzed by Moran and Baumann (1994) and discussed earlier. Phylogenetic trees of the bacteria and their aphid hosts are perfectly congruent. This suggests that speciation of the aphids has lead to speciation of the bacteria—they have cospeciated. Molecular and fossil evidence suggests that this association is incredibly ancient—between 160 and 280 million years. The *Buchnera* bacteria are thus a vital and remarkably reliably inherited developmental resource.

Vertical Transmission

The lack of strict vertical (parent to offspring) inheritance is another potential problem for the ability of certain forms of extended inheritance to lead to cumulative selection. Again Sterelny (chapter 23) has discussed the problem with considerable insight:

Transmission [of ecological and cultural differences] is not vertical. Indeed, it is not even individual. It is diffuse. Groups of trees engineer their soil structures or a fire-prone understory; individual trees do not make their microenvironments for themselves and their descendants. In most cases groups of animals make warrens, trackways, track-and-bowl systems, beaver lodges, termite mounds and other structures ultimately taken over by the next generation. If this is transmission at all, it is diffuse and development is holistic. (p. 344).

It is likely that even quite small departures from 100 percent vertical transmission could undermine the evolutionary coherence of some forms of extended inheritance with the rest of the developmental system. For example, in Whitehead's (1998) simulation of the effects of cultural selection on mtDNA diversity in whales, if horizontal transmission was much greater than 0.5 percent, then there was little reduction in genetic diversity. One possibility is that ecological and cultural inheritance, while diffuse and horizontal at the individual level, might be vertically inherited in larger units such as families and local

populations. Linguistic inheritance may operate like this (Gray and Jordan 2000). The inheritance of species-specific strains of *Wolbachia* bacteria in some species is also extremely unreliable for individuals while being highly reliable for the species as a whole. Ecological and cultural inheritance could then play an important evolutionary role by providing the basis for higher level trait groups (see Wilson 1997 and the discussion on lineages that follows). In fact, a group selective explanation has been offered for the individually unreliable nature of transmission in *Wolbachia* (Vines 1999: 47). However, as Sterelny notes, the conditions that allow group selection are quite restrictive. An alternative possibility is that although changes in ecological and cultural inheritance might not provide the heritable basis for cumulative selection, they could play important evolutionary roles both in opening up new sets of adaptive possibilities and by facilitating the dynamics of evolutionary change. It should also be emphasized that the problem of diffuse horizontal inheritance does not apply to all forms of expanded inheritance. Sterelny notes that cytoplasmic factors, symbiont systems, and ant nest types are all likely to be inherited in a highly vertical fashion.

Modularity and One Reason Why Selectionists Cannot Ignore Development

At the end of his classic paper on adaptation, Lewontin (1978) notes that adaptive evolution requires quasi-independence. By quasi-independence he means that selection must be able to act on a trait without causing deleterious changes in other aspects of the organism. If all the features of an organism were so closely developmentally integrated that quasi-independent variation did not exist, then "organisms as we know them could not exist because adaptive evolution would have been impossible" (Lewontin 1978: 169). This means that we must add a caveat to Lehrman's slogan "Nature selects for outcomes" (Lehrman 1970: 28). Development does matter.

The reliable reoccurrence of an advantageous variant is not enough. The developmental route which produces the variation must be quasi-independent if it is to be the basis for cumulative selection. A common finding in artificial selection experiments is that although many traits initially respond rapidly to selection, the response will often slow and reach a plateau (Lerner 1970). One interpretation of these results is that developmental links mean that many traits can only be changed to a certain degree without having deleterious effects on other aspects of the phenotype. Further directional change would thus require some kind of developmental reorganization.

There has been considerable recent interest in the extent to which the organization of development really is modular. For example, Halder, Callerts, and Gehring (1995) demonstrated the modularity of eye formation in *Drosophila* by successfully inducing eyes on the antennae, wings, and legs of *Drosophila*. The targeted misexpression of the "eyeless" gene produced structures in these unusual locations that contained a cornea, bristles, and photoreceptors and were responsive to light. Wagner and Altenberg (1996) suggest that directional selection might act on developmental systems to reduce pleiotropic effects between characters with different functions, thereby enhancing the modularity and evolvability of these developmental systems. They speculate that there should be evolutionary trends towards increased modularity. Brandon (1999) goes so far as to suggest that developmental modules are the units of selection. The study of developmental modularity is still in its infancy and the extent of modularity far from resolved. However, from a DST perspective the exact extent of modularity is not a pivotal concern. DST is agnostic on this question (Sterelny in press). Instead, we would emphasize three general implications that arise from the modularity issue. First, analyzing development itself is the key to understanding the ability of selection of act in a cumulative manner. Second, there is no reason to think that extended forms of inheritance such as symbiont transmis-

sion or cultural traditions will be any less modular in their developmental consequences than genetic factors, and thus extended forms of inheritance cannot be brushed off as lacking an essential requirement for cumulative selection. Third, because extended inheritance must be taken seriously rather than brushed aside, the range of phenomena that can and should be given selectionist explanations is considerably increased. This is something that we as Darwinian biologists should be excited about. DST expands the scope of Darwinian explanation, and that is exactly the general conclusion of this section on natural selection. From a DST perspective there is lots more work to be done—there are exciting new questions that we have, at best, only partial answers to. As students of cumulative selection, we really need to know the extent to which extended forms of inheritance fulfill the requirements of longevity, fidelity, and vertical inheritance, and we really need to investigate the modularity and evolvability of developmental systems.

Adaptation and Niche Construction

In three seminal papers Richard Lewontin criticized the metaphors that have traditionally been used to represent the process of adaptation by natural selection (Lewontin 1982; Lewontin 1983a; Lewontin 1983b). The metaphorical conception that Lewontin criticized is the so-called lock and key model of adaptation. Adaptations are solutions (keys) to the problems posed by the environment (locks). Organisms are said to be adapted to their ways of life because they were made to fit those ways of life. In place of the traditional metaphor of adaptation as "fit" Lewontin suggested a metaphor of construction. Organisms and their ecological niches are co-constructing and codefining. Organisms both physically shape their environments and determine which factors in the external environment are relevant to their evolution, thus assembling such factors into what biologists describe as a niche. Organisms are adapted to their ways of life because organisms and their way of life were made for (and by) each other. Lewontin also revised the popular metaphor of a "fitness landscape." In this image, populations occupy a rugged landscape with many fitness peaks and evolve by always trying to walk uphill. But because organisms construct their niches, the landscape is actually much like the surface of a trampoline. As organisms climb the hills they change the shape of the landscape. Lewontin's metaphor of construction is not merely a new way to describe the same evolutionary process. It is the public face of a substantially revised model of the actual process of natural selection, redefining the causal relationships which ecology and evolutionary biology must seek to model.

Lewontin's ideas challenge one of the central elements of contemporary neo-Darwinism, the idea that the source of the selective pressures that explain adaptive evolution can be sought in a relatively independent science of ecology.[5] Darwin himself was well aware of the reciprocal influence of organism and environment, and conducted pioneering studies on the role of earthworms in the formation of soils (Darwin 1881). Despite this, many presentations of Darwinism treat the environment as a source of fixed problems which every organism must solve or die. Environments for which there is no fossil or other direct evidence can be reconstructed by "reverse engineering" the organisms that those environments shaped to fit themselves. Some even suggest that evolutionary research can proceed by first identifying the niches in an environment and then predicting how organisms will evolve to fill them— so-called adaptive thinking.[6] But as Lewontin has pointed out, there are indefinitely many overlapping niches in an unoccupied physical landscape. Until the organisms that occupy the niches are specified, the concept of the niche is completely unhelpful. Of course, there is a sense in which every possible niche that an organism could construct in an area of space and time "exists." This

sense becomes still more tenuous, however, once it is recognized that occupied landscapes owe much of their physical structure to the activities of the organisms that occupy them. In this tenuous sense there were niches for species requiring high rainfall in the Amazon basin before the biota which make it a high-rainfall region had evolved. So a region of space and time contains not only the niches that can be defined using its existing features, but all those that could be defined using the features induced by the action of all the species that could evolve so as to make a niche in that region!

Sterelny and Griffiths have argued that the concept of a "vacant niche" makes sense only in an ecosystem that has already been structured by a collection of organisms which are part of it (Sterelny and Griffiths 1999). It may then be possible to determine that an organism of a specified type not present in that ecosystem could maintain itself were it introduced. The move from this idea to the idea of an unoccupied landscape with a determinate niche structure is an illicit idealization. It idealizes away from precisely those factors that create the possibility of identifying vacant niches. Similarly, the idea that knowing the "shape" of the vacant niches allows us to predict how organism will evolve to fit them ignores the fact that different organisms will construct different niches. Some eucalypt species can establish and sustain "islands" of dry sclerophyll forest in rainforest by facilitating bushfires. Once this process is understood it is possible to identify a vacant niche for these species in many other landscapes, despite the fact that they would not even germinate if simply planted there (Mount 1964). We would argue that the idea of a vacant niche for *E. delegatensis* in all non-Australian rainforests is simply perverse in comparison to the idea of niche construction.

The most detailed attempt to develop the new metaphor of construction is that of F. J. Odling-Smee and his collaborators (for an overview, see chapter 10 of this volume). The current prominence of the term *niche construction* is due to this

Table 16.1
Three pictures of the dynamical equations for evolution

Traditional Neo-Darwinism	Lewontin's Constructionism	Odling-Smee's General Coevolution
$dO/dt = f(O, E)$	$dO/dt = f(O, E)$	$dO_{pop}/dt = f(O_{pop}, E_{pop})$
$dE/dt = g(E)$	$dE/dt = g(O, E)$	$dE_{pop}/dt = g(O_{pop}, E_{pop})$
		$d(O_{pop}, E_{pop})/dt = h(O_{pop}, E)$

E, Environment; O, organism; E_{pop}, organism-referent environment of a population; O_{pop}, population of organisms. These variables are related by functions f, g, h. See text for explanation.

group. The first two columns in table 16.1 give the traditional neo-Darwinian model of adaptation as "fit" and the model of adaptation as construction as these two models are described by Lewontin (1982, 1983a). In the traditional picture, change in organisms over time is a function of the state of the organism and its environment at each the previous instant. The environment acts on the existing state of organisms by selecting from the pool of variation those individuals best fitted to the environment. The environment itself changes over time too, but as the bottom equation shows, these changes are not a function of what organisms are doing at each previous instant. In Lewontin's alternative picture, shown in the center column of table 16.1, organisms and their environments play reciprocal roles in each other's change. Change in the environment over time is a function of the state at each previous instant of both the environment and the organisms evolving in that environment.

The right-hand column of table 16.1 shows Odling-Smee's model of evolution as the co-construction of organism and environment. Odling-Smee's general coevolutionary model differs from Lewontin's in two ways (Odling-Smee 1988). First, Odling-Smee hoped to generate a common framework in which to represent both development and evolution. This explains why the terms E_{pop} and O_{pop} occur in the equations in table 16.1. Evolution is a process in which *populations* and their environments co-construct one another over time. If the terms were E_i and O_i, then in Odling-Smee's notation the equations would describe the co-construction of an individ-

ual organism and its developmental environment as the organism's life cycle unfolds. By introducing these indices, Odling-Smee is making explicit what was already implicit in the explanation of Lewontin's equations given in the previous paragraph—the term O in those equations refers to *populations* of organisms, not to some individual organism. Earlier versions of DST (e.g., Oyama 1985; Gray 1992) and some of Lewontin's writings are sympathetic to Odling-Smee's idea that there is a significant parallelism between the way populations of organisms and their environments reciprocally influence one another and way in which individual organisms and their developmental environments do so. But this is not the place to give this idea the attention it deserves.

The second way in which Odling-Smee's treatment differs from Lewontin's is that he is concerned not to represent the organism-environment system as a closed system, as the equations in the center column would seem to imply. Although the eucalypt-bushfire relationship, for example, is one of mutual construction, the changes in this system over time are externally driven by the progressive drying of the Australian continental climate. Organisms feel the impact of changes in the environment in the traditional sense of that term—their total biotic and abiotic surroundings—but they experience these impacts via the environment as it appears in relation to them, and thus different lineages experience "the same changes" quite differently. Odling-Smee tries to respect this situation by assigning separate roles to the environment of a particular lineage of organisms and what he calls the "universal physical environment." The former, organism-referent description of the environment is the source of evolutionary pressures on that organism, and the organism is the source of niche-constructing forces on that environment. The latter, the universal physical environment, is a source of exogenous change in the organism's environment.

Robert Brandon's theory of the role of the environment in adaptation is a useful complement to Odling-Smee's ideas. Brandon distinguishes three different senses of "environment" (Brandon 1990, 1992). His "external environment" corresponds to Odling-Smee's universal, physical environment. All organisms in a particular region of space and time share an external environment. The "ecological environment" must be described with reference to a particular evolving lineage. It consists of those environmental parameters whose value affects the reproductive output of members of the lineage. Finally, the "selective environment" is that part of the ecological environment which differentially affects the reproductive output of variant forms in the evolving lineage. It is this last which contains the sources of adaptive evolutionary pressures on the lineage. Brandon has used these ideas in the context of his own exploration of organism-environment coevolution (Brandon and Antonovics 1996). Organisms modify the selective and ecological environments in numerous ways. All these can potentially influence their evolution. Only some of these modifications of the selective and ecological environments also constitute modifications of the external environment, but whether they do or not is unimportant when determining their role in the future evolution of the organism-environment system. This point, which was touched on earlier, is one reason why the simple notion of (external) environment is inadequate. Many changes in the external environment do not constitute change from the point of view of the organism and, conversely, an organism can transform its environment without actually changing the universal physical environment (for example, by changing

a habitat association). It is the organism-referent description of the environment—the ecological environment—which captures the aspects of the environment that are relevant to the organism and defines what counts as a "change."

The developmental systems model of evolution (Gray 1992, 1997; Griffiths and Gray 1994a, 1997) can be clarified and improved by the insights of Odling-Smee and his collaborators. In particular, the insight that exogenous factors can affect the availability of developmental resources has not been sufficiently stressed in previous presentations. There remains, however, one major difference between DST and work on niche construction up to and including the present time. Niche construction is still a fundamentally dichotomous account of evolution (and, indeed, of development). There are two systems of heredity—genetic inheritance and environmental inheritance. There are, correspondingly, two causal processes in evolution—natural selection of the organism by the niche and construction of the niche by the organism. The niche-construction model could be modified to take account of recent work on narrow epigenetic inheritance, with a category like "intracellular inheritance" taking the place of genetic inheritance. This, however, would seem merely to substitute one rigid boundary for another. A central theme of the DST research tradition has been that distinctions between classes of developmental resource should be fluid and justified by particular research interests, rather than built into the basic framework of biological thought. Fundamentally, the unit of both development and evolution is the developmental system, the entire matrix of interactants involved in a life cycle. The developmental system is not two things, but one, albeit one that it can be divided up in many ways for different theoretical purposes. Hence we would interpret niche-construction models "tactically," as a method for rendering tractable some aspects of evolution. We would not interpret them "strategically" as a fundamental representation of the nature of the evolutionary process. This response is closely related to the comments about "multiple channel" models of inheritance given above.

The DST model of evolution can be represented in such a way as to make it directly comparable with the models in table 16.1. We can aptly represent the developmental system with the symbol Œ. We retain Odling-Smee's insight that evolutionary change in organism-environment systems is often exogenously driven by using E to represent the universal physical (external) environment. We end up with the equation:

$$dŒ_{pop}/dt = f\,(Œ_{pop},\ E)$$

Evolution is change in the nature of populations of developmental systems. This change is driven both endogenously, by the modification by each generation of developmental systems of the resources inherited by future generations, and exogenously, by modifications of these resources by factors outside the developmental system.[7]

Fitness and Adaptation

This representation of developmental systems evolution allows us to answer a persistent objection to DST. Since we claim that there is no distinction between organism and environment, where do evolutionary pressures on the developmental system come from?[8] What causes adaptation? To give a clear answer we must go back to the definition of the developmental system given in Griffiths and Gray (1994a). The developmental system of an individual organism contains all the unique events that are responsible for individual differences, deformities, and so forth. Just as a traditional model of evolution abstracts away from the unique features of individual phenotypes, developmental systems theory must abstract away from these features in order to tackle evolutionary questions. In evolutionary terms the developmental system contains all those features which reliably recur in each generation and which help to reconstruct the normal life cycle of the evolving lineage. Of course, many species have more than one normal life cycle, either because

there are different types of organism in a single evolving population, each reproducing its own differences (polymorphism) or because there are variations in the developmental matrix from one generation to the next (facultative development). For example, there are tall and short human families and heights also vary from one generation to another due to nutrition. These features are handled in the same way as in characterizations of "the" phenotype of an evolving lineage (Griffiths and Gray 1997). The resultant description of the idealized developmental system of a particular lineage at some stage in its evolution is highly self-contained. Because the focus is on how the complete life cycle is achieved, everything needed for that life cycle is assumed to be present. So everything that impinges on the process is an element of the system itself. It is this that creates the impression that all change in the system must be endogenously driven and creates the apparent puzzle about the source of selection pressures.

The puzzle is only apparent because to think about evolution we need to switch from describing the developmental system characteristic of an evolving lineage at a time to describing an evolving population of individual developmental systems. We need to look at the causes of variation, as well as how the characteristics of the lineage are reliably reconstructed. Hence we need to look at the causes of idiosyncratic development in particular individuals. These causes lie "outside" the description we have constructed of the typical developmental system of the lineage. A population of individual developmental systems will exhibit variation and differential reproduction for a number of reasons. Parental life cycles may fail to generate the full system of resources required to reconstruct the life cycle. Resources generated by the activities of an entire population (such as bushfires in eucalypt forest) may also be scarce, or patchily distributed, so that some individuals lack an important element of their developmental system. Finally, persistent resources—those developmental factors whose abundance is independent of the activities of the lineage—may be scarce or patchy and so some individuals may be unable to reestablish the relationship to these resources that is part of their life cycle. The external environment (E) can impinge on developmental systems by any of these routes. But this does not mean that we can go back to thinking of evolution as a response to the demands of the external environment. The effect of changes in the external environment on the evolution of a lineage can be understood only when those changes are described in terms of how they change the organism-referent environment (E_{pop}). "Changes" in parameters of the external environment that are developmentally equivalent are not changes from the point of view of the evolving system. People in different regions of Britain experience substantially different quantities of dissolved limestone in their drinking water, but this is generally of no ecological significance. Conversely, apparently trivial changes may seem momentous when described in terms of a particular developmental system. Far smaller changes in the concentration of lead from one region to another would have momentous consequences. This is, of course, the point already made by Lewontin, Odling-Smee's, and Brandon's work and by the concepts of ecological environment and organism-referent description of the environment.

So far we have concentrated on how failures of development can lead to evolutionarily significant variation. But positive innovations are possible as well. An individual difference in the system of developmental resources may allow some individuals to cope better when both are deprived of some developmental resource because of exogenous change. Alternatively, an individual difference may simply alter the life cycle in such a way that it gives rise to a greater number of descendants. The source of novelty can be a mutation in any of the developmental resources—parentally generated, population generated, or independently persistent. To make this discussion more concrete, imagine a typical population of hermit crabs. A key component of the developmental system in this lineage is a succession of discarded shells of other species. A dearth of shells would be an exogenous cause of selective pressure on the lineage. Variants with a beneficial set of behaviors or a beneficial habitat association that allowed them to continue to reliably reestablish their relationships to shells would be favored by selection. Shells will typically be an independently persistent resource, and the case in which an independently persistent developmental resource acts as a limiting resource has obvious resonance with traditional ideas of selection of the organism by an independent environment. But, to fictionalize the example slightly, suppose the crab life cycle includes disturbing the sand in such a way as to expose a greater supply of discarded shells. That would make shells a population-generated resource, but they might still act as a limiting resource. Or suppose a lineage evolves behaviors that allow crabs to bequeath shells to their offspring when they themselves seek a larger home. Shells would then be parentally generated, but exogenous change in the availability of shells might still leave some offspring without them, just as a shortage of a trace element in the parental diet may lead to a birth defect in a viviparous species.

One idea that really is missing from this picture is the notion that the external (universal physical) environment poses definite problems that lineages must seek to solve. Instead, the lineage helps to define what the problems are. A dearth of shells is a feature of the ecological environment of a hermit crab and a problem for the hermit crab, but it is completely invisible to a blue-swimmer crab. The number of discarded shells per square meter is a feature of the external environment of both species, but it is only a feature of the ecological environment of one of them. So it is true that the developmental systems treatment of evolution does not incorporate Darwin's original, intuitive idea of fitness as a measure of the match between an organism and an independent environment (e.g., Darwin 1859/1964: 472). But this is a feature which the developmental systems treatment shares with conventional neo-Darwinism. Adaptation is no longer defined intuitively, as the sort of organism/environment relationship that a natural theologian would see as a sign of God's beneficent plan. Darwin set out to explain the fact that the biological world is full of adaptations in this sense, but as so often happens in science, the phenomenon to be explained got redefined in the process of explaining it. In modern usage, an adaptation is whatever results from natural selection, even when what results is intuitively perverse and inefficient. Gould and Lewontin once described a mutation in a bird which doubles clutch size when the population is at the limits of the carrying capacity of the environment. The mutation sweeps to fixation with a consequent doubling of chick mortality. They described this as a case of selection operating without producing adaptation (Gould and Lewontin 1978). They were using "adaptation" in its original sense and consequently swimming against the neo-Darwinian tide. To most of their critics it seemed obvious that this trait was a paradigmatic adaptation. The first example of this shift in the concept of adaptation is, of course, Darwin's own idea of sexual selection. In at least some of his moods Darwin still saw this as a separate force that could act in opposition to natural selection. That suggests that he was limiting the idea of natural selection to processes that produce adaptations in the sense of William Paley's "contrivances"—features that suggest the world was designed by a beneficent creator. But it has been a long time since sexual selection seemed anything other than one kind of natural selection.

We hope it is now clear how DST can explain adaptation, in the modern sense of that term. Change over time in the developmental system of a lineage is driven by the differing capacity of variant developmental systems to reconstruct themselves, or, in a word, differential fitness. What is fitness? In contemporary evolutionary theory fitness is a measure of the capacity of a unit of evolution to reproduce itself (Mills and Beatty 1979). Fitness differences are caused by physical and behavioral differences between the individuals in the population. So fitness can be translated on a case-by-case basis into a detailed causal explanation of evolutionary success. Fitness in general, however, does not correspond to any single physical property (Rosenberg 1978). The only general account of fitness describes its role as a parameter in population dynamic equations. It is clear that this orthodox account of fitness applies equally well to the developmental systems theory. There is no puzzle about how developmental systems that incorporate the whole range of resources that reconstruct the life cycle could come to vary in their success in reconstructing themselves and be selected on that basis.

Individuals, Lineages, and the Units of Evolution

A coherent theory of evolution requires an accurate conception of its fundamental units. According to DST an evolutionary individual is one cycle of an complete developmental process—a life cycle. We have shown that natural selection can act on populations of developmental systems and give rise to adaptation, but in doing so we have assumed that developmental systems are the sort of things that can be counted, that they have clear boundaries and that they do not overlap so much that they cannot be distinguished from one another. We now turn to justifying this assumption. Developmental systems include much that is outside the skin of the traditional phenotype. This raises the question of where one developmental system and one life cycle ends and the next begins There is an enormous amount of cyclical structure in most biological lineages. As well as the life cycles associated with traditional physiological individuals there are "repeated assemblies" (Caporael 1995) within a single individual, such as cells or morphological parts like the leaves of a tree. There are also repeated assemblies of whole individual organisms, such as the characteristic mother/child dyad or the ephemeral dyad formed of a buyer and seller in a market. It has recently been suggested that repeated assemblies of human individuals like these can themselves be units of evolution (Wilson and

Sober 1994). In previous publications we have tried to identify what makes a repeated assembly a developmental system in its own right, as opposed to a part of such a system or an aggregate of several different systems (see especially Griffiths and Gray 1997). Our focus has been on identifying developmental systems that could be evolutionary individuals. We have also been concerned to examine the extent to which DST can support a hierarchical model of natural selection. On the one hand, there is no reason why natural selection should not operate at different levels of biological organization (Brandon 1988; Sober and Lewontin 1982). On the other hand, not every repeated assembly is the focus of a selection process. So we need criteria to identify which developmental systems count as evolutionary individuals. While we see some merit in our previous suggestions, we have learnt a great deal from the work of David Sloan Wilson and Elliott Sober on trait-group selection and the concept of a superorganism, and also from Kim Sterelny's work on higher level selection (Sober and Wilson 1994, 1998; Sterelny 1996; Wilson 1997; Wilson and Sober 1994). In this section we revise our previous account of the individual in the light of this work.

In earlier work we suggested that an evolutionary individual was distinguished from a collection of such individuals by the strength of the evolutionary association between its components: "We argue that the eukaryotic cell should be seen as a single life-cycle because its constituents are obligate symbionts and there are strong barriers to their evolving back to free-living forms. Strongly obligate symbioses like this one should be regarded as a single evolutionary lineage" (Griffiths and Gray 1997; 478). The inspiration for this idea was the now almost universal acceptance that symbiotic lineages sometimes merge into a single lineage, as in the origins of eukaryotic cells (Margulis 1970). The descendants of these symbiotic organisms, the cell nucleus and the cell organelles, are replicated with different periodicities and have different patterns of inheritance, just like the various developmental resources that go to make up the developmental system of an organism. Nevertheless, unicellular eukaryotes are regarded as individuals, not collections of individuals. In contrast, the ants and acacia trees of the ant/acacia symbiosis are normally regarded as two separate evolutionary lineages. We suggested that the distinction between one and many turns on whether one element of the symbiosis can give rise to new cycles of itself that are not coupled to the other members of the symbiosis in the characteristic way. The meaning of "cannot" here is that the free-living form is very distant in the space of biological possibility. There are strong barriers to the components of the eukaryotic cell evolving back to free-living forms. The barriers for the ant and the acacia tree are weaker. We defended the vagueness of this answer by arguing that there is no clear line on the continuum between strong symbioses, facultatively colonial organisms like slime-molds and obligate colonial organisms like sponges and metazoans at which individuality springs into existence. We suggested that Maynard-Smith and Szathmáry's concept of "contingent irreversibility" of an evolutionary development in a lineage came close to the concept of biological (im)possibility we required (Maynard Smith and Szathmáry 1995). Our proposal was intended to assist in defining an evolutionary lineage, as well as an evolutionary individual. If two life cycles become coupled in a way that is contingently irreversible, then the evolutionary lineages of which they are representatives have also merged.

We now see these ideas as strongly convergent with Wilson and Sober's idea of "shared evolutionary fate" (Wilson and Sober 1994). In our 1997 paper we wrote: "Two lineages whose evolutionary fates were previously separable (though interacting) are now inseparably bound together" (Griffiths and Gray 1997: 478). However, we had not really assimilated Wilson and Sober's proposal, one of whose advantages is the substance it gains from its relationship to Wilson's concept of trait-group selection (Wilson 1983). A trait group is a set of organisms relative to which some adaptation is, in economic terms, a public good. The beavers that share a lodge form a trait group with respect to dam-building adaptations because it is not possible for one beaver to increase its fitness by dam building without increasing the fitness of its lodge mates. The water will be 1 cm deeper for everyone. Trait group selection can occur when there is a correlation between having an adaptation and being part of the relevant trait group. If the beavers that share a dam are closely related, shared descent will produce such a correlation—a mechanism long recognized as kin selection. Another long-accepted evolutionary mechanism—reciprocal altruism—produces the same effect. If individuals differentially associate with those who return their favors, then there will be a correlation between giving and receiving favors. The advantage of the trait group-selection model is that it brings out the underlying unity of the various models for the evolution of cooperation and reveals the whole landscape, only a few peaks of which had previously poked up through the fog of the group-selection debate.

The general phenomenon of which kin selection and reciprocal altruism are special cases is population structured evolution. Organisms do not interact with equal probability with every other member of the population. Population structure creates opportunities for trait group selection. The most obvious form of population structure is geographic structure. Robert Triver's famous paper on the evolution of reciprocal altruism gives as an example a symbiotic relationship between small cleaner fish and the large predators whose parasites these fish remove (Trivers 1971). The larger fish do not eat the small cleaners after they have finished cleaning, as a simple game-theoretic model would suggest. Trivers argues that this is because the large fish find it hard to locate cleaners. They solve this problem by returning to the same cleaning site each time. By eating the cleaners they would reduce their probability of being cleaned next time. Trivers treats this as a case of reciprocal altruism, but it is more common to define reciprocal altruism as requiring recognition and memory of

individuals. What matters is that both the cleaner-fish example and more standard cases of reciprocal altruism are examples of population structured evolution and trait group selection. In the case of the fish, it is the patchy distribution of cleaners on the reef that creates a correlation between not eating cleaners and getting cleaned. In reciprocal altruism, the source of population structure is refusal to associate with those who do not dispense benefits. There are many other possibilities. Richard Dawkins has described a bizarre fictional example. He imagines some people with green beards who are disposed to help anyone else with a green beard (Dawkins 1976: 96–97). Given suitable cost-benefit ratios, individuals with this trait complex could outcompete those who neither dispense nor receive benefits. Dawkins uses this example to demonstrate that it is not kinship that is important in kin-selection models, but shared genes. He postulates that his green-bearded people are unrelated but share a pleiotropic gene that confers the trait complex upon them. However, the example would work just as well if different green-bearded individuals had many different genes that produced the same effects. The different genes could all impair the same normal biochemical pathway—many "genetic diseases" work this way. Dawkins's example would also work if the trait complex were produced by cultural inheritance—a subculture keen on hair dye and cooperation. The particular mechanism of inheritance is irrelevant. What the hypothetical green-beard example actually demonstrates is not that kin selection is a special case of gene selection, but that it is a special case of trait-group selection. It is not necessary that individuals who selectively benefit one another be related or that they share a gene. The essential feature needed to get models like this to work is population structure. Something has to create a statistical association between dispensing benefits and associating with other individuals who dispense benefits.

Wilson and Sober argue that trait groups are units of evolution. More specifically, trait groups are interactors *sensu* David Hull (Hull 1980).

That is to say, it makes sense to assign fitnesses to trait groups and to track the evolution of adaptations due to the differential reproduction of their associated replicators. In one respect, however, a trait group is very different from the kinds of evolutionary individuals we tried to define in previous work. Each trait potentially defines a different trait group. The beavers in a dam are not a trait group with respect to foraging behaviors. In humans, each cooperative behavior may define a different trait group. Paying for your shout in the bar does not benefit the same group as helping with the housework. This has caused Dawkins to deny that there is any room for the replicator/interactor distinction in the selection processes described by Wilson and Sober (Dawkins 1994). According to Dawkins, "vehicles" (interactors) only exist in a special subset of selection processes where the fate of a collection of replicators is strongly linked by their joint investment in a complex, physiologically interdependent collection of adaptations like organism. In other selection processes, there are only individual replicators and their impact on their own replication prospects. This criticism seems to us both right and wrong, and to point to the need to clearly distinguish the concept of an interactor as it figures in the replicator/interactor distinction from the interactor as a generalization of the concept of the organism. If we concentrate on the replicator/interactor framework for thinking about natural selection then Dawkins's criticism is clearly misguided. Replicator and interactor are two aspects of the process of selection. In classic cases of gene selection, such as meiotic drive, the replicator and the interactor are the same physical object—a stretch of DNA. Nevertheless, this object is playing two separate roles. The fact that it plays the replicator role is supposed to explain the stable replication of form from one generation to the next. The fact that the same stretch of DNA plays the interactor role explains the selection of those replicators associated with the most efficient interactors. In paradigm cases of phenotypic evolution, the two

roles are played by different physical objects—traditionally one or more genes and one or more phenotypic traits. Proponents of the replicator/interactor framework believe that identifying the occupants of these two roles is essential to understanding how natural selection is operating in a domain. We think the replicator/interactor framework has fatal flaws as a representation of evolution, embodying as it does the gene/environment dichotomy rejected by developmental systems theory (Griffiths and Gray 1994a, 1994b). But be that as it may, the arguments in favor of the framework have as much application in the case of trait group selection as elsewhere. In this sense, Dawkins is wrong and there are interactors in the selection processes Wilson and Sober describe.

The concept of an interactor has a double life, however. It also serves to generalize the notion of an organism. Sometimes, when we look for the interactor in a potential domain for natural selection we are looking, not for something that can play the abstract role of interactor in selection theory, but for something that corresponds to the organism. In earlier group selection models a spatially cohesive local population of a species—a deme—was assumed to correspond to the organism. In that case the analogy with paradigm cases of individual selection seemed clear enough. The new trait-group selection models show that the interactor role can be fulfilled in such a piecemeal way that nothing at this new, higher level of selection corresponds to the organism in paradigm cases. In this sense, Dawkins was right that in some trait-group selection models there is no interactor. Sterelny and Griffiths have argued that trait-group selection sometimes gives rise to functionally organized units with many adapted traits and sometimes does not. The old term *superorganism* is a useful one to replace the ambiguous *interactor* for these higher-level evolutionary units.

What makes a group into a superorganism? At an intuitive level, an ant nest is a much more convincing superorganism that a lodge full of beavers. There are a number of features that seem to underlie this intuition, such as the functional differentiation of parts and the dependence of parts on the whole for their viability. Sterelny and Griffiths argue that what is fundamental to a superorganism is that very many traits of its organisms are selected with respect to a single trait group. That trait group is the superorganism (Sterelny and Griffiths 1999: 172–177; see also Wilson 1997). The ants in a nest and the cells in a human body have a shared fate not just with respect to one part of their activities, but with respect to all of them. A liver cell does not have some adaptations with respect to the whole body and other with respect to the liver alone. This is because the only way the liver cell can reproduce itself is via the success of the whole organism. Similarly, the only way an ant can contribute to its own reproduction is via the success of the nest as a whole. The general phenomenon that we see in these cases is the existence of evolved features that suppress competition between the component parts of the superorganism. The best known of these is the segregation of the germline. In most animals a particular cell lineage is physiologically isolated relatively early in development as the source of all future gametes. This means that other cell lines that do not contribute to successful functioning of the whole organism are, like cancer cell lineages, doomed to extinction. Segregation of the germ line is an important mechanism. Organisms that do not have this feature, such as slime-molds, rapidly lose their multicellularity when selection for it is relaxed and become once again a population of free-living individuals. One species of aphid seems to segregate a proportion of its inherited endosymbionts as the exclusive source of founder populations of endosymbionts for its offspring, presumably in order to keep the remaining endosymbionts at work (Frank 1996). However, it is easy to overstate the importance of this particular mechanism. Plants typically do not have germline segregation, so it cannot be a prerequisite for complex multicellular life. Leo Buss has explored

some of the very different mechanisms that are used to bind the interests of cell lineages together in plants and fungi (Buss 1987). In bee nests, the queen marks her eggs with a pheromone that inhibits workers from eating them. Eggs laid by workers are eaten by other workers, so the only realistic way for workers to bring about the reconstruction of their life cycle is via the larger, colony life cycle (Ratnieks and Visscher 1989). The worker bee is reduced to a part of a larger cycle as effectively as the cell of a metazoan body is by segregation of the germline or an equivalent mechanism.

The idea that evolutionary individuals are trait groups thus converges on our older idea that an individual is a life cycle whose components cannot reconstruct themselves when decoupled from the larger cycle. Much trait-group selection does not give rise to new levels of individuality, but only to transitory interactors. However, when new features evolve, presumably by trait-group selection, which link together the members of a group in a way that is contingently irreversible, a new kind of individual emerges. At this point, the population structure on which trait-group selection depends is no longer just a cause of trait-group selection, but an effect of trait-group selection. An individual is a system in which the parts form a trait group with respect to most future evolutionary processes. This account of the evolution of individuality can actually explain why the distinction between a colony of organisms or a symbiotic association and an individual organism is not a sharp one. The mechanisms that bind the trait group together can be more or less effective. They may also keep the evolutionary interests of the same group aligned across a wider or narrower range of traits. The metazoan organism and the unicellular eukaryotic cell are clearly individuals. Jellyfish, lichens, eusocial insect colonies, and the ant/acacia symbiosis are less clearly so. Each of these has a life cycle and a developmental system that feed into its development. But in most of these latter cases, it is possible to describe evolutionary pressures with respect to which the smaller life cycles nested within the larger cycle do not form a trait group. The more forced and implausible these scenarios, the less theoretical role there is for a description in which these cycles are treated as independent and not as parts.

Conclusion

The aim of this chapter is quite simple. The fear that DST will lead to wholesale rejection of current evolutionary theory is not well founded. DST is not anti-Darwinian, nor does it render the basic explanatory terms of evolutionary theory incoherent. To the contrary, DST expands the scope and power of adaptive/historical explanation. The core Darwinian concepts of inheritance, natural selection, adaptation, individual, and lineage can be productively reworked in DST terms as follows:

• Developmental system—the interactants and processes that produce a life cycle.

• Evolutionary developmental system—the interactants and processes that produce those developmental outcomes that are reliably reproduced in a lineage.

• Inheritance—the reliable reproduction of developmental resources down lineages.

• Natural selection—the differential reproduction of heritable variants of developmental systems due to relative improvements in their functioning.

• Adaptation—the product of natural selection.

• Individual—the most inclusive sequence of developmental events (life cycle) such that the smaller repeated cycles nested within it form a trait group with respect to most plausible evolutionary scenarios.

• Lineage—a causally connected sequence of similar individual life cycles.

• Evolution—change over time in the composition of populations of developmental systems.

The benefits of this reconceptualization of evolution in DST terms are considerable. Phenomena that are marginalized in current genocentric conceptions of evolution, like expanded inheritance, niche construction and developmental organization, are placed center stage. Here are some suggestions for the kind of research questions we think DST encourages.

1. Treat all claims about instincts, genetic programs and other black boxes as potential research questions for developmental analysis, that is, how does this trait actually develop, what resources does its reliable development depend upon, are there many developmental routes to this outcome or only one, over what range of parameters is this developmental outcome stable, how does the "environment" change as a function of initial development differences that produce this trait?

2. Study expanded forms of inheritance. Conduct studies to investigate the longevity and fidelity of extended inheritance. Are there physiological and developmental mechanisms that enable these forms of inheritance to be vertically reproduced down lineages with fidelity and longevity? If extragenetic inheritance is an adaptive developmental resource, then developmental systems that reliably and efficiently pass on that resource would be at an advantage. Test adaptive hypotheses about extended inheritance using comparative methods—for example, is species diversity greater or rates of evolution higher in lineages with certain forms of extragenetic inheritance? Develop mathematical models of the impact of different types of extragenetic inheritance and their coevolution with genetic change. How tightly coupled do different developmental resources have to be to influence each other's evolutionary dynamics? What role have expanded forms of inheritance played in major evolutionary transitions?

3. Study "niche construction." Conduct field experiments to assess the fitness impact of niche construction. Develop models to investigate the

evolutionary consequences of the ways in which organisms select and modify their environments. Test predictions from these models using comparative methods.

4. Investigate the extent and functional basis of developmental modularity. How modular is development? Are functionally linked traits also linked together in development? Are there evolutionary trends in the degree of modularity? Does the modularity only hold over a restricted range of parameters? Can these parameters be changed by selection? Are extended forms of inheritance more or less modular than genetic inheritance in their developmental consequences?

Neo-Darwinism was the result of the union of Darwin's theory of natural selection with a particular view of heredity. The new view of heredity transformed Darwin's vision and gave rise to a wide range of research questions. In rejecting the narrowly gene-centered view of heredity and bringing developmental processes back into our account of evolution, we are not rejecting the theory of natural selection but are attempting to unite it with the developmental systems account of heredity and thus to reveal new and promising research agendas.

Acknowledgments

We would like to thank Kendall Clements, Fiona Jordan, Susan Oyama, Kim Sterelny, and Debbie Waldron for comments on drafts of this chapter.

Notes

1. Here is a random selection: "Evolution 1. Process by which organisms come to differ from generation to generation. 2. Change in the gene pool of a population from generation to generation" (Arms and Camp 1987: 1121). "Evolution is the result of accumulated changes in the composition of the gene pool" (Curtis and Barnes 1989: 989).

2. Although the definition of evolution as change in gene frequencies is widely accepted, its significance is disputed. In G. C. Williams's influential view, when we measure change in gene frequencies we are getting at the heart of the process of evolution. It is genes that compete and are selected (Williams 1966). In contrast, phenotypic or hierarchical views of evolution accept that evolution can be represented by change in gene frequencies, but locate the processes of competition and selection at higher levels of biological organization.

3. In the mathematical theory of information (Shannon and Weaver 1949) and its relatives (Dretske 1981), a signal sender conveys information to a receiver when the state of the receiver is correlated with the state of the sender. The conditions under which this correlation exists constitute the "channel" between sender and receiver. Changes in the channel affect which state of the receiver corresponds to which state of the sender. The information conveyed by a particular state of the receiver is as much a function of the channel, the context, as it is of the sender.

4. In local enhancement the model directs the subject's attention to salient features of environment and the subject then develops the appropriate behavior through individual trial-and-error learning. In emulation the subject learns about cause and effect relations rather than the behavior itself.

5. The traditional neo-Darwinian explanation of adaptation is an "externalist" explanation. Explanation flows from the environment to the organism and not vice versa. For a good discussion of "externalist," "internalist," and "constructionist" explanatory strategies in general, see Godfrey-Smith (1996).

6. For example, Dennett (1995) and Pinker (1997). The idea is popular in the evolutionary psychology movement (Barkow, Cosmides, and Tooby 1992). For an attempt to integrate the idea that each trait has its own selection of relevant environmental parameters into evolutionary psychology, see Irons (1998).

7. Susan Oyama (personal communication) is less happy than we are with the idea of the external or universal physical environment. The next section makes clear that our formulation actually preserves the DST insight that a full description of the developmental systems of an individual organism, as opposed to a description of the typical developmental system of a population, will include all the causal factors that influence that individual's development. Oyama is legitimately concerned that the need for a concept of endogenous sources of change in evolution will create the impression that there is a need for sources of change outside

the developmental system in the individual case. That would be exactly the sort of dichotomous account of development that DST seeks to avoid, albeit with the boundary in an unusual place.

8. We are not aware of any published version of this criticism, but it was first suggested to Griffiths in conversation by Lindley Darden in 1994 and has also been raised by Alexander Rosenberg (personal communication).

References

Arms, K., and P. S. Camp. (1987). *Biology.* (3d ed.). New York: CBS College Publishing.

Barkow, J. H., L. Cosmides, and J. Tooby (Eds.) (1992). *The Adapted Mind: Evolutionary Psychology and the Generation of Culture.* Oxford: Oxford University Press.

Boyd, R., and P. J. Richerson. (1996). Why culture is common, but cultural evolution is rare. *Proceedings of the British Academy* 88: 77–93.

Brandon, R. (1988). The levels of selection: A hierarchy of interactors. In H. Plotkin (Ed.), *The Role of Behavior in Evolution,* pp. 51–71. Cambridge, MA: MIT Press.

Brandon, R. (1990). *Adaptation and Environment.* Princeton, NJ: Princeton University Press.

Brandon, R. N. (1992). Environment. In E. Fox Keller and E. A. Lloyd (Eds.), *Keywords in Evolutionary Biology,* pp. 81–86. Cambridge, MA: Harvard University Press.

Brandon, R. (1999). The units of selection revisited: The modules of selection. *Biology and Philosophy* 14: 167–180.

Brandon, R., and J. Antonovics. (1996). The coevolution of organism and environment. In R. Brandon (Ed.), *Concepts and Methods in Evolutionary Biology,* pp. 161–178. Cambridge: Cambridge University Press.

Buss, L. (1987). *The Evolution of Individuality.* Princeton, NJ: Princeton University Press.

Caporael, L. (1995). Sociality: Coordinating bodies, minds and groups. *Psycoloquy.* http://www.princeton.edu/pub/harnad/Psycoloquy/1995.volume.6.

Cavalli-Sforza, L. L., A. Piazza, P. Menozzi, and J. Mountain. (1988). Reconstruction of human evolution: Bringing together genetic, archaeological and linguistic data. *Proceedings of the National Academy of Sciences* 85: 6002–6006.

Curtis, H., and N. S. Barnes. (1989). *Biology.* (5th ed.). New York: Worth Publishers.

Darwin, C. (1859/1964). *On the Origin of Species: A Facsimile of the First Edition.* Cambridge, MA: Harvard University Press.

Darwin, C. (1881). *The Formation of Vegetable Mould, Through the Action of Worms, with Observations on Their Habits.* London: Murray.

Dawkins, R. (1982). *The Extended Phenotype.* Oxford: Freeman.

Dawkins, R. (1994). Burying the vehicle. *Behavioral and Brain Sciences* 17: 616–617.

Dawkins, R. (1996). *Climbing Mount Improbable.* London: Viking.

Dennett, D. C. (1995). *Darwin's Dangerous Idea.* New York: Simon and Schuster.

Depew, D. J., and B. H. Weber. (1995). *Darwinism Evolving: Systems Dynamics and the Genealogy of Natural Selection.* Cambridge, MA: Bradford Books/MIT Press.

Dretske, F. (1981). *Knowledge and the Flow of Information.* Oxford: Blackwells.

Frank, S. A. (1996). Host control of symbiont transmission: The separation of symbionts into germ and soma. *American Naturalist* 148: 1113–1124.

Gilbert, S. F., J. M. Opitz, and R. A. Raff. (1996). Resynthesizing evolutionary and developmental biology. *Developmental Biology* 173: 357–372.

Godfrey-Smith, P. (1996). *Complexity and the Function of Mind in Nature.* Cambridge: Cambridge University Press.

Goodwin, B. C. (1984). Changing from an evolutionary to a generative paradigm in biology. In J. W. Pollard (Ed.), *Evolutionary Theory: Paths into the Future,* pp. 99–120. New York: Wiley.

Gould, S. J., and R. Lewontin. (1978). The spandrels of San Marco and the Panglossian paradigm: A critique of the adaptationist programme. *Proceedings of the Royal Society of London* 205: 581–598.

Gray, R. D. (1992). Death of the gene: Developmental systems strike back. In P. E. Griffiths (Ed.), *Trees of Life,* pp. 165–210. Dordrecht: Kluwer.

Gray, R. D. (1997). "In the belly of the monster": Feminism, developmental systems, and evolutionary explanations. In P. A. Gowaty (Ed.), *Evolutionary Biology and Feminism*, pp. 385–413. New York: Chapman and Hall.

Gray, R. D. (2001). Selfish genes or developmental systems? Evolution without interactors and replicators? In R. Singh, C. Krimbas, J. Beatty, and D. Paul (Eds.), *Thinking about Evolution: Historical, Philosophical, and Political Perspectives*, pp. 184–207. Cambridge: Cambridge University Press.

Gray, R. D., and F. M. Jordan. (2000). Language trees support the express train sequence of Austronesian expansion. *Nature* 405: 1052–1055.

Griffiths, P. E., and R. D. Gray. (1994a). Developmental systems and evolutionary explanation. *Journal of Philosophy* 91(6): 277–304.

Griffiths, P. E., and R. D. Gray. (1994b). Replicators and vehicles—or developmental systems? *Behavioral and Brain Sciences* 17: 623–624.

Griffiths, P. E., and R. D. Gray. (1997). Replicator II: Judgement day. *Biology and Philosophy* 12(4): 471–492.

Griffiths, P. E., and R. D. Knight. (1998). What is the developmentalist challenge? *Philosophy of Science* 65(2): 253–258.

Halder, G. P., P. Callerts, and W. J. Gehring. (1995). Induction of ectopic eyes by targeted expression of the eyeless gene in *Drosophila. Science* 267: 1788–1792.

Harlow, H. F., and M. K. Harlow. (1962). Social deprivation in monkeys. *Scientific American* 207(5): 136–146.

Hull, D. L. (1980). Individuality and selection. *Annual Review of Ecology and Systematics* 11: 311–332.

Irons, W. (1998). Adaptively relevant environments versus the environment of evolutionary adaptedness. *Evolutionary Anthropology* 6: 194–204.

Jablonka, E., and E. Szathmáry. (1995). The evolution of information storage and heredity. *TREE* 10(5): 206–211.

Johnston, T. D. (1987). The persistence of dichotomies in the study of behavioural development. *Developmental Review* 7: 149–182.

Keller, L., and K. G. Ross. (1993). Phenotypic plasticity and "cultural transmission" of alternative social organisations in the fire ant *Solenopsis invicta. Behavioural Ecology and Sociobiology* 33: 121–129.

Lansbury, P (1997). Yeast Prions: Inheritance by seeded protein polymerisations? *Current Biology* 7: R617.

Lehrman, D. S. (1970). Semantic and conceptual issues in the nature-nurture problem. In L. R. Aronson, E. Tobach, D. S. Lehrman, and J. S. Rosenblatt (Eds.), *Development and Evolution of Behavior: Essays in Memory of T. C. Schneirla*, pp. 17–52. San Francisco: Freeman.

Lerner, I. M. (1970). *Genetic Homeostasis*. New York: Dover Publications.

Lewontin, R. (1970). The units of selection. *Annual Review of Ecology & Systematics* 1: 1–14.

Lewontin, R. C. (1978). Adaptation. *Scientific American* 239 (September): 156–169.

Lewontin, R. C. (1982). Organism & environment. In H. Plotkin (Ed.), *Learning, Development, Culture*, pp. 151–170. New York: John Wiley.

Lewontin, R. C. (1983a). Gene, organism and environment. In D. S. Bendall (Ed.), *Evolution: From Molecules to Men*, pp. 273–285. Cambridge: Cambridge University Press.

Lewontin, R. C. (1983b). The organism as the subject and object of evolution. *Scientia* 118: 65–82.

Lindquist, S. (1997). Mad cows meet psi-chotic yeast: the expansion of the prion hypothesis. *Cell* 89: 495.

Lynch, A., and A. J. Baker. (1986). Congruence of morphological and cultural evolution in Atlantic island chaffinch populations. *Canadian Journal of Zoology* 64: 1576–1580.

Maynard Smith, J. (2000). The concept of information in biology. *Philosophy of Science* 67: 177–194.

Maynard Smith, J., and E. Szathmáry. (1995). *The Major Transitions in Evolution*. Oxford: W. H. Freeman.

Mills, S., and J. Beatty. (1979). The propensity interpretation of fitness. *Philosophy of Science* 46: 263–286.

Money, J. (1992). *The Kaspar Hauser Syndrome of "Psychosocial Dwarfism": Deficient Statural, Intellectual, and Social Growth Induced by Child Abuse*. Buffalo, NY: Prometheus Books.

Moran, N., and P. Baumann. (1994). Phylogenetics of cytoplasmically inherited microorganisms of arthropods. *Trends in Ecology and Evolution* 9: 15–20.

Mount, A. B. (1964). The interdependence of the eucalypts and forest fires in southern Australia. *Australian Forestry* 28: 166–172.

Newcomb, R. D., P. M. Campbell, D. L. Ollis, E. Cheah, R. J. Russell, and J. G. Oakeshott. (1997). A single amino acid substitution converts a carboxylesterase to an organophosphorous hydrolase and confers insecticide resistance on a blowfly. *Proceedings of the National Academy of Science* 94: 7464–7468.

Odling-Smee, F. J. (1988). Niche-constructing phenotypes. In H. C. Plotkin (Ed.), *The Role of Behavior in Evolution,* pp. 73–132. Cambridge, MA: MIT Press.

Oyama, S. (1985). *The Ontogeny of Information: Developmental Systems and Evolution.* Cambridge: Cambridge University Press. (2d rev. ed., Durham, NC: Duke University Press, 2000.)

Pal, C., and I. Miklos. (1999). Epigenetic inheritance, genetic assimilation and speciation. *Journal of Theoretical Biology* 200: 19–37.

Penny, D., E. E. Watson, and M. A. Steel. (1993). Trees from languages and genes are very similar. *Systematic Biology* 42: 382–384.

Pinker, S. (1997). *How the Mind Works.* New York: Allen Lane.

Ratnieks, F. L. W., and P. K. Visscher. (1989). Worker policing in honeybees. *Nature* 342: 796–797.

Rosenberg, A. (1978). The supervenience of biological concepts. *Philosophy of Science* 45: 368–386.

Shannon, C. E., and W. Weaver. (1949). *The Mathematical Theory of Communication.* Urbana: University of Illinois Press.

Sober, E., and R. C. Lewontin. (1982). Artifact, cause & genic selection. *Philosophy of Science* 49: 157–180.

Sober, E., and D. S. Wilson. (1994). A critical review of philosophical work on the units of selection problem. *Philosophy of Science* 61(4): 534–555.

Sober, E., and D. S. Wilson. (1998). *Unto Others: The Evolution and Psychology of Unselfish Behavior.* Cambridge, MA: Harvard University Press.

Sterelny, K. (1996). Explanatory pluralism in evolutionary biology. *Biology and Philosophy* 11(2): 193–214.

Sterelny, K. (in press). Development, evolution and adaptation. *Philosophy of Science.*

Sterelny, K., M. Dickison, and K. Smith. (1996). The extended replicator. *Biology and Philosophy* 11(3): 377–403.

Sterelny, K., and P. E. Griffiths. (1999). *Sex and Death: An Introduction to the Philosophy of Biology.* Chicago: University of Chicago Press.

Szathmáry, E., and J. Maynard Smith. (1997). From replicators to reproducers: The first major transition leading to life. *Journal of Theoretical Biology* 187: 555–571.

Tomasello, M., A. C. Kruger, and H. H. Ratner. (1993). Cultural learning. *Behavioral and Brain Sciences* 16(3): 495–552.

Trivers, R. L. (1971). The evolution of reciprocal altruism. *Quarterly Review of Biology* 46(4): 35–57.

Vines, G. (1999). Gendercide. *New Scientist* 158: 44–47.

Wagner, G. P., and L. Altenberg. (1996). Complex adaptations and the evolution of evolvability. *Evolution* 50: 967–976.

Whitehead, H. (1998). Cultural selection and genetic diversity in matrilineal whales. *Science* 282: 1708–1711.

Whiten, A., J. Goodall, W. C. McGrew, T. Nishida, V. Reynolds, Y. Sugiyama, C. E. G. Tutin, R. W. Wrangham, and C. Boesch. (1999). Culture in chimpanzees. *Nature* 399: 682–685.

Williams, G. C. (1966). *Adaptation and Natural Selection.* Princeton, NJ: Princeton University Press.

Wilson, D. S. (1997). Biological communities as functionally organized units. *Ecology* 78: 2018–2024.

Wilson, D. S., and E. Sober. (1994). Reintroducing group selection to the human behavioral sciences. *Behavioral and Brain Sciences* 17: 585–608.

17

Generative Entrenchment and the Developmental Systems Approach to Evolutionary Processes

William C. Wimsatt

Developmental Systems Theory as a Methodological Theory

DST is not a theory in the sense of a specific model that produces predictions to be tested against rival models. Instead DST is a general theoretical perspective on development and evolution at the same level of generality as genic seletionism. (Gray 1999)

DST is a broad research perspective rather than a specific model. Hence to succeed as a research program it must generate or entrain other theory, that is, specific models, data gathering, simulations, and experiments. We all have to eat, and theories or research perspectives without specific issue will not be pursued indefinitely. For reconceptualizing relations, and reanalyzing data, particularly to compromise facile "nothing-but" style claims about where "ultimate" explanations lie (somehow always microlevel and "genetic"), DST has done a lot already. (See Oyama's 1985 groundbreaking critiques of traditional nativist theories, and Griffiths and Gray's 1994 particularly compact and illuminating presentation of DST.) We will all benefit from these reconceptualizations. But we should work for more.

I have long felt my ideas consilient with DST. Perhaps this is self-selection; it is certainly also due to early extended associations with Richard Lewontin and Richard Levins. (See Wimsatt 1986 on development and nativism, and Wimsatt 1980b, 1981b for early analyses supporting Gray's attacks on gene selectionism and reductionism.) Nonetheless, I had tended not to count myself as a DST theorist because I sought more specific models.[1] Now I see from Gray's paper that I am a DST theorist after all—at least if DST is a family resemblance concept. (Some others in this book are more "specific" than I!) Here I will try to bridge the gap between theoretical perspective and specific models. Generative entrenchment (or GE) reflects factors of great generality having natural resonance with DST—a class of developmental and evolutionarily significant factors that can be modeled, generate predictions, and yield new understandings. If so, GE provides an important engine for evolutionary processes on the DST perspective—or perhaps a transmission coupling developmental and evolutionary processes to the same effect. In what is becoming a tradition for DST, this also involves recognizing what we already do and know in a new light.

How can generative entrenchment—a source of developmental constraints—be an engine? Blades in a water turbine constrain the flow of water and provide motive power—acting as an engine even though they are not the source of the energy—by transforming the potential energy or rectilinear motion of the water stream into the angular momentum and torque of the turbine shaft. They are levers, one of the five "fundamental machines" of high school physics.[2] Similarly, GE affects and constrains which kinds of variation get incorporated in processes of evolutionary change and elaboration. Things that look like constraints on processes on a shorter time scale can look like engines of that process in a larger context or on longer time scales. The main effects of GE constraints are to control or modulate the relative rates of different kinds of change processes (Schank and Wimsatt 1988; Wimsatt and Schank 1988). That is not only an engine, but potentially a cybernetic engine—a potential source of control structures. But unlike the picture painted by gene-selectionists, the loci of control here are distributed across system and environment, and up and down the levels of organization.

Neo-Darwinism ignored development in "the major synthesis" bringing population genetics to the theoretical core of evolutionary biology. In the last two decades this has emerged as a serious error. Perhaps at mid-century there was not the work (in developmental genetics and other work on development from neurophysiology to animal ethology and social psychology) to force a productive synthesis. But times have changed. "Evodevo" is one of the hottest research areas in both evolutionary biology and development. (Raff 1996 is a synthetic work by probably the major spokesman of this new field.) But we must be careful lest this come to be seen as just claiming the larger playing field that genes deserve.[3] Progress and balance require that the other relevant fields continue to develop in analyzing the causal efficacy of processes at their levels of organization. An extended multilevel population genetics surely must be part of an adequate picture, but this will not suffice. As DST urges, the whole ecologically embodied and embedded life cycle, or for multiple levels, interdigitated life courses or cycles—Caporael's (1997) "repeated assemblies"— must be treated as relevant units of analysis.[4] DST must play an active role to leave its mark in evo-devo and elsewhere as it continues to develop. We need more "top-down" conceptual and theoretical work to properly situate the significance of the emerging results. I have long argued against "nothing-but" style genetic reductionism.[5] We also need room to study the same systems from the top down. Both approaches are legitimate.

But this is not and should not be a one-way street. DST advocates should also be prepared to learn from the bottom up: we will learn as much in new evidence and ideas from developmental genetics and more broadly from evo-devo as we have to give to it. We have all benefited from careful analyses at the genetic level in the work of Jablonka (see chapter 9), Moss (1992, 1998; see chapter 8), Keller (1995), Neumann-Held (1998; see chapter 7), and back to work by Lewontin (1983) in criticism of the genes as self-activating and self-expressing. These works are as much about development as about embodiment. Griesemer's move (1999a, 2000) to reconceptualize and replace replicator-talk with developmentally embodied "reproducers" promises notions of inheritance more friendly to DST. Schaffner's (1998) rich exploration of developmental genetics

(by one traditionally more reductionistic) reveals his own surprise at the developmental complexities he found in the supposedly simple and determinate nematode soil worm *Caenorhabditis elegans*. We must deny the total hegemony of the genetic perspective, but at the same time we must not be afraid to use some of its products.

Darwin's Principles Embodied: The Evolution and Entrenchment of Generative Structures

Where should we start with our top-down theorizing to put developmental genetics in context? Why not with the only other general biological theory we have—evolution. Any evolving systems must meet what Lewontin (1970) has called "Darwin's principles." they must:

1. have descendants that differ in their properties (*variation*),

2. some of which are heritable (*heritable* variation), and

3. have varying causal tendencies to have descendants (heritable variation *in fitness*).

These three principles—all met simultaneously for the same processes or properties of the same entities—are widely acknowledged as core requirements for evolution to occur. Any population of entities meeting all three of these conditions will undergo an evolutionary process, and none which fail to meet even one of them can do so.[6] They might be thought of as the *logical* or *conceptual* conditions for an evolutionary process because they key into the fundamental requirements of evolution by natural selection.

But there are two further conditions of great generality. These conditions recognize development's central role in the evolutionary process. I know of no interesting evolutionary process whatsoever (physical or conceptual) that does not meet them.[7] With these DST becomes not just an additional perspective on evolution, but fundamental to it.[8] The same entities which meet the first three conditions must also be:

4. structures which are generated over time so they have a developmental history (*generativity*), and

5. some elements that have larger or more pervasive effects than others in that production (*differential entrenchment*).

Then different elements in the structures characteristically have downstream effects of different magnitudes. The *generative entrenchment* (GE) of an element is the magnitude of those effects in that generation or life cycle. Elements with larger degrees of GE are *generators*. This is a degree property. The GE of an element in an evolutionary unit has multiple deep consequences for its evolutionary fate and character, and that of systems impinging on it. I return to this later.

But first, why are these deep principles, and what kinds of principles are they? Exceptions are *logically possible*, so they cannot be logical principles. And we can imagine physical systems which fail to meet them, so they are *physically possible*, though, as we will see, rather unlikely. But there are weaker but still very strong kinds of principles producing universal or near-universal behavior. I characterize them and then raise some qualifications:

1. *Physically generic.* Kauffman (1993) urges us to build theory for the emergence of complexity around properties which are stochastically inevitable (or stochastically self-organizing)—motivated by his belief that there are things too complex to be maintained by selection (or by selection alone).[9] Consider an ensemble of systems defined by a set of constructional constraints but free to vary in other defined directions. A generic property for that ensemble is one characterizing virtually all systems in it. Thus (1) a system picked at random from the ensemble will probably have the property. Further, dynamical processes which change systems from one ensemble state to another will probably (2) conserve generic properties and (3) produce generic properties in systems lacking them. If degree of genericity is quantifiable, it will behave very

much like entropy, with dynamical changes in the system tending to increase genericity (Schank and Wimsatt 1988). These properties are physically robust just as entropy increase is on the statistical form of the second law of thermodynamics: it allows reversibility and reductions of entropy *in principle* but de facto guarantees irreversibility and increasing entropy if the system is not already at an entropic maximum!

2. *Evolutionarily generic.* Schank and I (1988) suggested that Kauffman's treatment of generic traits as monadic properties of the system was too narrow. We proposed a second complementary class of generic properties that are invariant across different selection regimes. An example of this is discussed further below: The fitness loss due to loss of a deeply GE'd trait is profound, and, because it affects so many things is almost context-independent. There is almost nothing you can do to help! This situation will therefore be almost universally selected against. *If physically generic means that a property is realized for nearly all contexts, evolutionarily generic means that it is selected for (or against) in nearly all contexts.*

3. *Rate-dominant generic.* Simon's (1962) argument for the evolution of hierarchical organization via the formation of (nearly decomposable) stable subassemblies suggest that hierarchically organized complex structures will evolve much faster than other alternatives. And as evolutionary biologists know, those that get there first get to eat the others as they arrive! So the evolutionary systems that we see should be hierarchical and nearly decomposable.[10] These could also be thought of as deep design principles of "meta-engineering" (Dennett 1995).

4. *Symmetry-breaking generic*: those properties of interest emerge or *self-organize through self-amplifying deviations from conditions of homogeneity or symmetry*. Most discussed cases are not principles but examples of pattern formation, like that of the famous Zhabotinski-Belusov reagent, in which spiral colored patterns of activity

emerge from local density fluctuations in the reactants, and expand to cover and entrain the whole system. I will argue that condition (5)—differential entrenchment—is such a principle.

Finally, there are results derivable from some combination of the above with possibly broad initial conditions and other assumptions. Apparent conflicts among these kinds of principles may arise when the conditions for deriving one of them are violated, or when expectations derived from application of them to simpler systems are overgeneralized.

So How Do Our Two New Principles Stand?

Any physical process (including reproduction) takes time, so part of (4) is trivially satisfied. But is any temporally extended process a developmental process?[11] Surely not. Lineages are not sequences of arbitrary and dissimilar developmental trajectories: Developmental structures or processes have characteristic patterns or *life cycles*. This is very important—enough to be added as a separate condition if we had to. But we don't: it already follows from the rest of (1) thru (5), *via* the evolutionary conservatism of entrenched elements of the life cycle. *That there are recognizable life cycles*—though not necessarily all of their details[12]—*is a consequence of generative entrenchment*—(and thus of 5 in the context of 1–4). We get them "for free"! Not only traits but their pattern of appearance is heritable, and may itself be treated as a trait. And *phenotypic* characters, which depend for their form of expression upon a prior relatively specific sequence of events, or more generally, on developmental timing, depend substantially upon it.

Darwin recognized age-specific patterning as "a rule governing inheritance" in the first chapter of *The Origin of Species*:

A much more important rule ... is that, at whatever period of life a peculiarity first appears, it tends to appear in the offspring at a corresponding age.... I believe this rule to be of the highest importance in explaining the laws of embryology. (Darwin 1859: 13–14)

Darwin considered this a "rule governing inheritance" though it would not appear as such (indeed, not at all!) in any genetics text in the post-Mendelian era.[13] It was not about combinatorial heredity, but about characters in relation to development. If we apply this rule in sequence to peculiarities (and to normalities!) appearing throughout the life *course* (not thereby implying a *repeated* trajectory), then if the organism has offspring, repeated application of this rule for the different characters predicts that there will be (at least roughly), a life *cycle*—thereby implying a repeated trajectory.

The "laws of embryology" Darwin mentions are presumably those of von Baer—in post-Darwinian guise, that earlier developmental traits tend to be more evolutionarily conservative, so more taxonomically widespread, than later ones (Gould 1977). These would produce life cycles with less recognizable variation at earlier developmental stages than at later ones—not exactly similar (for developmental as well as for hereditary reasons), but easily recognizable as life cycles. Von Baer's laws are a straightforward prediction of the simplest models of the action of generative entrenchment in evolution (Wimsatt 1986). So condition (4) is established as central, unavoidable, and having an honorable history.

Condition (5) is equally fundamental—and inevitable, though *not* widely recognized as such. It is inevitably satisfied for heterogeneous structures: Try to imagine a machine or system whose breakdowns are equally severe for any kind of failure, in any part, under all conditions.[14] (There aren't any!) Any differentiated system—biological, cognitive, or cultural—exhibits various degrees of generative entrenchment among its parts and activities. Condition (5) is essentially a generic property (*sensu* Kauffman 1993) of differentiated systems. Almost all possible systems of a given kind will meet it. But there is more. If you *could* find and begin with a system violating (5), so everything had effects of the same magnitude, mutational fluctuations and selection processes acting on it are self-amplifying, pushing evolution toward systems which increasingly satisfy it. So (5) is doubly guaranteed both as a generic property of mechanisms, and as a evolutionary generic property of arbitrary selection regimes acting on structures which (for whatever improbable reasons) fail to meet it. These surprising points were first made in Wimsatt and Schank 1988. The argument for the latter goes as follows:

Evolution is opportunistic—it uses what is already there, with new additions to modify existing or to make new adaptive systems. (1) Consider a system in which all elements make equal contributions, and thus engender equal fitness losses if they are knocked out. (2) Assume they all have equal probability per unit time of sustaining such a decrement. (3) Assume also a given probability per unit time that any element (gene, aspect of the phenotype, behavior, or phenotype-environment relational structure) will acquire a dependent adaptation—one adaptive only in the presence of that factor. (When genetic, these are called "modifier genes".) (4) Elements that stay around longer have a higher probability of acquiring such adaptations, and a higher expected number of such acquisitions.

Assumptions 1–3 are conditions of the idealization setting up the symmetry argument. (4) is entailed by (3) and the laws of probability. Continuing: (5) Elements that acquire dependent adaptations have, in consequence, greater generative entrenchment. (6) Degradational mutations in them or in genes affecting them would now cause greater fitness losses than before they acquired their dependent adaptations. (7) Analytical results of population genetic theory (Wright 1931) show that they then have increased chances of staying around longer, and indeed, the effect is nonlinear. By a reapplication of (4), they have also increased their chances of acquiring still other dependent adaptations—"downstream modifiers." This basic process is the developmental expression of what happens in much micro-

evolutionary change—stated in a way that fits also exaptive or macroevolutionary change.

This is a self-amplifying process. *Mutation and selection should act over long evolutionary periods to break symmetrical or homogeneous distributions of functions and selection coefficients, so that systems with equal or nearly equal selection coefficients for their parts would be driven by selection to greater variability in the distribution of the parts' selection coefficients.*[15] Stochastic fluctuations in selection coefficients due to environmental fluctuations (ecological drift) can also break symmetries.

So, in summary, why add conditions (4) and (5) to the widely accepted Darwin's principles? Because *any nontrivial physical or conceptual system satisfying (1–3) will do so via causal (phenotypic) structures satisfying (4) and (5).* So any evolutionary systems will satisfy (4) and (5). They will thus have a development, and if they can reproduce and pass on their set of generators, will have a heredity. This order is expository, not causal: without a minimally reliable heredity, they cannot evolve a complex developmental phenotype, but developmental architecture can increase the efficacy and reliability of hereditary transmission. Heredity and development thus bootstrap each other, as emerging genotype and phenotype, through evolution. This provides a symmetric view for the nature and origin of life, requiring only that generators retain their generative powers under a sufficient fraction of accessible small changes in their structure. Then they will also show phenotypic variations which (by the logic of Darwin's argument) yield fitness differences, natural selection, and evolution.

Genes or Generators?

"Darwin's principles" never mention genes. Neither do the new additions. *This expanded list of conditions gives heredity and development without ever introducing the usual replicator/interactor distinction,* widely adopted since Hull (1980) as the appropriate generalization of the genotype-phenotype distinction for the informational generalization of "genes." It is too early to tell whether we can eliminate it in all contexts, but it seems quite clear that here Griesemer's account of *reproduction* fits more naturally. Griesemer (1999, 2000) roots a conception of heredity within an account of evolution as a materially continuous lineage of developments to facilitate the central reintegration of development into our accounts of the evolutionary process.[16] Genes are agents in these stories (in biology) not the privileged bearers of information, but coactors with (the rest of) the developing phenotype and its environmental resources as bearers of relationally embodied information.[17]

I proposed in 1981b (see also Callebaut 1993: 425–429) that generative entrenchment could be used to individuate genes in terms of their heterocatalytic role by their phenotypic activity, rather than talk of copying (an abstraction of their autocatalytic role) as replicator accounts try to do. A "heterocatalytic" account of (biological) genes consilient with this is nicely elaborated by Eva Neumann-Held (1998). Sterelny, Smith, and Dickison (1996) present a rich and insightful friendly critique of DST, but in doing so defend a notion of "replicator" far broader than most (and closer both to Griesemer's contrasting "reproducer" concept, and to a heterocatalytic one)—one that if fleshed out is not ultimately at odds with DST. There is no reason however why the heterocatalytic roles picked out by generative entrenchment criteria must correspond to well-delineated genes—the context dependence and distributed character of the associated activities may be so great as to make that seem unwise (see chapter 7).

Heterocatalytic gene-like things picked out by the GE criterion in biology would include some but not all (traditional) genes, some things which are not genes, and most often, heterogeneous complexes of both.[18] In some domains (like cultural evolution) gene-like things may be picked out by GE criteria where there is arguably *nothing* picked out by autocatalytic criteria, or more

commonly, where autocatalysis is such a distributed, complex, and diffuse process (with identities or similarities determined more by context than by content), that there seems no point to try to track compact lineages through it. Paradoxically, if I am right, the DST approach may provide in that case (in at least some ways) a simpler and more compact account than a memetic one—as if genetics and development were changing places (Wimsatt 1999a). (How does a scientific theory make a copy of itself? *Very* indirectly!) Economist Kenneth Boulding used to quip that "a car is just an organism with an exceedingly complicated sex life." Like a technological virus, it takes over a complex social structure and redirects the resources of a large fraction of it to reproduce more of its own kind.[19] Indeed, our economic system has fostered an environment—a "culture dish" in which the invention, mutation, and expansion of such cultural viruses is encouraged until, many environmentalists would say, it has assumed cancerous proportions (see also Sperber 1995).

The Consequences of Differential Generative Entrenchment

If a structure meeting (1–5) is even minimally adapted to its environment or task, then modifications of more deeply GE'd elements will have higher probabilities both of being maladaptive, *and* of being *more seriously* maladaptive. These become more extreme—in the simplest models, exponentially so—either for larger structures (e.g., if they grow by adding elements downstream), or as one looks to more deeply entrenched elements in a given structure. Either change increases the degree of "lock in" of entrenched elements.[20]

Selection acts on the structure as a whole, so parts of an adaptive structure are inevitably coadapted to each other, and to different components of the environment. Larger changes in this structure will have to meet more design constraints. Fewer changes can do so. Ever larger changes have a rapidly decreasing chance of being adaptive. Mutations in deeply GE'd elements will have large and diverse effects, so are much more likely to be severely disadvantageous or lethal. Simple analytical models (Wimsatt 1986) and more realistic structures and simulations (Rasmussen 1987; Schank and Wimsatt 1988; Wimsatt and Schank 1988) all show that system elements with greater GE tend to be much more evolutionarily conservative. Changes accumulate elsewhere while these deeper features appear relatively "frozen" over evolutionary time. This is the basis of von Baer's "law"—revealed in our models as a probabilistic generalization. While things can appear early without being GE'd, and or appear late and be GE'd, there are strong stochastic associations between earliness in development and increasing probabilities and degrees of generative entrenchment.

We have ways for modulating or weakening generative entrenchment for some purposes so we can (occasionally) make deeper modifications and get away with it (Wimsatt 1987). Crucial for cultural evolution, most of these ways do not apply or apply only in weaker forms for biology.[21] For this and other reasons (notably the horizontal transmission of practices and parasites[22])—cultural evolution commonly proceeds much faster than biological evolution.[23]

These exceptions aside, one can predict which parts of such structures are more likely to be preserved, or to change—and over broader time scales, their relative rates of change—in terms of their generative entrenchment. What kinds of structures? They could be quite diverse: propositions in a generated network of inferences; laws or consequences in a scientific theory; experimental procedures or pieces of material technology; structures or behaviors in a developing phenotype; relationships between organism and niche resources; cultural institutions or norms in a society; or the dynamical structures—biological and cognitive—driving cognitive development. GE necessitates and provides a reconceptualization and replacement of the traditional innate/

acquired distinction (Wimsatt 1986, 1999b). Turner (1991) employed generative entrenchment to analyze the distinction between literal and figurative meaning. Griffith (1996) uses GE in a related way to suggest an analysis for natural kinds. It has powerful applications to cultural evolution in general, and scientific change in particular (see Callebaut 1993: 331–334, 378–383, 425–429; Griesemer and Wimsatt 1989; Wimsatt 1995). Entities with larger generative roles are more foundational (in role and properties), more likely to persist to be observed and, (for some fraction of them) to grow. If they are more robust than alternatives, they are more likely to have been there from the beginning, and to apparently have an almost unconditional necessity.

Returning to biology, Arthur (1997) provides the broadest current review but does not cover everything. In developmental genetics, Rasmussen (1987) explicitly drew on my 1986 to predict the broad architecture of the developmental program of *Drosophila melanogaster* from effects of genetic mutants and data from comparative phylogeny. Even emerging before most of the HOX "explosion" (which yield a strikingly rich GE structure), the vast majority of its predictions remain sound. Schank and I have applied GE models to problems in biological evolution and development ranging from the architecture of gene control networks (1988), and the role of modularity in development (1998), to the evolution of complexity (Wimsatt and Schank 1988).

The generative form of adaptive structures affects the foci and relative rates of their evolution. Evolution acts on that form. So one should be able to trace feedbacks from developmental pattern to evolutionary trajectory and back again, identifying pivot points where relative stasis or elaboration can cause major changes in evolutionary direction. (Generative entrenchment as an engine!) These are *second-order* effects. They give GE theories greater explanatory power than one might at first suppose. Campbell's (1974) "vicarious selectors" have demonstrated abilities to create or facilitate new effectively autonomous higher level dynamics.[24] Thus perception (a vicarious selector) plus mate-choice can create runaway sexual selection processes leading natural selection in new directions (Todd and Miller 1998). Cultural evolution (Boyd and Richerson 1985; Wimsatt 1999a) is similarly open-ended—partly for similar reasons. These processes build upon and interact richly with each other. Partial products of GE, they provide both new opportunities for its action and occasions for its use as a tool of analysis. Applied to the evolution of cognitive and cultural systems, generative entrenchment extends evolutionary epistemology with new predictions and explanations. Differences among these processes in different areas affect how the models must be developed and applied and the kinds of results expected. These differences, and the character, consequences of, and interactions among these elaborating second-order effects will provide many areas for further development of theory and models.

The biological theory of evolution is not properly seen as an eliminative reductive one, but as an integrative theory articulating a diversity of other theories about biological structures and processes. So DST must be capable of articulating the insights and assessing the consequences of the processes, forces, groups, and objects studied in the various biological and human sciences. Generative dependency structures and differential entrenchment are found widely there, and should therefore provide a widely useful tool in their analysis.

Generative Entrenchment, Satisficing Design, and Historical Contingency

History matters to evolution. It is not too far wrong to say that everything interesting about adaptation is a product of selection for improvements in design, or of history, or their interaction.[25] Gould (e.g., 1989) has emphasized the role of contingency in evolutionary processes, arguing that minor unrelated "accidents" or "incidents"

can massively change evolutionary history.[26] It seems plausible—indeed (as I will argue) almost inescapable—that a successive layered patchwork of contingencies has affected not only the detailed organic designs we see, and variations among conspecific organisms, but also much deeper things—the very configuration and definition of the possible design space, and the regions in it they occupy. Deep accidents from the distant past not only define the constraints of current (so-called) optimizations, but constraints on these constraints, and so on, moving backward through a history of the deposition of exaptive dependencies (Gould and Vrba 1982) which become framing principles for the design of successively acquired and modified adaptations.

GE provides an explanation, perhaps the only possible explanation, for how and why this is possible. In reproducing heritable systems, GE and selection may provide sufficient conditions for the incorporation and growing importance over time of contingency, and of history, in the explanation of form. Not only that: Because older contingencies become more general and more entrenched, we have an explanation through GE for the broader and deeper (contingent) regularities of nature—usually not for the details of the particular regularities, but for why there should be *any* such regularities.[27] In this GE is like "chaotic" dynamics—there can be sensitive dependence on initial conditions, but unlike that case, the interest of GE arises only after the impact of these conditions has become relatively entrenched, by which time further small changes lose their capacity to affect those aspects of the outcome.

Because of the dominant role of population genetics in evolutionary biology, stochastic genetic events and processes—point mutations, tandem duplications, inversions, segregation events, independent assortments, and other recombinations in inheritance—are the most commonly cited sources of contingency in evolution. Equally important are chance ecological events: meetings leading to matings, migrations, symbioses, exclusions, parasitisms, and predations. As George Williams (1966) quipped, "To a plankton, a great blue whale is an act of God." Better design as a plankton will not help if it is in the wrong place at the wrong time.

But accidental or contingent events do not by themselves make history. They *need not* leave distinguishable historical traces much later. Most do not. They may be:

1. Not heritable, or

2. even if locally heritable, are averaged out (or erased) in changing drift processes in space and time—lost in multiple intersecting entropic processes.

3. Damped out *via* "noise tolerant" design in phenotype or genotype. Biological processes are paradigmatically noise tolerant at all levels: synonymies in the genetic code, diploidy, alternative metabolic pathways for many critical functions, up through developmental canalization, growth allometries, bilaterally symmetric organs, and macroscopic regulatory features of individual physiology, a hierarchy of learning and generalization processes, to various mechanisms in social groups, breeding populations, ecosystems, and trophic levels. Multiple realizeability, so favored by philosophers of psychology, has its primary significance as noise tolerance to evolutionists. Noise tolerance produces neutrality or near-neutrality among evolutionary variants, allowing drift among them and increased probability of loss of any particular variant. (Most "neutral" mutations are not intrinsically so, but owe their neutrality to this class of mechanisms.) So things captured here feed to (2). Even if neutrality doesn't lead to loss, it may lead us to miss the variation—or to *decide* not to notice it (as unimportant) in our accounts.

4. Optimization (or selection) also erases history on evolutionary time scales if the optima do not stand still. Usually they do not. Over time, changing adaptations to changing circumstances gradually erode and reconfigure *everything that is changeable* (this point is robust, and arguable in different ways: contrast Lewontin's 1966 "capricious evolution" with Van Valen's 1973 "Red Queen"). But if the optima were static, the closer you approach an adaptive peak and the longer you remain there, the less obvious is the path you took to get there—or at least so it would seem.

The only systematic means of escape from the fluctuations of fleeting optima over time is through generative entrenchment *via* an important property of the latter: The more deeply generatively entrenched something becomes, the more *context-independent* and the larger is the fitness loss if it is disturbed. Deeply generatively entrenched things become really established—the deepest become *functional necessities*. This says in another way that deeply GE'd things will be preserved.[28] If the fitness decrement from their loss gets larger and more invariant, so do they.

To mark history, an event must cause cascades of dependent events that affect evolution. Some contingent events are massive, immediately marking diverse biotic and geophysical processes, like the "large body" impacts recorded at the K-T boundary extinguishing the dinosaurs and most other species, giving small mammals a chance. No surprise: such a massive cause *should* have had far-reaching effects.

But most evolutionary "contingencies" start small—the luck of the draw for George Williams's plankton or single-base mutations initiating selective cascades of layered exaptations with divergent consequences. Oxygen production—metabolic byproduct in ancient plants—presumably started small, but hardly any contingency has had broader or greater consequences for evolution, by spreading as these plants succeeded, and becoming a much larger process. As this atmospheric poison rose in concentration, oxygen was initially adapted to, then became utilized almost universally throughout the animal kingdom, driving an energetically richer metabolism which now depends on it. This is probably the best possible single case for Levins and Lewontin's (1983) argument that organisms construct their environments as well as conversely. Small contingencies that leave a mark in evolutionary history do so by becoming larger—amplified by recurrent processes—for organisms, by reproduction.[29]

Cascading sequential dependencies also occur in each individual during development—commonly, but not exclusively through descent processes in cellular reproduction.[30] Not all of these are "contingencies." "Generic" events (Kauffman 1993) or forces reflecting laws of nature (e.g., gravity or surface tension) can become GE'd—deeply utilized or "presumed" in the design of adaptive structures (Wimsatt 1986; Schank and Wimsatt 1988). D'Arcy Thompson (1917) saw in the rich similarity of animate and inanimate natural forms the direct action of laws of nature, and thought he could dispense with selection. But in most cases on the organic side he was seeing the interplay of selection with properties of matter and deep entrenchment—the generation of what Griffiths and Gray (1994) would classify as highly heritable nongenetic resources. Thus it is an error to suppose that generative entrenchment and self-organization must be alternative competing explanations for deep regularities of form. They are inextricably interdigitated (Wimsatt 1986).

But GE can also happen to arbitrary contingencies, rendering them more or less context-dependent adaptive necessities—all the more striking for their seeming arbitrariness. These reflect an evolutionary history of contingencies, of exaptation layered upon exaptation, a history unique to, characteristic of, and divergent in different lineages. This is the architecture of adaptation. This creation of layered dependencies produces GE. Laws of nature, and generic properties of ensembles of systems can be expected to show through a lot of noise—developmental noise and evolutionary noise. But the vast majority of arbitrary small contingencies vanish without a trace, through one of the four mechanisms given earlier.

So GE is essentially the *only* way that such smaller contingencies have a reasonable chance to be preserved over long stretches of macroevolution. Drift alone will not suffice. We see evolution as a contingent process primarily *because* of generative entrenchment. Or to put it another

way, it is probably not a bad first approximation to say: generative entrenchment is why history matters!

So these systematic effects—the structure of dependencies in development—can also have systematic evolutionary consequences. Developmental processes are the source of these patterns—for why contingencies can persist. So taking developmental systems seriously simultaneously explains two of the most striking features of organic life:

First, generative entrenchment explains the frequent deep similarities in organic architecture of varying degrees of generality roughly mapping phylogeny onto causal depth in developmental process in the life cycle of organisms. A cluster of related heuristic principles utilizing GE play major roles in inferences from development to phylogeny, and conversely. Haeckel's "Ontogeny recapitulates phylogeny" is dead, but these similar inferential principles are alive and well, though not unrestricted (Raff 1996; Wimsatt 1998; see Rasmussen 1987 for some of the principles and how they work; Nelson 1998 considers issues of GE in phylogenetic inference in detail).

Second, GE explains why some features that seem bizarre, inefficient, or only arbitrary should show so much persistence, so much evolutionary inertia after chance or selection have kept them for a time, and other things have come to depend upon them. It does not explain why they occur in the first place. These marks of contingency are more quickly lost—never visible (as in the quantum transitions through which an ionization might lead to a base substitution), or local and fleeting (the predator from which it zigged instead of zagged and got away), or not so local and not so fleeting but still not saved (changing selective optima).

An important part of seeing how GE can act as an explanation here is to see that differential GE is a *generic* feature of phenotypes—of life cycles, and repeated assemblies. Differential GE is lawlike in character—both in its widespread applicability, and in the near inevitability of its consequences. As something gets more deeply GE'd, selection forces opposing changes in it become—stochastically, but inevitably—more unconditionally negative, more context independent. This is because of the embeddedness of GE'd parts, the richness of their connections. The diverging consequences lead to less localized (more holistic) effects. And things with very different degrees of GE in the same repeated assembly can reliably and robustly—even if not certainly—be expected to change at different rates. That is a crisp prediction. In these important ways any holism attributable to DST is *not* totally symmetric, *not* uninformative, *not* uninvestigateable, *not* experimentally intractible.

People have pointed to the importance of contingency in evolution, argued for it, presupposed it, but not tried to explain it—not as a property of individual cases (left- versus right-coiling shells) or periods (the great loss of Cambrian Bauplans)—but as a generic property of systems that can fix and build up richly textured tapestries of accidents. GE does so. It does not explain their origin, or early preservation. But the longer the contingencies persist, becoming more deeply generatively entrenched, and commonly, more widely distributed among descendant species, the larger a mark they have left on evolution, and the larger the probability they will continue to leave marks. GE is also a sound and robust explanation for why evolution should be so much marked by contingency—and for why the contingencies persist. Unlike the leveling and homogenizing processes attributed to the second law of thermodynamics, evolution seems to leave an ever more complex and filigreed history over time. Generative entrenchment is the primary explanation for this difference. Note that in this sense, evolution can be cumulative without being progressive.

A Parting Shot Promising Later Broadsides

It is time to reclaim something for DST that has been long and proudly held (but only with squat-

ters' rights) by population genetics. Some of the most powerful explanations in evolutionary theory arise from observations or assumptions about reproductive rates, body size, and generation times of different species. Fitness is closely connected with reproductive rates, and evolutionary rates with generation time, and generation times with body size. On our time scale we can confidently predict that insects will rapidly evolve pesticide resistance, bacteria will generate antibiotic resistance even faster, that there will be new influenza viruses each year that threaten humans (even those who had it last year), that large fluctuations in prey numbers will render comparably sized predators at least locally extinct. (Predictions of this sort are strikingly robust and independent of detail for such complex phenomena: unlike the hypothetical and usually falsified predictions of idealized models.)

Because population genetic models abstract away from phenotypic process—body size, reproductive rate, and generation time must enter by assumption. This escapes notice because such information is close to our experience, and we do not derive it from our models. (Even a rocket scientist untutored in biology would know that flies reproduce like hell and that there are lots of them.) So flies have lots of mutations, lots of generations, and lots of offspring for every one of each that we can muster. So they can always run faster than we can—evolutionarily speaking. These properties are close to and flow naturally from developmental models. (Body size, reproductive rate, and generation time are all related phenotypic properties, and show strong correlations across species which are relatively insensitive to all kinds of empirical details (see Bonner 1965, chap. 2). Those who say that they don't need DST do not realize how much they have filched from it already or from common knowledge of developmental processes.

To see what kinds of things we might get from development, let's pick an area that has undergone massive recent elaborations without benefit of developmental insights. This is cultural evolution viewed through the lens of the selfish meme. The closest meme theorists come to metazoan development (or any sort of multiunit complexity) is vague talk of "meme-complexes." These references are so little elaborated that one cannot tell whether meme-complexes are supposed to correspond to multilocus genotypes, developmental complexity, ecosystems, all of the above, or none of the above. One of many problems with "memetics" (if there is any such beast) is that in focussing on autocatalytic ideas of self-replication, intuitions from development have been almost totally left out.

I will not provide such a theory here (though I am working on one). I sketch a simple model of cultural transmission, (Wimsatt 1999a; Callebaut 1993: 425–429, figure 9.2). Assume we are developing individuals who acquire cultural traits, practices, knowledge systems, and skills over time, embedding them in and relating them to other generative things we have acquired. Usually (save for very simple things) we pass on these generative structures not by passing on the whole developed thing (the adult phenotype as it were), in one bolus, but by passing on only a few generators, (heterocatalytically characterized gene-like things). We provide—as individuals, and through our institutions—the resources to interact with what we have already passed on, acquiring further generators, to develop a mature understanding and working knowledge in the recipients. So imagine a just-so story greatly truncated, collapsing into a stage what would be months of sequential assignments, a high school physics teacher gives his already substantially prepared students the following extended assignment:

So copy down Newton's three laws, and work in groups to solve the simple problems next to them on the board. You may ask me any questions. I'll help, but I won't tell you the answers. Then go home and buy an annual car magazine for this year—one with all of the specifications in it. In terms of the horsepower and weights for each, estimate the ratios of the 0–60 times for all new cars, and what proportion of the ideal values their actual acceleration times are. Given their stopping times

from that distance, estimate their deceleration rates. In terms of their cornering coefficients, estimate how fast they will be able to go before spinning out on a 100-yard radius track. Finally, assuming that they are all made of iron, run into a concrete wall at 60 MPH., and all of their kinetic energy is turned into heat, how much hotter do they get? How much should their front bumpers collapse? List any assumptions you make as you go along. For cases in which the published answers are different from your calculated ones (essentially all of them), can you give plausible reasons for the differences? You may need additional information for some problems, so tell me what it is and how you would find out. How will you evaluate your sources? Which problems can't be solved with plausibly available information at all? Invent a problem of your own of comparable richness utilizing Newton's laws (automotive design has many others) and solve it or tell why you couldn't for as many parts as possible. For two weeks from today, design a lab to investigate and illustrate to others a related phenomenon that you found particularly interesting, and think of what assignments you would give them to take home.

This made-up time slice has predecessors and descendants in the same lineage, starting with arithmetic and the lever, and ranging up to particle physics and chaotic dynamics, and is loaded with sequential dependencies for necessary skills. Pick the "gene-like" generators as Newton's laws in this simple story of a developing condensed-matter physicist (there are many other generators as well). It is clear how many and varied the conceptual, material, and social resources these students will have to draw upon to complete the assignment, and how much richer their knowledge of physics will be than the mere statement of these laws, even though many textbooks act as if all of the information "followed from" by the axioms. Consider how much development and integration of the resources is required for each successful answer, explained anomaly, or analysis of why no answer is possible. If this seems to have parallels with the gene-selectionist story, it should. But let me make some predictions about cultural evolution from a DST perspective—turning only on reproductive rate and generation time, and not significantly utilizing the conservative properties of generative entrenchment, which has particularly rich application in this case (Wimsatt 1987):

1. Things that are not GE'd are paradigmatic "naked memes" and should be capable of the highest horizontal transmission rates consistent with the channel characteristics. These are cases which r-selective meme theorists seem to like the most. (They also reflect—in themselves—the least structure, and are likely to most anger social scientists when offered as examples of "culture".)

2. The more complex the cultural trait, the more slowly it should spread (it is more complex to learn and to teach, and there may be increased complexity in interfacing it with other elements of the culture.)

3. If the trait is entirely new to the culture (so that no one knows much of what they need to master it), there should also be a longer lag before it can spread (because it takes time to teach and to learn) and

4. it will be learned by a smaller number of individuals (who specialize for it) unless it becomes "socially required."

5. All of these will tend to slow down the reproductive rate and decrease the fraction of the population picking it up. The chance of "falling off the track" will increase even for those intending to complete it.

These are heavy reproductive burdens for complex cultural traits. How can we escape these limitations? Consider a complex cultural trait or related set of traits as a conceptual phenotype. If "cell lineages" in the phenotype differ in length, and gene-like generators (transmissible generative parts) can be placed on shorter lineages, then transmission can be accomplished before the phenotype achieves full maturity—(defined as being able to transmit the generators for the full complex theory or set of practices). So then it is possible to transmit generators to produce immature or partial phenotypes earlier, and one can have conceptual overlapping generations, ranging

from high school and undergrad tutors through assistant professors to emeriti. (Or brown belts through third-degree black belts running Tae Kwon Do clubs with members of different degrees of sophistication, and skill-ranked tournaments.) Using a series of immature individuals to transmit simpler parts before they could transmit the mature theory or practice can substantially increase the reproductive rate (a "master" multiplies his presence with younger assistants—see the workshops of major Renaissance artists) *and* shorten the effective generation time by finding simpler tasks through which an individual can produce useful labor. In addition, apprentices learn by teaching, and any teacher is likely to remain more interested if they are optimally challenged, teaching not too far below (or above!) their competency level. (See Csikszentmihalyi 1990, on "flow.") So more are likely to complete the course of instruction, more usefully, and with greater competence.

In addition, having more individuals who know parts of the complex practice introduces redundancy—not just as multiple copies might, but because they are intelligent agents who together may be able to reconstruct, correct, or improve pieces that they would not have otherwise as mere passive redundant copiers. Thus are born scientific communities, craft guilds, and priesthoods.

Is this too far from biology? Notice that we have just reinvented population growth with overlapping generations and social structure—for memes. (With an important difference, however: The memotype transmitted by immature parents is also less "mature." That's why "population memetics" would be such a mess: We don't just track genes or genotypes regularly assembled from pairs of parents, but all of the developmental stages of phenotypes—variously assembled in different proportions from variable numbers of "parents.") For all that, the teaching of younger apprentices parallels the "helping" behavior of brown jays in Costa Rica. (Lawton and Guindon 1981; Lawton and Lawton 1985, 1986; Williams,

Lawton, and Lawton 1994). The brown jays who are older siblings capable of building nests and laying eggs do not make very good nests and do not have the complexities of parental care down too well, so their expected yield of successfully fledged offspring is not much above zero. But if they stay at their parents' nest they learn how "on the job" (those who have helped for a year do better than those of the same age who didn't) (Lawton 1985), while their parents raise more related young than they would be able to otherwise. Helpers sometimes "cheat" by laying an egg of their own in their parents' nest—getting a "free ride". But maybe that happens in cultural evolution too—like borrowing the family car for a date!

I leave this speculative example to suggest more general principles of adaptive design for cultural evolution which come out of a developmental perspective.

6. If a cultural trait has a structure which is GE'd, complex, and important, more elaborate scaffolding should evolve to facilitate learning it. And if the scaffolding can be used to learn other things, the scaffolding in turn may become more entrenched than the original. (This latter is exaptive elaboration.)

7. Evolution will structure the organization of developmental programs (dispositions of resources) so that assembly of GE structures will become (and look) self-organizing, and self-maintaining reliably across the range of environments normally encountered. (So GE and self-organization interact richly; see chapter 18.)

8. Judgments of "importance" for most cultural traits will derive in large part from their generative entrenchment in the production of other cultural traits.

Principle (6) roughly reflects the situation for learning mathematics, physics, or other complex "acquired" traits. And mathematics—though a difficult and GE'd subject matter in its own right—becomes scaffolding for all kinds of development, learning and empowering of theory in all of the "quantitative" sciences. By contrast, principle (7) reflects the situation for first-language acquisition, and for second-language acquisition if it is sufficiently similar to the first, and begun sufficiently early. And, if so, the self that gets organized should not be just the language ability but the cluster of competencies through which it emerges. These probably include a variety of cognitive, conative, social, affective, and motor skills. In this sense, language ability is probably no more *self*-organizing than genes are *self*-replicating. (It is a common reductionistic bias to draw the boundaries of the system in the wrong place—always too small; see Wimsatt 1980b, 1997b.) Finally principle (8) is an observation that follows when the organization of any produced system increases in complexity and is structured or maintained at least partially along functional lines. And that is—taken as a whole—many more parts of a functionally organized structure are means than ends (Wimsatt 1997b). This is likely to be the most controversial thing I've said. But to make sense of it, we need to remember that activities can become autotelic through learning and exposure—we can acquire tastes defining things as good-in-themselves. And that does not mean that they may not nonetheless also serve cultural ends.

Finally, are any of the predictions of this section surprising? Perhaps not. Most are already familiar. We had other ways of getting them already, or they were parts of our embodied common sense—save perhaps for the last four. (We've had examples of all of them staring us in the face!) What is more surprising is where the predictions came from. And these will not be the last.

Acknowledgments

Some of this paper is elaborated from my 1999a and 1999b. In thinking about developmental systems, I have benefited from long collaborations with Jim Griesemer and Jeff Schank, more recent discussions with Julio Tuma and Jason Roberts, and timely feedback from Susan Oyama and Paul Griffiths. Schank's work has been particularly pivotal in developing many of these ideas into richer and more testable forms.

Notes

1. See Glassman and Wimsatt (1984); Rasmussen (1987); Schank and Wimsatt (1988, 1998); Wimsatt and Schank (1989); Wimsatt (1981: 178–181, 1986, 1999a, 1999b); and Callebaut (1993: 331–334, 378–382, 385–386, 425–427).

2. Indeed, a closer look at any internal combustion engine would show nothing different: structures yielding patterns of constrained motion which, with the release of imported sources of energy, are converted into the ability to do work.

3. This would be the wrong lesson to draw from Schaffner (1998). Another careful exploration of this territory is Sterelny, Smith, and Dickison (1996). But both are more gene-centered in flavor than I prefer.

4. Sterelny, Smith, and Dickison (1996) see this thicket of overlapping boundaries as a problem, which they apparently think is special to DST. I disagree. Some kinds of systems, and problems require recognizing and dealing with multiple boundaries simultaneously, and multiple relevant ways of decomposing larger systems into parts. See Wimsatt (1974, 1994) for analysis and McHarg (1971) for lovely examples.

5. See Wimsatt (1976b, 1980b, 1981b, 1994, 1997a, 2001). The best recent review of this enormous literature is Sarkar (1998).

6. For qualifications see Brandon (1990: 6–9). See also Campbell's (1974) "Blind variation and selective retention." And substitute "number of" for "causal tendencies to have" in (3), and drift is also covered.

7. Focusing on replication as copying abstracts away from both of these processes. The focus is on the information transmitted, and not on the process which Griesemer calls "reproduction" (2000a, 2000b).

8. Arguments from and about fundamentality are almost inevitably misused. We must learn to deal with complex systems simultaneously through a variety of theoretical perspectives and at various levels of organization—usually without fundamentally privileging any of them (Wimsatt 1994). But these five conditions give a richer and more powerful view of evolution than that

of "the selfish gene," though it would be foolish not to use the rich structure arising from genetic specifications of the problem as well, *when that is genuinely available*. (Most commonly it is not.) So called multilocus traits are multidimensionally environmental too! An adequate genetic specification without genomic and environmental context-dependencies seems (just contingently?) *never* to be at hand.

9. Before (1993), e.g., in (1985), Kauffman presented self-organization and selection as alternative causes. In 1993, he adopted a weaker view, with selection and self-organization as complements (the parenthetic qualifications)—a view I have urged in print since 1986 (and in conversation much earlier). Kauffman had been treating *genericity* as a dichotomous variable, suggesting the former. But it is a degree property (Schank and Wimsatt 1988) so the latter seems inevitable.

10. I argued (1974) that it is not that simple. Simon's argument ignores coevolution of subassemblies and must be transformed to utilize development. See also Lewontin's "quasi-independence" (additivity or near-decomposability for small or moderate changes in a system that is globally epistatic), and "evolvability" (Lewontin 1978; Wimsatt 1981b; Dawkins 1989; Kauffman 1993; Dennett 1995; Schank and Wimsatt 1998; Brandon 1999).

11. I do not give a sufficient analysis of a developmental process; see Griesemer (2000a).

12. Some of these details are of crucial importance though—such as that metazoans go through a single-cell stage. Without this, GE structures could look radically different. So there is still a lot of work to do to get to life-cycles as we know them. But there is some hope; see Grosberg and Strathmann (1998).

13. Darwin opens (p. 13) by saying that "the laws governing inheritance are quite unknown," but then goes on to list a number of regularities variously described as "rules" or "facts" about inheritance. They appear not to be described as laws because (1) they have exceptions or (2) may be described in terms of tendencies, or (3) the conditions under which they work is unclear.

14. This is a kind of unavoidable axiom of design anchored in a combination of physically and evolutionarily generic principles which I will discuss elsewhere.

15. This argument works only from a symmetric starting point—and depending on conditions and the resulting organization of the phenotype, for some distance around it. It does not show that for *any* distribution of

selection coefficients over functional elements, "the rich will get richer, and the poor, poorer." Indeed, sufficiently extreme distributions could be expected to get less extreme.

16. Griesemer and I (1989) argue that misinterpretation of E. B. Wilson's diagram of Weismannism earlier this century contributed to the emergence of the modern genetic reductionist interpretation of evolutionary theory.

17. Griesemer's story (and those of Moss and Neumann-Held) are all richer than my proposal (1981b) to define genes in terms of their heterocatalytic function. Griesemer sees material transmission across generations as a crucial feature of reproduction—which may differentiate it from cultural transmission, with further interesting consequences. He also discusses the distorted representations of the biology induced by the "informational" or "replicator" conception of the gene.

18. In cases where either genes or environmental factors alone were picked out, it would usually be because explanatory contexts seemed to call for a focal object or process—which would never constitute the total cause.

19. From notes of Boulding's lecture at Max Black's seminar in the Humanities, Science, and Technology program at Cornell University, 1974–75.

20. "Lock in" is Brian Arthur's (1994) term for the same process; in economic models of developing technologies, city formation, and cases where adoptions of an action or technology reduces the cost of doing so to others—generating other adoptions. His simple models are in effect microevolutionary frequency-dependent selection models. They complement our microevolutionary simulations and macroevolutionary discussions of larger structures with bigger differences in GE.

21. "Deep" modifications in biology are a key theoretical problem in paleontology and developmental genetics, where GE arguments are commonly made. The problem is how to explain "Raff's hourglass"—the large amount of early developmental variation in closely related species of amphibia, echinids, and some other kinds of organisms. (Raff's 1996 famous "direct developing" sea urchins are evolutionarily recent but skip an entire and normally important larval stage.) The simplest GE models would predict monotonically declining variation at earlier developmental stages, a "cone" (Arthur 1997). So what is entrenched, what is not, and how does what is not escape? Some early things are obviously deeply entrenched—the genetic code, machinery for cell division, and the famous HOX clusters,

while others are not—synonymous code variants for the same protein, and any pathways which are not limiting in a redundant network. These are both discussed even in Wimsatt (1986). But there are many other possibilities. Michael Wade and I will review them with particular reference to Raff's hourglass in the near future.

22. Hull urges that from the practice's (or parasite's) point of view, the transmission is vertical not horizontal (so that differences between cultural and biological inheritance have been overstated). While correct, transmission is still "horizontal" relative to the host's biological descent. This situation arises with multiple intersecting channels or lineages of heredity. (We will tend to treat lineages of the larger units as vertical, and the smaller horizontal, especially if the smaller things also replicate faster.)

23. But biology often outruns culture, even if *our* biology does not. Bacteria may change properties (their virulence) affecting their fitness and ours in days or weeks in the course of an epidemic—much faster than insects evolve pesticide resistance, in turn notoriously faster than our response. Not only can biology be fast, but culture can be slow: Boyd and Richerson note (1985: 60) that Olduwan and Aechulian toolmaking practices were transmitted essentially unchanged for tens of thousands of generations—through major environmental changes and migrations.

24. Conditions for "dynamical autonomy" of higher-level phenomena consistent with reductive explanation of those same phenomena are provided in Wimsatt (1981, 1994).

25. Whether satisfied or optimized (I favor the former for theoretical and empirical reasons) design is subject to (usually unspecified or underspecified) constraints. History includes any particularities of initial conditions in space and time. These could seem like two capacious wastebaskets that together capture almost anything. Perhaps, but *evolutionary* history imposes additional constraints of generational repetition, and size and time scales which are used in the argument. And as we will see, GE gives history structure—a life on multiple scales simultaneously.

26. Evolutionary biologists use *contingency* differently from modern metaphysicians, but not unlike Aristotle's distinction between necessity versus chance or accident: The opposition is not between "necessary" or "analytic" versus "contingent" (as possibly false), but law-bound or highly probable versus unlikely or arbitrary.

27. Human history need not turn on the GE of contingencies, though invocation of cultural norms or institutions (which are characteristically GE'd) would bring it in. Narrative explanations can serve other ends. In contexts of fixing responsibility we can pick any event, large or small, and ask what lead up to it. If small, it need not lead to or through other events with diverging consequences. (But if it did we would likely use them to provide organizing patterns—likely particular and dated—for our story.)

28. Preserved, and also (*re* functional necessities) that our very conception of that kind of system now includes that feature—and returns to Griffith's (1996) use of GE in his nonessentialist gloss on natural kinds. It also has obvious implications for teleosemantics.

29. This applies also to their detection. A found fossil, even the only known instance of its type, was likely of an organism much more common in its time and place.

30. Cellular descent is an important process for GE, but only one of many. This was a flaw in Arthur's earlier presentations of GE-style theories. He drops this in his (1998), recognizing, as we have urged—see Rasmussen (1987); Schank and Wimsatt (1988)—that GE works as well for biochemical pathways (see Morowitz 1992) and "horizontal" interactions. Thus circulating hormones produced by localized tissues spread consequences to other cellular lineages, to the organism as a whole, and even to the social group (Stern and McClintock 1999).

References

Arthur, B. W. (1994). *Increasing Returns and Path Dependence in the Economy.* Ann Arbor: University of Michigan Press.

Arthur, W. (1997). *The Origin of Animal Body Plans: A Study in Evolutionary Developmental Biology.* Cambridge: Cambridge University Press.

Bonner, J. T. (1965). *Size and Cycle.* Princeton, NJ: Princeton University Press

Boyd, R., and P. Richerson. (1985). *Culture and the Evolutionary Process.* Chicago: University of Chicago Press.

Brandon, R. (1990). *Adaptation and Environment.* Princeton, NJ: Princeton University Press.

Brandon, R. (1999). The units of selection revisited: The modules of selection. *Biology & Philosophy* 14(2): 167–180.

Callebaut, W. (1993). *Taking the Naturalistic Turn: How to Do Real Philosophy of Science.* Chicago: University of Chicago Press.

Campbell, D. T. (1974). Evolutionary epistemology. In P. A. Schilpp (Ed.), *The Philosophy of Karl Popper* vol. 2, pp. 413–463. LaSalle, IL: Open Court.

Caporael, L. (1997). The evolution of truly social cognition: The core configuration model. *Personality and Social Psychology Review* 1: 276–298.

Csikszentmihalyi, M. (1990). *Flow: The Psychology of Optimal Experience.* New York: Harper and Row.

Darwin, C. (1859). *The Origin of Species.* London: John Murray.

Dawkins, R. (1989). The evolution of evolvability. In C. Langton (Ed.), *Artificial Life,* pp. 201–220. Redwood City, CA: Addison-Wesley.

Dennett, D. (1995). *Darwin's Dangerous Idea.* New York: Simon and Schuster.

Glassmann, R. B., and W. C. Wimsatt. (1984). Evolutionary advantages and limitations of early plasticity. In R. Almli and S. Finger (Eds.), *Early Brain Damage* vol. 1, pp. 35–58. Orlando: Academic Press.

Gould, S. J. (1997). *Ontogeny and Phylogeny.* Cambridge, MA: Harvard University Press.

Gould, S. J. (1989). *Wonderful Life: The Burgess Shale and the Nature of History.* New York: W. W. Norton.

Gould, S. J., and E. Vrba. (1982). Exaptation—a missing term in the science of form. *Paleobiology* 8: 4–15.

Gray, R. (2000). Selfish genes or developmental systems? Evolution without replicators and vehicles. In R. Singh, C. Krimbas, J. Beatty, and D. Paul (Eds.), *Thinking about Evolution: Historical, Philosophical, and Political Perspectives,* pp. 184–207. Cambridge, Cambridge University Press.

Griesemer, J. R. (2000a). Reproduction and the reduction of genetics. In P. Beurton, R. Falk, and H-J. Rheinberger (Eds.), *The Concept of the Gene in Development and Evolution,* pp. 240–285. Cambridge: Cambridge University Press.

Griesemer, J. R. (2000b). Development, culture and the units of inheritance. In *Proceedings of the Philosophy of Science Association 1998* vol. 2 (*Philosophy of Science,* 67: S348–S368.

Griesemer, J. R. (forthcoming). *Reproduction in the Evolutionary Process.*

Griesemer, J. R., and W. C. Wimsatt. (1988). Picturing Weismannism: A case study in conceptual evolution. In

M. Ruse (Ed.), *What Philosophy of Biology Is,* pp. 75–137. Dordrecht: Martinus-Nijhoff.

Griffiths, P. (1996). Darwinism, process structuralism and natural kinds. *Proceedings of the Philosophy of Science Association 1996* vol. 1 pp. 1–9.

Griffiths, P., and R. Gray. (1994). Developmental systems and evolutionary explanation. *Journal of Philosophy* 91: 277–304.

Grosberg, R. K., and R. R. Strathmann. (1998). One cell, two cell, red cell, blue cell: The persistence of a unicellular stage in multicellular life histories. *TREE* 13(3): 112–116.

Hull, D. L. (1980). Individuality and selection. *Annual Review of Ecology and Systematics* 11: 311–332.

Hull, D. L. (1988). *Science as a Process.* Chicago: University of Chicago Press.

Kauffman, S. A. (1993). *The Origins of Order.* Oxford: Oxford University Press.

Keller, E. F. (1995). *Refiguring Life: Metaphors of Twentieth-Century Biology.* New York: Columbia University Press.

Lawton, M. F., and C. F. Guindon. (1981). Flock composition, breeding success and learning in the brown jay. *The Condor* 83: 27–33.

Lawton, M. F., and R. O. Lawton. (1985). The breeding biology of the brown jay in Monteverde. *The Condor* 87: 192–204.

Lawton, M. F., and R. O. Lawton. (1986). Heterochrony, deferred breeding and the evolution of avian sociality. In R. Johnston (Ed.), *Current Ornithology* vol. 3, pp. 187–224. New York: Plenum Press.

Lewontin, R. C. (1966). Is nature probable or capricious? *Bioscience* 16: 25–27.

Lewontin, R. C. (1970). The units of selection. *Annual Review of Ecology and Systematics* 1: 1–18.

Lewontin, R. C. (1978). Adaptation. *Scientific American* 239(3): 156–169.

Lewontin, R. C. (1983). The organism as the subject and object of evolution. *Scientia* 118: 63–82.

Lorenz, K. Z. (1965). *Evolution and Modification of Behavior.* Chicago: University of Chicago Press.

McHarg, I. (1971). *Design with Nature.* New York: Natural History Press.

Moss, L. 1992. A kernel of truth? On the reality of the genetic program. In D. Hull, A. Fine, and M. Forbes (Eds.), *Proceedings of the Philosophy of Science Asso-ciation 1992* vol. 1, pp. 335–348. East Lansing, MI: The Philosophy of Science Association.

Moss, L. (1998). *What Genes Can't Do.* Doctoral dissertation, Northwestern University.

Morowitz, H. (1992). *The Origins of Cellular Life.* New Haven: Yale University Press.

Neumann-Held, E. M. (1999). The gene is dead—long live the gene! Conceptualizing genes the constructionist way. In Peter Koslowski (Ed.), *Sociobiology and Bioeconomics—The Theory of Evolution in Biological and Economic Theory,* pp. 105–137. Berlin: Springer.

Nelson, P. (1998). *Common Descent, Generative Entrenchment, and the Epistemology of Evolutionary Inference.* Doctoral dissertation, University of Chicago.

Oyama, S. (1985). *The Ontogeny of Information.* Cambrdige: Cambridge University Press. (2d. rev. ed., Durham, NC: Duke University Press, 2000.)

Raff, R. A. (1996) *The Shape of Life: Genes, Development and the Evolution of Animal Form.* Chicago: University of Chicago Press.

Rasmussen, N. (1987). A new model of developmental constraints as applied to the *Drosophila* system. *Journal of Theoretical Biology* 127(3): 271–301.

Sarkar, S. (1998). *Genetics and Reductionism.* Cambridge: Cambridge University Press.

Schaffner, K. F. (1998). Genes, behavior and developmental emergentism: One process, indivisible? *Philosophy of Science* 65: 209–252.

Schank, J. C., and W. C. Wimsatt. (1988). Generative entrenchment and evolution. In A. Fine and P. K. Machamer (Eds.), *Proceedings to the Philosophy of Science Association 1986* vol. 2, pp. 33–60. East Lansing, MI: The Philosophy of Science Association.

Schank, J. C., and W. C. Wimsatt. (2000). Evolvability: adaptation, and modularity. In R. Singh, C. Krimbas, J. Beatty, and D. Paul (Eds.), *Thinking about Evolution: Historical, Philosophical, and Political Perspectives,* pp. 322–335. Cambridge: Cambridge University Press.

Simon, H. A. (1962). *The Architecture of Complexity.* (3d ed. 1996.) Cambridge, MA: MIT Press.

Sperber, D. (1995). *Explaining Culture.* London: Blackwell.

Sterelny, K., K. Smith, and M. Dickison. The extended replicator. *Biology and Philosophy* 11: 377–403.

Stern, K., and M. K. McClintock. (1998). Regulation of ovulation by human pheromones. *Nature* 392: 177–179.

Szathmáry, E., and Maynard Smith, J. (1997). From replicators to reproducers: The first major transitions leading to life. *Journal of Theoretical Biology* 187: 555–571.

Thompson, D. W. (1917). *On Growth and Form.* Cambridge: Cambridge University Press.

Todd, P., and G. Miller. (1998). Biodiversity through sexual selection. In C. Langton (Ed.), *Artificial Life V.* Cambridge, MA: MIT Press.

Turner, M., (1991). *Reading Minds: The Study of English in the Age of Cognitive Science.* Princeton, NJ: Princeton University Press.

Van Valen, L. (1973). A new evolutionary law. *Evolutionary Theory* 1: 1–30.

Williams, D., M. F. Lawton, and R. O. Lawton. (1994). Population growth, range expansion, and competition in the cooperatively breeding brown jay, *Cyanocorax morio. Animal Behavior* 48: 309–322.

Williams, G. C. (1966). *Adaptation and Natural Selection.* Princeton, NJ: Princeton University Press.

Wimsatt, W. C. (1974). Complexity and organization. In K. F. Schaffner and R. S. Cohen (Eds.), *Proceedings of the Philosophy of Science Association 1972* vol. 20, pp. 67–86. Dordrecht: Reidel.

Wimsatt, W. C. (1976). Reductive explanation: A functional account. In A. C. Michalos, C. A. Hooker, G. Pearce, and R. S. Cohen (Eds.), *Proceedings of the Philosophy of Science Association 1972* vol. 30, pp. 671–710. Dordrecht: Reidel.

Wimsatt, W. C. (1980). Reductionistic research strategies and their biases in the units of selection controversy. In T. Nickles (Ed.), *Scientific discovery—Vol. 2: Case Studies,* pp. 213–259. Dordrecht: Reidel.

Wimsatt, W. C. (1981). Units of selection and the structure of the multi-level genome. In P. D. Asquith and R. N. Giere, (Eds.), *Proceedings of the Philosophy of Science Association 1980* vol. 2, pp. 122–183. Lansing, MI: The Philosophy of Science Association.

Wimsatt, W. C. (1986). Developmental constraints, generative entrenchment, and the innate-acquired distinction. In W. Bechtel (Ed.), *Integrating Scientific Disciplines,* pp. 185–208. Dordrecht: Martinus-Nijhoff.

Wimsatt, W. C. (1987). Generative entrenchment, scientific change, and the analytic-synthetic distinction. (Unpublished manuscript.)

Wimsatt, W. C. (1994). The ontology of complex systems: Levels, perspectives and causal thickets. *Canadian Journal of Philosophy* Supp. vol. 20, pp. 207–274.

Wimsatt, W. C. (1995). The analytic geometry of genetics: The evolution of punnett squares. (Unpublished manuscript.)

Wimsatt, W. C. (1997a) Aggregativity: Reductive heuristics for finding emergence. In L. Darden (Ed.), *Proceedings of the Philosophy of Science Association 1996* Vol. 2, pp. S372–S384.

Wimsatt, W. C. (1997b). Functional organization, functional analogy, and functional inference. *Evolution and Cognition* 3(2): 102–132.

Wimsatt, W. C. (1998). Simple systems and phylogenetic diversity. *Philosophy of Science* 65: 267–275.

Wimsatt, W. C. (1999a). Genes, memes, and cultural inheritance. *Biology and Philosophy* 14: 279–310.

Wimsatt, W. C. (1999b). Generativity, entrenchment, evolution, and innateness. In V. Hardcastle (Ed.), *Where Biology Meets Psychology: Philosophical Essays,* pp. 139–179. Cambridge, MA: MIT Press.

Wimsatt, W. C. (2001). *Re-engineering Philosophy for Limited Beings: Piecewise Approximations to Reality.* Cambridge: Harvard University Press.

Wimsatt, W. C., and J. C. Schank. (1988). Two constraints on the evolution of complex adaptations and the means for their avoidance. In M. Nitecki (Ed.), *Evolutionary Progress,* pp. 231–273. Chicago: University of Chicago Press.

Wright, S. (1931). Evolution in Mendelian populations. *Genetics* 16: 97–159.

18

Developmental Systems, Darwinian Evolution, and the Unity of Science

Bruce H. Weber and David J. Depew

We take it as a presumption of this chapter that we are living at a time when the central problem in theoretical biology is the integration of developmental biology with genetics and evolutionary theory (Gilbert, Opitz, and Raff 1996). Even its founders admit that developmental biology was set aside for the time being when the Modern Evolutionary Synthesis was shaped (Mayr and Provine 1980).[1] Surprisingly, orthodox neo-Darwinians (who assume that organisms are the primary targets and beneficiaries of selection) have made little progress on this issue since then. Accordingly, the field has been left to contestations between molecular reductionists, who assume that the problem of development is simply the problem of turning structural genes on and off, and those who identify in one way or another with the contemporary "developmentalist challenge," who are confident that what genes do is far from the whole story.[2]

A tendency to drive this contestation to extremes manifests itself whenever proposed revisions in received evolutionary theory are brought into play. At one extreme, molecular reductionists favor "selfish" genes as units and beneficiaries of Darwinian selection (Dawkins 1989). To do so, they blur the distinction between molecular and evolutionary gene concepts. At the other extreme, some developmentalists suspect that integration between developmental biology and evolutionary theory cannot take place until evolutionary theory has been wrested away not simply from reductionistic versions of genetic Darwinism, but from the Darwinian tradition as a whole. Authors of the latter stripe are keenly aware that the Modern Evolutionary Synthesis was not in fact the first evolutionary synthesis, and so might not be the last. Post-Weismannian Darwinism, of which the Modern Synthesis is a development, broke with the assumed parallelism between ontogeny, ecology, and phylogeny that was the common coin of evolutionists of all stripes at the end of the nineteenth century. Some contemporary developmentalists, especially those

who appreciate the ecological orientation of the older movement, would like to see evolutionary theory put back into the developmentalist matrix from which statistical Darwinism broke free in the twentieth century (Salthe 1993; Webster and Goodwin 1996; Ulanowicz 1997; see Depew and Weber 1995 and Gilbert 1991 for an historical discussion of this topic). For their part, Darwinians of every persuasion predictably resist these proposals as falsely implying that evolution over phylogenetic time is progressive in some meaningful sense. They are keenly sensitive to the fact that most public discourse about evolutionary matters shows ill-digested traces of the "old" evolutionary synthesis, many of which are far from benign in their consequences.

Developmental systems theory (henceforth DST) is a welcome intervention in this unhappy dialectic. Although those associated with it typically exhibit a marked animus against genocentric versions of Darwinism, DST (which should be construed as part of a more encompassing "developmentalist challenge") has in recent years shown that it can present itself as a genuine Darwinian research program (Griffiths and Gray 1994). DST's advocates begin by arguing that if we take as seriously as we should the fact that organisms develop we must regard them as epigenetically constructed in each generation from a large array of developmental resources, some heritable, some not, rather than as read out or printed out from a causally primary, quasi-preformationist genetic program (Oyama 1985). In consequence, genes are regarded by DST as only one of a number of developmental resources. Natural selection is construed as able to act in principle on variations in any or all of these resources. The differential retention of these variants will manifest itself as fitness-enhancing changes in aspects of the life cycle in a particular environment.

From the point of view of philosophers of biology, among whose major functions is to identify conceptual biases that retard the course of scientific work and to recommend new conceptual frameworks calculated to get things moving again, DST is perhaps the most interesting development since the nineteen seventies, when philosophers of biology intervened in the so-called units of selection controversy in order to suggest that Darwinism is not formally committed to selection at only one level. In part, that is because DST itself intervenes in this very controversy in a way that is intended to put an end to it. For DST's champions infer from the presumptive causal parity of all developmental resources that the replicator/interactor distinction, on which the units of selection debate has been predicated, is ill conceived (Griffiths and Gray 1994; Griffiths and Knight 1998). When phenotypic traits are construed as developmental resources it can be seen that they are as much replicators as interactors. Conversely, when the full extent of the dependence of genes on other developmental resources comes into view, they lose their supposedly privileged status as replicators.

In this essay, our aim is to discuss an aspect of DST that has not, to the best of our knowledge, been fully explored by its most prominent advocates. It is better able, we maintain, to forge closer links with other sciences than the versions of Darwinism that it seeks to displace. In particular, we believe that, in addition to being able to solve the specific problems within evolutionary theory to which it overtly addresses itself, DST allows evolutionary biology to be more substantively connected to the physical and chemical sciences below and to the much-contested evolutionary behavioral sciences above. In this sense, DST is able to support reasonable intuitions about the "unity of science" by maintaining a productive mean between the extremes of autonomy for evolutionary biology advocated most vociferously by Ernst Mayr (see Mayr 1988, for example) and the reductionist impulses of genocentric Darwinians (see E. O. Wilson 1998)—both of which seem to us generally more effective as means of freezing existing science into place than of encouraging new bursts of cutting-edge problem-solving. In the course of making this

argument, we offer some putative extensions of DST's main claims, some of which might well be contested by DST's advocates. We hope that our remarks will be taken in the purely exploratory spirit in which they are offered.

DST as a Darwinian Research Program

Susan Oyama's *The Ontogeny of Information* is a seminal text for developmental systems theorists (Oyama 1985). In the process of deconstructing the nature/nurture dichotomy, Oyama articulates a view in which each ontogenetic cycle employs a set of largely, but not exclusively, heritable developmental resources. These are reconstructed in each generation in what Maturana and Varela call an "autopoietic" process—a self-organizing process that does not rely on a central information source (Maturana and Varela 1980). DST, according to Paul Griffiths and Russell Gray, emphasizes the self-organizing properties of the system of physical resources which occupy a stretch of space-time as a result of the activities of past generation. Self-organization reconstructs the life-cycle (Griffiths and Gray 1997). To be sure, the resources needed to complete each stage of ontogenesis include genes. But other chromosomal, cytoplasmic, and metabolic structures (some of which are heritable in their own right), as well as behavioral, environmental, and social factors, play a causal role that should in no case be subordinated a priori to genes. On this view, functional information is the result of ontogeny rather than ontogeny being the result of the transmission of information stored in a genetic program. Thus it becomes inappropriate to say that genes "contain" information.

We may place research programs that flow from this guiding insight within a much longer lived research tradition, namely the epigenetic view of embryogenesis that by the eighteeenth century was vigorously contesting the preformationism that had flourished in the seventeenth. Following a suggestion of Evelyn Fox Keller, we are even tempted to call Oyama's proposal, and the DST research programs that it has inspired, "the new epigenesis" (Keller 1995). Reasonable Darwinians, such as Mayr, have always recognized that there are traces of preformationism in post-Weismannian Darwinism. The whole organism may not be antecedently encased within the sperm or egg, but the information for producing the organism is. These preformationist traces are most visible in genocentric forms of Darwinism like that of Dawkins, with its stress on "immortal replicators." Viewed in the light of DST's strongly epigeneticist approach, however, one can find traces of preformationist rhetoric even in Mayr, against whose notion of a genetic program Oyama protests. In rejecting all preformationist elements, DST reverts, in fact if not in name, to the original meaning of the term epigenesis, which was lost under the impact of post-Weismannian thinking, when it came to refer to all processes in forming the phenotype other than genetic transcription and translation. The older meaning of epigenesis, which runs from Aristotle through Harvey to an array of later eighteenth-century biologists, holds that an organism develops through a process in which the proximate cause of each step is the total set of interactions at the immediately preceding state, starting with the procreative act, moving through the differentiation and articulation of physical and psychological traits, and ending with the initiation of another life cycle. It is to just such a robustly epigenetic conception that DST adverts.

The antiepigenetic notion that genes play a causally dominant role in evolution has been encouraged by the view that other aspects of ontogeny are not in fact heritable. This is, however, false. Prions, for example—proteinacious particles that are responsible for "mad cow" and other scrapie diseases—are heritable, but are in no way coded for by DNA or RNA, as the Nobel laureate Stanley Prusiner has demonstrated. One may be loath to call a prion a developmental resource. But a comprehensive inventory of heritable, nongenetic factors that are indeed developmentally relevant, including cytoplasmic inheritance, has been collected and assessed in Jablonka and Lamb's *Epigenetic Inheritance and Evolution: The Lamarckian Dimension* (Jablonka and Lamb 1995). We will note some of their results below.

The claim that DST can be construed as a Darwinian research program achieves its greatest salience against the background of renewed interest in nongenetic forms of heritability (Griffiths and Gray 1994, 1997). Griffiths and Gray implicitly ask why we should believe that the existence of heritable developmental resources other than genes should in any significant way conflict with the theory of natural selection as such, even if they do conflict with certain versions of that theory. After all, heritability is simply "a measure of how well the state of the parent predicts the state of the offspring" (Sterelny and Griffiths 1999: 35). If you hold an environment constant enough you can increase the heritability of traits no matter what is happening with the genes. Working from this simple recognition, Griffiths and Gray do not criticize standard versions of genetic Darwinism in the manner of Darwinian "pluralists," who multiply the units and levels at which natural selection can work, or those who limit its adaptive power by assigning a large role to phylogenetic constraints, pure chance, or mechanisms of nonadaptive natural selection. On the contrary, they appear to be rather strong adaptationists, working for the most part at the traditional organismic level. With a little effort, their view can even be read as a defense of orthodox organism-level selection against both genocentric versions of natural selection and the fashionable "expansion" of Darwinism to countenance both higher and lower units and levels of selection. For the suborganismal and superorganismal traits that have led others rather promiscuously to postulate the existence of "superorganisms" above and "selfish genes" below are, whenever possible, reconceived as resources for the generation-by-generation construction of organisms. Variation in these resources drives adaptive natural selection.[3]

This is possible in part because of an ontological dimension of DST. Organisms are viewed by

DST's proponents as self-organizing processes rather than as discrete, hard entities on which "forces" impinge. Developmental resources often lie beyond the traditional boundary of the ontologically hardened organism—in the environment, for example, and especially in that part of the environment of some species that can be called cultural. In consequence, DST's proposed extermination of the preformationist elements latent within post-Weismannian Darwinism suggests that the centrality of organismic selection in traditional Darwinism can be preserved partly because the boundaries of what counts as a "part" become productively blurred when ontogeny is viewed as a process.

Developmental Systems as Autocatalytic Dissipative Structures

When we first read Griffiths and Gray's 1994 paper we had the experience Huxley reports himself as having had when he first encountered Darwin's theory: How very stupid of us not to have thought of that. For in *Darwinism Evolving: Systems Dynamics and the Genealogy of Natural Selection* we had, like Oyama, Griffiths, and Gray, insisted that organisms are developmental processes (Depew and Weber 1995). Like Oyama, Griffiths, and Gray too, we had thought of ontogenesis as self-organizing rather than as causally manipulated by genes. Moreover, we had expressed regret that the developmental and self-organizational nature of organisms had not yet been well integrated with Darwinian theory, leaving the field to antiselectionist developmentalists.

Our own response to this unsatisfactory situation was to argue in some historical detail that in its various stages and incarnations Darwinism had always benefited from reconceiving natural selection in terms of models drawn from various sorts of systems dynamics, and to express our hope (based on this weak induction from the past) that new models of natural selection, in which that phenomenon comes to be viewed as

a predictable property of certain kinds of self-organizing systems, would eventually allow Darwinism and developmentalism to be brought into a new unity. We also suggested that the developmental and self-organizational aspects of organisms are rooted in the prebiological fact that organisms are autocatalytic chemically dissipative systems before they are anything else, and that autocatalytic chemically dissipative systems are by their very nature self-organizing developmental systems. What we did not have, however, was an argument showing how Darwinism should be construed in a way that is sensitive to the self-organizing developmental nature of living systems. And what we especially did not have was an argument showing that the most problematic aspects of contemporary Darwinian theory could be resolved by just such a reconstruction. That is what we found by reading Griffiths and Gray and rereading Oyama. In sum, what we have learned from DST's proponents is what follows from the fact that organisms are self-organizing developmental systems. What we hope to offer them is some sense of why organisms are self-organizing developmental systems in the first place—and why, in our view, natural selection properly so called can arise only in such systems.

Our answers to these questions reflect the fact that one of us is a biochemist who has worked on the evolution of proteins and the origins of life. Living things, whatever else they may be, are open systems, pumping matter and energy into themselves and dissipating it in degraded form to their environments. As such, organisms are what Prigogine called "dissipative structures" (Prigogine 1980). The inherent tendency of dissipative structures to increase, even to maximize, their dissipative rate is linked with their ability to build better dissipative pathways, in the form of more efficient internal structures, which enable them to make other entities pay their entropic debts. The most effective way of building such structures in chemical systems is by means of autocatalysis. A chemical reaction that produces a substance that can facilitate the production of more of the original reactant will show a rapid amplification of the concentration of that substance. This is called an autocatalytic cycle. Organisms are chock full of autocatalytic cycling—indeed, within the environments to which they are deeply coupled, they simply *are* autocatalytic systems of a certain sort. Moreover, not only are cells highly complex autocatalytic cycles of energy use and dissipation, but, in their developmental trajectories, the organisms that are composed of such cells show precisely the predictable patterns of bifurcation and differentiation that we observe in all complex dissipative structures, such as the Belousov-Zabotinsky reaction. One can see the traces of such patterns in the developmental trajectories of organisms like slime molds, whose pattern of aggregation and dispersal proceeds along, and takes adaptive advantage of, dynamical paths already marked out by self-organization (Bonner 1993).

The fact that living things are autocatalytic dissipative systems constitutes, we believe, the root cause of the fact that organisms are self-organizing developmental systems. Accordingly, although we agree with advocates of the autonomy of evolutionary biology that some physical and chemical aspects of organisms are more or less irrelevant to their evolutionary histories (Burian and Richardson 1991), we do not think that all of them are. On the contrary, we think the chemical and biochemical underpinnings of organisms, which in turn rest on the thermodynamical parts of physics, are highly relevant to the evolutionary process.

The potential relevance of chemical self-organization and physical dissipation to selective processes arises because, in addition to spontaneous differentiation and bifurcation, chemical autocatalytic cycles themselves already exhibit a certain sort of selection—not natural selection for fitness, to be sure, but what we will call "chemical selection" for autocatalytic efficiency. The differential rates at which an autocatalytic cycle is completed, for example, entail differential

proliferation. In earth's early environment, accordingly, an enormous competitive advantage would have accrued to entities that could make use of templates, however crude, to enhance the reliability, speed, and redundant pathways of their autocatalytic activity. At the dawn of life this probably occurred through an intense coevolution between proteinoids, which have long been known spontaneously to form in chemical environments such as those that obtained early in earth's history, and nucleic acids, which may have formed either separately or, alternatively, in the presence of polypeptides within amphiphilic vesicles. Energy-capture mechanisms would have driven polymerization of amino acids and nucleotides in a chemically bounded space of amphiphiles to produce "generic" proteins and nucleic acids. These generic macromolecules would have had weak, but broad catalytic activity such that a complex array of chemical reactions would be possible within such protocells (Weber 1998).

An ensemble of protocells so characterized can be said to exhibit weakly heritable variation, even in the absence of genetic mechanisms. As Stuart Kauffman's simulations have suggested, increased coupling of such a system of autocatalyzed reactions would have led to catalytic closure (Kauffman 1993). Catalytic closure means that every member of an autocatalytic set has at least one of the possible last steps of its formation catalyzed by some other member of the set. There would have been increased rewards for catalytic efficiency, and thus selection pressure for a better fit between the sequence space of a given macromolecule and the catalytic task space with which it roughly overlaps. Under selection pressure for more efficient autocatalytic cycling, a better fit between the sequence space of a macromolecule and its catalytic function space would have occurred. That step might have taken place in the following way: A tightening of the interactions between nucleic acids and proteins would have led to a division of labor between them, one helping more reliably to ensure generation by generation reconstruction, the other carrying out the bulk of the catalytic tasks. As patterns of increasingly reliable and robust autocatalysis emerged, a transition from protocells to true cells, and hence to life properly so called, would have ensued. Thus life would have emerged by way of a process we call "chemical selection," a process that, in its very conception, is inseparably linked to the self-organizational and differentiating tendencies of autocatalytic dissipative structures and to the thermodynamical imperatives that underlie them.

This approach to the origins of life bears on a number of contested issues in both theoretical biology and the philosophy of biology. For one thing, it constitutes a critique of "magic molecule" thinking, according to which life emerged once, and by accident, in the form of a statistically fluky RNA molecule that thereupon proceeded to decorate itself out with survival machines. Because that sort of thinking lies at the basis of "selfish gene" theory, our rejection of "magic molecule" thinking is intended as a critique of "selfish gene" theorizing generally.

We see the basic fallacy of this approach as its tendency to push the notion of natural selection down into the very sorts of processes from which natural selection itself must arise. Thinking further along these lines, we noticed that our account of the emergence of life by way of chemical selection also constitutes a critique of the rather promiscuous, substrate-neutral conception of natural selection that currently flourishes among Darwinians. We believe that natural selection, properly so called, is not, at least in the first instance, an explanatory model, or a universal mechanism, or an "algorithm," as Dennett calls it (Dennett 1995). Nor do we believe that it is more or less indifferent to the kinds of materials and entities on which it works. On the contrary, natural selection is itself an emergent natural phenomenon[4]—a phenomenon that arises in, and properly applies only to, the chemically autocatalytic and physically dissipative systems of which we been speaking (Depew and Weber 1998).

It is one thing to say, with Lewontin, that natural selection, formally considered, ranges over all and only entities that exhibit differential retention of heritable variation (Lewontin 1980). But, as Lewontin himself admits, natural selection, formally defined by the three conditions of variation, heritability, and differential retention, is not sufficient for producing an adaptation. What is needed in addition is continuity and directionality maintained over a good deal of transgenerational time in an environment made constant enough (in part through the activities of organisms themselves) to add up to something useful (Sterelny and Griffiths 1999: 31–38). We agree with DST's advocates, who, perceiving this, argue that transgenerational continuity and directionality of the required kind can be assured only when ontogeny is construed as the reliable reconstruction in each generation of a range of developmental resources. But we are also inclined to think that this process itself depends on the self-organizational aspects of ontogenesis, and that these in turn rely substantially on autocatalytic processes at the level of matter and energy transformation. From this "bottom-up" point of view, in which it is construed as an emergent phenomenon, natural selection is presumptively adaptive. For adaptive natural selection is what realistically accounts for the emergence of the phenomenon of natural selection in the first place; merely formal natural selection, which may seem primary from a purely definitional point of view, and only contingently related to adaptation, appears from a more realistic view as a departure from the adaptive norm.

It was at this point that our thinking began to link up more substantively, if also more problematically, with that of DST's advocates. For it would seem that we are making a stronger claim than Griffiths and Gray. They contend that developmental systems can be, and indeed are, produced by natural selection. We contend in addition that anything that comes into being by natural selection, considered as a real phenomenon, must be a system that develops. We maintain, that is to say, that only the developmental systems identified by DST contain a rich enough notion of heritability to sustain the kind of selection that can properly be called adaptive natural selection, as distinct from chemical-autocatalytic selection below and certain conceptions of cultural selection above. It does not, of course, follow from this that natural selection is reducible to chemical selection. What follows is that natural selection presupposes, and is uniquely predicable of, systems in which chemical selection, as we have described it, has already been at work—and that cultural selection does not depend on a separate process working on cultural replicators ("memes"), but is an integral part of the coupling between organisms and aspects of their environments.

This way of situating the argument provides additional support for one of DST's most salient criticisms of the selfish gene hypothesis. DST's advocates have noted a strange anomaly in theories that treat genes not only as the units over which the selective "algorithm" ranges, but as the chief *beneficiaries* of this iterative process. Because on this account they are merely repositories of transgenerational information, genes seem on this view to have no biological function at all (Griffiths and Knight 1998; Agar 1999)! The absurdity of this becomes especially clear when one views organisms in the way we do—as self-organizing, autocatalytic dissipative systems. For in this case one can readily enough surmise that the biological function of protein-coding sectors of molecules like RNA and DNA is to stabilize and buffer developmental cycles by enhancing their ability to repeat more reliably in new autocatalytic cycles what has been effective in earlier ones. Genes are, metaphorically at least, catalysts of catalyzing reactions.

Bearing this picture in mind, we defined organisms not only as autocatalytic dissipative systems, but as "bounded and informed" autocatalytic dissipative systems (Weber and Depew 1996). For the catalytic effect of genes results in, and is adaptive for, enhanced reliability in the generation-by-generation construction of highly integrated autocatalytic systems, and hence is responsible for the increased correlation between the traits of parents and offspring, ancestors and descendants.

Our use of the terms "bounded" and "informed" has caused some squinting on the part of DST's proponents, who suspect that talk about the boundedness of organisms encourages too entitative a view of the ontology of living things, and hence obscures the processive, constructive view of ontogeny that DST supports (Griffiths, personal communication). Similar worries arise in connection with our use of the term "informed." For it might suggest the encoding of heritable information in a genetic program.

We are reasonably sure, however, that this worry can be assuaged by clarifying the point of view from which we take organisms to be bounded and informed. Organisms are bounded in contrast to more loosely coupled, prebiotic autocatalytic systems. For the latter are not sufficiently buffered from aspects of their environment to count as genuinely living systems.[5] This does not in the least mean or imply that all of the developmental resources organisms use to reconstruct themselves are located within their monadic boundaries—boundaries that are erected in the received view of natural selection less by empirical facts than by an inappropriately atomistic conceptual apparatus. Nor does our use of the term "informed" imply that genes "contain" information that is supposed to be semantically packaged within DNA and RNA, which is then "opened up" and "expressed." Genes, on our view, perform a stabilizing function, as noted above. This increases the degree of correlation between one generation and the next. The information associated with these correlations is of necessity distributed over the entire process, not encoded in informational packets and read out into matter by way of a ghostly computerlike program.

It is possible that DST's advocates will greet our proposal to root natural selection, development, and self-organization in prebiotic processes as imposing unnecessary baggage on them in their efforts to gain a foothold against the received view of genetic Darwinism. It is difficult to argue at one and the same time for a new view of the "units" on which natural selection works and against the autonomy of biology stance that often accompanies the standard picture, especially because the latter also serves to mark off the turf of evolutionary biologists in the discursive environment of academic life. Accordingly, it might be thought that our proposal holds DST's arguments about the units of selection hostage to quite independent arguments against the autonomist stance. We certainly hope this is not the case. We cannot, however, repress the passing thought that treating developmental systems as emergent from autocatalytic dissipative systems by way of a linked series of types of selection might actually help DST's advocates make their views about the units of selection controversy more persuasive. For when one looks at the topics of boundedness and information from the more "bottom-up" perspective we are advocating, DST will not appear merely to be negating the received Darwinian view by heavily policing the use of terms like *informed* and *bounded*. The best defense is a good offense; and we suggest that in this case the best offense is a more physically and chemically rooted conception of organisms as developmental processes, a view that will reveal clearly the biological functions that genes play in relatively bounded and reliably reconstructable developmental systems. In the following section, we will try to support this claim by exploring the potential relevance of our view to nongenetic forms of inheritance.

Epigenetic Inheritance and the Darwinian-Lamarckian Divide

We stated above that DST's ability to present itself as a Darwinian theory has been stimulated by empirical recognition that quite concrete resources other than genes vary and are heritable. Much of the evidence for these forms of heritable variation has been summarized by Jablonka and Lamb (Jablonka and Lamb 1995; see also Jablonka and Lamb 1998). These include cytoplasmic factors, organizing centers for basal bodies and microtubules, DNA methylation pat-

terns, membranes, and organelles, as well as extracellular factors such as the cellular matrix, developmental and environmental signals, and behavior. Taken collectively, Jablonka and Lamb call these "epigenetic inheritance systems" (EISs). The net effect of EISs is to provide for natural selection a wider range of heritable variation than that provided by genetic variation based on nuclear DNA alone.

It might be said by proponents of strongly genocentric versions of Darwinism, who assume that information must be stored, read out, and indeed be the chief beneficiary of natural selection, that these phenomena are too weakly and transiently heritable to have much of an effect on natural selection and evolution. It is just here, however, that one must guard against one's conceptual framework and ontology getting in the way of the facts. Admittedly, such variations might not be taken up by natural selection when natural selection is construed as occurring in systems that do not in any marked or essential way exhibit strongly nonlinear forms of feedback, including autocatalysis, within a self-organizing developmental system. In the standard view of natural selection positive feedback is more or less absent. Thus one will be tempted to cry out for an immortal, self-creating replicator, as Dawkins does (Dawkins 1989), or for mechanisms that can "pump up" the otherwise plodding pace of adaptation, as Dennett does (Dennett 1995). But, as we have argued above, it is precisely in systems that do exhibit self-organization through autocatalytic positive feedback that natural selection is at work. That in turn makes it much easier for otherwise weakly heritable variants to be taken up and amplified by natural selection. For such variants to count in the process of adaptive natural selection, all that is required is that they be reliably reconstructed in cooperation with a host of other interacting resources in each epigenetic cycle. Accordingly, the potential relevance to evolutionary theory of epigenetic variation increases, at least theoretically, when DST is assumed as a background theory; and it perhaps increases even more when our proposal to situate DST within a context that highlights the self-organizing nature of ontogeny through its essential connection to chemical autocatalysis is brought into play.

We can be more concrete about this. The best-documented example of an EIS is DNA methylation, which is one of the mechanisms of chromatin marking. Chromatin marking is important for cell memory. It ensures that during development and differentiation a cell lineage that has become a kidney cell, for example, continues upon somatic reproduction to produce kidney cells and not some other type of cell. This example serves to remind us that genic reductionists too often forget that chromosomes themselves are not exclusively made up of the base sequences of DNA. Other molecular structures, such as various nucleoproteins and chemical modifications of the DNA bases like methylation, may affect gene expression. Admittedly, the control of the pattern of methylation during development is not yet well understood. But what is clear is that such a pattern is preserved during mitotic division through the action of the methyl transferase enzyme. Methylation has been shown to alter the gene expression of tissue-specific genes, stage-specific genes, and silent genes, whereas the expression of genes needed for basic metabolism of any cell is not so controlled. Differences in DNA methylation have led to different chromatin structures, and in this respect they can be considered alternative phenotypes of a gene. Kermicle has suggested that these alternative phenotypes of a particular locus be called "epialleles" (Kermicle 1978). Epialleles arise during normal chromatin alterations that occur during somatic differentiation. Although all the DNA sequences at that locus are identical, there will be a number of epialleles within an individual organism and in a population.

This source of epigenetic variability is expressed during development. It has been demonstrated in several cases, moreover, that epigenetic variation can be transmitted to offspring. Flavell and O'Dell found seven epialleles for wheat high-molecular-weight glutinin that were heritable (Flavell and O'Dell 1990). Sano and colleagues have shown that changes of methylation pattern in rice result in heritable phenotypic change that remained stable over multiple generations, even though the relevant DNA sequence was unchanged (Sano et al. 1990). Thus variation in methylation patterns can, in principle, be subject to the action of natural selection, and can lead to evolutionary adaptation by means of various mechanisms, of which Waddington's genetic assimilation is one of the best known (Waddington 1940). Admittedly, plants, which have late to nonexistent germline segregation, seem to have a higher probability of having epigenetic inheritance play a direct role in adaptive evolution. Epigenetic factors in animals are more likely to affect early development or germ-cell lineages, when segregation of a germ line from somatic processes is far less rigid. Among animals with complex behaviors, however, the presumptive irrelevance of epigenetic variation can be compensated for by increased sensitivity to environmental signals and by behavioral variations.

The subtitle of *Epigenetic Inheritance and Evolution*, in which Jablonka and Lamb report results such as those just discussed, is *The Lamarckian Dimension*. Their use of the term *Lamarckian* is not intended to deny the evolutionary power of natural selection (Jablonka, personal communication). It suggests instead opposition to forms of Darwinism that attempt to assimilate the evolutionary gene concept to the molecular gene concept, thereby fetishizing and exaggerating the causal role of naked DNA and RNA in such a way that the heritability of variants that do not fit this model, such as epialleles, goes unrecognized or underestimated in its potentially adaptive effects. So, although they are not necessarily rendered heritable in the stereotypical ways that have led to the ill repute of Lamarckian evolutionary mechanisms by twentieth century Darwinians, it is useful to call attention to these phenomena by using the term *Lamarckian*.

We suspect, however, that this decision is not rhetorically cost-free. Even such a benign use of the term *Lamarckian* testifies to the power of genocentric Darwinians to monopolize the term *Darwinism* in contemporary discourse, a power that has increased remarkably in recent years through the popularity of works on sociobiology and evolutionary psychology. For in this discursive context, the term *Lamarckian* is more likely to suggest violations of the Central Dogma of Molecular Biology, which requires that information must never run directly from proteins to DNA and RNA, than to call attention to instances of epigenetic inheritance and natural selection that do not quite fit the genocentric Darwinian stereotype. Jablonka's and Lamb's primary purpose is to dispute genetic Darwinism's empirical marginalization of nongenetic forms of inheritance, and not to argue for stronger, more problematic, versions of Lamarckism that puts it into direct opposition with the Central Dogma (although results by Steele, Lindley, and Blanden that might suggest this have been reported [Steele, Lindley, and Blanden 1998]). This distinction becomes a good deal clearer when it is linked to the strongly epigenetic view of ontogeny that Jablonka and Lamb clearly assume. Seen in this light, their evidence for epigenetic inheritance supports DST's effort to expunge from evolutionary theory all traces of preformationism, and so to combine a pre-Weismannian conception of epigenesis with a post-Synthesis reconstruction of Darwinian natural selection. DST offers a framework for the notion of epigenetic inheritance that puts these phenomena in the best light.

DST, Developmental Psychology, and the Unity of Science

In this essay, we have suggested that developmental systems, as DST describes them, can best be viewed as emergent from autocatalytic dissipative systems. On this account, DST's proposed reunion of Darwinism with developmental biology will lead not only to a closer relationship between evolutionary theory and developmental

biology, but between a newly unified biology and the thermodynamics and chemical kinetics of dissipative structures.

This proposal can easily provoke fears of reductionism. Our general response to that fear is to distinguish between the unification of science and programs for reductionism. Reductionism failed over and over again as the twentieth century unfolded, provoking a strong reaction that prevails today in favor of incommensurability, pluralism, and even "the disunity of science" (Dupre 1993). Yet it remains true that one of the most salient characteristics of science is its increasing unity. This fact has been ill served by philosophers of science who, by means of the deductive-nomological account of explanation, too readily universalize the quest within mathematical physics for greater and greater reduction under a smaller and smaller range of laws. In actual fact, science as a whole becomes unified by an ongoing series of discoveries in which, as we come to understand a given phenomenon better, we also come to see more clearly its relation to other phenomena. We wish to go on record, accordingly, as supporting the unity of science in this wide sense without supporting reductionism or the philosophy of science that gives it undue importance in actual processes of scientific inquiry.

Unfortunately, genetic Darwinism has not yet exhibited this pattern. In its well-motivated efforts to steer clear of reductionism, especially after the rise of molecular biology, the Modern Evolutionary Synthesis has attempted to preserve its autonomy from physics and chemistry by categorically denying the relevance of the physical and chemical sciences to specifically biological problems (Mayr 1988). Our support for proposals to reformulate Darwinism in terms of DST's conception of ontogeny, and in turn to refer this conception to autocatalytic and dissipative processes, contains an implicit plea to relax this stance. For our hope is that this conception of ontogeny will lead to hitherto unsuspected links between biology and certain aspects of chemistry and physics without in the least suggesting that evolutionary biology is reducible to these sciences in the sense that biological processes are "nothing but" lower-level phenomena operating under certain initial and boundary conditions.

It is not only with sciences that on a conventional ordering lie below evolutionary biology, however, that we wish to urge a unifying stance, but also with some sciences above it. We are especially interested in this respect in evolutionary developmental psychology, which is at present in a state of intense ferment that we suspect can be clarified by Darwinian versions of DST. For in treating aspects of the environment as developmental resources, DST allows us to see what genetic reductionism obscures, namely that features of cultural environments, rather than being sites at which selection over memic units takes the place of natural selection, are, in many cases, aspects of the process of organismic natural selection itself.

Organocentric Darwinians like Mayr insist, quite correctly, that natural selection is a two-stage process (Mayr 1988; see Brandon 1985, 1990). Even the best genes have no future unless the organisms carrying them are successful enough in dealing with their environments to pass them on. Thus what Robert Brandon calls "environmental selection" in each generation must precede natural selection of genotypes over multiple generations (Brandon 1990). From this account, it is easy to see that, all other things being equal, there will exist selection pressure for behavioral innovations that help organisms respond to contingencies in their environments. Lewontin in particular has insisted that populations commonly evolve in a fitness-enhancing way by taking advantage of the considerable variation available in gene pools to evolve adaptations that enable organisms to construct and reconstruct their environments so that existing genotypes may remain as nearly as possible adapted to stabilized conditions (Lewontin 1983). It is precisely because the environment has been stabilized by the agency of organisms, in fact, that the traits whereby it comes to be so stabilized *can* be heritable. From here it is not much of a leap

to see that learned behaviors, and the learning mechanisms that support them, are extremely effective ways of assuring matches between organisms and otherwise shifting environments. These will, all other things being equal, be favored by adaptive natural selection. Learning mechanisms that rebuild behaviors in each generation by using mimetic capacities to better match organisms to environments are clearly evident among primates, including ourselves. One consequence is the pressure we are able to put on other species and the diminished pressure they are able to put on us.

By stressing the role of learning capacities and the behaviors they generate as developmental resources, DST provides additional conceptual backing for sometimes overlooked aspects of the orthodox, organocentric view of Darwinism. This approach also sounds a greater note of naturalism, and so far forth the unity of science. For it tends to break up the long-standing agreement that cultural evolution is "Lamarckian" (and all the speedier for that), while biological evolution, properly so called, is under the control of adaptive natural selection as measured by changing gene frequencies. The distinction between "memic" and "genic" selection is merely a version of the latter understanding. For DST, on the other hand, culture does not "arise from" nature, nor is it "constrained" by it. In view of the role cultural elements can play in the generation-by-generation construction of organisms with reliably heritable traits, culture is *part* of nature.

A relatively uncontroversial way in which DST-oriented Darwinians challenge standard understandings of the nature/nurture divide is to urge against most versions of "evolutionary psychology" that generation-by-generation *maintenance* of the genetic structures that facilitate learning should count as producing adaptations no less than the selection pressure that was *first* responsible for the emergence of the trait in question (Sterelny and Griffiths 1999: 219). On this view, too strong a distinction between the adaptations by which the learning abilities of our hunting and gathering ancestors first arose and the subsequent "exaptation" of these capacities for the uses of modern life should be resisted. Many of the later uses of earlier capacities are certainly adapt*ive*. But, in view of the power of natural selection to maintain, as well as to select for, genotypes that foster current fitness, there is no reason to refuse to call these adaptive traits adaptations, and far less reason to call them "exaptations." On this view, the sociobiological notion that our genes "hold us on a leash" more or less evaporates. With it too evaporate many of the motives that have provoked Darwinian opponents of sociobiological thinking to rely too heavily on strong forms of group selection to circumvent what wrongly are seen to be the limitations of individual-level selection.

We wish to end this essay on a note of query about this issue, posing two questions that suggest how open the question of cultural evolution still is, as well as our confidence that this disputed topic can be illuminated by using a DST interpretive strategy.

The first question has to do with the recent revival of talk about the so-called Baldwin effect. First proposed in the 1890s as a way of reconciling Weismannian neo-Darwinism with some aspects of Lamarckism, the Baldwin effect generally asserts that if an organism chances to exhibit a behavior that permits more effective interaction with its environment, and can pass that behavior along mimetically, then descendents of such individuals and populations will on average do better than their competitors, in turn making it more probable that germline factors that promote the behavior, if they independently arise, will be more quickly and effectively taken up and moved in an adaptive direction (Baldwin 1896). Even though, properly interpreted, the Baldwin effect does not suggest, or at least require, that multigenerational phenotypic continuity will induce the kind of genetic change necessary to support it, this notion was downplayed by the makers of the Modern Synthesis until it became very recessive indeed, killed for the most part by the usual method of accepting it in theory but denying in practice. In part, that was, in our

view, because this putative phenomenon became too closely linked to debates about the general efficacy of Waddington's genetic assimilation. But it was also because the Modern Evolutionary Synthesis likes to think of fit genotypes as being sifted from the process of phenotypic selection *in each generation*. By contrast, the Baldwin effect asserts that reconstruction of learned behaviors at the phenotypic level can occur over multiple generations before any genotypic change begins to follow suit (which presumably does not happen at all in a large number of cases). The latter notion seems to us quite congenial to DST. Because in the context of contemporary interest in the evolution of language and mind the Baldwin effect has recently made a big comeback as a putative evolutionary "driver" or "lifter" among thinkers of widely varied stripe (Dennett 1995; Michel and Moore 1995; Deacon 1997; Hinton and Nowlan 1996), we wish to call upon DST's advocates to help clarify the theoretical structure of Baldwin-type phenomena as well as their actual effects, if any. When, if reliably reconstructed in each generation, should we begin to talk about adaptations?

A related question has to do with cases that call into question the standing assumption that adaptations produced by natural selection must always involve changes in nuclear gene frequencies. A case that raises this question has been reported in matrilineal species of whales—species whose members live with female relatives and form new groups when females and their young fission off from larger groups (Whitehead 1998). These species, it seems, have five times less diversity in their mitochondrial DNA than humpback whales or dolphins, which do not have such a social structure, while nuclear DNA, the presumptive site for standard Darwinian models of evolutionary change, remains unchanged. To explain this, the investigator who has reported these results hypothesizes that cultural learning over many generations—the imparting of songs, migration strategies, foraging techniques, or baby-sitting tactics—may become adaptive in ways that are evidenced by, but not caused by, a reduction in genetic diversity in mitochondrial DNA, which hitchhikes on these changes (Whitehead 1998). Recall that mitochondrial DNA is passed exclusively through the female line and is not involved in adaptations except for the biochemical functions carried out by mitochrondia themselves. Where, then, are the expected gene frequency changes that underlie this presumptive adaptation?

Standard versions of Darwinism can readily answer this question by assigning to whales the so-called Lamarckian, or cultural, adaptive evolution that has hitherto been exclusively attributed to humans, or at least primates. That is not unreasonable considering the communicative traits of whales and the connective power of their brains. Assuming that it is a real phenomenon, however, we think it merely dogmatic not to call this a case of natural selection, especially because the process presumably goes on through a differential retention of behavioral variants, either at individual or trait-group levels. Its status as a genuine adaptation, built up over generations by fixation of heritable variation, can, we suspect, be affirmed by DST's approach to natural selection. But if this is the case, DST's innovative intervention in the quarrels that have always attended the multifaceted and still vital Darwinian tradition might have to confront a potential clash between its constructivist view of ontogeny and the assumption that adaptive natural selection is always measured by gene frequency changes in nuclear DNA. At the very least, that should be interesting.

Notes

1. Some think it had already been aside when T. H. Morgan separated genetics from embryology (see Gilbert, Opitz, and Raff 1996). Gottlieb has a more charitable view of Morgan, in which the separation was intended tactically and only for a time; "eventually Morgan believed that the gene would be incorporated into the developmental process and its activity seen as reciprocally altered thereby" (Gottlieb 1992: 139). Gottlieb further comments that the current theories of differentiation, although informed by a wealth of data

from molecular genetics, have advanced little beyond the state that Morgan found them. (Gottlieb 1997: 98).

2. We take the term *developmentalist challenge* from Schaffner (1998). Schaffner mentions Gottlieb, Gray, Griffiths, Bronfenbreener and Cici, Oyama, and Lewontin among others as partisans of this challenge. The heterogeneity of the views of those listed suggests the pluralism of the developmentalist challenge.

3. DST Darwinians do not object to group selection. Rather, they tend to offer support for construing group selection in the sense articulated and defended by David Sloan Wilson. In effect, they recode Wilson's notion of trait groups in an interpretive framework that turns such traits into developmental resources of individuals, not ontological markers of entities (see Sterelny and Griffiths 1999: 160–177).

4. The concept of phenomenon in play here is roughly that of Bogen and Woodward 1988.

5. In this connection, our view departs from the wish of many anti-Darwinian developmentalists to extend the notion of living system to ecosystems, and indeed to the entire earth, as in the Gaia hypothesis.

References

Agar, N. (1996). Teleology and genes. *Biology and Philosophy* 11: 289–300.

Baldwin, J. M. (1896). A new factor in evolution. *American Naturalist* 30: 441–451.

Bogen, J., and J. Woodward. (1988). Saving the phenomena. *Philosophical Review* 97: 303–352.

Bonner, J. T. (1993). *Life Cycles: Reflections of an Evolutionary Biologist.* Princeton, NJ: Princeton University Press.

Brandon, R. N. (1985). Adaptive explanations: Are adaptations for the good of replicators or interactors? In D. J. Depew and B. H. Weber (Eds.), *Evolution at a Crossroads: The New Biology and the New Philosophy of Science,* pp. 81–96. Cambridge, MA: MIT Press.

Brandon, R. N. (1990). *Adaptation and Environment.* Princeton, NJ: Princeton University Press.

Burian, R., and R. Richardson. (1991). Form and order in evolutionary biology: Stuart Kauffman's transformation of theoretical biology. *Proceeding of the Philosophy of Science Association 1990* vol. 2, pp. 357–402. East Lansing, MI: Philosophy of Science Association.

Dawkins, R. (1989). *The Selfish Gene.* (2nd ed.) Oxford: Oxford University Press.

Deacon, T. (1997). *The Symbolic Species.* New York: W. W. Norton.

Dennett, D. C. (1995). *Darwin's Dangerous Idea: Evolution and the Meanings of Life.* New York: Simon and Schuster.

Depew, D. J., and B. H. Weber. (1995). *Darwinism Evolving: Systems Dynamics and the Genealogy of Natural Selection.* Cambridge, MA: MIT Press.

Depew, D. J., and B. H. Weber. (1998). What does natural selection have to be like in order to work with self-organization? *Cybernetics and Human Knowing* 5: 18–31.

Dupre, J. (1993). *The Disorder of Things: Metaphysical Foundations of the Disunity of Science.* Cambridge, MA: Harvard University Press.

Flavell, R. B., and M. O'Dell. (1990). Variation and inheritance of cytosine methylation patterns in wheat at high molecular weight glutinen and ribosomal RNA loci. *Development* 1990 Supplement: 15–20.

Gilbert, S. F. (1991). Induction and the origins of developmental genetics. In S. F. Gilbert (Ed.), *A Conceptual History of Modern Embryology,* pp. 181–206. New York: Plenum.

Gilbert, S. F., J. M. Opitz, and R. A. Raff. (1996). Resynthesizing evolutionary and developmental biology. *Developmental Biology* 173: 357–372.

Gottlieb, G. (1992). *Individual Development and Evolution: The Genesis of Novel Behavior.* New York: Oxford University Press.

Gottlieb, G. (1997). *Synthesizing Nature-Nurture: Prenatal Roots of Instinctive Behavior.* Mahwah, NJ: Lawrence Erlbaum Associates.

Griffiths, P. E., and R. D. Gray. (1994). Developmental systems and evolutionary explanation. *Journal of Philosophy* 91: 277–304.

Griffiths, P. E., and R. D. Gray. (1997). Replicator II: Judgement day. *Biology and Philosophy* 12: 471–492.

Griffiths, P. E., and R. D. Knight. (1998). What is the developmentalist challenge? *Philosophy of Science* 65: 253–258.

Hinton, G. E., and S. J. Nowlan. (1996). How learning can guide evolution. In K. Belew and M. Mitchell (Eds.), *Adaptive Individuals in Evolving Populations,* pp. 443–446. Reading, MA: Addison-Wesley.

Jablonka, E., and M. J. Lamb. (1995). *Epigenetic Inheritance and Evolution: The Lamarckian Dimension.* Oxford: Oxford University Press.

Jablonka, E., and M. J. Lamb. (1998). Bridges between development and evolution. *Biology and Philosophy* 13: 119–245.

Kauffman, S. A. (1993). *The Origins of Order: Self-Organization and Selection in Evolution.* New York: Oxford University Press.

Keller, E. F. (1995). *Refiguring Life: Metaphors of Twentieth-Century Biology.* New York: Columbia University Press.

Kermicle, J. L. (1978). Imprinting of gene action in maize endosperm. In D. B. Walsden (Ed.), *Maize Breeding and Genetics,* pp. 357–371. New York: Wiley.

Lewontin, R. C. (1980). Adattamento. *The Encyclopedia Einaudi.* Milan: Einaudi. Reprinted as Adaptation in E. Sober (Ed.), (1984), *Conceptual Issues in Evolutionary Biology: An Anthology,* pp. 234–251. Cambridge, MA: MIT Press.

Lewontin, R. C. (1983). Gene, organism and environment. In D. S. Bendall (Ed.), *Evolution from Molecules to Men,* pp. 273–285. Cambridge: Cambridge University Press.

Maturana, J., and F. Varela. (1980). *Autopoiesis and Cognition: The Realization of the Living.* Dordrecht: Reidel.

Mayr, E. (1985). How does biology differ from the physical sciences? In D. J. Depew and B. H. Weber (Eds.), *Evolution at a Crossroads: The New Biology and the New Philosophy of Science,* pp. 43–63. Cambridge, MA: MIT Press.

Mayr, E. (1988). *Toward a New Philosophy of Biology: Observations of an Evolutionist.* Cambridge, MA: Harvard University Press.

Mayr, E., and W. Provine. (1980). *Perspectives on the Unification of Biology.* Cambridge, MA: Harvard University Press.

Michel, G. F., and C. L. Moore. (1995). *Developmental Psychology: An Interdisciplinary Science.* Cambridge, MA: MIT Press.

Moss, L. (1992). A kernel of truth? On the reality of the genetic program. *Proceedings of the Philosophy of Science Association 1992* vol. 1, pp. 335–348. East Lansing, MI: Philosophy of Science Association.

Oyama, S. (1985). *The Ontogeny of Information: Developmental Systems and Evolution.* Cambridge: Cambridge University Press. (2d rev. ed., Durham, NC: Duke University Press, 2000.)

Prigogine, I. (1980). *From Being to Becoming: Time and Complexity in the Physical Sciences.* San Francisco: W. H. Freeman.

Salthe, S. N. (1993). *Development in Evolution: Complexity and Change in Biology.* Cambridge, MA: MIT Press.

Sano, H., I. Kamada, S. Youssefian, M. Katsumi, and H. Wabiko. (1990). A single treatment of rice seedlings with 5-azacitidine induces heritable dwarfism and undermethylation of genomic DNA. *Molecular and General Genetics* 220: 441–447.

Schaffner, K. F. (1998). Genes, behavior, and developmental emergentism: One process, indivisible? *Philosophy of Science* 65: 209–252.

Steele, E. J., R. A. Lindley, and R. V. Blanden. (1998). *Lamarck's Signature: How Retrogenes are Changing Darwin's Natural Selection Paradigm.* Reading, MA: Perseus Books.

Sterelny, K., and P. Griffiths. (1999). *Sex and Death: An Introduction to the Philosophy of Biology.* Chicago: University of Chicago Press.

Ulanowicz, R. (1997). *Ecology, the Ascendent Perspective.* New York: Columbia University Press.

Waddington, C. H. (1940). *Organizers and Genes.* Cambridge: Cambridge University Press.

Weber, B. H. (1998). Emergence of life and biological selection from the perspective of complex systems dynamics. In G. Van de Vijver, S. N. Salthe and M. Delpos (Eds.), *Evolutionary Systems: Biological and Epistemological Perspectives on Selection and Self-Organization,* pp. 59–66. Dordrecht: Kluwer Academic Publishers.

Weber, B. H., and D. J. Depew. (1996). Natural selection and self-organization: Dynamical models as clues to a new evolutionary synthesis. *Biology and Philosophy* 11: 33–65.

Webster, G., and B. Goodwin. (1996). *Form and Transformation: Generative and Relational Principles in Biology.* Cambridge: Cambridge University Press.

Whitehead, H. (1998). Cultural selection and genetic diversity in matrilineal whales. *Science* 282: 1708–1711.

Wilson, E. O. (1998). *Consilience.* New York: Alfred A. Knopf.

19 From Complementarity to Obviation: On Dissolving the Boundaries between Social and Biological Anthropology, Archaeology, and Psychology

Tim Ingold

The roots of what I have to say in this chapter go back to concerns that originally led me, as an undergraduate at Cambridge in the late 1960s, to take up the study of anthropology. I had just completed my first year as a student of natural science and was profoundly disillusioned. It was not that I was any less fascinated by the phenomena of nature. My disenchantment stemmed rather from a dawning realization that the scientific establishment was so heavily institutionalized, internally specialized, and oppressively hierarchical that the most one could achieve as a professional scientist would be to become a very small cog in a huge juggernaut of an enterprise—one moreover that seemed to have lost touch both with its sense of social responsibility and with its original mission to enlarge the scope of human knowledge, and to have become largely subservient to the military-industrial complex. Looking around for something else to study, I wanted a discipline that would help to reconnect the sense of intellectual adventure associated with scientific inquiry with the realities of human experience in a world increasingly ravaged by massive technological intervention. Anthropology seemed, at the time, to fit the bill. Indeed, the reasons why I took up anthropology then are still the reasons why I continue to study it, though I might now express them in rather different terms. I believe that the discipline has a critical contribution to make to the way we understand the process of human being-in-the-world which is badly needed in an intellectual, political, and economic climate that has always tended to divorce human affairs from their bearings in the continuum of organic life.

Since embarking on my studies in anthropology I have never looked back. I have, however, often looked from side to side, observing with mounting despair how it has been torn apart by the very divisions I thought it existed to overcome. These divisions ultimately seem to derive from a single, master dichotomy that underpins the entire edifice of Western thought and science—namely, that between the "two worlds" of humanity and nature. For this is what has given us the overriding academic division of labor between those disciplines that deal, on the one hand, with the human mind and its manifold linguistic, social, and cultural products, and on the other, with the structures and composition of the material world. And it also cleaves anthropology itself into its sociocultural and biophysical divisions, whose respective practitioners have less to say to one another than they do to colleagues in other disciplines on the same side of the academic fence. Social or cultural anthropologists would rather read the work of historians, linguists, philosophers, and literary critics; biological or physical anthropologists prefer to talk to colleagues in other fields of biology or biomedicine.

I am not content to live with this situation. It was, in part, the challenge of closing the gap between the arts and humanities on the one hand, and the natural sciences on the other, that drew me to anthropology in the first place, and I still believe that no other discipline is in a better position to accomplish it. In this article, I present a program for how this might be done. My argument, in a nutshell, hinges on a distinction between two approaches to thinking about the relations between those aspects of human existence that have conventionally been parceled out between different disciplines (or, in the case of anthropology, subdisciplines) for separate study. For convenience, I call these the complementarity and obviation approaches, respectively. The first regards every aspect as a distinct, substantive component of being. It admits that the study of each component is bound to yield only a partial account, but promises that by putting these accounts together it should be possible to produce a synthetic account of the whole. These syntheses are characteristically denoted by such hybrid terms as *biosocial*, *psychocultural*, or even *biopsychocultural*. The obviation approach, by contrast, is intent on doing away with the boundaries by which these components have been distinguished. It claims that the human being is not a composite entity made up of separable but mutually complementary parts, such as body, mind, and culture, but rather a singular locus of creative growth within a continually unfolding field of relationships. In what follows, I argue for an obviation approach.

Before proceeding further, I should add a note about the terms I use for the different fields of anthropology. For a start, I do not deal here with the distinction between social and cultural anthropology: I believe this distinction is already widely regarded as obsolete, and I have no intention of reinstating it. So when I place the word social before anthropology, I mean it as a shorthand for social-cultural. Likewise, I am not concerned with the distinction between biological and physical anthropology. To my ear, the latter designation has a rather archaic ring, suggesting a preoccupation with measuring skulls and excavating for fossil bones. I prefer the designation biological, since it suggests a more rounded concern with the conditions of human life, both now and in the past. Finally, I shall make no attempt to distinguish between archaeology and prehistory, and will use the first term indiscriminately to cover both.

Social and Biological Anthropology

It is notoriously difficult to explain, to those new to the subject, what anthropology is all about. What, they might ask, is this being, this *anthropos*, from which our discipline takes its name? It is one thing, it seems, to ask what is *a* human being, quite another to ask what is human *being*. The first question is an empirical one, the second is a question of ontology. A modern evolutionary biologist, for example, might describe a human being as an individual of a species with a suite of built-in characteristics that owe their origin to a process of variation under natural selection. To

this, however, the philosopher might respond with the observation that the very possibility of such a description is only open to a creature for whom being is knowing, one that can so detach its consciousness from the traffic of its bodily interactions with the environment as to treat the latter as the object of its concern. It is in this transcendence over nature, our philosopher might isist, that the essence of our humanity resides. In short, the human being can only appear as a naturally selected, empirical object in the eyes of the rationally selecting epistemic subject.

This paradox, that accounting for our existence in nature means taking ourselves out of it, runs like a thread through the entire history of Western thought and science. And it lies at the root of the idea that humans—uniquely among animals—exist simultaneously in two parallel worlds, respectively of nature and society, in the first as biological individuals (organisms), in the second as cultural subjects (persons). As organisms, human beings seem inescapably bound to the conditions of the natural world. Like other creatures, they are born, grow old and die; they must eat to live, protect themselves to survive and mate to reproduce. But as persons, humans seem to float aloof from this world in multiple realms of discourse and meaning, each constitutive of a specific historical consciousness. From this exalted position they are said to transform nature, both ideationally through the imposition of schemes of symbolic representation and practically through the application of technology, thereby converting it into the object of relations among themselves, relations that are taken to make up the distinct domain of society.

Now a complementarity approach would accept this division between the organism and the person, and would aim to put together the partial accounts of human life obtainable on each of the two planes, of nature and society, to produce a complete "biosocial" picture. The obviation approach, by contrast, would reject the complementarity assumption, that human existence can be neatly partitioned into its biophysical and sociocultural components, not, however, by simply collapsing one side of the dichotomy into the other as in the more extreme forms of sociobiology and social constructionism, but by doing away with the dichotomy itself. Whereas an advocate of complementarity might assert that the human being is not merely a biological organism nor merely a social person, but the compound of one thing plus the other, the obviation approach asserts that humans are indeed all organism, as indeed they are all person, for in the final analysis organism and person are one and the same, and there is nothing *mere* (that is, residual or incomplete) about either (Ingold 1990: 220). By the same token, this approach would reject the idea that there is an essence of humanity that sets us radically apart from all other creatures whose lives are wholly contained within the world of nature, and with it the possibility of a purely objective account of the human being as a naturally existing, evolved entity. We may of course imagine ourselves to be suspended in a world of intersubjective meaning, over and above that of our material life, but such imaginings can only be carried on by a being who is already positioned in the world and, by virtue of that fact, already committed to relations with determinate components of the environment. You have to be in a world to imagine yourself out of it, and it is through this being-in-the-world that you become what you are.

Let me briefly compare the two approaches as they might be applied to one of the classic fields of anthropological inquiry, namely kinship. The complementarity approach would reject both the radical sociobiological thesis, that kinship can be reduced to a calculus of genetic relatedness, and the equally radical humanistic alternative, that it is an arbitrary social construct that bears no relation to genetic connection at all. Rather, it would suggest that for a complete understanding of human kinship we need to recognize not only how individuals may be innately predisposed to behave in certain ways toward those to whom they have a close genetic link,

but also how such ways of behaving are channeled, evaluated and made meaningful, and the persons to whom they are directed categorized, in terms of culturally specific, representational schemata. An obviation approach to the study of kinship, on the other hand, would begin by recognizing that behavioral dispositions are neither preconstituted genetically nor simply downloaded onto the passively receptive individual from a superior source in society, but are rather formed in and through a process of ontogenetic development within a specific environmental context. Kinship is about the ways in which others in the environment contribute—through their presence, their activities and the nurturance they provide—to this process.[1] Thus, insofar as it concerns the growth of the organism-person within a field of ongoing relations, kinship is indissolubly biological and social. But the biology pertains to development, not genetics; and the social to the domain of lived experience rather than its categorical representation.

The contrast between the two approaches may be illustrated by way of one other example. Bipedal locomotion, the capacity to walk on two feet, is generally assumed to be one of the hallmarks of our species, and as such to form part of an evolved human nature. Yet as we all know, and as Mauss famously observed in his essay of 1934 on body techniques (Mauss 1979: 97–123), people in different cultures are brought up to walk in very different ways. These ways are acquired, or as Mauss put it (p. 102), "there is no 'natural way' [of walking] for the adult." How would a complementarity approach deal with this? It would argue—very much, in fact, as Mauss did—that although the body is innately predisposed to walk, it is also educated by a received social tradition, transmitted orally or by other means, consisting of certain ideal rules and conventions that lay down standards of propriety, perhaps specific to age, sex or gender, that walkers are enjoined to follow, and in terms of which their performance is evaluated and interpreted. Thus, while the capacity to walk is a biological universal, particular ways of walking are expressive of social values.

But the obviation approach can readily find fault with this argument. For a start, human babies are not born walking; rather, the ability to walk is itself an acquired skill that develops in an environment that includes walking caregivers, a range of supporting objects, and a certain terrain. How, then, can one possibly separate learning to walk from learning to walk in the approved manner of one's society? Surely, the development of walking skills is just one aspect of the growth of the organism-person within a nexus of environmental relations, and as such is closely bound up with kinship. Walking is certainly biological, in that it is part of the modus operandi of the human organism, but it is also social—not because it is expressive of values that somehow reside in an extrasomatic domain of collective representations, but because the walker's movements, his or her step, gait and pace, are continually responsive to the movements of others in the immediate environment. It is in this kind of mutual responsiveness or "resonance" (Wikan 1992), not in the subjection of behavior to categorical rules, that the essence of sociality resides.

Body, Organism, and Development

Clearly, the problem with the complementarity thesis is that it is unable to offer a coherent account of ontogenetic development. Human beings are supposed to be in part preconstituted genetically, in part moulded through the superimposition (through enculturation or socialisation) of ready-made structures. Real humans, however, *grow* in an environment furnished by the presence and activities of others. It is precisely because the dynamics of development lie at the heart of the obviation approach that it is able to dispense with the biological/social dichotomy. And it leads, naturally, to a focus on issues of embodiment. By this I do not mean that the human body should be understood as a site or medium for the inscription of social values. I would rather use the term to stress that throughout life, the body undergoes processes of growth and decay, and that as it does so, particular skills, habits, capacities, and strengths, as well as debilities and weaknesses, are enfolded into its very constitution—in its neurology, musculature, even its anatomy. To adopt a distinction suggested by Connerton (1989: 72–73), this is a matter of incorporation rather than inscription. Thus walking, for example, is embodied in the sense of being developmentally incorporated through practice and training in an environment. The same, indeed, goes for any other practical skill.

Having said that, however, I must admit to a growing unease with the fashion for the "body" in current social anthropology, and indeed with the very notion of embodiment. Advocates of the "paradigm of embodiment," such as Csordas (1990), have drawn inspiration from the philosophy of Merleau-Ponty (1962) in treating the body as the form in which the human person, qua cultural subject, is intentionally present as a being-in-the-world. One of their aims in doing so is to break away from the Cartesian bias, still dominant in mainstream psychology, toward treating the body as the executive arm of a disembodied mind that, sheltered from direct contact with the external world, is presumed to organize the data of experience and to be the ultimate source of all meaning and intention. I sympathize with this aim, but I am not sure that the best way to overcome the troublesome mind/body dichotomy is by dropping the former term and retaining the latter. It would seem just as legitimate to speak of enmindment as of embodiment, to emphasize the immanent intentionality of human beings' engagement with their environment in the course of perception and action. The distance between a Merleau-Pontyan phenomenology of the body and what Bateson (1973) christened the "ecology of mind" is not as great as might first appear.

Perhaps this is merely an issue of semantics. Behind it, however, there lies a more fundamental question. How, if at all, are we to distinguish the body from the organism? One answer might be that the body is a discrete object composed of organs and tissues, like as not dead or at least anesthetized, as it might appear before the surgeon in the operating theatre, whereas the organism is a living being, situated and functioning in its proper environment. But neither Merleau-Ponty nor those who have followed his lead mean the body in this sense. They are rather referring to "the living body ... with feelings, sensations, perceptions and emotions" (Ots 1994: 116), or what is known in German as *Leib* (as opposed to *Körper*). Yet in their determination to treat the *leib*ly body as the *subject* of culture, anthropologists such as Jackson (1989: 119) and Csordas (1990: 5) cannot avoid the implication that there exists some kind of biological residuum that is objectively given, independently and in advance of the cultural process.[2] Culture and biology remain as far apart as ever, only the body has been repositioned: formerly placed with the organism on the side of biology, it has now reappeared with the person on the side of culture. Hence the body as subject is split off from the organism as object, leaving the latter bodiless, reduced to an inchoate mass of biological potential. The embodiment of culture leads to nothing less than the disembodiment of the organism!

It seems to me that the theoretical gains brought by the paradigm of embodiment will be more apparent than real, so long as we fail to take one final, and crucial step, which is to recognize that the body *is* the human organism, and that the process of embodiment is one and the same as the development of that organism in its environment. Once this step is taken, then one or other of the two terms, body and organism, becomes effectively redundant. Given the choice of which term to retain, I would opt for the latter, since it better conveys the sense of organized process, of movement, connectivity, and relationality, that I take to be fundamental to life. Substituting life for mind, and organism for body, the notion of a mindful body may be replaced by that of living organism, a substitution that has the effect both of restoring human beings to their proper place within the continuum of organic life, and of laying the Cartesian dualism finally to rest. Most social anthropologists, however, even

those committed to phenomenological or ecological approaches, are markedly reluctant to go this far. Their hesitation may be attributed in part to the continuing influence of dualistic thinking, but in part—too—to a certain nervousness about the implications of the position set out earlier for the distinction between culture and biology.

These implications are indeed radical. If, as I have suggested, those specific ways of acting, perceiving, and knowing that we have been accustomed to call cultural are incorporated, in the course of ontogenetic development, into the neurology, musculature, and anatomy of the human organism, then they are equally facts of biology. Cultural differences, in short, *are* biological. Now of course, it was precisely on the premise that cultural variation is independent of biology that anthropologists could claim to have refuted the raciology of the early decades of this century (Wolf 1994). In 1930, no less an authority than Boas had declared that "any attempt to explain cultural form on a purely biological basis is doomed to failure" (Boas 1940: 165). From then on, the biophysical and sociocultural divisions of anthropology have proceeded along markedly divergent paths. It is no wonder that contemporary social anthropologists should be fearful of going back on such a fundamental tenet of disciplinary integrity.

I believe this is an issue that has to be confronted. How can we rest secure in the conviction that raciology has long since been expurgated from the discipline, now that the premises on which this was done seem increasingly shaky, if not downright incoherent? Evidence of this incoherence is not hard to come by, for example in the "statement on race" recently endorsed by the International Union of Anthropological and Ethnological Sciences. Article 10 of this statement begins with the well-worn claim that "there is no necessary concordance between biological characteristics and culturally defined groups," and ends by asserting that "it is not justifiable to attribute cultural characteristics to the influence of genetic inheritance" (IUAES 1996: 19–20). What is striking here is the implicit attribution of "biological characteristics" to "genetic inheritance"—and this despite the recognition, elsewhere in the document, that "biological differences ... are strongly influenced by nutrition, way of life and other aspects of the environment" (Article 4). To return to my earlier example: consider a culturally specific way of walking. Is this not a property of the organism, the outcome of a process of development, and hence fully admissible as a "biological characteristic"? Despite Boas's strictures, there is nothing wrong with accounting for this or any other aspect of cultural form on a "purely biological basis," so long as the biology in question is of development, not genetics.

Evidently, the real source of the problem is not the identification of the social or cultural with the biological, but the assignment of the biological to the genetic.[3] For it is the latter assumption, which still lies unquestioned at the heart of much anthropological theory as well as in the discipline's public pronouncements, that forces us to choose between treating, say, a locally specific way of walking either as basically nonbiological or extrasomatic, governed in its bodily execution by a scheme of acquired mental representations, or as biological but genetically inherited. The first alternative reinstates the Cartesian antinomies of mind and body; the second takes us right back to raciology. Breaking the link between biological form and genetic inheritance, however, is easier said than done, for this link underpins the entire edifice of modern evolutionary theory and justifies the fundamental precept on which it rests, namely that the life history of the individual organism, its ontogenetic development, forms no part of the evolution of the species to which it belongs.[4]

The Myth of the Genotype

In brief, what is supposed to evolve is not the organism itself or its manifest capabilities of action, but rather a formal design specification for the organism known as the genotype. The evolution of this specification takes place over numerous generations, through changes brought about by natural selection in the frequency of its information-bearing elements, the genes. Development is then understood to be the process whereby the genotypic specification, by definition context-independent, is translated within a particular environmental context into the manifest form of the phenotype. In this standard account, the genotype is privileged as the locus of organic form, while the environment merely provides the material conditions for its substantive realisation. To be sure, an organism may develop different features in changed environments, but these differences are regarded as no more than alternative phenotypic expressions of the same basic design. Only when the design itself changes does evolution occur.

Let me return for a moment to the example of walking. According to orthodox evolutionary biology, bipedal locomotion is one of a suite of anatomical and behavioral characteristics that have emerged in the course of human evolution. It—or, rather, a program for its development—must therefore form part of the species-specific genotypic endowment that each one of us receives at the point of conception. It is in this sense that human beings are said to be universally equipped, as part of their evolved makeup, with an innate capacity to walk on two feet, regardless of how they walk in practice, or of whether they walk at all—or go everywhere by car! Specific ways of walking have not themselves evolved, they are just alternative phenotypic realizations of an evolved, genotypic trait. By the same token, we should all be genotypically endowed with the capacity to rest for long periods in a squatting position, yet this is something that I (along with fellow Westerners) am quite unable to do, since I have been brought up in a society where it is normal to sit on chairs. As this example shows, the notion of capacity is almost totally vacuous unless it refers back to the overall set of conditions that must be in place, not only in the individual's genetic constitution but also in the surrounding environment, to make the subsequent development of the characteristic or capability in question a realistic possibility (Ingold 1996a). One would otherwise have to suppose that human beings were genotypically endowed, at the dawn of history, with the capacity to do everything that they ever have done in the past, and ever will do in the future—not only walk and squat but also swim, ride bicycles, drive cars, fly airplanes, carry out scientific research, and so on (the list would be endless).

What this means, in general terms, is that the forms and capacities of human and other organisms are attributable, in the final analysis, not to genetic inheritance but to the generative potentials of the *developmental system* (Oyama 1985), that is, the entire system of relations constituted by the presence of the organism in a particular environment. This is not to deny that every organism starts life with—among other things—its complement of DNA in the genome. Orthodox evolutionary theory has it that this DNA encodes the formal design specification. Because, however, there is no reading of the genetic code that is not itself part of the process of development, it is only within the context of the developmental system that we can say what any particular gene is *for*. It follows that there can be no specification of the characteristics of an organism, no design, that is independent of the context of development. The genotype simply does not exist. And so too, in the case of human beings, there is no such thing as "bipedal locomotion" apart from the manifold ways in which people actually learn to walk in different communities (Ingold 1995a).

Now if, as I have argued, organic form is a property not of genes but of developmental systems, then to account for its evolution we have to understand how such systems are constituted and reconstituted over time. This conclusion has three major implications. First, far from being a tangential offshoot of the evolutionary process, ontogenesis is the very crucible from which it unfolds. Second, because organisms, through their activity, can influence the environmental conditions for their own future development and that of others to which they relate, they figure not as passive sites of evolutionary change but as creative agents, producers as well as products of their own evolution. Third, and most crucially for my present purposes, this applies equally to human beings. "Our basic image of human ontogeny," as Robertson (1996: 595) insists, "should therefore be that of a lifespan set between an ascendent and a descendent generation, linked by the process of begetting and being begotten." Human lives overlap: fashioned within contexts shaped by the presence and activities of predecessors, they in turn affect the conditions of development for successors. There is nothing strange about this idea; on the contrary it sums up the process we are used to calling history. So conceived, however, history is not so much a movement in which, as Maurice Godelier puts it (1989: 63), human beings "produce society in order to live," as one in which, in the course of their social lives, they *grow* one another, establishing by their actions the conditions for each other's development. But taken in this sense, history is no more than a continuation, into the field of human relations, of a process that is going on throughout the organic world. That process is one of evolution. The distinction between history and evolution is thus dissolved (Ingold 1995a: 210–211).

Anthropology and Archaeology

Between Evolution and History

This is where prehistoric archaeology comes in. Do archaeologists study human history or human evolution? So long as the distinction remains in place, archaeology seems to fall awkwardly between the two stools. An indicator of this predicament is the fact that while there has long been a strongly held view in social anthropology that there is little to distinguish it from the discipline of history, and despite the rather obvious links between history and archaeology (in that both study the lives of people in the past), the majority of social anthropologists insist that their subject has little or nothing to do with archaeology. Of course, this was not always so. The evolutionary anthropologists of the nineteenth century were keen to study allegedly primitive peoples because it was thought that their present existence could illuminate the earlier conditions of humankind in the spheres of social and intellectual life, just as archaeology could reveal the early stages of material culture. But the subsequent rejection of this kind of progressive evolutionism broke the link between social anthropology and archaeology, and at the same time ruled out of order any suggestion that humans might be *more* or *less* cultural, or that they might be further along or lag behind in the course of history.

So far as most contemporary social anthropologists are concerned, living beings either inhabit a historically constituted world of cultural meaning or they do not: all human beings do, other animals do not. There are no differences of degree. Yet the very idea that humans inhabit separate, cultural worlds implies that at some point, the history of culture must have lifted off from a baseline of full-blown, evolved human capacities. Short of supposing some kind of unfathomable quantum leap, there is no alternative but to imagine a historical trajectory that rises inexorably from a point of emergence or origin. Figure 19.1 shows an early example of just such a view, taken from Kroeber's classic paper of 1917 on the superorganic (Kroeber 1952: 50). Here, the history of culture is seen taking off from organic evolution at point B; by point C we have the rudimentary culture of "primitive man," while by point D (the present) the origins of culture have been left far behind. Present-day social anthropologists may well frown at this picture, and scoff at its invocation of progressive development, but they themselves have nothing better to offer. And one of the reasons why they tend to steer clear of prehistoric archaeology, I suggest, is that it throws the spotlight on just those awkward questions

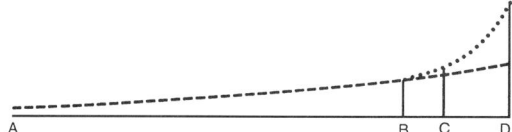

Figure 19.1
Organic evolution and the history of culture (after Kroeber 1952: 50). "In this illustration," Kroeber explains, "the continuous line denotes the level inorganic; the broken line the evolution of the organic; the line of dots, the development of civilisation. Height above the base is degree of advancement." A marks "the begining of time on this earth," B "the first human precursor," C "primitive man," and D "the present moment."

that they would rather not have to think about. If human history has a point of origin, what could it mean to have been living close to that point, or even at the crucial moment of transition itself? Were such people semicultural, gearing up for history? How can one conceivably distinguish those actions and events that carried forward the movement of human history from those that set it in motion in the first place?

In recent years archaeologists have expended a great deal of effort in revealing the origins of culture and history in what has come to be called the "human revolution" (Mellars and Stringer 1989). This is now supposed to have taken place during the Upper Palaeolithic, though archaeologists remain perplexed by the apparent fact that so-called modern humans—that is, beings equipped with the full suite of evolved capacities needed to set the cultural ball rolling—arrived on the scene a good hundred thousand years before we find any evidence for the sorts of things with which culture is usually associated: burials, art, complex and regionally diverse toolkits, language, and so on. Indeed, the alleged revolution seems to have taken about twice as long as the fifty-thousand-year history it is supposed to have inaugurated! Be that as it may, the argument I have set out above suggests that the entire project of searching for the genesis of some essential humanity is seriously misguided. We look in vain for the evolutionary origins of human capacities for the simple reason that these capacities continue to evolve in the very historical unfolding of our lives.

Of course, even an orthodox evolutionary theorist would have to admit that the evolution of humankind did not exactly stop once history and culture were underway. However the conventional view, exemplified in Kroeber's diagram (figure 19.1) and reiterated by countless authors ever since, is that by comparison with the rate of historical change, this evolution has continued at snail's pace, so that to all intents and purposes, contemporary human beings may be regarded as not significantly different from their predecessors of the Upper Palaeolithic.[5] They are equipped with the same basic morphology, capacities and dispositions that, packaged in the genotype, have been passed on down the generations for tens of thousands of years. To be sure, the amount of genetic change in human populations over this period may have been relatively small. My contention, however, is that in their dispositions and capacities, and to a certain extent even in their morphology, the humans of today are not at all like their predecessors. This is because these characteristics are not fixed genetically but emerge within processes of development, and because the circumstances of development today, cumulatively shaped through previous human activity, are very different from those of the past.

It is, I believe, a great mistake to populate the past with people like ourselves, equipped with the underlying capacities or potentials to do everything we do today, such that history itself appears as nothing more than the teleological process of their progressive realization. Indeed the very notion of an origin, defined as the point at which these capacities became established, awaiting their historical fulfilment, is part of an elaborate ideological justification for the present order of things and, as such, but one aspect of the intense presentism of modern thought. In so far as the task of archaeology is to illuminate the past rather than legitimate the present, archaeologists

should be foremost in combating the pretensions of the origin-hunters. And they should help us to recognize that our humanity, far from having been set for all time as an evolutionary legacy from our hunter-gatherer past, is something that we continually have to work at and for which we ourselves must bear the responsibility. There is, in short, no way of saying what a human being is apart from the manifold ways in which human beings become. "Modern humans" have not originated yet, and they never will.

Landscape and Environment

This recognition that the forms of human being, and the capacities they entail, are continually evolving as life goes on, helps to put paid to another dichotomy which has been particularly troublesome for archaeology. This is between the natural and the artificial, and it lies at the source of the idea that archaeologists study artifacts. Now the very notion of artifact implies the working up of some raw material to a finished form, corresponding to a preconceived design in the mind of the artisan. Only once it has first been *made*, in this sense, can it be brought into play in the ordinary business of life, in the course of which it is *used*. This distinction between making and using is fundamental to what I have elsewhere (Ingold 1995b) called the "building perspective": the idea that life goes on within structures that have been constituted in advance, rather than these structures arising within the life process itself. Adopting such a perspective, it is easy to imagine that the forms of objects recovered from archaeological sites correspond to designs that were originally in the heads of their one-time makers.

However as Davidson and Noble (1993: 365) have pointed out, it is a fallacy—and one that is found very frequently in archaeological writing—to suppose that objects are ever finished in this sense. For one thing, their forms are not imposed by the mind, but arise within the movement of the artisan's engagement with the material; for another, in the course of being used for one purpose, objects may undergo further modification that make them peculiarly apt for another. Whether, at any moment, we say the object is being used or made depends entirely on whether the reference is to a present or future project. Although at a certain point, the artisan may claim to have completed his work, that is certainly not the end of the object he has produced. Indeed, an artifact can never really be said to be finished until it is of no further use to anyone and is finally discarded. The lesson to be learned from this is that the objects around us have histories which, in certain respects, are not unlike the life histories of persons. Just as persons continually come into being through their involvement in relationships with other persons and objects in their environments, so the forms and meanings of objects are generated within the contexts of their involvement in the diverse life-projects of the beings (human and nonhuman) with which they are surrounded. In this respect they are never made but always in the making.[6]

We cannot, then, make a hard and fast distinction between one class of things that are ready-made in nature, and another class of things that have been made through the shaping of a naturally given raw material into a finished artefactual form. Nor can we adopt an analogue of the complementarity thesis, and suppose that objects, like persons, are in part naturally preconstituted, and in part molded through the imposition of cultural design. Just such an analogue is implicated in the unfortunate designation of artifacts as objects of "material culture," suggesting as it does that to make an artifact you first take an object with certain intrinsic material properties and then add some culture to it. I noted earlier that a critical weakness of the complementarity thesis, applied to persons, is that it cannot offer a realistic account of ontogenetic development. In precisely the same way, an approach that stresses the complementarity of natural and artificial—or built—components of the environment cannot begin to grasp the ways in which environments

enfold, into their very formation and constitution, the lives and works of their inhabitants (Ingold 1993b: 156–157). To appreciate what is going on here, we need to adopt a different perspective, one that recognizes that the forms people make or build, whether in their imagination or on the ground, arise within the current of their involved activity, in the specific relational contexts of their practical engagement with their surroundings. Building, in short, is encompassed by dwelling, making by use. I call this the "dwelling perspective" (Ingold 1995b).

Alongside the relatively recent anthropological recognition, discussed earlier, that the human intentional presence in the world is an embodied presence, there has been a remarkable upsurge of interest in the landscape. This is already proving to be a particularly fruitful area of collaboration between anthropologists and archaeologists (for example, Bender 1993; Tilley 1994). Though the precise meaning of "landscape," as that of "body," continues to be the subject of intense controversy, there is a clear connection between the two concerns. For if persons inhabit intentional worlds, and if bodies inhabit landscapes, then to reunite persons with their bodies is also to restore their intentional worlds to the landscape. But this raises a parallel problem, too. How, if at all, are we to distinguish landscape from environment? Just as the body has come to be identified with the cultural subject, and the organism with the residual biological object, so there is a temptation to treat the landscape as an intersubjectively constituted, existential space, while reducing the environment to a mere substrate of formless materiality. In this vein, Weiner (1991: 32) speaks of how the bestowal of place names intentionally transforms "a *sheer physical terrain* into a pattern of historically experienced and constituted space and time," thereby creating "existential space out of a *blank environment*" (my emphases).

Once again, the division between the sociocultural and the biophysical is reproduced rather than dissolved. This, in my view, is a retrograde step. The environment of persons is no more reducible than is their organic existence to pure molecular substance. It is not merely physical, and it is certainly not blank. For example, the ground I walk on is surely a part of my environment, but in a physicalist description the ground, as such, does not exist; there are only packed molecules of carbon, nitrogen, silicon, and so on. As Reed has eloquently put it, "it is the earth on which we walk, and the soil in which we plant, that is relevant for us as perceiving and acting creatures; not the molecules discovered by scientists" (Reed 1988: 111). The environment, in short, is not the same as the physical world; that is to say, it is not describable in terms of substance. Rather, the environment is the world as it exists and takes on meaning in relation to the beings that inhabit it (Gibson 1979: 8). As such, its formation has to be understood in the same way that we understand the growth of organisms and persons, in terms of the properties of dynamic self-organization of relational fields. But precisely because environment does not stand as material substance to the immaterial forms of landscape—because it undergoes a continual process of formation with, and around, its inhabitants—I see no basis on which the two terms, environment and landscape, may be distinguished.

Earlier, I suggested that the concept of the "mindful body" should be replaced by that of "living organism." But the formulation remained incomplete, since neither the body nor the organism, however conceived, can exist in isolation. We can now complete it with the proposition that the mindful body in a landscape be replaced by the living organism in its environment. And this, to conclude, offers the basis for a real synthesis of archaeology and anthropology. Instead of being separated by contrived divisions between past and present, or between artifacts and bodies, we might say that where anthropology studies the conditions of human living in the environment—or of what phenomenological philosophers call "being-in-the-world"—archaeology studies the formation of the environment of our living-in,

our dwelling. In practice, of course, one cannot do one without the other, nor can either be done without regard to the inherent temporality of the processes both of ontogenetic development and environmental formation. I have already shown how we can dispense with the distinction between social and biological anthropology. It is now possible to see how the anthropology/archaeology distinction might be thrown out as well.

Psychology and Anthropology

One of the reasons why, up to now, it has proved so difficult to effect a reintegration of the subdivisions of anthropological inquiry, and particularly of its social and biological components, lies in the fact that in the conventional complementarity approach, the necessary link between the individual organism and the cultural subject can only be established by way of a third term, namely what is called the "human mind." The discipline that exists to study the human mind is, of course, psychology. Thus for advocates of the complementarity approach, psychology would have to be along in any complete, synthetic account of human existence. The synthesis would not be "biosocial" but "biopsychosocial." As Mauss put it, an exclusive focus on the relations between the biological and the sociological "leaves but little room for the psychological mediator." An account of walking, for example, that rested solely on an anatomical or physiological base, or even on a psychological or sociological one, would be inadequate. "It is the *triple viewpoint*, that of the 'total man,' that is needed" (Mauss 1979 [1934]: 101, my emphasis). In what follows I propose to argue, to the contrary, that the human mind—conceived as some kind of structured entity—is as much an invention of modern science as is the human genotype. Mind, as I have already suggested with acknowledgment to Bateson (1973), is not in the head rather than out there in the world, but immanent in the active, perceptual engagement of organism-person and environment.

As the study of the conditions of such engagement, psychology should be no different from anthropology.

To begin, we have to consider why a psychological mediator should be deemed necessary at all. The reason lies in the fact that the classical division between the biological and the social is not based on one opposition but is a compound of two: between body and mind, and between the individual and the collectivity. Thus psychology has traditionally shared with biological anthropology an exclusive focus on the individual, and with social anthropology a focus on mental rather than bodily states. I have already addressed the problem of the mind/body dichotomy, but it remains to deal with that of individual versus collectivity. This latter dichotomy rests on a hierarchical conception of the relations between parts and wholes that is very deeply embedded in the structure of our thought. Anthropologists have always professed their commitment to a holistic approach, but they have tended to take this to mean a focus on wholes—conceived as total societies or cultures—as opposed to their parts or members, individual human beings. Following principles set out by Durkheim a century ago, it has generally been conceded that as the whole is more than the sum of its parts, so "society is not the mere sum of individuals, but . . . a specific reality which has its own characteristics" (Durkheim 1982 [1895]: 128–199).

Now the very logic of summation invoked here entails that every part is a self-contained, indivisible, naturally bounded unit whose integrity and constitution are already given, independently and in advance of any relations it may enter into with others of its kind. These relations, in short, have no bearing upon the constitution of the individual parts themselves, but are rather constitutive of a distinct entity, namely society, located at a higher level of abstraction. This Durkheimian view has long underwritten the academic division of labor between psychology and social anthropology: whereas the former is said to study the mind of the individual, the latter is concerned with the collective mind of society. Much recent work in social anthropology, however, has pointed to the inadequacy of the classical individual/society dichotomy (Strathern 1996). We have begun to recognize (see, for example, Toren 1993) that those capacities of conscious awareness and intentional response normally bracketed under the rubric of mind are not given in advance of the individual's entry into the social world, but are rather fashioned through a lifelong history of involvement with both human and nonhuman constituents of the environment. We have realized, too, that it is through the situated, intentional activities of persons, not through their subjugation to the higher authority of society, that social relationships are formed and reformed.

With this, the hierarchical conception of part/whole relations simply collapses. Every particular person, in so far as it enfolds in its constitution the history of its environmental relations, gathers the whole into itself.[7] But that whole, so conceived, is not an entity but a movement or process: the process of social life. Persons come into being, with their specific identities, capacities, and powers of agency, as differentially positioned enfoldments of this process, and in their actions they carry it forward. Consciousness and social existence, though they appear at any particular moment to offer alternative perspectives on the person, respectively inward-looking and outward-looking (Ingold 1983: 9), turn out in their temporal unfolding to be one and the same, like the single surface of a Möbius strip. Taking this view, I can see no further intellectual justification for continuing to uphold the boundary that has traditionally divided psychology from social anthropology. The discipline that will be brought into being through the dissolution of this boundary, whatever we choose to call it, will be the study of how people perceive, act, feel, remember, think, and learn within the settings of their mutual, practical involvement in the lived-in world. In the following paragraphs I should like to review some of the consequences of this perspective in three areas that have traditionally been central to psychological inquiry: perception, memory, and learning.

Perception

Why do people perceive the world in the particular ways that they do? Mainstream psychology has long regarded perception as a two-step operation: in the first, sensory data are picked up from the environment by means of the receptor organs of the body; in the second these data are processed by a range of devices in the mind, to generate images or representations, internal models of an external reality. This processing is known as cognition. By and large, psychologists have been concerned to discover universals of cognition, which are attributed to structures established in the course of human evolution. Anthropologists, by contrast, have wanted to explain why people from different cultural backgrounds perceive the world in different ways. They have done so by suggesting that human cognized models are constructed on the basis of programs or schemata that are acquired as part of a tradition, and vary from one culture to another. What people see will therefore be relative to their particular framework for viewing the world. At first glance, the universalistic claims of psychology seem incompatible with the relativistic stance adopted by social anthropology. But as several authors have pointed out (e.g., D'Andrade 1981; Sperber 1985; Bloch 1991), the two perspectives are, in fact, perfectly complementary. For unless innate processing mechanisms are already in place, it would not be possible for human beings to acquire the programs for constructing their culturally specific representations from the data of experience.

I will spell out the logic of this argument later on, because it bears on the issue of learning. My present concern is with the way in which the approaches outlined above, both in psychology and anthropology, reproduce the Cartesian duality of mind and body, removing the former from the contexts of human engagement with the environment while treating the latter as no more than a kind of recording instrument, converting the stimuli that impinge upon it into data to be processed. One of the most powerful critiques of this view has come from advocates of so-called ecological psychology, who have drawn inspiration above all from the pioneering work of Gibson on visual perception (Gibson 1979). Ecological psychologists reject the information-processing view, with its implied separation of the activity of the mind in the body from the reactivity of the body in the world, arguing instead that perception is an aspect of functioning of the total system of relations constituted by the presence of the organism-person in its environment. Perceivers, they argue, get to know the world directly, by moving about in the environment and discovering what it affords, rather than by representing it in the mind. Thus meaning is not the form that the mind contributes, by way of its acquired schemata, to the flux of raw sensory data, but is rather continually being generated within the relational contexts of people's practical engagement with the world around them.

It follows from this approach that if people raised in different environments perceive different things, this is not because they are processing the same sensory data in terms of alternative representational schemata, but because they have been trained, through previous experience of carrying out various kinds of practical tasks, involving particular bodily movements and sensibilities, to orient themselves to the environment and to attend to its features in different ways. Modes of perception, in short, are a function of specific ways of moving around—of walking, of sitting or squatting, of tilting the head, of using implements, and so on, all of which contribute to what Bourdieu (1977: 87) would call a certain "body *hexis*." And as we have already seen, these forms of motility are not added to, or inscribed in, a preformed human body, but are rather intrinsic properties of the human organism itself, developmentally incorporated into it modus operandi through practice and training in a particular environment. Hence capacities of perception, as of action, are neither innate nor acquired but undergo continuous formation within processes of ontogenetic development. This result is clearly in line with the conclusions to be drawn from an obviation approach to the relation between social and biological phenomena. In their rejection, on the one hand, of the Cartesian view of action as the bodily execution of innate or acquired programs, and on the other hand, of the cognitivist view of perception as the operation of the mind upon the deliverance of the senses, the obviation approach in anthropology and the ecological approach in psychology find common cause. Both take the living-organism-in-its-environment as their point of departure. This is why (*contra* Bloch 1991) I believe that an anthropology that sets out from this point has more to gain from an alliance with ecological psychology than from an alliance with cognitive science.

Memory

Another way of expressing the difference between cognitivist and ecological approaches is in terms of a contrast suggested by Rubin (1988). One may understand what is going on, he writes, in terms of one or other of two alternative metaphors. The first is a complex structure metaphor, the second a complex process metaphor. The former, which is dominant in cognitive psychology, works by converting what is observed in the world into a formal account, whether envisaged as a script, schema, grammar, program, or algorithm, and then has that account copied into the mind so that the observed behavior can be simply explained as the expression of this mental blueprint. The latter, the dominant metaphor in ecological psychology, imputes little or no structured content to the mind. Instead, behavior is explained as the outcome of a complex process set in train by virtue of the immersion of the practitioner, whose powers of perception and action have been fine-tuned through previous experience, within a given environmental context. Let me clarify the contrast by means of a simple analogy. Suppose I play a record of one of Bach's suites for unaccompanied cello. An exceedingly complicated pattern is engraved on the otherwise blank surface of the disc, but the mechanical processes—of rotation and amplification—involved in the operation of the record player could hardly be more simple. Now suppose I pick up my cello to perform the suite myself. In this case, the music issues directly from my own movement, a movement that involves the whole of my being indissolubly coupled with the instrument. The process of playing a musical instrument like a cello is enormously complex, and calls for embodied skills that take years to acquire. But whether the music exists at all as a structure in the head or mental score, independently of the activities of practice and performance, is a moot point.

Now in introducing his distinction between complex structure and complex process metaphors, Rubin was actually concerned with the psychological study of memory. His point was that in mainstream cognitive psychology, it is usual to regard memory as a kind of mental store, in which past experiences and received information are engraved and filed, as on the grooves and bands of a record (or, to adopt a more contemporary analogy, a computer disc). Remembering is then a rather simple process of searching or scanning, across a complexly structured cognitive array. It is, moreover, a purely mental, inside-the-head operation. Once a particular memory is retrieved, it may or may not be expressed in overt, bodily behavior. But every behavioral expression, like every playing of a record, is no more than a replica run off from a preexisting template. With a complex process model, by contrast, remembering is itself a skilled, environmentally situated activity. It is in playing the Bach suite that I remember it; the processes of remembering and playing are one and the same. It follows that every performance, far from being a replica, is itself an original movement in which the music is not so much reproduced as created anew. More generally, remembering is a matter not of discovering structures in the attics of our minds, but of generating them from our movements in the world.

Armed with this contrast, let me now turn to the role of remembering in social life. It is remarkable that the two pioneering figures in the study of social memory, Halbwachs and Bartlett, took opposite sides on the issue. Halbwachs, a committed Durkheimian, identified memory with the very framework of collective representations that are supposed to give order and meaning to the otherwise chaotic influx of raw sensation. If memory is social rather than individual, it is because the complex structures that underwrite the human capacity for recollection have their source in a collective tradition. "Our recollections," Halbwachs wrote, "depend on those of all our fellows, and on the great frameworks of the memory of society" (1992: 42). For Bartlett, to the contrary, what counted was not the structure of memory, but the process of remembering. This process, he argued, depends upon an organization of what he called "schemata." Ironically, though it was Bartlett who introduced the concept of schema into psychology, he did not like it, and warned explicitly against regarding schemata as static, maplike structures—which is precisely how they are understood by most cognitive psychologists and cognitive anthropologists today. According to Bartlett, the schema is an *active* organization of past reactions or experiences, which is continually brought to bear, and at the same time continually evolves, in the complex process of our engagement with the environment (Bartlett 1932: 201).[8] And it is because this is largely an environment of other persons that remembering is social.

Clearly, without the ability to remember, human beings would be unable to learn anything at all. But there is a world of difference between learning as adding more to one's internal, representational structure, and learning as the development of a skill (Rubin 1988: 379–380). Bartlett preferred the approach of skill (one of his examples was of strokes in tennis or cricket). However as Connerton has pointed out (1989: 28), the cognitivist emphasis on looking for structures in the mind, and the concomitant reduction of action to a simple process of mechanical execution, has left no conceptual space for the investigation of bodily enskilment, or what he calls "habit memory." It is true that most social anthropological work on memory has actually been about commemoration—the present reenactment of past events in ritual practice, storytelling, writing, and the like. And commemoration needs to be distinguished from memorization: the developmental incorporation of specific competencies (such as playing a musical instrument) through repeated trials. While the relation between memorization and commemoration has yet to be fully unraveled (Ingold 1996b: 203), the essential point to recognize is that the one cannot occur without the other. To commemorate the music of Bach, for example, it must be possible to perform it, one cannot perform it without skill, and the development of skill implies memorization (Connerton 1989: 5).[9]

Learning

This is the point at which to return to the psychological version of the complementarity thesis, namely that the acquisition of culture is possible thanks to innate mental processing devices. It is perfectly true that if culture consisted of a corpus of transmissible knowledge, or in the words of Quinn and Holland (1987: 4), of "what [people] must know in order to act as they do, make the things they make, and interpret their experience in the distinctive way they do," then the mind would have to be pre-equipped with cognitive devices of some kind that would allow this knowledge to be reassembled inside every individual head through a processing of the raw input of sensory data. In other words, the programs or schemata that enable people to construct their culturally specific representations of the world, and to deliver appropriate plans of action, would themselves have to be constructed from the elements of experience, on the basis of certain rules and principles. So how were these acquired? Perhaps in the same way, through the processing of experiential input according to yet another program. "You can learn to learn," Johnson-Laird explains, "but then that learning would depend on another program, and so on. Ultimately, learning must depend on *innate* programs that make programs" (Johnson-Laird 1988: 133). Whence, then, comes the information that specifies the construction of the innate devices, without which, it would seem, no learning could take place at all?

By and large, in the literature of cognitive psychology, the postulation of innate structures is taken to require no more justification than vague references to genetics and natural selection. Thus it is assumed that the design specifications for what is often called the mind's "evolved architecture" (Cosmides, Tooby and Barkow 1992: 5) must form one component of the human genotype. I have already shown, however, that it is impossible to derive a design specification for the organism from its genetic constitution alone, independently of the conditions of its development in an environment. For cognitive psychology this problem is further compounded, for if the theory of learning as the transmission of cultural information is to work, the requisite cognitive devices must already exist, not merely in the virtual guise of a design, but in the concrete hardwiring of human brains. Somehow or other, in order to kick-start the process of cultural transmission, strands of DNA have miraculously to transform themselves into data processing mechanisms. This is rather like supposing that merely by replicating the design of an aircraft, on the drawing board or computer screen, one is all prepared for takeoff.

Attempts in the literature to resolve this problem, insofar as it is even recognized, are confused and contradictory. To cut a rather long and tangled story short, they boil down to two distinct claims. One is that the concrete mechanisms making up the evolved architecture are reliably constructed, or wired up, under all possible circumstances. The other is that these universal mechanisms proceed to work on variable inputs from the environment to produce the diversity of manifest capabilities that we actually observe. Consider the specific and much-vaunted example of language acquisition. Here, the alleged universal mechanism is the so-called language acquisition device (LAD). It is assumed that all human infants, even those (hypothetically) reared in social isolation, come equipped with such a device. During a well-defined stage of development, this device is supposed to be activated, operating on the input of speech sounds from the environment so as to establish, in the infant's mind, the grammar and lexicon of the particular language spoken in his or her community. It would thus appear that language acquisition is a two-stage process: in the first, the LAD is constructed; in the second, it is furnished with specific syntactic and semantic content. This model of cognitive development is summarized in figure 19.2. Notice how the model depends on factoring out those features of the environment that are constant, or reliably present, in every conceivable developmental context, from those that represent a source of "variable input" from one context to another. Only the former are relevant in the first stage (the construction of "innate" mechanisms); only the latter are relevant in the second (the acquisition of culturally specific capabilities).

For comparative analytic purposes, it is sometimes helpful to sift the general from the particular, or to establish a lowest common denominator of development. But real environments are not partitioned in this way. Let me continue for a moment with the example of language learning. From well before birth, the infant is immersed in a world of sound in which the characteristic patterns of speech mingle with all the other sounds of everyday life, and right from birth it is

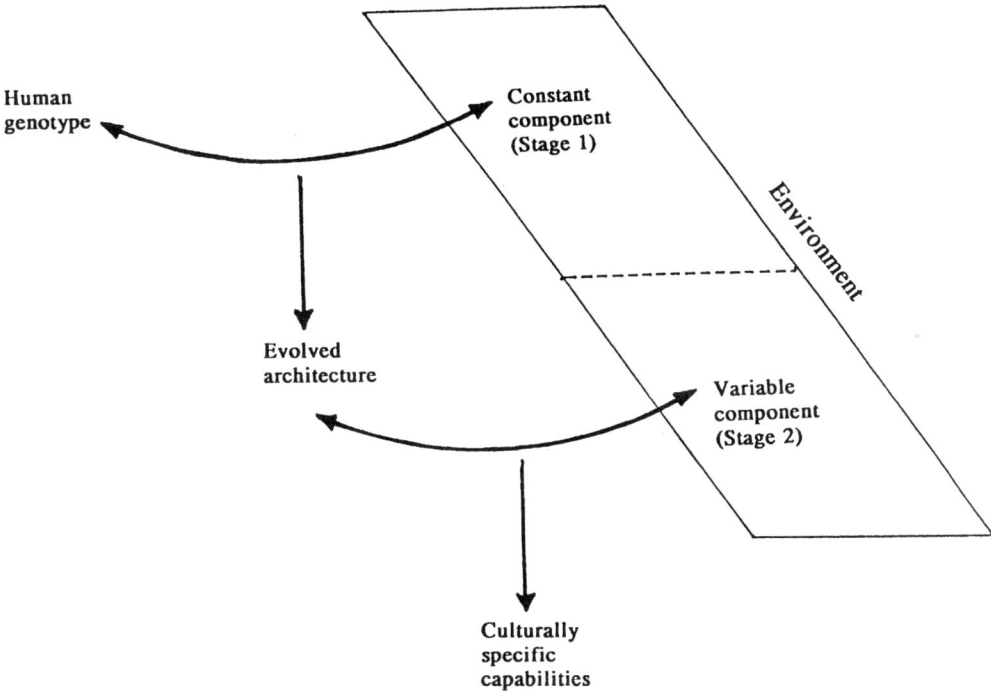

Figure 19.2
The two stages of cognitive development according to the complementarity model. In the first stage the human genotype interacts with the constant component of the environment to produce the universal mechanisms of the mind's evolved architecture. In the second, this architecture operates on variable environmental inputs to produce culturally specific capabilities.

surrounded by already competent speakers who provide support in the form of contextually grounded interpretations of its own vocal gestures. This environment, then, is not a source of variable input for a preconstructed device, but rather furnishes the variable conditions for the growth or self-assembly, in the course of early development, of the neurophysiological structures underwriting the child's capacity to speak. As the conditions vary, so these structures will take manifold forms, each differentially tuned both to specific sound patterns and to other features of local contexts of utterance. These variably attuned structures, and the competencies they establish, correspond of course to what appear to observers as the diverse languages of the world. In short, language—in the sense of the child's capacity to speak in the manner of his or her community—is not acquired. Rather, it is continually being generated and regenerated in the developmental contexts of children's involvement in worlds of speech. And if language is not acquired, there can be no such thing as an innate language-learning device.

What applies specifically in the case of language and speech also applies, more generally, to other aspects of cultural competence. Learning to walk in a particular way, or to play a certain musical instrument, or to practice a sport like cricket or tennis, is a matter not of acquiring *from* an environment representations that satisfy the input conditions of preconstituted cognitive devices, but of the formation, *within* an environment, of the necessary neurological connections, along with attendant features of musculature and anatomy, that underwrite the various skills involved. This conclusion is once again concordant with the obviation approach developed earlier, and it undermines one of the key ideas of the complementarity thesis—that cultural learning is like filling a universal, genetically specified container with culturally specific content. The notion that culture is transmissible from one generation to the next as a corpus of knowledge, independently of its application in the world, is untenable for the simple reason that it rests on the impossible precondition of a ready-made cognitive architecture. In fact, I maintain, nothing is really transmitted at all. The growth of knowledge in the life history of a person is a result not of information transmission but of guided rediscovery, where what each generation contributes to the next are not rules and representations for the production of appropriate behavior but the specific conditions of development under which successors, growing up in a social world, can build up their own aptitudes and dispositions.

The process of learning by guided rediscovery is most aptly conveyed by the notion of *showing*. To show something to someone is to cause it to be made present for that person, so that he or she can apprehend it directly, whether by looking, listening, or feeling. Here the role of the tutor is to set up situations in which the novice is afforded the possibility of such unmediated experience. Placed in a situation of this kind, the novice is instructed to attend to this or that aspect of what can be seen, touched or heard, so as to get the feel of it for him- or herself. Learning in this sense is tantamount to what Gibson (1979: 254) called an "education of attention." Gibson's point, in line with the principles of his ecological psychology, was that we learn to perceive by a fine-tuning or

sensitization of the entire perceptual system, comprising the brain and peripheral receptor organs along with their neural and muscular linkages, to particular features of our surroundings. Through this process, the human being emerges not as a creature whose evolved capacities are filled up with structures that represent the world, but rather as a center of awareness and agency whose processes resonate with those of the environment. Knowledge, then, far from lying in the relations between structures in the world and structures in the mind, mediated by the person of the knower, is immanent in the life and consciousness of the knower as it unfolds within the field of practice set up through his or her presence as a being-in-the-world.

The three topics I have reviewed above—of perception, memory, and learning—are of course closely connected. All of them could be addressed in terms of a complex structure metaphor, by imagining the world of our experience to be decomposed into a myriad of ephemeral fragments, unit events, samplings of which the mind has then to piece together into some coherent pattern by means of totalizing frameworks of social rather than individual provenance. I have argued, by contrast, for an approach that starts from relations and processes rather than structures and events. Whether our concern be with perceiving, remembering, or learning, the workings of mind are to be found in the unfolding relations between organism-persons and their environments. There is no way of saying what the human mind is, or of specifying its essential architecture, outside of this unfolding. For the forms of human knowledge do not stamp themselves upon the substance of human experience, but themselves arise within the complex processes of people's engagement with their surroundings. In short, the phenomena of mind are as much ecological and social as they are psychological. To conclude this section I should like to show why the approach adopted here promises to shed an entirely fresh light on one of the most neglected areas of anthropological inquiry, namely the knowledge and activities of children.

Children

In a paper presented some twenty years ago, Theodore Schwartz spoke of his "sudden and belated realization ... that anthropology had ignored children in culture while developmental psychologists had ignored culture in children" (Schwartz 1981: 4). There are signs, today, of a change of heart in both disciplines. The reasons for the anthropological neglect of children, however, do not lie merely in a certain observational blindness—the failure of ethnographers in the field to notice children, or to pay attention to their activities and what they have to say. Nor do they stem from the real difficulties, practical as well as ethical, of collaborating with children in ethnographic research. To bring children back to where they belong—at the center of our inquiries, just as they are at the center of social life—will require more than just a different attitude on the part of ethnographers. For what is at stake is the very framework of theory and concepts that we bring to our scientific project. Once again, the source of the problem lies in the thesis of complementarity. Developmental psychologists could afford to ignore culture, so long as they concerned themselves with supposedly universal mechanisms of acquisition, whose structure and functioning were conceived to be indifferent to the specificities of the acquired content. But conversely, social anthropologists could afford to ignore children, so long as they were regarded as incomplete adults whose personhood was not yet fully formed and who had still to take on the total complement of cultural knowledge from their predecessors.

In a sense, anthropology would have rather not had to deal with children for the same reason that it has shunned inquiry into human origins. In both cases, the received theoretical wisdom implies a transition from an initial state of biological existence, defined in terms of naturally evolved potentials, to a final state of full-blown cultural life, but nevertheless cannot countenance the possibility of a form of life that is semicultural, betwixt and between nature and history. Substitute "ancestral hominid" for "infant," and the following characterization of childhood offered by Goldschmidt (1993: 351)—"the process of transformation of the infant from a purely biological being into a culture-bearing one"— would serve equally well to define the so-called human revolution. And just as it is difficult to see how the events of this prehistoric revolution can possibly be distinguished from those of the history it is alleged to have inaugurated, so too, there seems to be no obvious way of telling apart the experiences, supposedly constitutive of childhood, that make a human being ready for history, from those that belong to the historical process itself. An obviation approach, however, enables us to dispense with such troublesome distinctions. The infant, who admittedly starts life as a "purely biological being," remains so for the rest of his or her life. Yet right from the moment of conception, this being is also immersed at the center of a world of other persons—a social world—and participates in the historical process of its unfolding (Toren 1993: 470). Surrounded by its entourage of adults, the infant contributes—by way of its presence and activities— to the latter's growth and development, just as they contribute to its own.

To be sure, children *are* different. For one thing, they are physically smaller, so that the environment affords them possibilities of action that are not available to grownups, and of course constrains what they can do as well—especially if it is full of structures built to adult dimensions. It is reasonable, too, to distinguish degrees of maturity in the life histories of organism-persons. Neurophysiologically, the brain of the adult human is more complex than that of the small child. It is not reasonable, however, to equate smallness or immaturity with a state of incompletion. Persons, as I have shown, are never complete, never finished, but undergo continual development within fields of relationships. The image of the child as an incomplete person has its source in the complementarity thesis, with its assumption that humans come into the world with their capacities already in place, waiting to be filled up with cultural content. A classic statement to this effect comes from Geertz (1973: 50): "Between what our body tells us and what we have to know in order to function, there is a vacuum we must fill ourselves, and we fill it with information (or misinformation) provided by our culture." The implication is that children's ability to function in the world is at best imperfect. Yet as Toren has rightly observed, "children have to live their lives in terms of their understandings just as adults do; their ideas are grounded in their experience and thus equally valid" (1993: 463). Adults and children may, then, function differently from one another, but no better or worse.

The complementarity approach, in effect, hides children from view behind a category of childhood which marginalizes them, or even excludes them altogether, from full participation in social life. The obviation approach, by contrast, brings children out into the open, but it does so by dissolving the categorical distinction between childhood and adulthood. Children and adults are no longer conceived to stand on either side of a boundary between becoming a person and being one, between undergoing socialization and participating in social life, between acquiring cultural knowledge and applying it in practice, or in short, between learning and doing. Children are persons just as adults are, and their knowledge and skills are likewise developed through participation both with other children and with adults in the joint practical activities of social life. This is not to say, however, that children and adults are the same. It is possible to speak of children without a special category of childhood, simply in recognition of the inherent temporality of human life, of the fact that organism-persons grow older, increasing in skill and maturity—and in that sense also in knowledge—as they do so. In the course of this aging process, one grows

out of certain ways of doing things, and grows into others. But no one has ever grown out of biology, nor has anyone grown into society or culture.

Conclusion

Throughout this chapter, I have argued against the idea that human beings participate concurrently in two distinct worlds, of nature and society, figuring as biological individuals in the former and as cultural subjects in the latter. Instead, I propose that we consider humans as indistinguishably organisms and persons, participating not in two worlds but in one, consisting of the entire field of their environmental relations. Figure 19.3 illustrates schematically the contrast between these two views. Needless to say, the environment of a person will include beings of many kinds, both human and nonhuman, to which that person will relate in different ways depending on their particular qualities and characteristics, and on the project in hand. As one passes from relations with humans to relations with nonhuman animals, plants, and inanimate objects, there is no Rubicon beyond which we can say of any relation that it is directed toward things in nature rather than persons in society. For as the edge of nature is an illusion, so too is the image of society as a sphere of life that exists beyond it (Ingold 1997: 250). But by the same token, in the project of scholarly research, there can be no absolute division of method and objective between studying the lives and works of humans and of nonhumans. Why, then, should the participatory and interpretative approaches of the arts and humanities be limited to the study of human subjects? And why, conversely, should the observational and explanatory approaches of science be limited to the domain of nonhuman "nature"? Why, indeed, should these approaches be separated at all?

Ever since its relatively recent inception, the credibility of "social science" has been compromised by the recognition that the observer of hu-

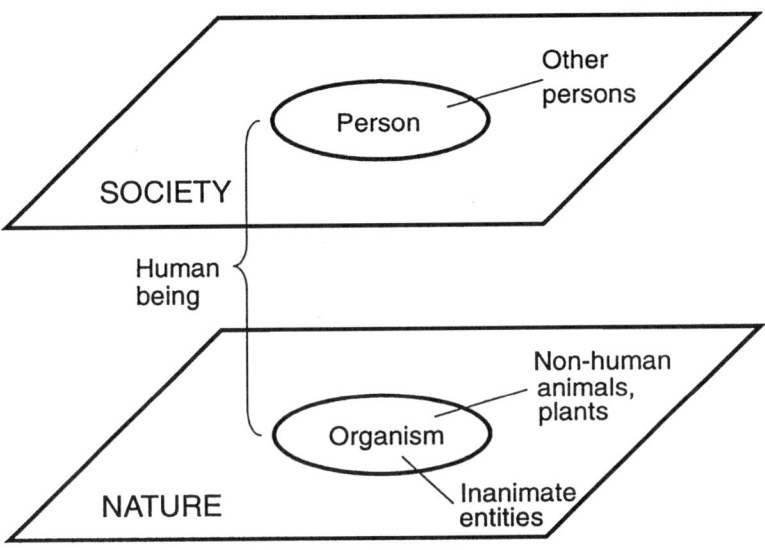

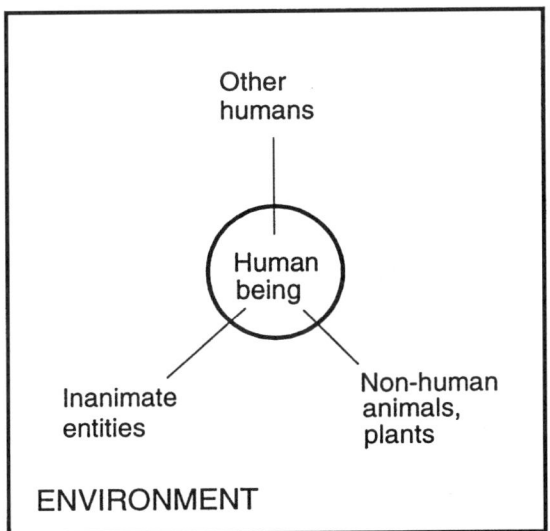

Figure 19.3
A schematic comparison of the complementarity and obviation approaches (after Ingold 1996c: 127). In the complementarity approach (upper diagram), every human being is, in part, a person in society and, in another part, an organism in nature. In the obviation approach (lower diagram), the human being is a person-organism situated in an environment of human and nonhuman others.

man behavior is necessarily a participant in the field of observation. In this vein, numerous critics have pointed out that participant observation, the methodological crux of social anthropological inquiry, is a contradiction in terms. To participate, it is said, is to swim with the current, to observe is to stand on the bank: how can one possibly do both at once? Now it is doubtless true that scientific inquiry of any kind depends upon observation. But there is more to observation than mere spectating. A disinterested bystander who did not, in some way, couple the movement of his or her attention to the surrounding currents of activity, who failed to watch what was going on, would see much, but observe nothing. Observation, in short, is itself an environmentally situated activity that requires the observer to place himself or herself, in person, in a relation of active, perceptual engagement with the object of attention. It is from this kind of sensory participation, proceeding against the background of involved activity in the wider environment of human and nonhuman others, that all scientific knowledge grows.

Thus, whether our concern be with humans or nonhumans, there can be no observation without participation, no explanation without interpretation, no science without engagement. As one such science, I believe that anthropology is destined to take its place as part of a broader ecological study of the relations between organism-persons and their environments, premised on the inescapable fact of our involvement in the one world in which we all live (Ingold 1992: 693–694). Any divisions within this field of inquiry must be relative rather than absolute, depending on what is selected as one's focus rather than on the a priori separation of substantive, externally bounded domains. I hope, in this chapter, to have given some idea of how we might proceed with reconstructing the discipline along these lines.

I would like to end, however, with a word about the teaching of anthropology. I have shown that the forms of human knowledge are not made by society, and handed down to its in-

dividual members for them to use in their everyday lives, but are rather generated and sustained within the contexts of people's engagements with one another and with nonhuman components of the environment. If this applies to knowledge in general, in must apply to anthropological knowledge in particular. In the field, anthropologists learn; in the classroom they teach. This does not mean, however, that they are receiving knowledge in the first case and transmitting it in the second. For in both, whether with local people or with students, they collaborate in the dialogic processes of its creation. It is through bringing the two dialogues, in the field and the classroom, into a productive interplay that anthropological knowledge is generated. It follows that the dialogue in the classroom is as important, and as integral to the anthropological project, as the dialogue in the field. Belatedly, we have begun to recognize the contribution that local collaborators—erstwhile "informants"—have made to the advance of our subject. It is high time we recognized the contribution of students as well.

Notes

1. Because these others may be nonhuman as well as human, there is nothing strange about the extension of kinship relations across the species boundary that is commonly taken for granted among non-Western peoples.

2. This position is beautifully epitomized, and parodied, in the title of a recent article by Morton, "The organic remains" (Morton 1995).

3. In this vein, for example, Goldschmidt (1993: 355) writes of the "dynamic relation between the genetic and the cultural, between biology and anthropology." His equation of biology with genetic programming leads him to the bizarre thought that even the human embryo, to the extent that its development is affected by environmentally specific "intra-uterine experiences," could not be "purely biological" since it would already have acquired a modicum of culture (1993: 357, n. 19).

4. While recognizing that these processes, of individual ontogeny and evolutionary phylogeny, are distinct, biologists do not deny that there are connections between

them. Thus the circumstances of ontogenetic development, insofar as they affect genetic replication, may exert an influence on evolution; conversely the evolved genetic specification is supposed to establish a schedule for development (Hinde 1991: 585).

5. As Kroeber wrote, "All evidence directs us to the conviction that in recent periods civilization has raced at a speed so far outstripping the pace of hereditary evolution, that the latter has, if not actually standing still, afforded all the seeming, relatively, of making no progress" (1952: 51).

6. The "finished artifact fallacy" has its precise counterpart in standard notions of socialization or enculturation as the working up of human raw material into finished forms, ready for entry into social life.

7. I am compelled here to use the neuter pronoun "it," with regard to persons, rather than "he/she," in recognition of the fact that gendering is itself an aspect of the way in which relations are enfolded in the consciousness and identity of the self.

8. "What is very essential to the whole notion," Bartlett writes, is "that the organised mass results of past changes of position and posture are actively *doing* something all the time; are, so to speak, carried along with us, complete, though developing from moment to moment" (Bartlett 1932: 201). There are, I think, striking similarities between Bartlett's notion of *schema* and Bourdieu's (1977) of *habitus*. Both terms suggest an active, dynamic organization of past experience, rather than a passive, static framework for accommodating it.

9. Connerton prefers the notion of habit to that of skill, arguing that as a habit becomes established, "awareness retreats," leading ultimately to bodily automatisms (1989: 93–94). I do not think it is right to describe the movements of the musician or craftsman as habitual in this sense. In such skilled activity, awareness does not retreat but becomes one with the movement itself. This movement, far from being automatic, carries its own immanent intentionality (Merleau-Ponty 1962: 110–111).

References

Bartlett, F. C. (1932). *Remembering: A Study in Experimental and Social Psychology.* Cambridge: Cambridge University Press.

Bateson, G. (1973). *Steps to an Ecology of Mind.* London: Fontana.

Bender, B. (Ed.) (1993). *Landscape: Politics and Perspectives.* Oxford: Berg.

Bloch, M. (1991). Language, anthropology and cognitive science. *Man* 26: 183–198.

Boas, F. (1940). *Race, Language, and Culture.* New York: Free Press.

Bourdieu, P. (1977). *Outline of a Theory of Practice.* Cambridge: Cambridge University Press.

Connerton, P. (1989). *How Societies Remember.* Cambridge: Cambridge University Press.

Cosmides, L., J. Tooby, and J. H. Barkow. (1992). Introduction: Evolutionary psychology and conceptual integration. In J. H. Barkow, L. Cosmides, and J. Tooby (Eds.), *The Adapted Mind: Evolutionary Psychology and the Generation of Culture,* pp. 3–15. New York: Oxford University Press.

Csordas, T. (1990). Embodiment as a paradigm for anthropology. *Ethos* 18: 5–47.

D'Andrade, R. G. (1981). The cultural part of cognition. *Cognitive Science* 5: 179–195.

Davidson, I., and W. Noble. (1993). Tools and language in human evolution. In K. R. Gibson and T. Ingold (Eds.), *Tools, Language and Cognition in Human Evolution,* pp. 363–388. Cambridge: Cambridge University Press.

Durkheim, E. (1982). *The Rules of Sociological Method.* S. Lukes (Ed.), W. D. Halls (Trans.). London: Macmillan. (Original work published 1895.)

Geertz, C. (1973). *The Interpretation of Cultures.* New York: Basic Books.

Gibson, J. J. (1979). *The Ecological Approach to Visual Perception.* Boston: Houghton Mifflin.

Godelier, M. (1989). Incest taboo and the evolution of society. In A. Grafen (Ed.), *Evolution and its Influence,* pp. 63–92. Oxford: Clarendon Press.

Goldschmidt, W. (1993). On the relationship between biology and anthropology. *Man* 28: 341–359.

Halbwachs, M. (1992). *On Collective Memory.* Chicago: University of Chicago Press. (Original work published in 1950).

Hinde, R. A. (1991). A biologist looks at anthropology. *Man* 26: 583–608.

Ingold, T. (1983). The architect and the bee: Reflections on the work of animals and men. *Man* 18: 1–20.

Ingold, T. (1990). An anthropologist looks at biology. *Man* 25: 208–229.

Ingold, T. (1992). Editorial. *Man* 27: 693–696.

Ingold, T. (1993a). Technology, language, intelligence: A reconsideration of basic concepts. In K. R. Gibson and T. Ingold (Eds.), *Tools, Language and Cognition in Human Evolution*, pp. 449–472. Cambridge: Cambridge University Press.

Ingold, T. (1993b). The temporality of the landscape. *World Archaeology* 25: 152–174.

Ingold, T. (1995a). "People like us": The concept of the anatomically modern human. *Cultural Dynamics* 7: 187–214.

Ingold, T. (1995b). Building, dwelling, living: How animals and people make themselves at home in the world. In M. Strathern (Ed.), *Shifting Contexts*, pp. 57–80. London: Routledge.

Ingold, T. (1996a). The history and evolution of bodily skills. *Ecological Psychology* 8: 171–182.

Ingold, T. (1996b). Introduction to the 1992 debate, "The past is a foreign country." In T. Ingold (Ed.), *Key Debates in Anthropology*, pp. 201–205. London: Routledge.

Ingold, T. (1996c). Hunting and gathering as ways of perceiving the environment. In R. Ellen and K. Fukui (Eds.), *Redefining Nature: Ecology, Culture and Domestication*, pp. 117–155. Oxford: Berg.

Ingold, T. (1997). Life beyond the edge of nature? Or, the mirage of society. In J. D. Greenwood (Ed.), *The Mark of the Social*, pp. 231–252. Lanham, MD: Rowman and Littlefield.

IUAES (International Union of Anthropological and Ethnological Sciences). (1996). The IUAES and the new Unesco statement on race. *Human Peace* (IUAES Quarterly News Journal of the Commission on the Study of Peace) 11(1): 18–20.

Jackson, M. (1989). *Paths toward a Clearing: Radical Empiricism and Ethnographic Inquiry*. Bloomington: Indiana University Press.

Johnson-Laird, P. N. (1988). *The Computer and the Mind: An Introduction to Cognitive Science*. London: Fontana.

Kroeber, A. L. (1952). *The Nature of Culture*. Chicago: University of Chicago Press.

Mauss, M. (1979). Body techniques. In *Sociology and Psychology: Essays by Marcel Mauss* (B. Brewster, Trans.), part IV, pp. 97–123. London: Routledge & Kegan Paul. (Original work published in 1934).

Mellars, P., and C. Stringer. (Eds.). (1989). *The Human Revolution: Behavioural and Biological Perspectives on the Origins of Modern Humans*. Edinburgh: Edinburgh University Press.

Merleau-Ponty, M. (1962). *Phenomenology of Perception*. (C. Smith, Trans.) London: Routledge & Kegan Paul.

Morton, J. (1995). The organic remains: Remarks on the constitution and development of people. In J. Morton and M. Macintyre (Eds.), *Persons, Bodies, Selves, Emotions*, pp. 101–118. Special issue of *Social Analysis*, no. 37. Department of Anthropology, University of Adelaide.

Ots, T. (1994). The silenced body—the expressive *Leib*: On the dialectic of mind and life in Chinese cathartic healing. In T. J. Csordas (Ed.), *Embodiment and Experience: The Existential Grounds of Culture and Self*, pp. 116–136. Cambridge: Cambridge University Press.

Oyama, S. (1985). *The Ontogeny of Information: Developmental Systems and Evolution*. Cambridge: Cambridge University Press. (2d rev. ed., Durham, NC: Duke University Press, 2000.)

Quinn, N., and D. Holland. (1987). Culture and cognition. In D. Holland and N. Quinn (Eds.), *Cultural Models in Language and Thought*, pp. 3–40. Cambridge: Cambridge University Press.

Reed, E. S. (1988). The affordances of the animate environment: Social science from the ecological point of view. In T. Ingold (Ed.), *What Is an Animal?* pp. 110–126. London: Unwin Hyman.

Robertson, A. F. (1996). The development of meaning: Ontogeny and culture. *Journal of the Royal Anthropological Institute* 2: 591–610.

Rubin, D. C. (1988). Go for the skill. In U. Neisser and E. Winograd (Eds.), *Remembering Reconsidered: Ecological and Traditional Approaches to the Study of Memory*, pp. 374–382. Cambridge: Cambridge University Press.

Schwartz, T. (1981). The acquisition of culture. *Ethos* 9: 4–17.

Sperber, D. (1985). *On Anthropological Knowledge*. Cambridge: Cambridge University Press.

Strathern, M. (1996). For the motion. Contribution to the 1989 debate, "The concept of society is theoretically obsolete." In T. Ingold (Ed.), *Key Debates in Anthropology*, pp. 60–66. London: Routledge.

Tilley, C. (1994). *A Phenomenology of Landscape: Places, Paths and Monuments.* Oxford: Berg.

Toren, C. (1993). Making history: The significance of childhood cognition for a comparative anthropology of mind. *Man* 28: 461–478.

Weiner, J. (1991). *The Empty Place: Poetry, Space and Being among the Foi of Papua New Guinea.* Bloomington: Indiana University Press.

Wikan, U. (1992). Beyond words: The power of resonance. *American Ethnologist* 19: 460–482.

Wolf, E. (1994). Perilous ideas: Race, culture, people. *Current Anthropology* 35: 1–12.

V RESPONSES TO DEVELOPMENTAL SYSTEMS THEORY

20 On the Status and Explanatory Structure of Developmental Systems Theory

Peter Godfrey-Smith

What kind of theory is DST? Is it a scientific theory or a philosophical theory? Is it an empirical hypothesis, a suggested program of research, a philosophical gloss on our existing knowledge, or what? What difference does it make whether or not the central ideas associated with DST are true?[1]

DST advocates and critics alike have wondered about this issue. Within the movement, some, like Russell Gray, have tried to be very explicit about the difference made to concrete scientific research by accepting DST (Gray 2000; Griffiths and Gray 1994). Oyama, it seems to me, has been consistently more cagy about tying DST to some specific direction of empirical research, and has presented DST more as a very general and abstract "way of seeing" the biological world and our investigation of it. These two types of emphasis do not conflict, of course. DST can be a general way of seeing, and that way of seeing can suggest specific projects of empirical work. Still, it can be hard to sort through the collection of ideas associated with DST, and work out what sort of role each idea plays.

One critic of DST, Philip Kitcher, has complained that DST does not "offer anything that aspiring researchers can put to work" (Kitcher in press). Perhaps Kitcher's overall view would be more accurately expressed by saying that he thinks that mostly DST does not offer anything that aspiring researchers can put to work, and when it does offer something, what it offers is an insistence on taking every causal factor in biology equally seriously in every investigation, an insistence that would simply shut down scientific activity. Kitcher thinks that both DST and Lewontin's "dialectical biology" tend toward an extreme holism that is inconsistent with the practice of ordinary empirical research. The criticisms of DST in Sterelny, Smith, and Dickison's "The Extended Replicator" (1996) are much less damning of DST than Kitcher's, but Sterelny, Smith, and Dickison converge somewhat with Kitcher on the issue of holism (p. 382). Although Sterelny, Smith, and Dickison do not think the problem is insoluble, they think DST does face a problem from its apparent commitment to a holistic view of causation in development and evolution.[2]

I will outline one way for DST to steer a path through these problems. I will describe two intellectual projects that *need not* be distinct, but that *can* be distinct, and that are certainly consistent with each other. I will sort some of the main DST ideas into different categories associated with this distinction, and in doing so I hope to lay to rest (among other things) the problem of causal holism for DST. More generally, I will outline a way for DST advocates to conceive the place their program occupies within science and philosophy. So the first part of the paper is entirely constructive, from the DST point of view.

The rest of this chapter will be slightly more critical. Although there is no question that "DST" names a large and somewhat heterogeneous collection of ideas, I suggest that one helpful way to think about DST is to think of it as an assertion of very strong *antipreformationism* about development. This idea is the origin of much else in DST. In particular, it is helpful in understanding the DST critique of the "informational" perspective of the gene. For DST, the informational gene is the preformationist's last stand. No one can deny that many forms of preformationism turned out to be wrong. But I will suggest in the second part of this paper that DST might sometimes go too far in its rejection of the preformationist pattern of explanation.

DST as Research Program and as Philosophy of Nature

DST is a collection of ideas about development, causation, inheritance, and evolution. But is it a collection of scientific ideas, philosophical ideas, or both? I suggest that DST is a contribution to two different kinds of intellectual project. Firstly, DST can be regarded as a proposal for a scientific research program. DST contains a set of core negative and positive ideas about biological systems. These ideas do have the ability to steer biological research in particular directions, and they have the ability to be confirmed and disconfirmed through empirical testing.

The second project that some DST ideas contribute to is what I will call *philosophy of nature*.[3] When doing philosophy of nature in my sense, a writer comments on the overall picture of the natural world that science, and perhaps other types of inquiry, seem to be giving us. But this commentary does not have to use language in the same way that scientists find convenient for their own work. It can use its own categories and concepts, concepts developed for the task of describing the world as accurately as possible when a range of scientific descriptions are to be taken into account, and when a philosophical concern with the underlying structure of theories is appropriate. The claims made by a good philosophy of nature do have to be *consistent* with the claims made by science. But the concepts employed by a good philosophy of nature do not have to be the same as those used in the relevant science, and the organization and presentation of information in the two projects can be quite different.

I say the concepts "might" be different, and the presentation of information "can" be different— they *need not* be different. It might turn out that the concepts used every day by scientists are exactly the right ones for doing philosophy of nature as well. Any degree of overlap is possible. The key point is that a commentary on scientific knowledge that serves the purposes of philosophy of nature can, if necessary, fashion its own way of describing the structure that has been uncovered by science. The demands on a good scientific vocabulary can be quite different from the demands on a vocabulary used in philosophy of nature. Indeed, because a philosophy of nature typically tries to accommodate the claims made

by a number of different sciences, some nonscientific vocabulary will probably be needed in most cases. Also, when scientific ideas are relevant to other problems treated in philosophy—for example, ethical and political problems—the scientific ideas should be fed into such discussions in a philosophically processed form, not in the raw language of science.

The term *philosophy of nature* has a history that might be suspicious to many; it might suggest a project in which the philosopher erroneously tries to do science from the armchair, by deploying special logical tools of dubious value and arrogantly saying how the course of empirical science must necessarily go.[4] I stress that nothing like that is meant here. "Philosophy of nature" in my sense comes *after* empirical science and tries to redescribe structures in the world that have already been described by the sciences. A good philosophy of nature makes no empirical claims that are inconsistent with those made in the relevant sciences. Philosophers are at liberty to make novel empirical suggestions, of course. They are also sometimes able to help science by describing and exploring whole new *kinds* of theoretical structure, at a stage when these theoretical structures cannot yet be brought into much contact with empirical methods. In recent work, Michael Friedman has been making a very impressive case for the importance of this kind of "forward-looking" role of philosophy in relation to science (1999). I agree with Friedman on the importance of this role, but this is a different role from the one I am describing here.

So "philosophy of nature" in the present sense is not the same as the kind of "naturalistic philosophy" that conceives all philosophical work as completely continuous with science. That sort of philosophy, associated in particular with Quine (1969) but influential in various other forms, gives up the autonomy of philosophy with respect to its choice of questions, and hopes instead to make substantial contributions to the overall scientific project. While holding that philosophy requires constant input from science, I oppose the kind of naturalism that requires philosophy to give up any aspiration to formulate and address its own distinctively philosophical set of questions.

My claim that philosophy of nature comes "after" science should not be taken to deny an interaction between more philosophical and more empirical commitments within science. Lewontin (personal correspondence) expressed the suspicion that the separation I am describing between scientific description and philosophy of nature implies the view that science itself can proceed in a purely empirical way unaffected by philosophical doctrines and commitments. But I accept, with Lewontin, that empirical work within science is guided—in different ways and in different degrees—by philosophical ideas. As Kuhn (1970) stressed, abstract views about the proper relation between theory and evidence and the ultimate composition of the world are important parts of traditions of "normal science." Still, I hold that one valuable and distinct type of work that philosophy can do is work that does aspire to come "after" science. This is work that tries to give a careful philosophical redescription of the picture of the world that science seems to be delivering. This philosophical work might well come to have an effect on the science itself; it might change the hidden or overt philosophical commitments of the scientists. But the absence of such an effect on science does not rob the philosophical work of its value.

Turning back to DST, my suggestion is that existing DST writings tend to combine contributions to two projects—describing a scientific research program and outlining a philosophy of nature. As the two projects are associated with different goals, different types of assessment are relevant in each case. I will sort through some DST claims, and criticisms made of those claims, in this light of this categorization.

In what sense is DST a scientific research program? Is this a "research program" in the sense of Lakatos (1970) or Laudan (1977)? In this chapter I will not try to link DST with those existing philosophical descriptions of research programs (though some of my language will recall those discussions). In part this is because DST is something of a special case; it would be an error to say that DST is a research program in just the same sense as Newtonianism, Darwinism, or classical AI. But I do think there is some value in extracting a "research program" out of DST.

I think of the DST research program as including at least a set of core empirical claims, a set of suggestions about which sorts of scientific work are likely to pay off, and a set of concepts that are to be used in doing and describing the work. To this basic set of elements, some might want to add others, such as a set of habits of mind and a set of standards for good explanations. Different philosophies of science differ about the "thickness" of anything that deserves to be called a research program, tradition of normal science, or consensus practice. There is no need to take sides here on questions like that, as we can proceed with a simple and fairly "thin" concept of what a research program is like.

A research program has to contain suggestions for empirical work. What sort of work is supposed to be done? Here DST is in an unusual situation, as part of what it does is pick out a scattered body of existing scientific work, some of it quite old, and claim: "there should be more of *this*." In a footnote to his critical paper (2000, note 6), Kitcher says he would like to see what an example of DST-guided work in developmental biology would look like, or an example of dialectical population genetics. The right DST reply is to say that there is *already* a good deal of work in developmental biology and population genetics that does everything (or nearly everything) that DST could ask for. Some of it is reviewed in part 1 of this volume. This work exists, but in mainstream biological thinking it is often regarded as describing oddities, details, and exceptions—not as describing cases that provide models for thinking about development and evolution in general. In mainstream biological thinking, the developmental work that is supposed to provide a model is work on the expression of specific genes, and on how genes and their products exert sensitive control on developmental processes.[5] According to DST, the mainstream view erroneously holds that work focused on gene expression provides a general model, while work on such things as the inheritance of cytosine methylation patterns and endosymbionts merely describes interesting oddities.

Lewontin can reply to Kitcher in similar terms; there is lots of existing work in population genetics that does fit the framework of "dialectical biology" reasonably well, and lots that does not. For Lewontin, epistasis in fitness relations is to be regarded as the rule rather than the exception. A high degree of sensitivity of evolutionary predictions to the details of genetic systems and background parameters is to be expected. Lewontin also asserts the importance of "coupled" causal structures, in which variables affect each other symmetrically, rather than in an asymmetrical way where one variable is always cause and the other is effect.[6]

So part of the project in recent DST literature is an attempt to *draw together* a large number of existing pieces of scientific work and to suggest that when these are rightly interpreted and considered together, they suggest a project for further empirical work. For DST advocates, there is already a scattered collection of successful scientific work that provides a set of models and concepts to move ahead with.

Earlier I said that DST as research program is committed to a set of empirical claims. Which claims are these? Any suggestion will probably generate debate, but for purposes of illustration I will focus on one key example. DST the research program holds that extragenetic inheritance is more common and more causally important than mainstream views realize. Gray (in press) discusses a range of examples that he views as especially suggestive for DST, and these range from gut microorganisms through social behaviors to habitats. Jablonka's contribution to the present volume also contains a very well organized survey of these phenomena. In order to categorize some of these cases as real extragenetic inheritance, DST has to engage in some conceptual battles over what counts as inheritance, and over possible mainstream redescriptions of these in standard genetic terms.[7] Those questions are very difficult. At least part of DST's defence of its preferred way of categorizing cases of inheritance must be the claim that *if* the DST categorization of types of inheritance is used, new patterns will emerge and new insights will result. Perhaps there are some risks of circularity here, but I do not see these risks as worse than they are in other cases involving categorization in science. Here as elsewhere, a categorization is to be judged, at least in part, by its empirical fruits. And there are certainly some relevant phenomena that are agreed on all sides to constitute extragenetic inheritance. Either these cases are rare and make little difference to the working of the biological world, or they are not.

So DST claims that extragenetic inheritance is common, and that once it is demarginalized we will improve our understanding of the way traits appear and reappear across generations. This I take to be perhaps the clearest case of a DST idea that constitutes part of a scientific research program.[8] I stress it here for illustrative purposes, but there are others as well. Other empirical applications of DST, and explicit outlines of DST-based research programs, are discussed by scientists in other chapters of this book. (See, for example, the chapters by Nijhout and by Weber and Depew.)

I turn now to the second project that DST can be seen as undertaking, the project of "philosophy of nature." As I said, a philosophy of nature is an attempt to describe the world in a way that is closely informed by scientific theories, but that is free to reject the vocabulary and perhaps some of the classifications and interpretations of the world associated with the relevant sciences. I stress again that it is possible for someone to suggest that standard scientific descriptions already constitute an adequate philosophy of nature, or that scientific descriptions need only a minimal additional commentary before they will function as an adequate philosophy of nature. But the other view can be argued as well—the view that scientific descriptions of the world typically need a lot of massaging and interpretation before they can be seen as our best-possible description.

Why should this be? Why should there be a need to add to or modify the scientific description? One's answer to this will depend on general views about how science works and about scientific language. I will illustrate with a simple case that is also highly relevant in the present context. Suppose one holds that metaphorical and analogical language is essential to science, even in highly developed fields. When I say "essential" I mean something strong; I mean that ordinary scientific work requires, for communication between scientists and for creative thinking, the constant deployment of a rich set of metaphors when thinking about some particular system. Perhaps it is hard to imagine a case when the scientists will not be able to drop, perhaps with some struggle, the metaphors. But it might be the case in some particular field that the metaphors are never dropped, and hence that they become tightly fused to the nonmetaphorical language of that field. Then one task for philosophy of nature will be to come along afterward and sort through the scientific descriptions, distinguishing the parts that are properly taken literally from the parts that must be regarded as merely metaphorical.[9]

In the present context, the status of "genetic coding" is a relevant problem case. There has been a growing philosophical discussion of what exactly we should make of the language of genetic coding in mainstream biology (Sarkar 1996; Godfrey-Smith 2000a), and this is associated with discussions of the role of other semantic properties in biology, and the concept of a "genetic program." (See also Moss 1992 and Keller's contribution to this volume.) But as Maynard Smith (in press) stresses, the symbolic perspective on genes has become completely embedded in the theoretical language of biology. Terms like *transcription, translation, proof-reading*, and *synonymous* are all technical terms in biology, with quite precise conditions of use. It might be argued that these concepts are now essential to ordinary, productive, empirical research in genetics. Though derived at least semimetaphorically from the case of language, they are just as deeply embedded as any other specialized terms in the practice and thinking of geneticists.

DST has criticized the symbolic, informational perspective on the gene extensively. I suggest that there are two distinct criticisms DST makes here. First, there is no doubt that if DST is a research program, the program includes, as a negative heuristic, a rejection of that conceptual framework. DST makes an empirical bet that biology would be better off without this way of talking about genes—or at least that it would be better off with a drastically reduced role for such descriptions. Maybe DST is right about this. But suppose for a moment that DST is wrong, and for a particular reason. Someone like Maynard Smith might argue that it is no accident that the introduction of the symbolic perspective on genes coincided with an avalanche of scientific progress in this area. And DST does not deny the avalanche. Maybe the symbolic perspective on genes just happens to be a uniquely useful framework for guiding empirical work in this area. If that is true, it is true not just because of what genes are like, but also because of what our minds are like. A conceptual framework like this might, for a mixture of reasons, turn out to be ideal for human rumination and communication about genetics. That could be true even if genes have many properties that are not captured by the symbolic perspective. If genetics is complicated enough, and if everyday scientific work can only proceed with the aid of some simplifying framework or another, then the question becomes which is the simplifying framework best suited for us.

I am not saying that the possibility in the preceding paragraph is true; it might be totally wrong, and certainly DST bets that it is wrong. DST bets that genetics would benefit in many ways from giving up the special kinds of semantic and computational descriptions of genes that are now so common. But suppose the argument in the preceding paragraph is right. Then the *scientist* has no reason to drop the symbolic perspective on genes, when doing scientific work. But the fact that the scientist finds the framework indispensable does not stop the philosopher (or the scientist when wearing the hat of philosopher of nature) from taking an entirely different attitude. The philosopher of nature should ask: even if this way of talking has great practical value, what is its real status? Is this a good literal description of what genes are like, or is it a metaphor? Is it perhaps language that is being applied nonmetaphorically and being used to make literally false claims, but that nonetheless has heuristic value? In a situation when ease of scientific communication and spurring of creativity count for nothing—that is, in a situation when the science is done, for now—should we describe genes in these terms or not?

I said earlier that philosophy of nature also aims to "process" scientific descriptions prior to their public assimilation and utilization in other kinds of intellectual and practical decision-making. The semantic perspective on genes appears, when used in nontechnical discussions in newspapers and the like, to give support to genetic determinist views widespread in the general population. Clarifying and criticizing such connections is another role for philosophy of nature.

So on a question like the status of "coding for" descriptions, and other semantic talk in genetics, I understand DST as saying two things. DST as research program bets that science would be better off without such talk. But DST as philosophy of nature should recognize that all sorts of different factors might influence the usefulness of a conceptual framework in science. So DST as philosophy of nature should add that even if it turns out that genetics benefits from using the semantic

framework, when we try to give the most accurate possible description of what we know about genes, we should not use those terms. (The status of genetic coding will arise again below.)

During discussion of the first drafts of papers in this volume, two of the editors suggested that I compare DST with the "gene's eye view" on evolution, with respect to the duality between research program and philosophy of nature. To some extent the comparison is appropriate, as we do find a bifurcation of this kind in the literature on genic selection. But DST should be wary of the analogy.

The "gene's eye view" is sometimes presented as a specific program for empirical research. As such, it is intended to guide researchers away from looking for an organism-level benefit in all evolutionary explanations. Phenomena of meiotic drive and junk DNA are taken to be instructive models that tell us how genes can proliferate without helping organisms. This empirical orientation is one aspect of the literature on the gene's eye view, but this view coexists with Richard Dawkins's attempts to recast *all* of evolution in gene-selectionist terms, where this recasting is justified by appeal to abstract and general arguments about the role of "replicators" (1976, 1982). This latter type of argument for a gene's eye view makes no particular claims about how often there will be a familiar organism-level benefit involved in evolution; however common these cases are, they can be redescribed as selection on genes. Even phenomena previously categorized as group selection can be redescribed in this way. Dawkins's *The Selfish Gene* can be seen as outlining a philosophy of nature in which replicators are elevated to a position of primacy in evolution, and are regarded as the ultimate beneficiaries of everything natural selection produces. Replicators become the fundamental causal agents in the biological world.[10]

The analogy with the gene's eye view illustrates some moves that DST should avoid. Dawkins and others have sometimes blurred the two construals of gene-selectionism in misleading ways.

A separate question is whether the gene's eye view makes sense as a philosophy of nature. I think it does not. The topic is too big for discussion here, but one reason is that, contra Dawkins, replicators are not essential to evolution by natural selection (Godfrey-Smith in press b).

I turn now to the specter of holism, which has haunted DST in recent years and which has been the subject of vigorous discussion.

If one is engaged in philosophy of nature, the *unwieldiness* of a description is no objection to that description. The heuristic emptiness of a description is no objection to it either. Even if a way of seeing the world would be positively stultifying if adopted by the working scientist, that does not show anything wrong with that way of seeing things in the context of philosophy of nature.

So here we see one way of resolving the debates about holism. Critics of DST suggest that DST views about causation are holistic. And in some respects, they certainly are holistic. DST claims that various causal factors are relegated to the status of mere "support" or "background" by mainstream biological views, while other causal factors are regarded as primary or as the sources of form. But according to DST, this "privileging" of some causes over others makes no sense. "What I am arguing for here is a view of causality that gives formative weight to all necessary influences, since none alone is sufficient for the phenomenon or for any of its properties" (Oyama 1985: 15).[11] For Oyama and others, it is a crucial error to suppose that "some influences are more equal than others" in the explanation of biological form. Rather than disregard some causal factors in order to "privilege" others, all causal influences that bear on an event should be accorded comparable status. These are the sorts of claims to which the critics cry "holism!" and in some respects the label is entirely appropriate. But holism per se is only a bad thing in situations in which it puts impossible demands on empirical investigation, situations in which it is paralyzing to scientific work. In empirical work, it is prob-

ably inevitable that the researcher sort through causal factors and distinguish some as primary and others as background conditions, and to deny the researcher this strategy is to shut down research. But holism per se is not a bad thing in philosophy of nature, because it is an error to demand that a philosophy of nature be a useful tool in the laboratory, or a good heuristic for guiding research. In a philosophy of nature, holism is just one possible view about the causal structure of the world.

As I stressed earlier, to make this point it is not necessary to insist that the vocabularies and frameworks used in science and those used in philosophies of nature are distinct. Whether they are distinct or not will depend on many facts, including the flexibility of the language of scientific research. There is nothing stopping a scientist from first espousing a very holistic view of a system, and then going on to say he will relegate some causal factors to background status merely for immediate practical purposes. My separation of philosophy of nature from science is intended to show the *possibility* of a mismatch between the two, and the possibility of useful commentary on biology using a framework that is of little help to an empirical scientist.

Another way to put my point—a way more critical of DST—is this. One never knows in advance which heuristics and idealizations will be helpful in science. That is something only known after the fact, and it is an error to make strong predictions about such matters. Occasionally defenders of DST have done this, have said that such-and-such is bound to be a useful heuristic in research.[12] I suggest that we should never expect to know such things in advance. But a redescription of development or evolution does not have to inspire scientists for it to be valuable.

Maybe defenders of DST will not want to accept the strategy offered in this section; they might think that "philosophy of nature" in my sense is an empty exercise. Certainly some people will think that it is empty or at least dubious. I do suggest, however, that an appeal to the distinction I have outlined is an important option to have on the table.

Before leaving this topic, I will make a brief return to Kitcher's criticism of DST. It is hard for DST or any view like it to establish its superiority to a sparse and austere philosophy of nature. By "sparse" I mean a philosophy of nature that does not recognize the reality of the distinctions that DST makes at key places, and which treats some aspects of the DST description as mere colorful talk. The opposition will be difficult for the richer view because the proponent of the sparser view will always claim to be able to reexpress the only significant parts of the richer view in his or her own sparse terms. Then a challenge will be issued to the richer view to say why we should care about the remainder that cannot be sparsely expressed.[13] In the context under discussion, Kitcher has a sparse and austere view. Notably, he is skeptical about strong claims about the objective reality of causal relations. Kitcher is prepared to engage in causal talk, but I think it is fair to say he thinks that such talk should not be taken too metaphysically seriously. His own analysis of what is meant by causal talk is a sophisticated view, based on the concept of explanation, which at bottom is an unusual version of the regularity theory of causation (1989). In science we collect and classify patterns that permit us to unify phenomena. The "causal" structure of the world is no more than this structure of patterns and regularities, viewed by us in a certain way. A consequence of this is that when writing philosophically seriously about the nature of reality, for Kitcher a rather slim vocabulary is, in principle, always sufficient. All we need to do is to describe patterns in the flow of events—in this respect Kitcher is an heir to the positivist tradition.

If one has a view like this, one will be very resistant to views about biological systems (or anything else) that are driven by highly specific and elaborate claims about causation and causal dependence. DST is a view of that kind. DST, especially in Oyama's and Gray's versions, is concerned to make a number of subtle and controversial claims about what causation is like, and how causal relations are structured within biological systems. DST asserts strong views about causal connectedness, interpenetration, "reciprocal selectivity," the active and reactive properties of biological matter, and so on. These are all things that DST takes seriously which can only be expressed, let alone defended, if one conceives of causal facts in a rich and realist way. For someone like Kitcher, on the other hand, there is little or nothing at issue when someone asks whether it is true that causation is "multiple, interdependent and complex" (Oyama 1985: 32) and exhibits "reciprocal selectivity" (p. 15). To ask whether these things are true or not is to choose between different colorful glosses that one might put on the world. The colors are different but nothing substantial hangs on them.

DST is causally rich as a philosophy of nature. So is Lewontin's dialectical biology, and so are various other views quite unlike these two. Causally rich views must struggle hard, although they might in the end prevail, when exposed to the deflationary tendency of sparser philosophies, like Kitcher's.

DST as Extreme Antipreformationism

In this section I will discuss some DST views about the explanation of biological development. I suggest that it is helpful to see DST as, among other things, an assertion of a very strong form of antipreformationism.

Recall the eighteenth- and nineteenth-century debate between the preformationists and the defenders of epigenesis.[14] Does the adult form preexist, in some way, in the fertilized egg, or is development a process in which order really does come from disorder? If the latter, it was thought, this order must be brought about by a special organizing force acting on the inert matter.

Oyama (1985) discusses this debate, and sees her view as denying an assumption common to both sides, the assumption that matter cannot acquire biological form without there being some external source of this form. So she does not ally herself with the traditional "epigenetic" form of opposition to preformationism. That is reasonable enough, but it is possible to recast this old dispute within a more modern and naturalistic set of assumptions. Then when we ask the question of DST I think we get a different answer. Is development a process in which, despite appearances, a good deal of the structure of the organism-to-come preexists in the fertilized egg, or is it a process in which comparatively little structure is pre-given, and the active powers of biological matter, working through local causal interactions, give rise to complex eventual structures in reliable ways? Once expressed like this, it seems to me that much of the thrust of DST is to assert the latter option, the antipreformationist option.

The denial of preformationism is connected to DST's views about the attribution of informational and semantic properties to genes. For DST, the idea that genetic information is the source of form in development is a lapse back into preformationist error. According to DST, the informational gene is the preformationist's last stand; it expresses the idea that although the adult structure does not preexist in the egg, something just as good does. Against this, Oyama asserts that information is not something that preexists the biological interactions with which it is associated. Rather, information (as her 1985 title suggests) has its own ontogeny. Preformationism is denied even for information.

I suggest that DST might have a tendency to go too far on this point. The preformationist pattern of explanation did not turn out to be so completely misguided. To suggest this, I will first work through a series of different kinds of preformationist view, and then show the continuity between a reasonable interpretation of modern theories, and preformationist styles of explanation.

One extreme kind of preformationism is wholly false; this is the view in which a more or less complete adult organism exists in miniature within the fertilized egg. That view marks out one pole in the conceptual landscape. Consider next two ways to move away from this extreme view.

Option 1:   The zygote contains a blueprint or set of instructions for the adult organism.

Option 2:   The zygote contains a complete set of parts needed to build an adult organism, which have the ability to self-assemble after assuming the proper size and/or multiplying to the right numbers. Here "self-assembly" is conceived as a fairly trivial matter, once one has the parts.

Option 1 is familiar. Option 2 is a somewhat modernized version of a historical version of pre-formationism (Gould 1977: 20). Is there any truth in either of these views? Certainly many people write as if option 1 is true. Some who write this way mean it literally and others think it is a useful metaphor. I accept, with DST, that option 1 is not literally true, or close to true, and not a very helpful metaphor either.

Is there any truth in option 2? Here matters are a bit more complicated. Certainly a simple version of 2 is not at all true, when given as an explanation of development. The "parts" that historical defenders of this kind of view had in mind are certainly not present intact in the zygote. But there is an element of truth in option 2, and DST does not deny this. A zygote does contain structures that will also appear in the adult; the zygote contains intact *some* adult part types—chromosomes, mitochondria, membranes, and so on.

DST not only accepts this, but it also extends the list of inherited parts, via the extended DST view of inheritance (as Sterelny stressed to me). The "persistent resources" in an organism's environment are, for DST, genuine parts of DST's analog of the adult organism—the late stages in the life cycle. Resources such as habitats and energy sources are parts of later stages, and they are also present at the initial stages. This is not true of all parts of course, or of global features of the organic structure; those do not preexist, but are causal products of the interactions of various "resources" that exist at the earlier stages. So DST accepts that some key parts are present, either as tokens or as members of a closely copied lineage, at the earliest stages. DST denies that the existence of these parts vindicates preformationism about the eventual biological forms—the later stages of the life cycle.

So option 2 is generally false, according to DST and according to standard views, even though it has a few grains of truth. I now introduce another descendant of the original naive preformationist view, one which combines aspects of option 1 and option 2 above.

Option 3:   A large and crucial set of components of the adult (and of all developmental stages) are not materially present themselves in the zygote, but are coded for in structures within the zygote. These components make up most of the machinery by which the organism stays alive and controls its passage through the developmental sequence.

The relevant parts here are amino acid chains —which when folded and processed become functional protein molecules. So this view is preformationist about parts, not the whole, and the parts themselves are not physically present in the zygote but instead are coded for.

We have arrived, of course, at one aspect of the view found in mainstream modern genetics. This is not the *only* way to describe what modern genetics has taught us, but I claim that it is one reasonable way to describe the present picture. And this way of describing it shows the way in which current mainstream views are continuous with older preformationist ideas. The more common formulations, in terms of a genetically encoded "program" or genetic "instructions" for development, are not the only ways to express a continuity between preformationism and modern views.

Of course, here we must confront the question: Do the DNA sequences in the zygote really constitute a coded representation of proteins? I have

struggled with this elsewhere (2000a), and DST tends to oppose this type of description of genes. I do not think the exact status of genetic coding is crucial to the present point. At least it is correct to say that the DNA sequences contain structures that act as templates for the proteins produced.[15] That claim should suffice for the purposes of present discussion. In fact though, I hold that there are good reasons for claiming that genes code for proteins.

I have argued so far that it is possible to present modern genetics as having an explanatory structure that descends nontrivially from preformationism. Consider now Oyama's reaction to a related claim made by Stephen Jay Gould. Gould says "Modern genetics is about as midway as it could be between the extreme formulations [of preformationism and epigenesis] of the eighteenth century" (1977: 18; quoted in Oyama 1985: 24). Oyama rejects this "Golden Mean" resolution of the old debate, thinking the debate rested on false assumptions. But there is an important element of truth in Gould's claim. Our analysis of genetic causation should make sense of the empirical fact that preformationism turned out to contain an element of truth, of a perhaps unexpected kind. If a general analysis of causation, development and information denies this, it has probably overextended.

Does DST overextend? It is hard to say. Many strands of the DST literature bear on this point, and there is no simple resolution. At another point Oyama says: "What is transmitted [in reproduction] is *macromolecular form,* which, though it is necessary for the development of phenotypic form, neither contains it nor constitutes plans for it" (1985: 2). That is a good way to express her position. I may also differ from Gould on *how* an element of preformationism has been vindicated. Gould says that the modern partial vindication of preformationism is via our discovery of "coded instructions" in the genes. If Gould has in mind a set of instructions for the eventual adult form, then I agree with Oyama that this is a wrong turn. If all Gould has in mind here is the specification of proteins, then his Golden Mean is rightly expressed.

Another interesting passage suggests that Gould does have in mind a strong version of the "coded instructions" claim, which goes beyond the specification of proteins. He accepts, with the preformationists, that if the egg was truly unorganized, it could not develop into a complex adult without a directing entelechy. "[The egg] does, and can only do so, because the information—not merely the raw material—needed to build this complexity already resides in the egg" (1977: 21–22). Thus Gould squarely denies what Oyama calls the "ontogeny of information."

Both Oyama and the mainstream view (exemplified here by Gould) are very willing to express their views and conduct the dispute using the language of information: is developmental information preexisting or does it have an ontogeny? In my own statements of what is true in preformationism I avoid any formulation in terms of information. This is not because I think information is a useless concept in biology. Rather, it is because I suggest that the clear and useful way to handle informational terms here is to restrict the concept of information to the weak sense associated with the mathematical theory of information—the sense developed by Shannon (1948) and others, and sharpened up philosophically by Dretske (1981). Griffiths and Gray (1994) successfully argued that this concept cannot be used to describe a special kind of specificity that genes have and nongenetic factors do not have. So I hold that the partial modern vindication of preformationism lies in the fact that genes code for almost all the proteins—and hence much of the cellular machinery—that an organism will use during its life. The same sort of claim *can* also be expressed using the language of information; genes contain information specifying the proteins that the organism will use. But I hold that this formulation is more misleading, so I resist attempts to make the concept of information carry the main weight in philosophical discussions like the present one.

Although I have tried here to raise a suspicion, in this chapter I will not attempt to pin any definite error on DST regarding preformationism. Matters are just not as clear as that. I will, however, discuss another specific idea, associated with DST, that appears to be linked to a strongly antipreformationist message.

Advocates of DST, and similar views, have often been impressed by Gunther Stent's comparison of development with ecological succession (Stent 1981). Griffiths and Gray say this is "perhaps the best metaphor" for biological development (1994: 284; see also Oyama 1985: 177, note 4; Francis in press.)[16] What is the message of the ecological analogy? The central point is that contingent causal processes without any guiding plan or central controller can reliably produce specific outcomes. Each stage in the sequence gives rise to the next, but no part of the system governs the whole process in a centralized way.[17] DST sees the development of an organism as similar, and the ecological analogy is a good one for making this point.

Another point is contained in the analogy, however—an antipreformationist point. Ecological succession is a process in which the late stages owe their existence and structure to immediately prior stages, not to a special set of components present at the first stage. Illustrations of the succession phenomenon sometimes use cases where weedy plants and then pines colonize a disturbed area, to give way in turn to hardwoods. The hardwood forest (the analog of the adult organism) does not exist in miniature within the earlier stages, and neither does a representation of, or recipe for, the eventual forest.

The comparison is made complicated by the fact that seeds of the hardwood trees do have to exist prior to the trees, and these seeds have to come from somewhere. Many components of the eventual "climax" ecology will have been present all along in the area. Some aspects of the soil and its chemistry are more or less stable and preexisting, though other aspects are dependent on the process.[18] And some of the kinds of organisms present in late stages will have been present during the initial stages, perhaps in different numbers. (Other components, like oak trees, will have had their seeds brought in from neighboring areas.) So there are certainly some structures in the late stages that preexisted. Despite this, the ecological analogy is associated with a distinctive *pattern* of explanation, and this is a pattern of explanation stressing the way complex results can arise reliably from a long process in which elaborate final structures *need not* be present, in any sense, at the start. Structures present at time t arise from conditions at $t-1$, which arise from conditions at $t-2$, and so on. The transitions from one stage to another are not explained in terms of the continuing action of a key set of components that were present at the start.[19] In the case of genetics and development, however, the causal structure that we now recognize does give a special role, at every developmental stage, to protein molecules. These molecules are not all materially present at the start of the process, but they have their structures coded for (or templated for) by nucleic acid structures present at the start. This is the feature of contemporary views that does justify something like Gould's "Golden Mean" interpretation of the situation. And this is a point of disanalogy between development and ecological succession.

What does this tell us about DST? I have not given an argument against DST itself, or even against the overall DST enthusiasm for the ecological analogy. My suggestion is just that some aspects of the analogy might be misleading, because the analogy downplays a real continuity between modern biological knowledge and preformationist styles of explanation.

In correspondence about this issue, Richard Francis, one of the advocates of the ecological analogy, objected that the "continuity" that Gould and I allege between modern genetics and older kinds of preformationism is so slight as to be completely insignificant. If so, there has not been any Golden Mean resolution, and DST certainly cannot be faulted for failing to recognize it.

The issue is certainly difficult—what counts as a significant continuity with preformationism as opposed to an insignificant one? But a quotation might help to illustrate my view. The quote is from Albert Matthews, in 1924, given by Robert Olby (1994). Matthews's aim is to poke fun at protein-based versions of preformationism, and he does so in a way reminiscent of caricatures of more traditional preformationism.

In the author's opinion ... the chromatin of spermatozoa is nothing else than the chromatin of spermatozoa; and that of an egg cell is the chromatin of an egg cell. They are not nerve, muscle, epithelial chromatin, in masquerade.... And it must be remembered that the onus of proof is on those who assert that the chromosomes are ... museums containing samples of all the chromatin of all the cells of the body, not only all the chromatins which develop during life, but all that infinite collection of old masters inherited from the past, and all the infinite numbers of descendants yet to appear in the eons before us.... They are concealed no doubt in the chromosomal attic, ready to be produced when the occasion arises. (Quoted in Olby 1994: 80)

Matthews here takes chromosomes to contain nucleic acid and protein, with most biological specificity residing probably in the protein. He doubts that there is *any* sense in which the structure of biologically important parts of, say, muscle cells, are pre-given in egg or sperm. Matthews changed his mind within a decade. But Matthews's earlier critical position was coherently stated and false. The truth turned out to be not so distant from the view he was making fun of here. Rather than actual samples of muscle tissue, spermatozoa contain structures that code (or template) for the proteins specifically needed eventually for muscles. Of course, subsequent discoveries did not in any sense vindicate the picture of gametes containing "old masters" from past generations, and future generations as well. Still, the first part of the quote illustrates the unobviousness, the nontriviality, of the preformationist element in modern genetics.

Before leaving this point, I will note another possible objection to what I have said. Keller, in her contribution to this volume, argues that the constantly growing explanatory role within genetics for mechanisms of editing, regulation, splicing (and so on) is undermining the idea that "proteins are simply and directly encoded in the DNA" (p. 299; see also chapter 7; Sarkar 1996). Applied to the present point, this is an argument that even fewer traces of what I have identified as a preformationist pattern of explanation really remain within genetics. The DNA exists in the zygote, but the eventual protein products of that DNA depend on a great array of other components, many of which will only be built up in the course of development. So Keller might say that laying out a partial continuity between preformationism and modern genetics, as Gould did in 1977 and I do now, is *wholly* misleading. Certainly any view in this area must take account of the recently discovered complexity of gene regulation, and of post-transcriptional and post-translational modification of gene products. The question is whether these facts completely undermine the value of familiar generalizations about genetic specification of proteins—whether they completely undermine such claims as "almost all the proteins that will be used by an organism during its life are coded for in the DNA of the zygote." This objection, like that of Francis, is important and hard to assess; if true it would refute my argument over the last few pages. For now I will just leave this issue on the table.

Conclusion

This chapter has included a discussion of the status of DST and a rumination on one element of the DST perspective on biological development. How do the two discussions fit together? Is antipreformationism part of DST as research program or philosophy of nature? The issues involving coding, preformationism and explanation that I have tried to untangle are largely philosophical. To try to locate present biological knowledge in relation to a Golden Mean resolution of earlier extremes is to engage in a project of

interpreting and restating what we have learned via science. That project lies within philosophy of nature; the empirical outlines of present-day biology are accepted at face value, but the significance of the results and the fine details of the explanatory structure are up for debate. Though my own discussion here focused on the philosophy, the DST attitude to preformationist styles of explanation also plays a role within DST as research program. The DST resistance to preformationist ideas, however subtle, is part of its attempt to steer work toward some scientific options and causal possibilities, and away from others.

Acknowledgments

Thanks to all three editors for helpful comments on this material. Thanks also to Richard Francis, Kim Sterelny, Rasmus Winther, and Kritika Yegnashankaran for discussions and correspondence.

Notes

1. I use "DST" or "developmental systems theory" as a convenient label here; nothing much hangs on the use of "theory" rather than "viewpoint" or some other term. I take Oyama (1985) and Griffiths and Gray (1994) to be basic DST texts. I have given a simple outline of basic DST ideas in Godfrey-Smith (forthcoming).

2. See Griffiths and Gray (1997) for a rather forceful reply.

3. I introduced this concept briefly in Godfrey-Smith (2000), where it is applied to Lewontin's "dialectical biology."

4. *Naturphilosophie* was an ambitious late eighteenth- to early nineteenth-century approach to science and philosophy, influenced by German romanticism, which featured a good deal of unempirical argument about how the course of science must go. Naturphilosophie was based on strong doctrines of progressive development, and unity in nature's principles. Schelling and Oken were major figures. See Gould (1977) for a good outline, which also happens to be relevant to the second part of the present chapter. I use the term "philosophy

of nature" despite, not because of, this historical connection. Sometimes (as Oyama says elsewhere in this volume) the good words are already taken, and have to be reappropriated.

5. Mainstream biology advertises, as a particularly important recent triumph in the general field of development, the discovery of the "homeobox" system of gene regulation. This is a rather general-purpose mechanism for gene regulation that has been found in genetic systems affecting early development in a huge range of animals. Remarkably, homeobox-containing genes have often been found in a similar layout on chromosomes in different animals, where the anterior-posterior axis of structures affected by these genes is mirrored by the linear layout of the genes on the chromosome. Such work provides a good deal of basis for great optimism at present about the power of mainstream developmental genetics.

6. See Lewontin (1974), his two contributions to this volume, and his foreword to the second edition of Oyama (1985).

7. On this point see Griffiths and Gray (1994); Sterelny, Smith, and Dickison (1996).

8. The idea stressed here comprises two parts of a list of six empirical suggestions in Gray (in press).

9. That philosophical work might in turn affect the science, as scientists come to realize they have been using metaphors in certain ways—see above on the interaction between philosophy of nature and science.

10. The version of the gene's eye view outlined in Sterelny and Kitcher (1988) is best construed as showing how a gene-level redescription is possible that (1) makes philosophical sense if combined with an austere and partially conventionalist view about causation (see below in this section), and that (2) might (but also might not) have heuristic benefits for science via its unification of certain phenomena.

G. C. Williams's discussions of the gene's eye view themselves contain some of the philosophical/scientific duality I describe here (e.g., 1992), though he does not try to establish the kind of very ambitious philosophical view found in Dawkins (1976).

11. In the 2000 edition of her 1985 book, Oyama has changed "necessary" to "operative." This takes some, but not all, content out of the point. Certainly the earlier formulation was stronger.

12. Griffiths and Gray (1997), section 6, is one example. I think the suggestion is also made in more subtle

ways in other DST discussions. Note also that Griffiths and Gray's (1997) discussion was prompted by a rival set of claims about heuristic value in Sterelny, Smith, and Dickison (1996).

13. Quine, with his liking for "desert landscapes," is often associated with this kind of challenge.

14. A good outline is found in part 1 of Gould (1977).

15. Note that we could have had a more literal "protein preformationism" as the truth—that is the thought experiment in Godfrey-Smith (2000a), based on actual mid-century speculation. What if genes were little samples of all the proteins needed by the cell?

16. Lewontin (1983: 100 and elsewhere) likes to use ecological succession to stress another theme, the profound and systematic effects of organisms on their environments. Francis (in press) has a particularly detailed and picturesque discussion of ecological succession and its relation to development.

17. Stent's aim in his interesting 1981 discussion was to oppose the overuse of the "program" concept in developmental biology.

18. In some cases, such as recolonization of a recently deglaciated area, even the soil is largely a product of the early stages of the process.

19. Again, the point is made complicated by the fact that a lower-level story could be told in terms of the lives and deaths of individual organisms, and an element of preformationism is true at that lower level.

References

Dawkins, R. (1976). *The Selfish Gene.* Oxford: Oxford University Press.

Dawkins, R. (1982). *The Extended Phenotype.* Oxford: Oxford University Press.

Dretske, F. (1981). *Knowledge and the Flow of Information.* Cambridge, MA: MIT Press.

Francis, R. (in press). *Genes, Brains and Sex in the Information Age.* Princeton, NJ: Princeton University Press.

Friedman, M. (1999). Dynamics of reason: Kantian themes in the philosophy of science. Papers presented as the 1999 Kant Lectures at Stanford University.

Godfrey-Smith, P. (in press). Explanatory symmetries, preformationism and developmental systems theory. *Philosophy of Science* (supplement).

Godfrey-Smith, P. (2000). Organism, environment and dialectics. In R. Singh, C. Krimbas, J. Beatty, and D. Paul (Eds.), *Thinking about Evolution: Historical, Philosophical, and Political Perspectives*. Cambridge: Cambridge University Press.

Godfrey-Smith, P. (2000a). On the theoretical role of "genetic coding." *Philosophy of Science* 67: 26–44.

Godfrey-Smith, P. (in press b). The replicator in retrospect. *Biology and Philosophy*.

Gould, S. J. (1977). *Ontogeny and Phylogeny*. Cambridge, MA: Harvard University Press.

Gray, R. (1992). Death of the gene: Developmental systems strike back. In P. E. Griffiths (Ed.), *Trees of Life: Essays in Philosophy of Biology*, pp. 165–209. Dordrecht: Kluwer.

Gray, R. (in press). Selfish genes or developmental systems: Evolution without replicators and vehicles. In R. Singh, C. Krimbas, J. Beatty, and D. Paul (Eds.), *Thinking about Evolution: Historical, Philosophical, and Political Perspectives*. Cambridge: Cambridge University Press.

Griffiths, P., and R. Gray. (1994). Developmental systems and evolutionary explanation. *Journal of Philosophy* 91: 277–304.

Griffiths, P., and R. Gray. (1997). Replicator II: Judgment day. *Biology and Philosophy* 12: 471–492.

Kitcher, P. S. (1989). Explanatory unification and the causal structure of the world. In P. S. Kitcher and W. Salmon (Eds.), *Minnesota Studies in the Philosophy of Science: Vol. XIII. Scientific Explanation*. Minneapolis: University of Minnesota Press.

Kitcher, P. S. (in press). Battling the undead: How (and how not) to resist genetic determinism. In R. Singh, C. Krimbas, J. Beatty, and D. Paul (Eds.), *Thinking about Evolution: Historical, Philosophical, and Political Perspectives*. Cambridge: Cambridge University Press.

Kuhn, T. S. (1970). *The Structure of Scientific Revolutions*. (2d ed.). Chicago: University of Chicago Press.

Lakatos, I. (1970). Falsification and the methodology of scientific research programs. In I. Lakatos and A. Musgrave (Eds.), *Criticism and the Growth of Knowledge*. Cambridge: Cambridge University Press.

Laudan, L. (1977). *Progress and Its Problems*. Berkeley: University of California Press.

Lewontin, R. C. (1974). *The Genetic Basis of Evolutionary Change*. New York: Columbia University Press.

Lewontin, R. C. (1983). The organism as the subject and object of evolution.

Maynard Smith, J. (in press). The concept of information in biology. In *Proceedings of the 1999 International Congress for Logic, Methodology and Philosophy of Science*. Kluwer.

Moss, L. (1992). A kernel of truth? On the reality of the genetic program. *Proceedings of the Philosophy of Science Association 1992* vol. 1, pp. 335–348. East Lansing, MI: Philosophy of Science Association.

Olby, R. (1994). *The Path to the Double Helix*. (Rev. edition.) New York: Dover.

Oyama, S. (1985). *The Ontogeny of Information: Developmental Systems and Evolution*. Cambridge: Cambridge University Press. (2d rev. ed., Durham, NC: Duke University Press, 2000.)

Quine, W. V. O. (1969). Epistemology naturalized. In *Ontological Relativity and Other Essays*, pp. 69–90. New York: Columbia University Press.

Sarkar, S. (1996). Decoding "Coding"—Information and DNA. *BioScience* 46: 857–864.

Shannon, C. E. (1948). A mathematical theory of communication. *Bell System Technical Journal* 27: 379–423, 623–656.

Sober, E., and R. C. Lewontin. (1982). Artifact, cause and genic selection. *Philosophy of Science* 49: 157–180.

Sober, E., and D. S. Wilson. (1994). A critical review of philosophical work on the units of selection problem. *Philosophy of Science* 61: 534–555.

Stent, G. (1981). Strength and weakness of the genetic approach to the development of the nervous system. In W. M. Cowan (Ed.), *Studies in Developmental Neurobiology*. New York: Oxford.

Sterelny, K., and P. S. Kitcher. (1988). The return of the gene. *Journal of Philosophy* 85: 339–361.

Sterelny, K., K. Smith, and M. Dickison. (1996). The extended replicator. *Biology and Philosophy* 11: 377–403.

Williams, G. C. (1992). *Natural Selection: Domains, Levels and Challenges*. Oxford: Oxford University Press.

Beyond the Gene but Beneath the Skin

Evelyn Fox Keller

The past decade has witnessed an efflorescence of critical commentary deploring the excessively genocentric focus of contemporary molecular and evolutionary biology. Among philosophers, the best known work may well be that of developmental systems theorists (DST) Oyama, Griffiths, and Gray (and sometimes including Lewontin and Moss), but related critiques have also emerged from a variety of other quarters. Expressing diverse intellectual and philosophical preoccupations, and motivated by a variety of scientific and political concerns, these analyses have converged on a number of common themes and sometimes even on strikingly similar formulations.[1] Common themes include: conceptual problems with the attribution of causal primacy (or even causal efficacy) to genes;[2] disarray in contemporary uses of the very term *gene*; confusions and misapprehensions generated by use of the particular locution of "genetic program." Much of the impetus behind these critiques issues from long standing concerns, and indeed, many of the critical observations could have been (and in some cases have been) made long ago. Why then their particular visibility today? An obvious answer lies close at hand: Critiques of genocentrism have found powerful support in many of the recent findings of molecular biologists. Indeed, I would argue that it is from these empirical findings that the major impetus for a reformulation of genetic phenomena now comes.

Three developments (or findings) are of particular importance here: (1) the need for elaborate mechanisms for editing and repair of DNA to ensure sequence stability and fidelity of replication; (2) the importance of complex (and nonlinear) networks of epigenetic interactions in the regulation of transcription; (3) the extent to which the "sense" of the messenger transcript depends on highly regulated mechanisms of editing and splicing. The implication of the first is that the structure of the gene (or sequence of DNA) may be the (or, from the perspective of

DST, "a") locus of heredity constancy, but it can no longer be supposed to be its source: Particular genes (or sequences) persist as stable entities only as long as the machinery responsible for that stability persists. The dependence of gene function on complex epigenetic networks challenges (or at least seriously complicates) the attribution of causal agency to individual genes. Finally, the third finding radically undermines the assumption that proteins are simply and directly encoded in the DNA (indeed, it undermines the very notion of the gene as a functional unit residing on the chromosome).

On this much, one finds a certain general agreement. But differences—deriving in part from the different intellectual, scientific, and political perspectives of their authors—can also be found. Sometimes these are matters of emphasis, sometimes of focus, and sometimes of more substantive import. In this chapter, I want to focus on what I believe to be a substantive issue distinguishing my own perspective from that which tends to dominate the DST literature, and that issue can be put in the form of a question: Is there a place on our biological map for the material body of the organism, for that which lies beyond the gene yet beneath the skin? And if so, where is that place?

The Body in Question

I share with proponents of DST the conviction that the oppositional terms in which the nature/nurture debate has historically been framed are both artificial and counterproductive. But the particular question I pose here reflects an additional source of unease, and that is over the tacit elision of the body implied not only by the framing of the classical controversies, but at least partially continued in the solutions that have been thus far been put forth.

Without question, the most conspicuous roots of my concern are to be found in the history of genetics and neo-Darwinian evolutionary theory. With the emergence of genetics in the early part of the century, debates over the relative force of nature and nurture (first framed as such by Francis Galton in 1874) were recast, initially, in terms of heredity and environment (see, e.g., Barrington and Pearson 1909; Morgan 1911; Conklin 1915), and soon after, in terms of genes and environment (see, e.g., analyses of the relative importance of heredity and environment in Fisher 1918 and Wright 1920). This second reframing may have been an inevitable consequence of the terminological shift in the biological literature of the 1920s and later, in which the term *heredity* came to be replaced by the newer term *genetics*,[3] but the consequence of this shift was more than terminological: it amounted to a conceptual reduction of "nature" to "genes," and with that reduction, only one of two possible statuses for the extragenetic body: either its complete elision or its relegation to the category of "nurture" or "environment." Genetics further contributed to this relegation with its recasting of another and far older controversy, namely that concerning the relations between form (generally construed as active) and matter (construed as passive). That ancient discussion could now be (and was) reconceptualized in terms of genes as the agents of "action" (later, as sources of "information"), and of a cellular or extracellular environment that is simultaneously acted upon and informed, serving as passive material substrate for the development (or unfolding) of the organism.[4] But where many of the discussions of heredity and environment among geneticists focused on their relative force in individual development, elsewhere, such debates more commonly focused on their relative force in shaping the course of evolution. Here, the neo-Darwinian synthesis was of particular importance. In identifying genetic continuity and change as the sole fundament of evolution, it contributed powerfully to the polarization of debates over the relative force of genes and environment in such highly charged arenas as eugenics, or the "heritability" of in-

telligence and other behavioral attributes; it also helped pave the way for the mid-twentieth-century recasting of the nature-nurture debate in one of its crudest forms, that is, as a battle between advocates of "Darwinian" and "Lamarckian" evolution (see, e.g., Keller 1991; Jablonka and Lamb 1995).

To the extent that such debates imply a logical disjunction (form *or* matter; nature *or* nurture; genes *or* environment), they are clearly counterproductive. But my particular argument here is that replacing an implied disjunction by an explicit conjunction (nature *and* nurture; genes *and* environment) does little to ameliorate the particular problem of the role of the body that resides beyond the gene yet beneath the skin. To be sure, the disjunctive framing absolutely denies informational function to the material environment, whereas the conjunctive framing advocated by DST clearly does permit such a function.[5] But in both, there is discernible tendency to figure the organismic body qua environment (and qua nurture), and accordingly, to leave any distinctively informing role the organism's "internal environment" might play in development and evolution concealed from view.

Which Body, Which Skin? And, Anyway, Why Stop at the Skin?

Should the organismic body be singled out as having particular biological significance? And if so, which body, and which skin? Biology recognizes many bodies, corresponding to many skins: in higher organisms, there is the multicellular body contained within an outer integument; in all organisms, cellular bodies are contained by cell membranes; and in eukaryotic organisms, nuclear bodies are contained by nuclear membranes. To avoid some of this ambiguity, I choose to focus on that moment in the life cycle of higher (metazoan) organisms in which the outer integument *is* the cell membrane, and the organismic body *is* the cellular body—that is, in which the body in question is the fertilized egg or zygote. But there remains the question, why stop at the skin? Certainly, no biological integument provides an absolute divide between interior and exterior, and the cell membrane of a fertilized egg is, of necessity, more porous than most. Furthermore, because it regulates so much of the traffic between inside and out, the cell membrane is itself an active agent in shaping the body it contains, indeed, in determining the very meaning of interiority. These facts constitute a warning against conceptualizing the organism as an autonomous individual, sealed off from an exterior world by a static or preexisting boundary. Yet even so, the cell membrane, dynamic and permeable though it may be, defines a boundary which evolution has not only crafted into a cornerstone of biological organization but has endowed with absolutely vital significance. And given the dire effect the physical erasure of this boundary would have on the survival of the organism or cell, it scarcely seems necessary to elaborate upon the inappropriateness of its conceptual erasure.

In other words, the immediate and most obvious reason for taking this boundary seriously is grounded in its manifest indispensability for viability. But this said, we are still not any closer to understanding why is it so important. By way of addressing this last question, I would like to suggest that the primary function of the cell membrane (as of any other biological skin) is simply that it holds things together—more specifically, that it keeps in proximity the many large molecules and subcellular structures required for growth and development. Proximity is crucial, for it enables a degree of interconnectivity and interactive parallelism that would otherwise not be possible, but that is required for what I take to be the fundamental feature of the kind of developmental system we find in a fertilized egg, namely, its robustness. Prior to all its other remarkable properties—in fact, a precondition of these—is the capacity of a developmentally competent zygote to maintain its functional specificity in

the face of all the vicissitudes it inevitably encounters. This paradigmatic body may not be autonomous, but as embryologists have always known, it is far more tolerant of changes in its external environment than in its internal milieu. Indeed, were it not for their robustness, that is, for the tolerance of (at least many) early embryos to being moved from one environment to another, embryonic manipulation would not be possible, and much of what we know of as experimental embryology would never have come into being.

All of this may seem too obvious to need saying, but there are times when the obvious is what most needs saying. We have learned that no elision is innocent. Nor, for that matter, is any reminder of elision. Politics are everywhere. Just as there are important political dimensions to the history of debates over genes and environment,[6] so too, there are political dimensions to the elision of the body in genetical discourse,[7] as there also are, inevitably, to my insistence here on the boundary of the skin. In fact, the title of this essay contains its own elision: it reminds us of another title (Barbara Duden's *The Woman Beneath the Skin*), while at the same time suppressing the subject of that other title. There is, of course, a reason. This is not an essay in feminist theory, nor is it about women. The "woman" in my title is signified only by its absence—intended, by that absence, to evoke nothing more than a recognition of the trace of the woman beneath the skin that still lurks, if not in the body more generally, surely, in the reproductive body of the fertilized egg. Because, in sexual reproduction, the cytoplasm derives almost entirely from the unfertilized egg, it is no mere figure of speech to refer to it as the maternal contribution. Furthermore, the representation of that body as "genetic environment," as nothing more than a source of nurture for the developing organism, is a bit too reminiscent of conventional maternal discourse for at least this author's comfort. My title, in short, is deliberate in its allusivity: I want to indicate the possibility that gender politics has been implicated in the historic elision of the body in question, without, at the same time, reinscribing the woman in that or any other body. The primary aim of this essay is, finally, a biological one, by which I mean that it is to reclaim the possibility of finding biological significance and agency in that no man's land beyond the gene but beneath the skin. Contra Oyama (1992),[8] I want to argue that taking the cell rather than the gene as a unit of development *does* make a difference: not only does it yield a significant conceptual gain in the attempt to understand development, but also, it permits better conformation to the facts of development as we know them.

Is There a Program for Development? And If So, Where Is It to Be Found?

Many authors have taken issue with the concept of a *program* for development, noting its teleological implications, its metaphoric reliance on computer science, its implication of a unidirectional flow of information.[9] But my concern here is not with the concept of *program* per se: rather, it is with the more specific notion of a genetic program, especially in contradistinction to its companion notion, that of a developmental program. In other words, for my particular purposes here, I accept the metaphor of *program*, warts and all, and focus my critical attention instead on the implications of attaching to that metaphor the modifier "genetic." And I ask two questions: First, what is the meaning of a "genetic program"? Second, how did this concept come to be so widely accepted as an "explanation" of biological development?[10]

Taken as a composite, the meaning of the term *genetic program* simultaneously depends upon and underwrites the particular presumption that a "plan of procedure" for development is itself written in the sequence of nucleotide bases. Is this presumption correct? Certainly, it is almost universally taken for granted, but I want to argue that, at best, it must be said to be misleading, and at worst, simply false: To the extent that we may

speak at all of a developmental program, or of a set of instructions for development, in contradistinction to the data or resources for such a program, current research obliges us to acknowledge that these "instructions" are not written into the DNA itself (or at least, are not all written in the DNA), but rather are distributed throughout the fertilized egg. Indeed, if the distinction between program and data is to have any meaning in biology, it has become abundantly clear that it does not align (as had earlier been assumed) either with a distinction between "genetic" and "epigenetic," or with the precursor distinction between nucleus and cytoplasm. To be sure, the informational content of the DNA is essential—without it development (life itself) cannot proceed. But for many developmental processes, it is far more appropriate to refer to this informational content as data than as program (Atlan and Koppel 1990). Indeed, I want to suggest that the notion of genetic program both depends upon and sustains a fundamental category error in which two independent distinctions, one between "genetic" and "epigenetic," and the other between program and data, are pulled into mistaken alignment. The net effect of such alignment is to reinforce two outmoded associations: on the one hand, between "genetic" and active, and, on the other, between "epigenetic" and passive.

Development results from the temporally and spatially specific activation of particular genes, which in turn depends on a vastly complex network of interacting components including not only the "hereditary codescript" of the DNA, but also a densely interconnected cellular machinery made up of proteins and RNA molecules. Necessarily, each of these systems functions in relation to the others alternatively as data and as program. If development cannot proceed without the "blueprint" of genetic memory, neither can it proceed without the "machinery" embodied in cellular structures. To be sure, the elements of these structures are fixed by genetic memory, but their assembly is dictated by cellular memory.[11] Furthermore, one must remember that more than genes are passed from parent to offspring. To forget this is to be guilty of what Richard Lewontin calls an "error of vulgar biology." As he reminds us, "an egg, before fertilization, contains a complete apparatus of production deposited there in the course of its cellular development. We inherit not only genes made of DNA but an intricate structure of cellular machinery made up of proteins" (Lewontin 1992: 33).

Assuming one is not misled by Lewontin's colloquial use of the term *inherit* to refer to transmission over a single generation (as distinct from multigenerational transmission), none of this is either controversial or new, nor does it depend on the extraordinary techniques now available for molecular analysis. Yet, however surprisingly, it is only within the last decade or two that the developmental and evolutionary implications of so called "maternal effects" has begun to be appreciated.[12] Current research now provides us with an understanding of the mechanisms involved in the processing of genetic data that make the errors of what Lewontin calls "vulgar biology" manifest. Yet, even when elaborated by the kind of detail we now have available, such facts are still not sufficient to dislodge the confidence that many distinguished biologists continue to have in both the meaning and explanatory force of the genetic program. The question I want therefore to ask is, how come? What grants the "genetic program" its apparent explanatory force, even in the face of such obvious caveats as those above? To look for answers, I will turn to history, more specifically, to the history of the term itself.

"Programs" in the Biological Literature of the 1960s

The metaphor of a "program," borrowed directly from computer science, entered the biological literature in the 1960s not once, but several times, and in at least two distinctly different registers. In its first introduction, simultaneously by Mayr (1961) and by Monod and Jacob (1961), the locus of the "program" was explicitly identified as the genome, but, over the course of that decade, another notion of "program," a "developmental program," also surfaced, and repeatedly so. This program was not located in the genome, but instead, distributed throughout the fertilized egg (see, e.g., Apter 1966). By the 1970s, however, the "program" for development had effectively collapsed into a "genetic program," with the alternative, distributed, sense of a "developmental program" all but forgotten.

Francois Jacob, one of the earliest to use the concept "genetic program" contributed crucially to its popularization. In *The Logic of Life,* first published in French in 1970, Jacob describes the organism as "the realization of a programme prescribed by its heredity" (Jacob 1976; 2), claiming that "when heredity is described as a coded programme in a sequence of chemical radicals, the paradox [of development] disappears" (Jacob 1976: 4). For Jacob, the genetic program, written in the alphabet of nucleotides, is what is responsible for the apparent purposiveness of biological development; it and it alone gives rise to "the order of biological order." (Jacob 1976: 8) He refers to the oft-quoted characterization of teleology as a "mistress" whom biologists "could not do without, but did not care to be seen with in public," and writes, "The concept of programme has made an honest woman of teleology" (Jacob 1976: 8–9). Although Jacob does not exactly define the term, he notes that "[t]he programme is a model borrowed from electronic computers. It equates the genetic material of an egg with the magnetic tape of a computer" (Jacob 1976: 9).

However, equating the genetic material of an egg with the magnetic tape of a computer does not imply that that material encodes a "program"; it might just as well be thought of as encoding "data" to be processed by a cellular "program." Or by a program residing in the machinery of transcription and translation complexes. Or by extranucleic chromatin structures in the nucleus. Computers have provided a rich source of metaphors for molecular biology, but they cannot by themselves be held responsible for the notion of "genetic program." Indeed, as already indicated, other, quite different, uses of the program metaphor for biological development were already in use. One such use was in the notion of a "developmental program"—a term that surfaced repeatedly through the 1960s, and that stood in notable contrast to that of a "genetic program."

Let me give an example of this alternative use. In 1965, a young graduate student, Michael Apter, steeped in information theory and cybernetics, teamed up with the developmental biologist Lewis Wolpert to argue for a direct analogy not between computer programs and the genome, but between computer programs and the egg:

if the genes are analogous with the subroutine, by specifying how particular proteins are to be made ... then the cytoplasm might be analogous to the main programme specifying the nature and sequence of operations, combined with the numbers specifying the particular form in which these events are to manifest themselves.... In this kind of system, instructions do not exist at particular localized sites, but the system acts as a dynamic whole. (Apter and Wolpert 1965: 257)

Indeed, throughout the 1960s, a number of developmental biologists attempted to employ ideas from cybernetics to illuminate development, and almost all shared Apter's starting assumptions (see Keller 1995, chap. 3 for examples)—that is, they located the program (or "instructions") for development in the cell as a whole.

The difference in where the program is said to be located is crucial, for it bears precisely on the controversy over the adequacy of genes to account for development that had been raging among biologists since the beginning of the century. By the beginning of the 1960s, this debate had subsided, largely as result of the eclipse of embryology as a discipline during the 1940s and 1950s. Genetics had triumphed, and after the identification of DNA as the genetic material, the successes of molecular biology had vastly consolidated that triumph. Yet the problems of development, still unresolved, lay dormant. Molecular

biology had revealed a stunningly simple mechanism for the transmission and translation of genetic information, but, at least until 1960, it had been able to offer no account of developmental regulation.

James Bonner, a professor of biology at CalTech, in an early attempt to bring molecular biology to bear on development, put the problem well. Granting that "the picture of life given to us by molecular biology ... applies to cells of all creatures," he goes on to observe that this picture

is a description of the manner in which all cells are similar. But higher creatures, such as people and pea plants, possess different kinds of cell. The time has come for us to find out what molecular biology can tell us about why different cells in the same body are different from one another, and how such differences arise. (Bonner 1965: v)

Bonner's own work was on the biochemistry and physiology of regulation in plants, in an institution well known for its importance in the birth of molecular biology (see, e.g., Kay 1992). Here, in this work, published in 1965, like Apter and a number of others of that period, Bonner too employs the conceptual apparatus of automata theory to deal with the problem of developmental regulation. But unlike them, he does *not* locate the "program" in the cell as a whole, but rather, in the chromosomes, and more specifically in the genome. Indeed, he begins with the by then standard credo of molecular biology, asserting that "[w]e know that ... the directions for all cell life [are] written in the DNA of their chromosomes" (Bonner 1965: v). Why? An obvious answer is suggested by his location. Unlike Apter and unlike other developmental biologists of the time, Bonner was situated at a major thoroughfare for molecular biologists, and it is hard to imagine that he was uninfluenced by the enthusiasm of his colleagues at CalTech. In any case, Bonner's struggle to reconcile the conceptual demands posed by the problems of developmental regulation with the received wisdom among molecular biologists is at the very least instructive, especially given its location in time, and I suggest it is worth examining in some detail for the insight it has to offer on our question of how the presumption of a "genetic program" came—in fact, over the course of that very decade—to seem self-evident. In short, I want to take Bonner as representative of a generation of careful thinkers about an extremely difficult problem who opted for this (in retrospect, inadequate) conceptual shortcut.

Explanatory Logic of the "Genetic Program"

From molecular biology, Bonner inherited a language encoding a number of critical if tacit presuppositions. That language shapes his efforts in decisive ways. Summarizing the then current understanding of transcription and translation, he writes:

Enzyme synthesis is therefore an information-requiring task and ... the essential information-containing component is the long punched tape which contains, in coded form, the instructions concerning which amino acid molecule to put next to which in order to produce a particular enzyme. (Bonner 1965: 3)

At the same time, he clearly recognized that only the composition of the protein had thus been accounted for, and not the regulation of its production required for the formation of specialized cells, that is, cell differentiation remained unexplained. As he wrote, "Each kind of specialized cell of the higher organism contains its characteristic enzymes but each produces only a portion of all the enzymes for which its genomal DNA contains information" (Bonner 1965: 6). But, he continues: "Clearly then, the nucleus contains some further mechanism which determines in which cells and at which times during development each gene is to be active and produce its characteristic messenger RNA, and in which cells each gene is to be inactive, to be repressed" (Bonner 1965: 6).

Two important moves have been made here. Bonner argues that something other than the information for protein synthesis encoded in the

DNA is required to explain cell differentiation (and this is his main point), but on the way to making this point, he has placed this "further mechanism" in the nucleus, with nothing more by way of argument or evidence than his "Clearly then." Why does such an inference follow? And why does it follow "clearly"? Perhaps the next paragraph will help:

The egg is activated by fertilization.... As division proceeds cells begin to differ from one another and to acquire the characteristics of specialized cells of the adult creature. There is then within the nucleus some kind of programme which determines the property [sic] sequenced repression and derepression of genes and which brings about orderly development. (Bonner 1965: 6)

Here, the required "further mechanism" is explicitly called a "program," and, once again, it is located in the nucleus. But this time around, a clue to the reasoning behind the inference has been provided in the first sentence, "The egg is activated by fertilization." This is how I believe the (largely tacit) reasoning goes: If the egg is "activated by fertilization," the implication is that it is entirely inactive prior to fertilization. What does fertilization provide? The entrance of the sperm, of course, and unlike the egg, the sperm has almost no cytoplasm: it can be thought of as pure nucleus. Ergo, the active component must reside in the nucleus and not in the cytoplasm. Today, the supposition of an inactive cytoplasm would be challenged, but in Bonner's time, it would have been taken for granted as a carryover from what I have called "the discourse of gene action" of classical genetics (Keller 1995). And even then, it might have been challenged had it been made explicit, but as an implicit assumption encoded in the language of "activation," it was likely to go unnoticed by Bonner's readers as by Bonner himself.

Bonner then goes on to ask the obvious questions: "What is the mechanism of gene repression and derepression which makes possible development? Of what does the programme consist and where does it live?" (Bonner 1965: 6) And he answers them as best he can: "We can say that the programme which sequences gene activity must itself be a part of the genetic information since the course of development and the final form are heritable. Further than this we cannot go by classical approaches to differentiation" (Bonner 1965: 6).

In these few sentences, Bonner has completed the line of argument leading him to the conclusion that the program *must* be part of the genetic information, that is, to the "genetic program." And again, we can try to unpack his reasoning. Why does the heritability of the course of development and the final form imply that the program must be part of the genetic information? Because—and only because—of the unspoken assumption that it is only the genetic material that is inherited. The obvious fact—that the reproductive process passes on (or transmits) not only the genes but also the cytoplasm (the latter through the egg for sexually reproducing organisms)—is not mentioned. But even if it were, this fact would almost certainly be regarded as irrelevant, simply because of the prior assumption that the cytoplasm contains no active components. The conviction that the cytoplasm could neither carry nor transmit effective traces of intergenerational memory had been a mainstay of genetics for so long that it had become part of the "memory" of that discipline, working silently but effectively to shape the very logic of inference employed by geneticists.

Yet another ellipsis becomes evident (now, even to Bonner himself) as he attempts to integrate his own work on the role of histones in genetic regulation. Not all copies of a gene (or a genome) are in fact the same: Because of the presence of proteins in the nucleus, capable of binding to the DNA, "in the higher creature, if it is to be a proper higher creature, one and the same gene must possess different attributes, different attitudes, in different cells" (Bonner 1965: 102). The difference is a function of the histones. How can we reconcile this fact with the notion of a "genetic program"? There is one simple way, and

Bonner takes it—namely, to elide the distinction between genome and chromosome. The "genetic program" is saved (for this discussion) by just a slight shift in reference: now it refers to a program built into the chromosomal structure—that is, into the complex of genes and histones, where that complex is itself here referred to as the "genome."

But the most conspicuous inadequacy of the location of the developmental program in the genetic information becomes evident in the final chapter, in which Bonner attempts to sketch out an actual computer program for development. Here, the author undertakes to reframe what is known about the induction of developmental pathways in terms of a "master program," proposing to "consider the concept of the life cycle as made up of a master programme constituted in turn of a set of subprogrammes or subroutines" (Bonner 1965: 134). Each subroutine specifies a specific task to be performed. For a plant, his list includes: cell life, embryonic development, how to be a seed, bud development, leaf development, stem development, root development, reproductive development. Within each of these subroutines is a list of cellular instructions or commands, such as, "divide tangentially with growth"; "divide transversely with growth"; "grow without dividing"; and "test for size or cell number" (Bonner 1965: 137). He then asks the obvious next question: "[H]ow might these subroutines be related to one another? Exactly how are they to be wired together to constitute a whole programme?" (Bonner 1965: 135). Conveniently, this question is never answered. If it had been, the answer would have necessarily undermined Bonner's core assumption. To see this, two points emerging from his discussion need to be underscored: First, the list of subroutines, although laid out in a linear sequence (as if following from an initial "master program") actually constitute a circle, as indeed they must if they are to describe a life cycle. Bonner's own "master program" is in fact nothing but this composite set of programs, wired together in a structure exhibiting the characteristic cybernetic logic of "circular causality."

The second point bears on Bonner's earlier question, "Of what does the programme consist and where does it live?" The first physical structures that were built to embody the logic of computer programs were built out of electrical networks[13] (hence the term "switching networks"), and this is Bonner's frame of reference. As he writes, "[t]hat the logic of development is based upon [a developmental switching] network, there can be no doubt" (Bonner 1965: 148). But what would serve as the biological analogue of an electric (or electronic) switching network? How are the instructions specified in the subroutines that comprise the life cycle actually embodied? Given the dependence of development on the regulating activation of particular genes, Bonner reasonably enough calls the developmental switching network a "genetic switching network." But this does not quite answer our question; rather, it obfuscates it. The clear implication is that such a network is constituted of nothing but genes, whereas in fact, many other kinds of entities also figure in this network, all playing critical roles in the control of genetic activity. Bonner himself writes of the roles played by histones, hormones, and RNA molecules; today, the list has expanded considerably to include enzymatic networks, metabolic networks, transcription complexes, signal transduction pathways, and so on, with many of these additional factors embodying their own "switches." We could, of course, still refer to this extraordinarily complex set of interacting controlling factors as a "genetic switching network"—insofar as the regulation of gene activation remains central to development—but only if we avoid the implication (an implication tantamount to a category error) that that network is embodied in and by the genes themselves.

Indeed, it is this "category error" that confounds the very notion of a "genetic program." If we were now to ask Bonner's question, "Of what does the programme consist and where does it live?" we would have to say, just as Apter saw long ago, that it consists not of particular gene entities, and lives not in the genome itself, but of and in the cellular machinery integrated into a dynamic whole. As Garcia-Bellido writes, "Development results from local effects, and there is no brain or mysterious entity governing the whole: there are local computations and they explain the specificity of something that is historically defined" (1998: 113). Thus, if we wish to preserve the computer metaphor, it would seem more reasonable to describe the fertilized egg as a massively parallel processor in which "programs" (or networks) are distributed throughout the cell.[14] The roles of "data" and "program" here are relative, for what counts as "data" for one "program" is often the output of a second "program," and the output of the first is "data" for yet another "program," or even for the very "program" that provided its own initial "data." Thus, for some developmental stages, the DNA might be seen as encoding "programs" or switches which process the data provided by gradients of transcription activators, or alternatively, one might say that DNA sequences provide data for the machinery of transcription activation (some of which is acquired directly from the cytoplasm of the egg). In later developmental stages, the products of transcription serve as data for splicing machines, translation machines, and the like. In turn, the output of these processes make up the very machinery or programs needed to process the data in the first place. Sometimes, this exchange of data and programs can be represented sequentially, sometimes as occurring in simultaneity.

Into the Present

In the mid 1960s, when Bonner, Apter, and others were attempting to represent development in the language of computer programs, automata theory was in its infancy, and cybernetics was at the height of its popularity. During the 1970s and

1980s, these efforts lay forgotten: cybernetics had lost its appeal to computer scientists and biologists alike, and molecular biologists found they had no need of such models. The mere notion of a "genetic program" sufficed by itself to guide their research. Today, however, provoked in large part by the construction of hard-wired parallel processors, the project to simulate biological development on the computer has returned in full force, and in some places has become a flourishing industry. It goes by various names—Artificial Life, adaptive complexity, or genetic algorithms. But what is a genetic algorithm? Like Bonner's subroutines, it is "a sequence of computational operations needed to solve a problem" (see, e.g., Emmeche 1994). And once again, we need to ask, why "genetic"? Furthermore, not only are the individual algorithms referred to as "genetic," but also "in the fields of genetic algorithms and artificial evolution, the [full] representation scheme is often called a 'genome' or 'genotype'" (Fleischer 1995: 1). And, in an account of the sciences of *complexity* written for the lay reader, Mitchell Waldrop quotes Chris Langton, the founder of *Artificial Life*, as saying:

[Y]ou can think of the genotype as a collection of little computer programs executing in parallel, one program per gene. When activated, each of these programs enters into the logical fray by competing and cooperating with all the other active programs. And collectively, these interacting programs carry out an overall computation that is the phenotype: the structure that unfolds during an organism's development. (Waldrop 1992: 194)

Like their counterparts in molecular genetics, workers in Artificial Life are not confused. They well understand, and when pressed readily acknowledge, that the biological analogs of these computer programs are not in fact "genes" (at least as the term is used in biology), but complex biochemical structures or networks constituted of proteins, RNA molecules, and metabolites that often, although certainly not always, execute their tasks in interaction with particular stretches

of DNA.[15] Artificial Life's "genome" typically consists of instructions such as "reproduce," "edit," "transport," or "metabolize," and the biological instantiation of these algorithms is found not in the nucleotide sequences of DNA, but in specific kinds of cellular machinery such as transcription complexes, spliceosomes, and metabolic networks. Why then are they called "genetic," and why is the full representation called a "genome"? Is it not simply because it so readily follows from the usage the term "genetic program" had already acquired in genetics?

Words have a history, and their usage depends on this history, as does their meaning. History does not fix the meaning of words; rather, it builds into them a kind of memory. In the field of genetic programming, "genes" have come to refer not to particular sequences of DNA, but to the computer programs required to execute particular tasks (as Langton puts it, "one program per gene"); yet, at the same time, the history of the term ensures that the word *gene*, even as adapted by computer scientists, continues to carry its original meaning. And perhaps most importantly, that earlier meaning remains available for deployment whenever it is convenient to do so. Much the same can be said for the use of the terms *gene* and *genetic programs* by geneticists.

A Recapitulation

I have taken some time in examining Bonner's argument for "genetic programs," not because his book played a major role in establishing the centrality of this notion in biological discourse, but rather because of the critical moment in time at which it was written and because of the relative accessibility of the kind of slippage on which his argument depends. The very first use of the term *program* that I have been able to find in the molecular biology literature had appeared only four years earlier.[16] In 1961, Jacob and Monod, published a review of their immensely influential work on a genetic mechanism for enzymatic adaptation in *E. coli*, that is, the operon model. The introduction of the term *program* appears in their concluding sentence: "The discovery of regulator and operator genes, and of repressive regulation of the activity of structural genes, reveals that the genome contains not only a series of blue-prints, but a coordinated program of protein synthesis and the means of controlling its execution" (Jacob and Monod 1961: 354).

Three decades later, Sydney Brenner refers to the belief "that all development could be reduced to [the operon] paradigm"—that "It was simply a matter of turning on the right genes in the right places at the right times"—in rather scathing terms. As he puts it, "[o]f course, while absolutely true this is also absolutely vacuous. The paradigm does not tell us how to make a mouse but only how to make a switch" (Brenner et al. 1990: 485).[17] And even in the first flush of enthusiasm, not everyone was persuaded of the adequacy of this particular regulatory mechanism to explain development.[18] Lewis Wolpert was one of the early skeptics. In the late 1960s, he seemed certain that an understanding of development required a focus not simply on genetic information, but also on cellular mechanisms.[19] But by the mid 1970s, even Wolpert had been converted to the notion of a "genetic program" (see, e.g., Wolpert and Lewis 1975).

What carried the day? Certainly not more information about actual developmental processes. Far more than most histories of scientific terms, the history of *genetic programs* bears the conspicuous marks of a history of discourse and power. Initially founded on a simple category error, in which the role of genes as subjects (or agents) of development was unwittingly conflated with their role as objects of developmental dynamics, the remarkable popularity of this term in molecular genetics over the last three decades cries out for an accounting. Certainly, it provided a convenient gloss, an easy way to talk that rarely if ever trips scientists up in their daily laboratory work. But it does trip them up in their efforts to explain development; indeed, the term has proven remarkable effective in obscuring enduring gaps in our understanding of developmental logic. Arguably, it has also contributed to the endurance of such gaps.

So why did it prevail? If its popularity cannot be accounted for in strictly scientific or cognitive terms, we must look elsewhere. I suggest we look to the consonance of this formulation with the prior history of genetic discourse, particularly with the discourse of "gene action" that earlier prevailed. Fortifying the "genetic program" in the postwar era, with its easy and continuing elision of the cytoplasmic body, were an entirely new set of resources. Primary among these were the new science of computers, the imprimateur of Schrödinger, and the phenomenal success of the new molecular biology. Jacob's *Logic of Life* was of key importance in the popularization of the concept of "genetic program." Invoking the approval of both Schrödinger and Wiener, Jacob endows the transition from past to future metaphors with the stamp of authority.[20]

What's in a word? As I have already suggested, quite a lot. Words shape the ways we think, and how we think shapes the ways we act. In particular, the use of the term *genetic* to describe developmental instructions (or programs) encourages the belief even in the most careful of readers (as well as writers) that it is only the DNA that matters; it helps all of us to lose sight of the fact that, if that term is to have any applicability at all, it is primarily to refer to the *entities upon which instructions directly or indirectly act* and *not of which these instructions are constituted*. The necessary dependency of genes on their cellular context, not simply as nutrient but as embodying causal agency, is all too easily forgotten. It is forgotten in laboratory practice, in medical counseling, and perhaps above all, seduced by the promise of utopian transformation, in popular culture.

Notes

1. See, e.g., Griffiths and Neumann-Held (1999); Keller (1995, 1999).

2. See, e.g., Lewontin (1992); Moss (1992); Keller (1995); Strohman (1997).

3. Earlier, in the late nineteenth century, the term *heredity* had commonly been used far more inclusively, encompassing both the study of both genetics and embryology (see, e.g., Sapp 1987). Furthermore, in the 1920s, the term *genetics* was largely understood to refer solely to transmission genetics.

4. For further discussion, see Griesemer (forthcoming); Keller (1995).

5. Indeed, the brunt of much of this literature is to argue for symmetry between the role of genes and other developmental resources. Thus, for example, in arguing against the conventional view that genes code for traits, Griffiths and Gray suggest that "we can talk with equal legitimacy of cytoplasmic or landscape features coding for traits in standard genic backgrounds" (1994; 1998: 122).

6. See, e.g., Kevles (1986) and Paul (1995, 1998) for discussions of the politics of eugenics debates, and Sapp (1987) for a discussion of the impact of Lysenko's anti-genetic crusade in the Soviet Union on American genetics just before and during the cold war.

7. See my discussion of the "discourse of gene action" in Keller (1995, chap. 1).

8. I refer in particular to Oyama's discussion of Brenner's abandonment of the concept of a "genetic program" and his emerging conviction that the proper "unit of development is the cell." She writes, "Having given up genetic programs, [Brenner] now speaks of internal representations and descriptions. In doing so he is like many workers who have been faced with the contradictions and inadequacies of traditional notions of genetic forms and have tried to resolve them, not by seriously altering their concepts, but by making the forms in the genome more abstract: not noses in the genes, but instructions for noses, or potential for noses, or symbolic descriptions of them. This solves nothing" (Oyama 1992: 55).

9. See, e.g., Stent (1985), Newman (1988), Oyama (1989), Moss (1992), and de Chardarevian (1994).

10. The remainder of this paper is adapted from Keller (2000).

11. A vivid demonstration of this interdependency was provided in the 1950s and 1960s with the development of techniques for interspecific nuclear transplantation. Such hybrids almost always fail to develop past gastrulation, and in the rare cases when they do, the resultant embryo exhibits characteristics intermediate between the two parental species. This dependency of genomic function on cytoplasmic structure follows as well from the asymmetric outcomes of reciprocal crosses demonstrated in earlier studies of interspecific hybrids (Markert and Ursprung 1971: 135–137).

12. "Maternal (or cytoplasmic) effects" refers only to the effective agency of maternal (or cytoplasmic) contributions (such as, e.g., gradients). Because such effects need not be (and usually are not) associated with the existence of permanent structures that are transmitted through the generations, they should not be confused with "maternal inheritance."

13. In modern computers, such networks are electronic.

14. Supplementing Lenny Moss's observation that a genetic program is "an object nowhere to be found" (Moss 1992: 335), I would propose the developmental program as an entity that is everywhere to be found.

15. Executing a task means processing data provided both by the DNA and by the products of other programs—that is, by information given in nucleotide sequences, chromosomal structure, gradients of proteins and RNA molecules, the structure of protein complexes, and so on.

16. Simultaneously, and probably independently, Ernst Mayr introduced the notion of "program" in his 1961 article on "Cause and Effect in Biology" (adapted from a lecture given at MIT on Feb. 1, 1961). There he wrote, "The complete individualistic and yet also species-specific DNA code of every zygote (fertilized egg cell), which controls the development of the central and peripheral nervous system . . . is the *program* for the behavior computer of this individual" (Mayr 1961: 1504).

17. As Soraya de Chadarevian points out (1994), Brenner had taken a critical stance toward the use of the operon model for development as early as 1974 (see his comments in Brenner 1974).

18. Or even of the appropriateness of the nomenclature. Waddington, for example, noted not only that it "seems too early to decide whether all systems controlling gene-action systems have as their last link an influence which impinges on the gene itself," but also redescribed this system as "genotropic" rather than "genetic" in order "to indicate the site of action of the substances they are interested in" (Waddington 1962: 23).

19. For example, Wolpert wrote in 1969: "Dealing as it does with intracellular regulatory phenomena, it is not directly relevant to problems where the cellular bases of the phenomena are far from clear" (Wolpert 1969: 2–3).

20. He writes: "According to Norbert Wiener, there is no obstacle to using a metaphor 'in which the organism is seen as a message'" (Jacob 1976: 251–252). Two pages later, he adds, "According to Schrödinger, the chromosomes 'contain in some kind of code-script the entire pattern of the individual's future development and of its functioning in the mature state.... The chromosome structures are at the same time instrumental in bringing about the development they foreshadow. They are law-code and executive power—or, to use another simile, they are architect's plan and builder's craft all in one'" (Jacob 1976: 254).

References

Apter, M. J. (1966). *Cybernetics and Development*. Oxford: Pergamon Press.

Apter, M. J., and L. Wolpert. (1965). Cybernetics and development. *Journal of Theoretical Biology* 8: 244–257.

Atlan, H., and M. Koppel. (1990). The cellular computer DNA: Program or data. *Bulletin of Mathematical Biology* 52(3): 335–348.

Barrington, A., and Pearson, K. (1909). *A First Study of the Inheritance of Vision and of the Relative Influence of Heredity and Environment on Sight*. Cambridge: Cambridge University Press.

Bonner, J. (1965). *The Molecular Biology of Development*. Oxford: Oxford University Press.

Brenner, S. (1974). New directions in molecular biology. *Nature* 248: 785–787.

Brenner, S., W. Dove, I. Herskowitz, and R. Thomas. (1990). Genes and development: Molecular and logical themes. *Genetics* 126: 479–486.

Conklin, E. G. (1915). *Heredity and Environment in the Development of Men*. Princeton: Princeton University Press.

de Chardarevian, S. (1994). Development, programs and computers: Work on the worm (1963–1988). Paper presented at the Summer Academy, Berlin, Germany.

Duden, B. (1991). *The woman beneath the Skin*. Cambridge, MA: Harvard University Press.

Emmeche, C. (1994). *The Garden in the Machine*. Princeton, NJ: Princeton University Press.

Fisher, R. A. (1918). The correlation between relatives on the supposition of Mendelian inheritance. *Transactions of the Royal Society of Endinburgh* 52: 399–433.

Fleischer, K. (1995). *A Multiple-Mechanism Developmental Model for Defining Self-Organizing Geometric Structures*. Doctoral dissertation, California Institute of Technology.

Galton, F. (1874/1970). *English Men of Science: Their Nature and Nurture*. London: Cass.

Garcia-Bellido, A. (1998). "Discussion." In *The limits of reductionism*. Chichester: John Wiley & Sons.

Griesemer, J. R. (forthcoming). The informational gene and the substantial body: On the generalization of evolutionary theory by abstraction. In N. Cartwright and M. Jones (Eds.), *Varieties of Idealization*. Poznan Studies (Leszek Nowak, series ed.). Amsterdam: Rodopi Publishers.

Griffiths, P. E., and R. Gray. (1994). Developmental systems and evolutionary explanation. *Journal of Philosophy* 91: 277–304.

Griffiths, P. E., and E. M. Neumann-Held. (1999). The many faces of the gene. *BioScience* 49: 656–662.

Hull, D., and M. Ruse. (1998). *The Philosophy of Biology*. Oxford: Oxford University Press.

Jablonka, E., and M. Lamb. (1995). *Epigenetic Inheritance and Evolution*. New York: Oxford University Press.

Jacob, F. (1976). *The Logic of Life*. New York: Vanguard.

Jacob, F., and J. Monod. (1961). Genetic regulatory mechanisms in the synthesis of proteins. *Journal of Molecular Biology* 3: 318–356.

Keller, E. F. (1995). *Refiguring Life*. New York: Columbia University Press.

Keller, E. F. (2000). Decoding the genetic program. In P. Beurton and R. Falk (Eds.), *Genes*, pp. 159–177. Cambridge: Cambridge University Press.

Kevles, Daniel J. (1986). *In the Name of Eugenics: Genetics and the Uses of Human Heredity*. Berkeley: University of California Press.

Lewontin, R. (1992). The dream of the human genome. *New York Review of Books*, May 28, pp. 31–40.

Markert, C. L., and H. Ursprung. (1971). *Developmental Genetics.* Englewood Cliffs, NJ: Prentice-Hall.

Mayr, E. (1961). Cause and effect in biology. *Science* 134: 1501–1506.

Monod, J., and F. Jacob. (1961). General conclusions: Teleonomic mechanisms in cellular metabolism, growth, and differentiation. *Cold Spring Harbor Symposia Quantative Biology* 26: 389–401.

Morgan, T. H. (1911). The influence of heredity and of environment in determining the coat colors in mice. *Annals of the New York Academy of Science* XXI: 87–118.

Moss, L. (1992). A kernel of truth? On the reality of the genetic program. *Philosophy of Science Association* 1: 335–348.

Newman, S. A. (1988). Idealist biology. *Perspectives in Biology and Medicine* 31(3): 353–368.

Oyama, S. (1989). Ontogeny and the central dogma: Do we need the concept of genetic programming in order to have an evolutionary perspective? In M. R. Gunnar and E. Thelen (Eds.), *Systems and Development,* pp. 1–34. Hillside, NJ: Lawrence Erlbaum Associates.

Oyama, S. (1992). Transmission and construction: Levels and the problem of heredity. In G. Greenberg and E. Tobach (Eds.), *Levels of Social Behavior: Evolutionary and Genetic Aspects,* pp. 51–60. Wichita, KS: T. C. Schneirla Research Fund.

Oyama, S. (2000). *Evolution's Eye: A Systems View of the Biology-Culture Divide.* Durham, NC: Duke University Press.

Paul, D. (1995). *Controlling Human Heredity.* Atlantic Highlands, NJ: Humanities Press.

Paul, D. (1998). *The Politics of Heredity.* New York: New York University Press.

Sapp, J. (1987). *Beyond the Gene.* Oxford: Oxford University Press.

Stent, G. S. (1985). Hermeneutics and the analysis of complex biological systems. In D. J. Depew and B. Weber (Eds.), *Evolution at a Crossroads,* pp. 209–225. Cambridge, MA: MIT Press.

Strohman, R. C. (1997). The coming Kuhnian revolution in biology. *Nature Biotechnology* 15: 194–200.

Waddington, C. H. (1962). *New Patterns in Genetics and Development.* New York: Columbia University Press.

Waldrop, J. M. (1992). *Complexity.* New York: Simon and Schuster.

Wolpert, L. (1969). Positional information and the spatial pattern of cellular differentiation. *Journal of Theoretical Biology* 25(1): 1–48.

Wolpert, L., and J. H. Lewis. (1975). Towards a theory of development. *Federation Proceedings* 34(1): 14–20.

Wright, S. (1920). The relative importance of heredity and environment in determining the piebald pattern of guinea-pigs. *Proceedings of the National Academy of Sciences* 6: 320–332.

22 Distributed Agency within Intersecting Ecological, Social, and Scientific Processes

Peter Taylor

Societies emerge as changing alignments of social groups, segments, and classes, without either fixed boundaries or stable internal constitutions.... Therefore, instead of assuming transgenerational continuity, institutional stability, and normative consensus, we must treat these as problematic. We need to understand such characteristics historically, to note the conditions for their emergence, maintenance and abrogation.
—Wolf 1982: 387

The anthropologist, Eric Wolf, proposes a conceptual inversion. Whenever theory has built on the dynamic unity and coherency of structures or units—in Wolf's case, societies or cultures—consider, instead, what would follow if those units were to be explained as contingent outcomes of "intersecting processes."[1] This broad "Wolfian" heuristic informs this essay's extensions of Developmental Systems Theory (DST) to cases in the sociology of mental illness, social-environmental studies, and social studies of science.[2] I link the three cases in a project of reconceptualizing human agents, in particular agents who are establishing knowledge and engaging in change. I show that viewing agents in terms of intersecting processes is also equivalent to teasing open their "heterogeneous construction," that is, their contingent and ongoing mobilizing of webs of diverse materials, tools, people, and other resources.

The importance for DST of reconceptualizing agency is indicated by a the section of Susan Oyama's *Ontogeny of Information* on "Subjects and Objects." Oyama describes our primary experience of ourselves as subjects maturing from dependence and passivity to independence and control—what I call "concentrated" agency. We come to experience temporal continuity and casual potency and are able to impart order according to prior knowledge and plan. This experience, however, "exaggerates our role as detached subjects and denies our object-like status" (Oyama 1985: 76). Accordingly, when we try to explain development, interaction, and perception, we tend to posit another subject inside ourselves—mental modules, optimizing or rational actors, or, most notably, genes. Similarly, to explain the order of the world people have traditionally posited a subject outside it, God, or, more recently, "the-forces-of-natural-selection."

In order to develop better explanations of development, interaction, and perception, we need, Oyama implies, metaphors and concepts that do not rely on the dynamic unity and coherency of agents, or on superintending agents within or outside those agents. And, to the extent that such patterns of thought persist because of their resonance with the experience agents have of their relations and actions in the material and social world, we need different experience. Or, better, we need to highlight submerged experience of ourselves as "object-like" or "distributed," that is, as agents dependent on other people and many, diverse resources beyond the boundaries of our physical or mental selves. After all, the primary experience of becoming an autonomous subject is not "raw" experience, let alone uniform and universal experience (Lebra 1984, cited in Kondo 1990: 32), but experience mediated through particular social discourse.

There are circles here to be wrestled with. New concepts and metaphors might emerge if we experienced ourselves differently, but what counts as our primary experience is mediated by prevailing conceptual schemes and shared metaphors. And in current Western social discourse, these highlight our autonomy as subjects. Conversely, when some of us seek to theorize developmental systems or, in my case, to highlight distributed agency, we foresake the facilitation afforded by prevailing concepts and metaphors of concentrated agency. To so distance ourselves from the dominant discourse, however, requires a strong sense of "independence" and "causal potency" in attempting to impose an order—on one's world and on one's audiences.[3] With these tensions acknowledged, but not resolved, let me move to the three cases.

Case I.   The Development of Severe Depression in a Sample of Working-Class Women

A body of research initiated by the British sociologists Brown and Harris in the 1960s has interpreted the social origins of mental illnesses in a way that undercuts the persistent dichotomization of genes versus environment. This "life events and difficulties" research, which is not well known in the United States, allows one to conclude that apportioning behavior to genes or environment is, at least for those seeking to reduce the incidence of mental illness, at best, not very informative or helpful.[4] To see how this follows, let me sketch their explanation of acute depression in working-class women in London (Brown and Harris 1978, 1989). I will also work in the extensions of their findings and generalized narrative contributed by Bowlby, a psychologist who focused on the long term effects of different patterns of attachment of infants and young children to their mothers (Bowlby 1988).

Four factors are identified by Brown and Harris as statistically more common in women with severe depression: a severe, adverse event in the year prior to the onset of depression; the lack of a supportive partner; persistently difficult living conditions; and the loss of, or prolonged separation from, the mother when the woman was a child (under the age of eleven). Bowlby interprets this last factor in terms of his and others' observations of secure versus anxious attachment of young children to caregivers. In a situation of secure attachment the caregiver, usually the mother, is, in the child's early years, "readily available, sensitive to her child's signals, and lovingly responsive when [the child] seeks protection and/or comfort and/or assistance" (Bowlby 1988: 167). The child more boldly explores the world, confident that support when needed will be available from others. Anxious attachment, on the other hand, corresponds to inconsistency in, or lack of, supportive responses. The child is anxious in its explorations of the world, which can, in turn, evoke erratic responses from caregivers,

and the subsequent attempt by the child to get by without the support of others.

The top three strands of figure 22.1 (class, family, psychology) combine the observations above to explain the onset of serious depression.[5] The factors are not separate contributing causes, like spokes on a wheel, but take their place in the multistranded life course of the individual. Each line should be interpreted as one contributing causal link in the construction of the behavior. The lines are dashed, however, to moderate any determinism implied in presenting a smoothed out or averaged schema; the links, while common, do not apply to all women at all times, and are contingent on background conditions not shown in the diagram. For example, in a society in which women are expected to be the primary caregivers for children (a background condition), the loss of a mother increases the chances of, or is linked to, the child's lacking consistent, reliable support for at least some period. Given the dominance of men over women and the social ideal of a heterosexual nuclear family, an adolescent girl in a disrupted family or custodial institution would be likely to see a marriage or partnership with a man as a positive alternative, even though early marriages tend to break up more easily. In a society of restricted class mobility, working-class origins tend to lead to working-class adulthood, in which living conditions are more difficult, especially if a woman has children to look after and provide for on her own. In many such ways these family, class, and psychological strands of the woman's life build on each other. Let us also note that, as an unavoidable side effect, the pathways to an individual's depression intersect with and influence other phenomena, such as the state's changing role in providing welfare and custodial institutions, and these other phenomena continue even after the end point, namely, depression, has been arrived at.

Suppose now, quite hypothetically, that certain genes, expressed in the body's chemistry, increase a child's susceptibility to anxiousness in attachment compared to other children, even

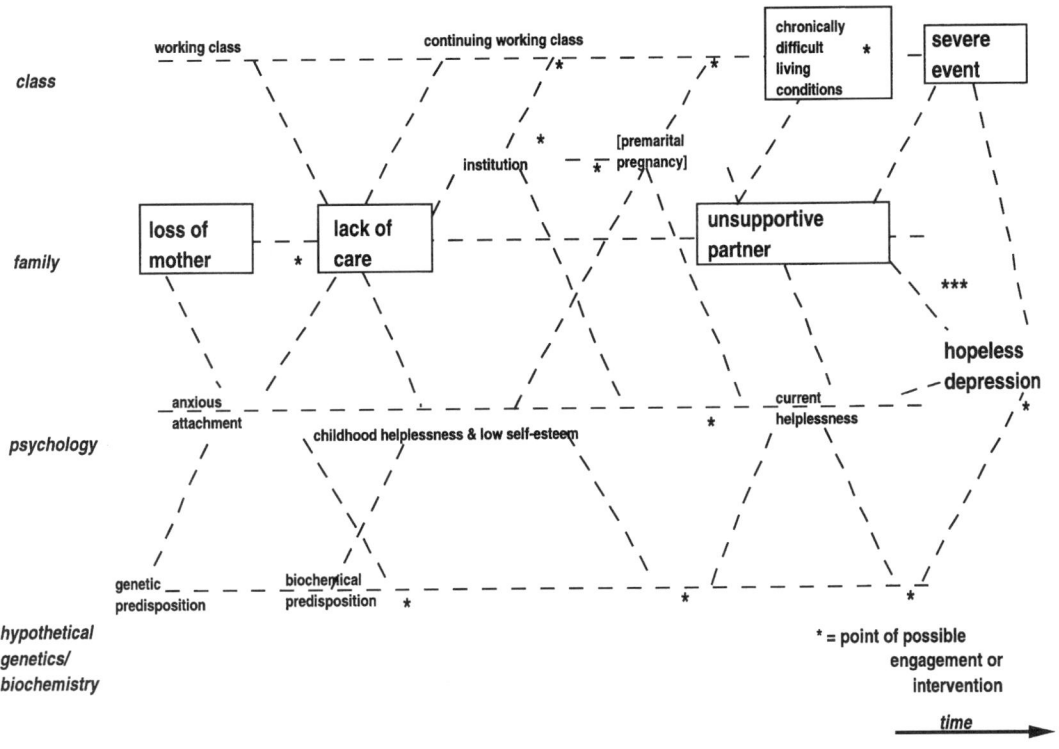

Figure 22.1
Pathways to severe depression in a study of working-class women. The dashed lines indicate that each strand tends to build on what has happened earlier in the different strands. See text for discussion and note 5 for sources.

those within the same family. Suppose also that this inborn biochemistry, or the subsequent biochemical changes corresponding to the anxiety, rendered the child more susceptible to the biochemical shifts that are associated with depression. (This hypothetical situation is given by the bottom strand of figure 22.1.) It is conceivable that early genetic or biochemical diagnosis followed by lifelong treatment with prophylactic antidepressants could reduce the chances of onset of severe depression. This might be true without any other action to ameliorate the effects of loss of mother, working-class living conditions, and so on. There are, however, many other readily conceivable engagements to reduce the chances

of onset of depression, for example, counseling adolescent girls with low self-esteem, quickly acting to ensure a reliable caregiver when a mother dies or is hospitalized, making custodial institutions or foster care arrangements more humane, increasing the availability of contraceptives for adolescents, increasing state support for single mothers, and so on. If the goal is reduction in depression for working-class women, the unchangeability of the hypothetical inherited genes says nothing about the most effective, economical, or otherwise socially desirable engagement— or combinations of engagements—to pursue. Notice also that many of these engagements have their downstream effect on depression via pathways that cross between the different strands. For example, if self-esteem counseling were somewhat effective, then fewer unwanted pregnancies and unsupportive partnerships might be initiated; both effects could, in turn, reduce the incidence of single parenthood and difficult living conditions.

These sequences of multiple causes, building on each other over the individual's life history, permit a number of conclusions about the nature/nurture debate:

1. Neither the unchangeability of genes nor the reliability of some gene- or biochemistry-based intervention, such as the hypothetical prophylactic antidepressants, would prove that the genes are the most significant cause of the acute depression that has been occurring in the absence of such treatment.

2. Critics of genetic explanations could dismiss the attribution of an individual's behavior to genes (or 50 percent or 80 percent to genes) as a technically meaningless partitioning of causes without placing themselves at the other pole from genetic determination.[6] That is, they would not have to make the counterclaim that the environment determines behavior or that, if the right environment were found, any desired behavior could be elicited. The Brown-Harris-Bowlby (BHB) account addresses malleability or immalleability of behavioral outcomes without ruling out genetic contributions.

3. Similarly, critics would not need to rest their case on demonstrations that behavioral genetics has been or still is methodologically flawed (Lewontin, Rose, and Kamin 1984), on textual deconstructions of the categories and rhetoric employed (Lewontin 1979), or on attributions of political bias to the supporters of behavioral geneticists. These are all interesting, but, in light of the BHB account of the behavior, not necessary for a conceptual critique of genetic determinism.

Over and above these conclusions, the BHB account of the origins of acute depression in working-class women also displays the following

features that I associate with the idea that something is "heterogeneously constructed," or an outcome of "intersecting processes." (Most of these have equivalents in DST.)

a. Without any superintending constructor or outcome-directed agent,

b. many heterogeneous components are linked together, which implies that

c. the outcome has multiple contributing causes, and thus

d. there are multiple points of intervention or engagement that could modify the course of development. In short,

e. causality and agency are distributed, not localized. Moreover,

f. construction is a process, that is, the components are linked over time,

g. building on what has already been constructed, so that

h. it is not the components, but the components in linkage that constitute the causes. Points (c) and (f–h) together ensure that

i. it is difficult to partition relative importance or responsibility for an outcome among the different types of cause (e.g., 80% genetic vs. 20% environmental). Generally,

j. there are alternative routes to the same end, and

k. construction is "polypotent" (Sclove 1995), that is, things involved in one construction process are implicated in many others. Engaging in a construction process, even in very focused interventions, will have side effects. Finally, points (f) and (k) mean that

l. construction never stops; completed outcomes are less end points than snapshots taken of ongoing, intersecting processes.

I am aware that there may be objections to the case I have chosen to make the preceding points. In discussing depression among working-class women, rather than in other groups, I could be seen as perpetuating a male, professional-class perspective. However, the politics of the case can be viewed quite differently. Although depressed working-class women are the focus, the intersecting processes account brings a range of other agents into the picture. While the account does not identify ways to cure the women studied, other girls and women that follow them might seek support from, or find themselves supported by—to pick up on the potential engagements mentioned earlier—counselors, hospital social workers, people reforming custodial institutions, family-planning workers, social policy makers, and so on. Moreover, these agents can view their engagement as linked with others, not as a solution on its own. For example, when women's movement activists create women's refuges as a step away from living in unsupportive households, this makes it possible for therapists who specialize in the psychological dynamics of the woman in her family to consider referring women to refuges as a critical disruption to the family's dynamic. The politics of highlighting different kinds of causes and their interlinkages can be seen as promoting such exchange among the distributed set of agents and contributing to the potential reformation of the social worlds intersecting around the development of any given focal individual or outcome.

Case II.   The History of Soil Erosion in a Region of Oaxaca, Mexico

In the mid-1980s resource economist Raúl García-Barrios and his ecologist brother Luis studied severe soil erosion in a mountainous agricultural region near San Andrés in Oaxaca, Mexico, and traced it to the undermining of traditional political authority after the Mexican revolution (García-Barrios and García-Barrios 1990). The soil erosion of the twentieth century is not the first time this has occurred in this region of Oaxaca. After the Spanish conquest, when the indigenous population collapsed from disease, the communities moved down from the high-

lands, abandoning terraced lands, which then eroded. The Indians adopted labor-saving practices from the Spanish, such as cultivating wheat and using plows. As the population recovered during the eighteenth and nineteenth centuries, collective institutions evolved that reestablished and maintained terraces and stabilized the soil dynamics. Erosion was reduced and soil accumulation was perhaps stimulated. This type of landscape transformation also needed continuous and proper maintenance, since it introduced the potential for severe slope instability. The collective institutions revolved around first the church and then, after independence from Spain, the rich Indians, *caciques*, mobilizing peasant labor for key activities. These activities, in addition to maintaining terraces, included sowing corn in work teams, and maintaining a diversity of maize varieties and cultivation techniques. The caciques benefited from what was produced, but were expected to look after the peasants in hard times—a form of moral economy (Scott 1976). Given that the peasants felt security in proportion to the wealth and prestige of their cacique and given the prestige attached directly to each person's role in the collective labor, the labor tended to be very efficient. In addition, peasants were kept indebted to caciques, and could not readily break their unequal relationship. The caciques, moreover, insulated this relationship from change by resisting potential labor-saving technologies and ties to outside markets.

The Mexican revolution, however, ruptured the moral economy and exploitative relationships by taking away the power of the caciques. Many peasants migrated to industrial areas, returning periodically with cash or sending it back, so that rural transactions and prestige became monetarized. With the monetarization and loss of labor, the collective institutions collapsed and terraces began to erode. National food-pricing policies favored urban consumers, which meant that in Oaxaca corn was grown only for subsistence needs. New labor-saving activities, such as goat herding, which contributes in its own way to erosion, were taken up without new local institutions to regulate them.

Although this synopsis of the García-Barrioses' account is brief and, like the first case, smoothed out, it allows me to reiterate and elaborate on the intersecting processes viewpoint in the context of social-environmental studies:

1. *Differentiation among unequal agents.* Sustainable maize production depended on a moral economy of cacique and peasants, and the inequality among these agents resulted from a long process of social and economic differentiation. Similarly, the demise of this agro-ecology involved the unequal power of the State over local caciques, of urban industrialists over rural interests, and of workers who remitted cash to their communities over those who continued agricultural labor.

2. *Heterogeneous components and inseparable processes.* As highlighted in figure 22.2, the situation has involved intersecting processes operating at different spatial and temporal scales, involving elements as diverse as the local climate and geomorphology, social norms, work relations, and national political economic policy. The processes are interlinked in the production of any outcome and in their own ongoing transformation. Each is implicated in the others, even by exclusion (Smith 1984), such as when caciques kept maize production during the nineteenth century insulated from external markets. No one kind of thing, no single strand on its own, could be sufficient to explain the currently eroded hillsides. In this sense, an intersecting processes account contrasts with competing explanations that center on a single dynamic or process, for example, climate change in erosive landscapes; population growth or decline as the motor of social, technical, or environmental change; increasing capitalist exploitation of natural resources; modernization of production methods; or peasant marginalization in a dual economy (Peet and Watts 1996).[7]

3. *Historical contingency of processes.* The role of the Mexican revolution in the collapse of nine-teenth-century agro-ecology reveals the contingency that is characteristic of history. The significance of such contingency rests not on the event of the revolution itself, but on the different processes, each having a history, with which the revolution intersected.

4. *Structuredness.* Although there is no reduction to macro- or structural determination in the above account, the focus is not on local, individual/individual transactions. Regularities, for example, the terraces and the moral economy, persist long enough for agents to recognize or abide by them. That is, structuredness is discernible in the intersecting processes.

5. *Distributed agency.* The agency implied in the account of the García-Barrios brothers was distributed, not centered in one class or place. In the nineteenth-century moral economy caciques exploited peasants, but in a relationship of reciprocal norms and obligations. Moreover, the local moral economy was not autonomous; the national political economy was implicated, by its exclusion, in the actions of the caciques that maintained labor-intensive and self-sufficient production. Although the Mexican revolution initiated the breakdown in the moral economy, the ensuing process involved not only political and economic change from above, but also from below and between—semiproletarian peasants brought their money back to the rural community and reshaped its transactions, institutions, and social psychology.

6. *Intermediate complexity.* The García-Barrios brothers include heterogeneous elements in their account, but, as my synopsis and figure 22.2 indicate, different strands can be teased out. The strands, however, are cross-linked; they are not torn apart. In this sense, the account has an intermediate complexity—neither highly reduced, nor overwhelmingly detailed. By acknowledging complexity, the account steps away from debates centered around simple oppositions, e.g., ecology/geomorphology versus economy/society. Similarly, by placing explanatory focus on the

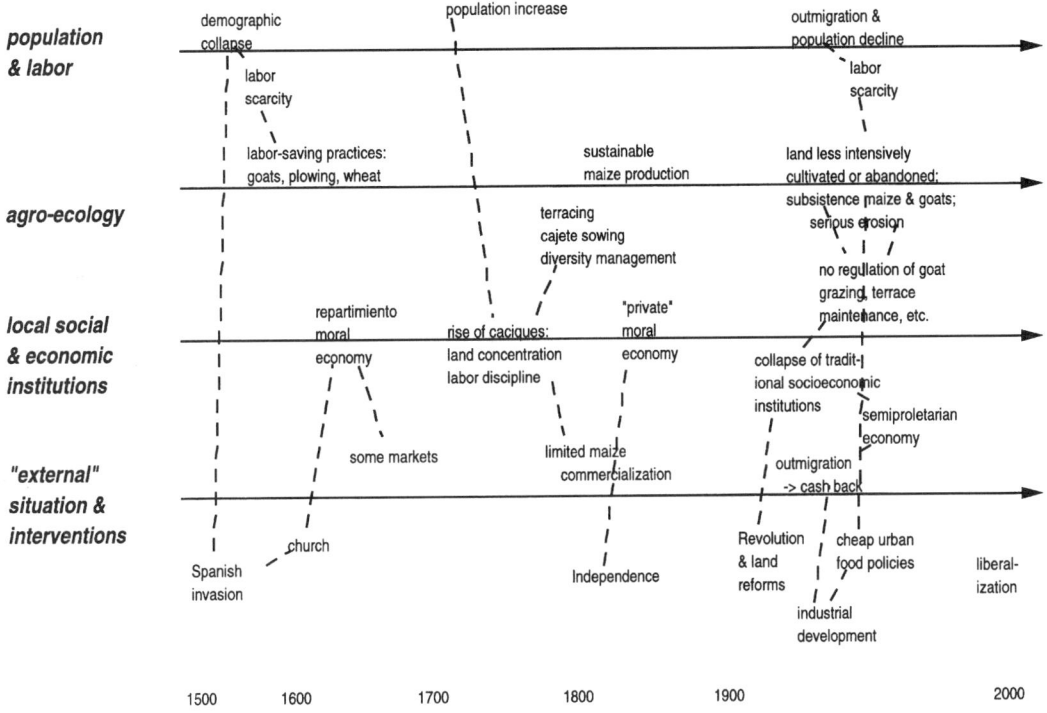

Figure 22.2
Intersecting processes leading to soil erosion in San Andrés, Oaxaca. The dotted lines indicate connections across the different strands of the schema. See text for discussion.

ongoing processes involved in the historically contingent intersections, the account discounts the grand discontinuities and transitions that are often invoked, for exmaple, peasant to capitalist agriculture, or feudalism to industrialism to Fordism to flexible specialization.[8]

7. *Multiple, smaller engagements.* Distributed agency, intermediate complexity, and the other features of intersecting processes have implications, not only for how environmental degradation is conceptualized, but also for how one responds to it in practice. Intersecting processes accounts do not support government or social movement policies based on simple themes, such as economic modernization by market liberalization, or sustainable development through promo-

tion of traditional agricultural practices. They privilege multiple, smaller engagements, *linked together* within the intersecting processes.[9]

This shift in how policy is conceived requires a corresponding shift in scholarly practice. On the level of research organization, intersecting processes accounts highlight the need, in brief, for transdisciplinary work grounded but not localized in particular sites. They do not underwrite the customary, so-called interdisciplinary projects directed by *natural* scientists, nor the economic analyses based on the kinds of statistical data available in published censuses. In all these different ways, representing intersecting processes is inseparably bound up with engaging or intervening[10] in a way that further extends the idea of distributed agency.

Case III.   The Simulated Future of a Salt-Affected Agricultural Region

The "Institute" is an economic and social research organization based in Melbourne, the major city of the southern Australian state of Victoria. The Kerang region, 240 kilometers north of Melbourne, is an agricultural region where farmers irrigate some pasture, which is grazed by beef or dairy cattle and sheep, and irrigate some crops. Soil salinization has been a chronic problem; during the middle 1970s, after some very wet years, the problem was acute. The rise in salinity, following a decline in beef prices, threatened the economic viability of the region. The "Ministry" of the state government overseeing water resource issues commissioned the institute in late 1977 to study the economic future of the region. An agricultural economist from the ministry and the principal investigator from the Institute formulated a project to evaluate different government policies, such as funding regional drainage systems, reallocating water rights, and raising water charges. This evaluation would take into account possible changes in farming practices, such as improvements in irrigation layout, drainage, and water management, and changes in the mix of farm enterprises. The analysis was to be repeated for different macroeconomic scenarios as projected by the institute's national forecasting models.

The central part of the project was the construction of what came to be known as the Kerang Farm Model (KFM). Using an optimization technique called linear programming the KFM would determine for each of four composite representative farms the mix of farming activities that produced the most income. Different factors, such as water allocation, could be changed and the effect on the income and mix of activities ascertained. The division of labor in the project was as follows. The principal investigator,

an econometrician, continued his work on the agricultural component of the institute's forecasting model. The agricultural economist conducted extensive surveys of farm operations for forty farms and acted as liaison with two senior agricultural extension officers in the region who helped screen the production relationships and parameters used in the KFM. I was hired for fifteen months as a statistician and modeler to analyze the farm surveys and to construct and operate the KFM. The ministry maintained oversight of the project through its agricultural economist and through regular meetings with the project team and an advisory committee.

The tangible products of the study included the survey and data analysis incorporated in one report to the ministry, the KFM and economic analysis making up the second report, a technical monograph documenting the KFM, papers presented at two national conferences of agricultural economists, and a public meeting in the Kerang region to explain the results of the study (Ferguson, Smith, and Taylor 1978, 1979; Taylor 1979). Although some refinements were omitted to meet the ministry's deadline, the KFM was sufficiently flexible to allow evaluation of the required range of factors, yet not so complex so as to be unmanageable.

At the public meeting to present the study's findings some local agricultural extension officers raised objections to the study's having endorsed irrigation of pasture over irrigation of crops. This ran contrary to the advice they had been giving to farmers ever since the decline in beef prices. Subsequent reanalysis, incorporating generous increases in crop yields into the KFM's parameters, was completed rapidly. This showed the result favoring pasture irrigation was robust and could be attributed to beef prices having recovered by this time in the late 1970s. The ministry, meanwhile, focused its attention simply on results indicating that water charges were not a primary limiting factor on farm enterprises or viability. These results eclipsed others concerning the larger range of options that the institute had been commissioned to analyze, which suggests that justifying an increase in water charges had been the ministry's primary concern all along.

This last outcome could engender or reinforce cynicism or fatalism about social impact studies commissioned by the authorities. However, if I were able to show the ways in which particular aspects influenced the results, I would be identifying how the research could have been done differently. The possibility of identifying sites for possible modification of similar research informs the analysis to follow.

My entry point for analyzing the project will be around the modeling because that was the part that I, as a participant, observed more closely. I refer to myself in the third person as "the modeler" to express some distance between my position and actions in 1978–79 and my interpretive role today. I do not want to discount my observations and understandings as a participant, but it would be misleading to imply that during the Kerang study I had in mind a later analysis in terms of the sociology of science.

Building and Probing the Kerang Farm Model

Diverse components went into the KFM: data on soil quality, expected crop yields, range of farm sizes, technical assumptions used in the linear program, the status of the different agents in the project, the geographical distance between the institute and the Kerang region, the computer packages available, the terms of reference set by the ministry, and so on. Moreover, many of these components span the different realms of action of the various agents—from the modeler to the farmers—who are implicated in the building of the KFM. I need to put some order into this heterogeneity of components and assess their relative importance. Let me use my observations as the modeler to unpack parts of the processes of model-building here.

Consider a central technical assumption in the KFM. The use of a linear program for economic analysis assumed that farmers operate to maximize one objective: in the KFM, this objective was income. Furthermore, the use of a linear program for policy formation assumed that if the optimal mix of farming activities according to the KFM were different from a farmers' existing mix, the farmer would change accordingly and immediately. Even though the economic future of the region obviously entailed the farmers' participation, the study did not investigate why and how farmers change, how directly and readily they respond to economic signs, or the extent to which any overriding economic rationality governed their actions.

The modeler questioned these assumptions. He expressed interest in techniques that incorporated more than one objective, but the principal investigator could not envisage modeling an alternative objective to income. In any case, software for multiobjective analysis was not available at the computer center used by the Institute. The modeler designed the KFM to allow examination of the course over time of new investments needed, but when the project approached its deadline, this part of the model development was halted. The modeler learned of the existence of a sociological study on the factors influencing Kerang farmers to change their practices. This study had not, however, been released at that time and the principal investigator lent no institutional support to obtaining advance access to it. These and other issues were, he maintained, outside the economic specialization of the Institute and best left for others to deal with.

In affirming the technical assumptions in the KFM in response to the modeler's questioning, the principal investigator drew variously on his senior and permanent position at the institute, the Institute's specialization in quantitative economic research, and the terms of reference and deadlines that the ministry had set. These assumptions, in turn, had several consequences. They eliminated certain issues from investigation, for example, farmer's objectives. They shaped the data that needed to be collected, for example, obviating the need to investigate how farmers change. And they colored the relationships put into the model, for example, the time course of investment became a secondary issue to locating the farming activities that optimized income. As an exercise in the authority of an experienced principal investigator over a young researcher, this was not at all extraordinary. Nevertheless, through such exchanges the principal investigator and the modeler were negotiating the different components of what would count as a representation of reality and a guide to policy formation.

Of course, there were parties other than the principal investigator and the modeler potentially involved in accepting or disputing the KFM. The farmers might have objected to the way their behavior was modeled. The KFM could also have been disputed by economists interested in multiobjective techniques, by sociologists interested in how people act, interact, and change, or by agricultural policymakers interested in having the study's results translated successfully into changes in the state of farming in the region. None of these potential disputes proved significant at the time. The farmers were separated from the formulation and operation of the KFM, and, conversely, the KFM was insulated from the farmers, by several considerations: by location (the modeling was performed in the city); through a chain of personnel (modeler—agricultural economist—senior agricultural extension officers—local agricultural extension officers—farmers); and by levels of abstraction and generalization. No one in the institute, the principal investigator in particular, had training in multiobjective economic analysis or ready access to suitable computer software. There were no sociologists included in the project team or advisory committee. The ministry, through the range of options established in the terms of reference for the study, indicated that change would be initiated by government policy based on economic and engineering criteria. The farmers were, in effect, to be instruments, more than coparticipants, in determining the future of the region. In short, the ministry did not dispute the KFM as a representation of reality; neither were any farmers, economists, or sociologists in a position to do so.

Six Heuristics Drawn from the Reconstruction of the Kerang Study

The description of the building of the KFM, although brief and clearly partial, is sufficient to introduce six propositions concerning the processes of science in the making and interpreting those processes. I begin with the observation that heterogeneous components from a range of realms of social action are being drawn on by the different agents involved in the KFM (proposition 1). Each of the other propositions follows more or less directly from the ones that have preceded it. These propositions are advanced heuristically, without expecting them to apply to all situations.

1. *Science-in-the-making depends on heterogeneous webs, not unitary correspondence.* From the description above, it is clear that diverse components were involved in building the KFM. Moreover, they were interconnected in practice, forming heterogeneous webs. The assumption that farmers were subordinate to economic rationality in the KFM facilitated the formulation of conclusions in the form of government policy options. The power of the government to enact its decisions rendered investigation of how farmers change less relevant, which shaped the data needing to be collected. Generalized agronomic data, rather than sociological insights, would suffice. This, in turn, conditioned the relationships that could appear in the model. Similarly, the modeler's mediated relationship with the modeled situation and his geographical separation from the region rendered it less relevant to model long-term options, such as selective reforestation and organic soil restoration. These possibilities, although potentially of economic and ecological benefit, would have required such things as experimental plots, publicity, education, advocacy, subsided loans for tree planting from the government, and other institutional changes before they could be adopted. With so many contingent fac-

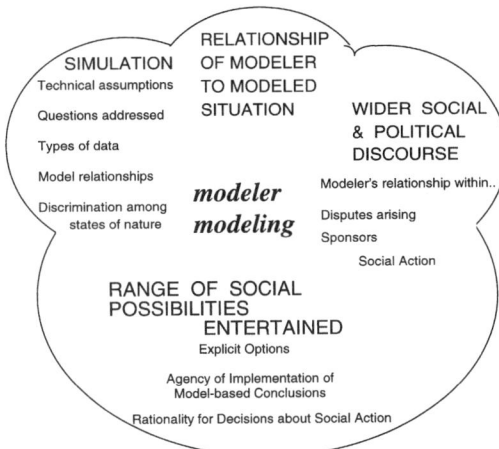

Figure 22.3
Different components of socio-environmental modeling. See text for discussion of their interconnections.

tors it was impossible even to estimate their costs. Omission of such options from the modeling, in turn, helped ensure that such aspects of the future reality would be less realizable, and the model's account more real. Figure 22.3 presents a schematic picture of diverse components interconnected in the making of the KFM.

"Technical" considerations, such as the assumption of income optimization, and "social" considerations, such as the separation of the modeler from the farmers, had implications in practice for each other. "Local" interactions were connected with activities at a distance. For example, the modeler and the principal investigator decided not to pursue sociological inquiry into how farmers change, which meant that the content of and conduct of the survey of farms and farmers could remain unchanged. No one component in the web stood alone in supporting the KFM as a representation of reality; in the actual intersecting processes of building the model, technical components could not be detached from social ones, nor local ones from those that spanned levels.

In this sense I would say that science is *constructed*; science-in-the-making is an ongoing process of building from diverse components, as in building a house from the ground up using concrete, bricks, cement, wood, nails, and so on. This is *social* construction, but not "merely" social construction. Moreover, the associations that social construction has with reflection are not apt here. It might be possible to say that the model *reflected* all the different social components, but it would be stretching the metaphor of reflection. The heterogeneity and interlinkage of the components make it difficult and uninformative to collapse science-in-the-making to a unitary idea of reflection of society in theory, or, similarly, to an issue of correspondence of theory to natural reality. In short, science, I would say, is *heterogeneously constructed* (Taylor 1995a).[11]

2. *Scientists represent-engage.* In the process of building the model, the modeler, principal investigator, and other agents linked together technical and social components in order to make a model that worked for them. These scientific agents tended to make the different components reinforce, not undermine, each other, rendering both the model and the ongoing scientific activity more difficult for others to oppose or modify *in practice* (see proposition 1).[12] This insight goes beyond the observation that representations of natural reality support interventions in different realms of social action, or the claim that repeatable interventions provide the basis for scientific representations (Hacking 1983). Through the model's heterogeneous construction, representations and engagements were being formed simultaneously, and, moreover, jointly. "Interaction" between "technical" and "social" considerations fails to capture this relationship. Let me instead speak of scientists *representing-engaging.*

3. *Scientists are practically imaginative agents.* The idea of representing-engaging implies that scientific agents are mindful both of nature and of the social worlds in which they act, and that they project continuously between these realms. This attention to their social situatedness is not

an accusation that scientists are corrupt, fallible, or lazily taking the path of least resistance. On the contrary, it is an affirmation of the view that all human activity is imaginative, that is, the result of a labor process that has to exist in the laborer's *imagination* before the process commences. Agents assess, not necessarily explicitly, the practical constraints and facilitations of possible actions in advance of their acting (Robinson 1984).[13]

Imagination in the sense I use it here is not at all like fantasy, in which worlds can be envisaged and mentally inhabited so as to escape from the practical difficulties of their realization. Achieving some result in the material world, in contrast to in fantasy, requires human agents to be engaged with materials, tools, and, usually, other people. The KFM modeler had to engage with pasture growth, government sponsorship, an agricultural extension system, and so on. Moreover, materials, tools, and other people confront scientists with their recalcitrance. So scientists project themselves into possible engagements out in the world in order to imagine what will work easily for them and what will not. These constant projected confrontations with the components that personal and collective histories make available lie behind all the actions people take, including scientists' representing-engaging. Through them people build up knowledge—not necessarily consciously articulated—about their changing capabilities for acting in relation to the conditions in which they operate.

4. *The agency of heterogeneously constructing agents is distributed.* If we focus on agents' contingent and ongoing mobilizing of webs of materials, tools, people, and other components, we can think of their psychology or agency as *distributed*, not *concentrated* mentally inside socially autonomous agents. That is, although agents work with mental representations of their worlds, the malleability of those representations should not be understood merely in internal mentalistic terms related to belief or rationality. In-

stead, we should inquire into the heterogeneity of resources that facilitate agents acting as if the world were like their representations of it. During the Kerang study, the principal investigator may well have *believed* deeply that economic decision-making was of primary importance in people's lives. However, he was able to sustain this belief against possible challenges by many *practical measures*, such as not securing access to the sociological study on how farmers change, and concentrating on his econometric investigations rather than developing skills in multiobjective analysis.

5. *Resources are causes.* Up to this point in my description of the construction of the KFM, I have used the neutral term *component* to refer to the diverse things that scientists link into webs to support their theories and ongoing scientific activity. But there are many components linked together through the construction process that have little significance in explaining the development of theories and activity. The modeler, for example, used baking soda to clean his teeth at that time. But let me reserve the term *resource* for components that make a claim or a course of action more difficult for others to modify. Resources make a difference; that is, when resources are deployed they function as *causes*. In this light, the term resource cannot be used descriptively without also implying a claim about causes, and such claims invite analysis (see Taylor 1995a, appendix A).

6. *Counterfactuals are valuable for exposing causes.* With the exception just now of the baking soda, the components of the construction process I have chosen to mention were significant resources in the building of the KFM. Or so my account of the KFM would imply. But how can I support the causal claims that I have thus structured into my account of the KFM? For a start, let me note that, to support the causal claim that something made a difference logically requires an idea of what else could have been if the resource in question had been absent. There are many

sources for ideas about what else could have been. Sociologists and historians of science listen to opposing parties in controversies (Collins 1981)—which include activists in movements for social change (Nelkin 1984)—undertake conceptual analysis or historical and cross-cultural comparisons (Harwood 2000), and give rein to their sociological imagination (Hughes 1971). Analyses of controversies have been popular; they provide the clearest, most concrete evidence of alternatives, because the agents themselves identify the resources they consider important.

There is no logical reason, however, why the resources explicitly exposed during a controversy constitute the full set used by a scientist. There are resources taken for granted and shared by opposing parties and, moreover, resources that must be mobilized even when there is no apparent controversy. In short, ideas of what else could have been should not be limited by whether anyone actually attempted to construct the alternative situation. For all these reasons, explicit use of counterfactuals may be needed in order to analyze a more inclusive array of resources used in the construction of science.

If we look back we can see that, although I began my account of the building of the KFM as a fairly neutral description, once I started to draw connections among the heterogenous components I began introducing counterfactuals. For example, in contrast to a single objective of maximizing income in the modeled farms, I mentioned the counterfactual possibility of multiobjective techniques. In explaining why this was not incorporated in the KFM, I mentioned that the principal investigator's training, his status relative to the modeler, software availability, and the institute's specialization were invoked during the course of the study. These were constraints for anyone wanting to construct a multiobjective model. By identifying them I was implying that the principal investigator's training and so on were resources for constructing a model with a single objective function. In this general fashion,

exploring the practical constraints on realizing *counterfactual* possibilities can, by a logic of inversion, expose the resources facilitating those who constructed what *actually* happened.

The emphasis on multiple, heterogeneous resources means that the relevant counterfactuals are multiple and particular. We could formulate an all-encompassing counterfactual, in which, for example, the Kerang study is replaced by a project that could not be used for top-down government policymaking. However, once we began to consider the practical implications of such a counterfactual, we would be challenged to identify specific sites for possible modification of the research. This would be all the more the case if we focused on the practical implications for the specific scientific agents involved. The modeler's ability to produce results based on sociologically realistic processes of change was constrained, as I observed earlier, by his distance from the farmers' realm of social action—distance given not only by location, but also by the chain of mediating personnel and degree of abstraction. The geographical and organizational distance was, in turn, related to the centralized character of government and intellectual activities in the one major city of each Australian state, something given by the previous two hundred years of development. Toward the end of the project the modeler considered a move counter to that centralization, namely, to live and work in the Kerang region as an agricultural consultant. He was aware that this would raise practical issues such as purchase and maintenance of a car, long-distance access to computer facilities and libraries, keeping abreast of discussions about the wider state of the rural economy, and other considerations of a more personal nature. The modeler's decision not to move meant the representation of the Kerang region he was able to produce facilitated the making of policy based on simple economic grounds. This outcome did not flow from a political or intellectual commitment to the economically based technocratic rationality; many practical, not only intellectual or ideological, considerations would have been entailed in producing a different result.

Conclusion: Persisting Tensions between Concentrated and Distributed Agency

For the three cases in this chapter I have provided overviews, in which the complexities have been smoothed or "disciplined." The emphasis, however, on heterogeneous resources and on intersections of processes at different scales and types highlights the range of agents whose different engagements jointly might contribute to modifying the focal outcomes and the social worlds intersecting around the development of those outcomes. As a result, the overviews do not privilege interventions from a superintending or master position.

I find the politics of distributed agency congenial, but recognize that a central question has been left open—what would lead any agent to engage so as to change the intersecting processes? Actually, no one can simply continue to mobilize the same resources as previously, because the contingent intersection of different processes ensures ongoing change and restructuring. So the open question becomes what would lead any agent to try self-consciously to steer the restructuring in certain directions over others? One kind of answer would return us to concentrated agency, in that we could point to the agent's goal, such as preventing illness, soil erosion, and the production of models whose results can be manipulated by policy-makers, or convincing readers of the virtues of an intersecting processes framework. A different kind of answer, more consistent with the spirit of this chapter, would stem from investigating the intersecting processes that have formed the particular agents in question (analogous to the first case, but without the focal outcome of a mental *illness*). Indeed, the Wolfian heuristic would have us subsume the first kind of answer into the second. That is, "goals" become discursive shorthand for the particular

intersecting processes of different agents. At the same time, to the extent that agents need to explain their actions to others—and to themselves —when they attempt to mobilize different resources or organize them in new directions, such discursive themes may be valuable resources.

The image that emerges is one in which agents are always "vibrating" among their experiences of concentrated and distributed agency. The challenge becomes to acknowledge the discursive impact of simple themes, but to strengthen the vibrations in the direction of agents attending to their dependency on other people and many, diverse resources beyond the boundaries of their physical or mental selves (Taylor 1999a).[14] The intersecting processes/heterogeneous construction framework introduced in this chapter clearly highlights distributed agency, but it would be better for my case if I could move beyond text and argument, to lead my audience into positive experiences of their distributed agency in establishing knowledge and engaging in change (Taylor 1990). To this end, the workshop processes developed by the Institute of Cultural Affairs (ICA) have become my model.[15] I could try to evoke the experience of ICA processes, but I am going to leave it for interested readers to gain this experience first hand.

Coda: Evolution in a Context of Unruly Ecological Complexity

Given this open-ended conclusion, my *argument* for pursuing and promoting the experience of our distributed agency cannot on its own be expected to move readers to change their work and ideas. Moreover, I have not explicitly bridged the gap between the immediate focus of DST on development and evolution and the areas addressed in this essay. I offer this coda, therefore, on theorizing ecological complexity to nudge DST theorists in the direction of exploring the kinds of intersecting processes I have highlighted. Indeed, my own interest in intersecting processes grew from a search for ways to theorize the complexity of eco-logical dynamics. Along the way I also observed that the structure and dynamics of this ecological context have not been well integrated into de-velopmental and evolutionary theory.[16] The challenge of doing so needs eventually to be ad-dressed—after all, *all* development and evolution occurs within a dynamic ecological context (Taylor 2001a).

During the last decade or so a reassertion of historical contingency, nonequilibrium formula-tions, local context and individual detail has subdued the ambitions many ecological theorists had in the 1960s and '70s for identifying general principles (Taylor 1992; Kingsland 1995). Ecolo-gists have become increasingly aware that situa-tions may vary according to historical trajectories that have led to them; that particularities of place and connections among places matter; that time and place is a matter of scale that differs among species; that variation among individuals can qualitatively alter the ecological process; that this variation is a result of ongoing differentiation oc-curring within populations (which are specifically located and inter-connected); and that interac-tions among the species under study can be arti-facts of the indirect effects of other "hidden" species.

In patch dynamic studies, for example, the scale and frequency of disturbances that create open "patches" is now emphasized as much as species interactions in the periods between distur-bances (Pickett and White 1985). Studies of suc-cession and of the immigration and extinction dynamics for habitat patches pay attention to the particulars of species dispersal and the habitat being colonized, and how these determine suc-cessful colonization for different species (Gray, Crawley, and Edwards 1987). On a larger scale, such a shift in focus is supported by biogeo-graphic comparisons which show that conti-nental floras and faunas are not necessarily in equilibrium with the extant environmental condi-tions (Haila and Järvinen 1990). From a different

angle, models that distinguish among individual organisms (in their characteristics and spatial location) have been shown to generate certain observed ecological patterns, such as patterns of change in size distribution of individuals in a population over time, where large scale, aggre-gated models have not (Huston, DeAngelis and Post 1988, Lomnicki 1988). And, the effects me-diated through the populations not immediately in focus, or, more generally, through "hidden variables," upset the methodology of observing the direct interactions among populations and confound many principles derived on that basis (Strauss 1991; Taylor 2001b).

To incorporate this new, or perhaps resurgent, emphasis, it has been suggested that ecology be conceived as an "historical" science (Ricklefs and Schluter 1993). Like the fields of epidemiology, psychoanalysis, structural geology, paleontology, and history proper, ecology faces the challenge of historical explanation: how to assemble a com-posite of past conditions sufficient for the subse-quent outcomes to have followed, while, at the same time, not obscuring the provisional quality such accounts have, their being subject to compe-tition from other plausibly sufficient accounts (Taylor 1987). The phrase "a composite of past conditions" could conjure up pure historical con-tingency, but I do not mean this. Like the ac-counts of intersecting processes in this essay, his-toricity in ecological thought should preserve a place for regularities or structuredness of ecolog-ical patterns and processes. To say that ecological structure has a history is to say that it changes in structure and is subject to contingent events, while at the same time it constrains and facili-tates the living activity that constitute any eco-logical phenomenon in its particular place. The challenge facing ecology then is to theorize par-ticularity and contingency intersecting with struc-ture, and of that structure changing in structure, being internally differentiated and, because of overlapping scales of different species' activities, having problematic boundaries—in short, to discipline, without suppressing this "unruly complexity" of ecological processes (Taylor 1992, 2001a, 2001b).

It is within such unruly ecological complexity that organisms, for almost four billion years, have constructed their living and "evolved," that is, given rise to descendants that differ from them. It ought not be assumed that the ecological context remains consistent, that is stable or repeatable, with respect to evolution occurring in populations of individuals or, as in DST, of life cycles of organisms and resources (Gray 2001). Although consistency of context may sometimes be the case, the relevant processes are not necessarily separable into "ecological" and "evolutionary" time scales (Taylor 2001a).[17] The challenge then for DST theorists is—as researchers conditioned by intersecting scientific and social processes—to make sense of the appearance of organisms as intersections of ecological, developmental, and evolutionary processes.

Acknowledgments

I acknowledge valuable comments from Susan Oyama and Russell Gray, which helped me link my thinking to DST.

Notes

1. Two comments on terminology: First, in other publications I use the term *system* or *strong system* to denote structures or units assumed to have dynamic unity and structure (e.g., Taylor 1988, 1992, 1998, 2001b), but, given the distinctions and arguments Oyama makes in this volume about the positive connotations of system in DST, I have chosen not to cast the term in a critical light in this essay. Second, I use the term *process* in the sense of sequences of events that persist or are repeated sufficiently long for us to notice them and need to explain them. This contrasts with an essentialist sense of process as a basic underlying causal structure that allows people to explain events as instances of the process or as noisy deviations from it. "Maturation," "modernization," and "population growth" are examples of the latter sense of process.

2. Portions of this essay are adapted from other publications, with acknowledgment of the respective publishers. Parts of Case I appeared in Taylor (1995a; University of Chicago Press); Case II, in Taylor (1999; Oxford University Press); Case III, in chapter 4A of Taylor (2001b; University of Chicago Press); and the coda, in Taylor (1997a; Taylor and Francis). Although I have developed these cases without explicit reference to DST, I have benefited from conversations with this volume's editors since the middle 1980s. I welcome this opportunity to let readers consider the convergences between DST and ideas arrived at along some different paths.

3. Indeed, when we search for new concepts and metaphors, or more generally, use words and text to make arguments and seek to convince others, we privilege three related and persistent "meta-metaphors": "1) metaphors are root, fundamental, underlying things that shape the surface layers; 2) mental things— thoughts, expectations, what we see—shape our actions; and 3) culture or society get into these thoughts (and so we can be taught [or argued into] how to conceive/perceive the world" (Taylor 1997b: 222, note 37). These meta-metaphors discount our experience of thought being constructed in practical activity from diverse resources.

4. Although associations between life events and difficulties are also studied in the United States, conventional quantitative epidemiology still dominates that research (Brown and Harris 1989: x–xi, 3–45). Associations are thought of in statistical terms, that is, as if causality were a matter of adding up separate "effects" (see note 6 below; chapter 15). In contrast, as the text and figure 22.1 indicate, Brown and Harris focus on the development of life histories and the contingencies involved.

5. Figure 22.1 is adapted from Bowlby (1988: 177). His schema is, in turn, adapted from Brown and Harris (1978: 265). The hypothetical genetics/biochemistry strand is my addition. Its significance will become clear in due course.

6. The nonpartitionability of different kinds of biological and social causes, given the interdependence of their effects, is demonstrated well by Lewontin (1974), when he argues that statistical partitioning of effects ("analysis of variance") does not constitute an analysis of causes. Of course, partitioning of biological and social causes does have ideological meaning (Lewontin, Rose, and Kamin 1984).

7. The combination of differentiation, historical contingency, and structuredness distinguishes this intersecting processes view of socio-environmental change from Vayda (1996). Although his approach shares many qualities with mine, he is more particularist and skeptical of theory based on social structures or structured processes.

8. Such discontinuities and transitions often rely on the sense of process that I want to avoid; see note 1.

9. In the Oaxacan case, the changes of past centuries cannot be undone, but a more fine-grained intersecting processes analysis focusing on recent decades would expose a range of potential engagements—from Non-Governmental Organizations promoting conservation of traditional cultivars to efforts to redirect international financial policies so as to support, rather than reduce, rural credit at the local level (de Janvry and García-Barrios 1989).

10. See Case III, heuristic 2.

11. If we think of construction in terms of sequences of diverse, multiple causes, we can reject the terms of the realism/relativism dichotomy persisting in explanations of the course of science in similar ways to the argument against nature vs. nurture from case I. Let us read genes as underlying, "mind-independent" reality and the environment as social influences on science. If the outcome (mental illness, or, by analogy, some aspect of science, e.g., an established theory) is the result of many heterogeneous components linked in a process, in which each step builds on the outcomes of the previous steps, then it is difficult to partition relative importance or responsibility for an outcome among the different contributing causes, or components in linkage. It becomes quite difficult to give meaning to determination by either nature or society, and not very helpful besides. Heterogeneous constructionism would, therefore, lead us to conclusions analogous to the three stated in Case I for the nature/nurture debate:

1. Suppose there are fundamental principles of nature that are difficult or impossible to modify (a tenet of scientific realism). This does not imply that this deep reality predominantly or ultimately governs the actions of scientific agents, in particular, their success in establishing some representation of this reality. Moreover, the reliability of certain science-based interventions in the world (also important to most scientific realists) is interesting and worth explaining, but it does not justify the belief that sound scientific method is the most efficacious route to exposing any unreliable knowledge or

eliminating problems in correspondence between theory and reality.

2. Critics of scientific realism do not need to claim that construction is entirely a matter of social influences, conventions or personal beliefs. Heterogeneous constructionism addresses the malleability or immalleability of scientific knowledge without entailing such relativism.

3. Sociology of science's analyses of methodology, interests, and rhetoric are illuminating, but not strictly necessary in the conceptual critique of scientific realism.

More needs to be said to argue these propositions, but not here (Taylor 1995a, 1995b); after all, analogies are meant to open discussions more than close arguments.

12. The work of Latour (1994) and Pickering (1995) shares with this essay an emphasis on process and scientific practice. Latour explores a Whiteheadean process metaphysics for science studies. Heterogeneous construction might be read as Pickering's "mangle" and "impure dynamics," and the imagination of scientists as his "modeling." In contrast, however, to my goals of explanation and spanning levels of social practice, Pickering develops a position opposed to social construction and causes and he theorizes practice mostly in terms of experimental practice. That is, it is mostly because scientists tinker with tangible objects, whose resistance requires accommodation, that their goals and interests are subject to ongoing revision. Studies of practice in that sense are reviewed by Golinski (1990); see also Pickering (1992).

13. Associating *imagination* and the *labor-process* is Marx's idea. See *Capital*, volume 1, part 3, chapter 7, section 1, reprinted, e.g., in Tucker (1978: 344–345). The convention in social studies of science has been to avoid reference to an agent's psychology for fear of shifting the terms of explanation from the social realm to an unobservable realm of the agent's mind. I find dubious both the equation of social with observable and the empiricist rejection of unobservables, but, in any case, notice that imagination relies on a distributed, not an internal, notion of mind and psychology. Furthermore, psychological or cognitive models of the scientist as social agent are implicit in every explanation of the outcome of scientific activity. For example, Latour (1987) depicts scientists building "networks" in response to the stimulus of others building competing networks, and assumes that scientists seek to accumulate resources, all of which results, if successful, in

"centers of calculation," "obligatory passage points" (Callon 1985), and their becoming macroactors (Callon and Latour 1981). Like the psychology of pigeons in the accounts of behaviorists, the psychology implied is both strong and minimal—the scientists are governed only by this egocentric metric of resource accumulation; they are not assumed to have multiple projects in their lives and work. This, like most other models of psychology and rationality implicit in social studies of science, is quite restrictive, even when rationalized as a methodological tactic to highlight the flexibility of agents' actions and network building.

14. In this project I am inspired by DST-like work on the development of the self in relationship to others (Fogel 1993) and to the "intentional scaffolding" others provide (Hendriks-Jansen 1996).

15. ICA workshops elicit insight from a large range of participants in analyzing a situation and usually lead to plans that no one participant has envisaged beforehand and that the participants are invested in carrying out. This is achieved by a neutral facilitator leading participants through four phases—objective, reflective, interpretive, and decisional—a structure best represented in "focused conversations" (Spencer 1989; Stanfield 1997). For an elaboration of the basic propositions of ICA facilitation and group process, see http://omega.cc.umb.edu/~ptaylor/ICApropositions. html, which is adapted from workshop materials of ICA Canada; see http://www.icacan.ca/.

16. In a sense, ecological dynamics are implicit in any evolutionary theory, but with "genetic (transmission)," "developmental," "ecological" and "evolutionary" time scales theoretically separated (Taylor 2001a). See note 17.

17. Laland and Odling-Smee (see chapter 10), who extend the important emphasis of Lewontin (see chapter 6) on organisms constructing the environments, recognize that the persistence of a constructed environment ("ecological inheritance") conditions subsequent evolution in the constructing species and others. Notice, however, that Laland and Odling-Smee do not otherwise theorize the dynamics of environmental change or explore the significance of those dynamics for the theory of natural selection (Taylor 2001a).

References

Bowlby, J. (1988). *A Secure Base.* New York: Basic Books.

Brown, G. W., and T. Harris. (1978). *Social Origins of Depression.* New York: Free Press.

Brown, G. W., and T. Harris. (1989). *Life Events and Illness.* New York: Guilford Press.

Callon, M. (1985). Some elements of a sociology of translation: Domestication of the scallops and the fishermen of St. Brieuc Bay. In J. Law (Ed.), *Power, Action, Belief: A New Sociology of Knowledge?* pp. 196–233. London: Routledge & Kegan Paul.

Callon, M., and B. Latour. (1981). Unscrewing the big Leviathan: How actors macro-structure reality and how sociologists help them to do so. In K. Knorr-Cetina and A. V. Cicourel (Eds.), *Advances in Social Theory and Methodology: Toward an Integration of Micro- and Macro-Sociologies,* pp. 277–303. Boston: Routledge & Kegan Paul.

de Janvry, A., and R. García-Barrios. (1989). Rural poverty and environmental degradation in Latin America: Causes, effects, and alternative solutions. Paper presented at the International consultation on environment, sustainable development, and the role of small farmers, Institute for Food, Agriculture and Development, Rome.

Ferguson, J., A. Smith, and P. Taylor. (1978). *Economic Aspects of the Use of Water Resources in the Kerang Region.* (Vol. 1). Melbourne: Institute of Applied Economic and Social Research.

Ferguson, J., A. Smith, and P. Taylor. (1979). *Economic Aspects of the Use of Water Resources in the Kerang Region.* (Vol. 2). Melbourne: Institute of Applied Economic and Social Research.

Fogel, A. (1993). *Developing through Relationships: Origins of Communication, Self, and Culture.* New York: Harvester Wheatsheaf.

García-Barrios, R., and L. García-Barrios. (1990). Environmental and technological degradation in peasant agriculture: A consequence of development in Mexico. *World Development* 18(11): 1569–1585.

Golinski, J. (1990). The theory of practice and the practice of theory: Sociological approaches in the history of science. *Isis* 81: 492–505.

Gray, A. J., M. J. Crawley, and P. J. Edwards. (Eds.), (1987). *Colonization, Succession and Stability.* Oxford: Blackwell.

Gray, R. D. (2001). Selfish genes or developmental systems? Evolution without replicators and vehicles. In R. Singh, C. Krimbas, J. Beatty, and D. Paul (Eds.), *Thinking about Evolution: Historical, Philosophical, and*

Political Perspectives, pp. 184–207. Cambridge: Cambridge University Press.

Griffiths, P. E., and R. D. Gray. (1994). Developmental systems and evolutionary explanation. *Journal of Philosophy* 91: 277–304.

Hacking, I. (1983). *Representing and Intervening.* Cambridge: Cambridge University Press.

Haila, Y., and O. Järvinen. (1990). Northern conifer forests and their bird species assemblages. In A. Keast (Ed.), *Biogeography and Ecology of Forest Bird Communities,* pp. 61–85. The Hague: SPB Academic Publishing.

Harwood, J. (2000). National academic cultures: Science and scholarship in interwar Germany and the United States. In C. Charle, J. Schriewer, and P. Wagner (Eds.), *Transnational Intellectual Networks and the Cultural Logics of Nations,* Oxford: Berghahn Books.

Hendriks-Jansen, H. (1996). *Catching Ourselves in the Act.* Cambridge, MA: MIT Press.

Hughes, E. C. (1971). *The Sociological Eye.* Chicago: Aldine Atherton.

Huston, M., D. DeAngelis, and W. Post. (1988). From individuals to ecosystems: A new approach to ecological theory. *Bioscience* 38: 682–691.

Kingsland, S. (1995). Afterword. *Modeling Nature: Episodes in the History of Population Biology.* (2d ed.). Chicago: University of Chicago Press.

Kondo, D. K. (1990). The eye/I. In *Crafting Selves: Power, Gender, and Discourses of Identity in a Japanese Workplace,* pp. 3–48, 309–312, 329–339. Chicago: University of Chicago Press.

Latour, B. (1987). *Science in Action: How to Follow Scientists and Engineers through Society.* Milton Keynes: Open University Press.

Latour, B. (1994). Les objets ont-ils une histoire? Recontre de Pasteur et de Whitehead dans un bain d'acide lactique. In I. Stengers (Ed.), *L'effet Whitehead,* pp. 197–217. Paris: Vrin.

Lebra, T. (1984). *Japanese Women: Constraint and Fulfillment.* Honolulu: University of Hawaii Press.

Lewontin, R. C. (1974). The analysis of variance and the analysis of causes. *American Journal of Human Genetics* 26: 400–411.

Lewontin, R. C. (1979). Sociobiology as an adaptationist program. *Behavioral Science* 24: 5–14.

Lewontin, R. C. (1982). *Human Diversity.* New York: W. H. Freeman.

Lewontin, R. C. (1983). The organism as the subject and object of evolution. *Scientia* 118: 63–82.

Lewontin, R. C. (1985). Adaptation. In R. Levins and R. C. Lewontin (Eds.), *The Dialectical Biologist,* pp. 65–84. Cambridge, MA: Harvard University Press.

Lewontin, R. C., S. Rose, and L. J. Kamin. (1984). *Not in our Genes: Biology, Ideology and Human Nature.* New York: Pantheon.

Lomnicki, A. (1988). *Population Ecology of Individuals.* Princeton, NJ: Princeton University Press.

Nelkin, D. (Ed.) (1984). *Controversy: Politics of Technical Decisions.* (2d ed.). Beverly Hills, CA: Sage.

Oyama, S. (1985). *The Ontogeny of Information: Developmental Systems and Evolution.* Cambridge: Cambridge University Press. (2d rev. ed., Durham, NC: Duke University Press, 2000.)

Oyama, S. (1988). Stasis, development and heredity. In M-W. Ho and S. Fox (Eds.), *Evolutionary Processes and Metaphors,* pp. 255–274. New York: John Wiley & Sons.

Peet, R., and M. Watts. (1996). Liberation ecology: Development, sustainability, and environment in an age of market triumphalism. In R. Peet and M. Watts (Eds.), *Liberation Ecologies: Environment, Development, Social Movements,* pp. 1–45. London: Routledge.

Pickering, A. (Ed.) (1992). *Science as Practice and Culture.* Chicago: University of Chicago Press.

Pickering, A. (1995). *The Mangle of Practice: Time, Agency and Science.* Chicago: University of Chicago Press.

Pickett, S. T. A., and P. S. White. (Eds.) (1985). *The Ecology of Natural Disturbance and Patch Dynamics.* Orlando, FL: Academic Press.

Ricklefs, R., and D. Schluter. (Eds.) (1993). *Species Diversity in Ecological Communities: Historical and Geographical Perspectives.* Chicago: University of Chicago Press.

Robinson, S. (1984). The art of the possible. *Radical Science Journal* 15: 122–148.

Sclove, R. (1995). *Democracy and Technology.* New York: Guilford.

Scott, J. C. (1976). *The Moral Economy of the Peasant: Rebellion and Subsistence in Southeast Asia.* New Haven: Yale University Press.

Smith, C. A. (1984). Local history in global context: Social and economic transitions in Western Guatemala. *Comparative Studies in Society and History* 26(2): 193–228.

Spencer, L. J. (1989). *Winning through Participation.* Dubuque, Iowa: Kendall/Hunt.

Stanfield, B. (Ed.) (1997). *The Art of Focused Conversation.* Toronto: Canadian Institute of Cultural Affairs.

Strauss, S. (1991). Indirect effects in community ecology: Their definition, study and importance. *Trends in Ecology and Evolution* 6(7): 206–210.

Taylor, P. J. (1979). *The Kerang Farm Model.* Melbourne: Institute of Applied Economic and Social Research.

Taylor, P. J. (1987). Historical versus selectionist explanations in evolutionary biology. *Cladistics* 3(1): 1–13.

Taylor, P. J. (1988). Technocratic optimism, H. T. Odum, and the partial transformation of ecological metaphor after World War II. *Journal of the History of Biology* 21: 213–244.

Taylor, P. J. (1990). Mapping ecologists' ecologies of knowledge. *Philosophy of Science Association* 2: 95–109.

Taylor, P. J. (1992). Community. In E. F. Keller and E. Lloyd (Eds.), *Keywords in Evolutionary Biology,* pp. 52–60. Cambridge, MA: Harvard University Press.

Taylor, P. J. (1995a). Building on construction: An exploration of heterogeneous constructionism, using an analogy from psychology and a sketch from socioeconomic modeling. *Perspectives on Science* 3(1): 66–98.

Taylor, P. J. (1995b). Co-construction and process: a response to Sismondo's classification of constructivisms. *Social Studies of Science* 25: 348–359.

Taylor, P. J. (1997a). Appearances nonwithstanding, we are all doing something like political ecology. *Social Epistemology* 11(1): 111–127.

Taylor, P. J. (1997b). Afterword: Shifting positions for knowing and intervening in the cultural politics of the life sciences. In P. J. Taylor, S. E. Halfon, and P. N. Edwards (Eds.), *Changing Life: Genomes, Ecologies, Bodies, Commodities,* pp. 202–224. Minneapolis: University of Minnesota Press.

Taylor, P. J. (1998). How does the commons become tragic? Simple models as complex socio-political constructions. *Science as Culture* 7(4): 449–464.

Taylor, P. J. (1999a). What can agents do? Engaging with complexities of the post-Hardin commons. In L. Freese (Ed.), *Advances in Human Ecology* vol. 8, pp. 125–156. Greenwich, CT: JAI Press.

Taylor, P. J. (1999b). Mapping the complexity of social-natural processes: Cases from Mexico and Africa. In F. Fischer and M. Hajer (Eds.), *Living with Nature: Environmental Discourse as Cultural Critique,* pp. 121–134. Oxford: Oxford University Press.

Taylor, P. J. (2001a). From natural selection to natural construction to disciplining unruly complexity: The challenge of integrating ecological dynamics into evolutionary theory. In R. Singh, K. Krimbas, J. Beatty, and D. Paul (Eds.), *Thinking about Evolution: Historical, Philosophical and Political Perspectives.* Cambridge: Cambridge University Press.

Taylor, P. J. (2001b). *The Limits of Ecology and the Re/Construction of Unruly Complexity.* Chicago: University of Chicago Press.

Taylor, P. J., and R. García-Barrios. (1995). The social analysis of ecological change: From systems to intersecting processes. *Social Science Information* 34(1): 5–30.

Tucker, R. C. (Ed.) (1978). *The Marx-Engels Reader.* (2d ed.). New York: W. W. Norton.

Vayda, A. P. (1996). *Methods and Explanations in the Study of Human Actions and Their Environmental Effects.* Jakarta: Center for International Forestry Research.

Wolf, E. (1982). *Europe and the People without History.* Berkeley: University of California Press.

23 Niche Construction, Developmental Systems, and the Extended Replicator

Kim Sterelny

Lewontin's World

In the world of evolutionary biology, Richard Lewontin, perhaps more than anyone else, has emphasized the complexity, the subtlety, and the variability of the relationship between genotype and phenotype (Lewontin 1974, 1991). But he is also famous for rejecting a certain conception of adaptation and adaptationism. Organisms, he maintains, do not just accommodate themselves to their environment. Organisms are active. They *select* their environment. For example, when an insect selects a crack in bark in which to shelter, it chooses to experience a range of temperature and humidity. These conditions might be quite different from the local norm. But they also transform their environment, affecting both themselves and the biological communities of which they are a part (Lewontin 1983, 1982, 1985). Just as environments alter organisms, organisms alter their environments. Thus Jones et al. (1997) write:

What does a tree do in a forest? Of course the living and dead tissues are eaten ... and the tree competes with other plants.... But the tree does much more than provide food and directly compete for resources. The branch, bark, root, and living and dead leaf surfaces make shelter, resting locations and living space. Small ponds full of living organisms form where throughfall gets channelled into crotches ... and the soil cavities that form as roots grow provide animals with places to live and cache food.... The leaves and branches cast shade, reduce the impact of rain and wind, moderate temperature extremes, and increase humidity for organisms in the understory and soil.... Root growth aerates the soil, alters its texture and affects the infiltration of water.... Dead leaves fall to the forest floor altering raindrop impact, drainage, and heat and gas exchange in the soil habitat, and make barriers or protection for seeds, seedlings, animals and microbes.... The roots can bind around rocks, stabilizing the substrate. (p. 1946)

Some of these impacts are mere effects; they are by-products of the organism's way of life. But

sometimes we should see the impact of organism on environment as the organism's *engineering* its own environment: The environment is altered in ways that are adaptive for the engineering organism. Thus some trees affect their local soil in ways that improve their own prospects. Pine needles, for example, tend to suppress the growth of other plants that might compete for nutrients. But organisms do not just engineer their own environments, thus in part constructing their own niches. They engineer their descendants' environments. Downstream engineering establishes a causal nexus between the generations additional to that of genetic transmission. Thus we can see Lewontin's critique of externalist adaptationism as adding to his rejection of gene-centered biology. For environmental engineering creates what Laland, Odling-Smee, and Feldman call "ecological inheritance" (Odling-Smee et al. 1996; see also chapter 10). So, for example:

1. Many animals make burrows, nests, shelters (like the beaver's dams), and other structures. These structures modify the impact of the environment on both engineers and their offspring. Thus beaver lodges provide shelter against the extremes of climate and protection against predators. Beaver lodges, rabbit warrens, the tunnel systems of naked mole rats, and some ant and termite nests are utilized and reengineered by those offspring as they become adults.

2. In selecting their microhabitat, organisms filter the way the world impinges on them. But they also often filter the way the world impinges on their offspring, for at least part of the life cycle of those offspring.

3. Organisms do not just engineer the physical world. They engineer *other organisms*, for themselves and their descendants. Parasites and parasitoids often engineer the behavior or morphology of their host, often very gruesomely. These effects on the host typically buffer or transform their offspring's environment. Thus many wasps

paralyze their prey without causing immediate death, leaving the host incapable of defense and preventing decay before the host is converted into wasp. Less gruesomely, the same is true of mutualisms. Leaf cutter ants are equipped with adaptive specializations of many kinds in the care and transmission of their fungal symbiont. There are many other examples of obligate associations of this kind. David Wilson discusses the mutual adaptation of the members of phoretic associations with one another. A phoretic association is a group of organisms with intrinsically poor powers of dispersal that rely on a winged host for transport to the ephemeral and widely scattered resources they need. Because their own reproduction relies on the continued success of their transportation system, Wilson notes, shared evolutionary interests can cause "some phoretic associations to evolve into functionally organised teams that both feed and protect their insect carrier" (Wilson 1997: 2021).

4. Sometimes plants change their environment but do not engineer it. They make it less rather than more suitable for themselves and their seeds, and hence promote a successional cycle. But we should not suppose that this is the typical impact of plants on their immediate environment. For trees and other plants do not just change their environment, they adapt it. They change it in ways that promote their own survival, growth, or reproduction. Perhaps the most spectacular examples of such engineering is the way fire-resistant trees like eucalypts engineer their environment for fire: their distinctive menu of litter makes their environment more fire-prone.

5. Animals sometimes engineer their epistemic environment. They do this for themselves when they probe their environments, manipulating it to access the information they need. For example, small fish like guppies sometimes deliberately inspect potential predators. Predators also inspect prey. Pursuit predators like African wild dogs are thought to probe the behavior of herds of potential prey, attempting to induce in them behaviors that reveal the weakest and most vul-

nerable among them. Engineering the epistemic environment of offspring is probably much more common. There is very little evidence of explicit teaching amongst animals (Caro and Hauser 1992), though female cats of various species provide opportunities for their young to learn to hunt by providing them with particularly easy prey. But there are many ways animals can engineer the epistemic environment of their young that do not depend on explicit teaching. What an animal does, where it goes, and what it eats will provide information of itself—will provide learning opportunities—especially if young animals are predisposed to pay special attention to the behavior of its parents. And, of course, many mammals monitor quite closely the exploratory activities of their young, making trial-and-error learning much safer that it would otherwise be.

This phenomenon is now, I think, widely recognized. But its analysis and explanation remain highly contested. This chapter proceeds as follows. In the next section I distinguish three takes on environmental engineering. One downplays the cross-generation significance of environmental engineering, not treating it as an inheritance mechanism. Another focuses only on inheritance, treating downstream ecological engineering as an aspect of cultural inheritance. Developmental systems theory fuses the two, treating environmental engineering both as part of the expanded view of inheritance, but also treating an expanded view of inheritance as a reason for focusing on organism/environment interactions.

A core feature of developmental system theory's expanded view of inheritance is the "parity thesis." Genes play an important role in development and evolution, but they do not play a unique role. The parity thesis is central to the third and fourth sections, where I develop a case for a substantially modified and qualified version of the thesis. The usual way of rejecting parity is to argue that genes play an informational role in development, and other developmental resources do not. That is not my strategy. Instead,

I argue that an inheritance system must have a particular suite of features if life is to be evolvable. Genetic inheritance does have these features, but much of what is lumped into ecological inheritance—though not all—does not. So although downstream environmental engineering is important, for the most part it should not be treated as an inheritance system. So parity is rejected, or accepted in only a qualified way.

Inheritance and Ecological Engineering

The Extended Phenotype

Dawkins (1982) reorganizes Lewontin's insight. He argues that the effects of environmental engineering should be seen as part of the phenotype of the engineering organism, for these are the effects through which the responsible replicators have been selected. I have argued elsewhere for the evolutionary significance of this conception. The single best reason for taking the gene/vehicle conception of evolution to be a genuine alternative to the standard genotype/phenotype conception is the fact that the systematic, evolutionarily significant phenotypic effects of genes are not confined to the body of the organism carrying them (Sterelny et al. 1996; Sterelny and Griffiths 1999, chap.3). For many extended phenotype effects—the construction of shelters, the effects of parasites and symbionts on their associates, the impact of plants on soil structure—are as developmentally stable as standard features of an organism's phenotype. Moreover, as Laland, Odling-Smee, (and Feldman chapter 10) note, extended phenotypic traits can lead to evolutionary cascades. The evolutionary invention of dam building will trigger other changes in beaver phenotypes, for life in the dam will modify the effects of natural selection on many beaver traits. But this conception can be at most part of the analysis of the phenomena illustrated by 1–5 earlier. For although it captures the way biological entities transform their own environments, it does not capture the cross-generational significance of

environmental engineering. It says nothing about ecological inheritance.

Developmental Systems Theory and Niche-Constructing Phenotypes

Developmental systems theorists wholeheartedly embrace Lewontin's world. As I understand it, developmental systems theory is characterized by three critical theses and a positive proposal. First, the critical theses: (1) We cannot simply assume that the organism/environment boundary is of theoretical significance for developmental and evolutionary biology. Here there is a point of intellectual contact with the extended phenotype perspective. (2) It may be legitimate to foreground genetic structure and genetic change for specific explanatory or predictive purposes. But in general, the genes an organism carries are just one set of developmental resources among many. Genes and gene changes are important both to development and to evolution, but they are not of primary or privileged importance. (3) Developmental systems theorists are skeptical about the project of explaining intergenerational similarity by appealing to the transmission of phenotype-making information across generations. In part, it is this scepticism about the idea of a preexisting program that endogenously sequences development that underpins their deflationary view of the evolutionary significance of genes.

The positive program of developmental systems theory is that the fundamental unit of evolution is the life cycle. In turn, the life cycle is the set of developmental resources that are packaged together and interact in such a way that the cycle is reconstructed. The most obvious life cycle is that of the organism plus its immediate environment, but developmental systems theorists are open to the idea that cycles will exist at both finer and coarser grains. For example, Griffiths and Gray (1997) argue that Keller and Ross's work on pheromone-mediated transmission of fire ant queen morphology is not just a good example of nongenetic inheritance; it is also a case where we should think of the life cycle as a nest life cycle, not an organism life cycle (Keller and Ross 1993). So in this respect, their view does depart from that of Lewontin. For he does seem to have an organism-centered view of evolution.

Laland, Odling-Smee, and Feldman have developed a view of the Lewontin phenomena that seems to me to be a cut-down or conservative version of developmental systems theory (see Odling-Smee 1994; Odling-Smee et al. 1996; Laland, Odling-Smee et al. 2000). In these papers, they develop two key ideas, one about evolutionary feedback loops, the second and stronger about inheritance. The first idea is that ecological inheritance causes evolutionary cascades. So in constructing their own niches, organisms do not just engineer their own environments and directly influence the environments of their own offspring. Ecological inheritance—the constructed niche—changes the way selection acts on genetic inheritance. There are feedback links between ecological and genetic inheritance. In chapter 10, Laland, Odling-Smee, and Feldman focus on this aspect of their model; that is, on the idea that ecological inheritance feeds back onto genetic inheritance. So, for example, they say, "We argue that through niche construction organisms not only shape the nature of their world but also in part determine the selection pressures to which they and their descendants are exposed" (p. 117). The impact of organisms on their environment is a product of evolution, but it is also a cause of evolution, acting as a lens through which selection effects further changes in the lineage.

But they also view niche construction as a bona fide inheritance system—as an independent channel through which intergenerational similarity is maintained. Thus, they write, "the effects of niche construction are frequently nontrivial, directional, accumulatory, pervasive and likely to change the nature of the evolutionary process" (p. 118) and "the problem is that ancestral organisms not only bequeath genes to their descen-

dants, relative to their selected environments in the standard way but also bequeath a legacy of modified natural selection pressures, as an ecological inheritance, relative to those same genes" (p. 125). Putting these ideas together, they argue:

When phenotypes niche-construct, they can no longer be thought of as simply "vehicles for genes," since they are now responsible for modifying some of the sources of selection in their environments that may subsequently feed back to select their genes. Moreover, there is no requirement for niche construction to result directly from genetic variation before it can influence the selection of genetic variation. (Odling-Smee et al. 1996: 645–646)

They think that this is most obviously true of human culture, for we are ecological engineers par excellence. But it is true of organisms in general. The result is a conception of evolution strikingly similar to that of developmental systems theory. Thus they claim:

The entities that are selected, and between which there are fitness differences, are not well-described as "vehicles" or even "interactors" . . . but rather are "organism-environment systems." (Laland et al. 2000: p. 143)

However, though their ultimate conception of evolution converges on developmental systems theory, in some respects their view is more conservative. They do not reject the information-flow conception of cross-generation similarity, and hence genetic inheritance remains central to their conception. Moreover, they have some tendency to focus on just two streams of cross-generation influence: genetic inheritance and cultural inheritance, with genetic inheritance being in some way primary, but modified by a secondary system. Other factors are important because they modify, and are modified by, the genetic inheritance system. I think this picture is particularly evident in Odling-Smee (1994). There he treats genetic inheritance as a flow of information, and other mechanisms as the provision of resources (p. 178). But it emerges in many other places in their argument; for example, in their discussion of Boyd and Richerson's model of cultural selection through individual imitation.

Thus the niche-construction view seems to come to something like this. Fundamentally, the history of evolution is the history of the evolution of organism/environment constructions. Nonetheless, we can usefully treat organisms as the units of selection (*qua* interactors), but they pass similarity across the generations by both genetic and culturally transmitted information. So though their ultimate conception of evolution seems almost the same as that of developmental systems theory, they think that dual inheritance models are useful simplifications, whereas defenders of developmental systems theory do not (Gray 1992; Griffiths and Gray 1994).

Nonetheless, developmental systems theory and the niche-construction conception are the only two approaches that identify an intrinsic connection between environmental engineering and a critical role for nongenetic inheritance. Thus, though the extended phenotype conception identifies an important role for ecological engineering, it was developed to sustain the theoretical primacy of genetic inheritance. Conversely, the expanded inheritance models that I shall shortly consider are *ex officio* silent on the significance of environmental engineering, though they are certainly compatible with recognizing its significance. Only developmental systems theory and its conservative cousin see these as two sides of the same coin.

Expanded Inheritance

The theory of the extended phenotype recognizes the pervasive ways in which genes and genotypes engineer their own environments. But though consistent with extragenetic inheritance, this view is silent on it. Dual inheritance theories are the inverse of the extended phenotype model. Dual inheritance theories explain intergenerational similarity by an appeal to a flow of information across the generations. But they recognize the existence of a cultural as well as a genetic channel.

These theories come in two versions. One version defines cultural inheritance rather narrowly. In our lineage, the flow of the skills of individuals and of groups to the next generation through imitation and explicit teaching constitute a cultural inheritance system. On this narrow view, there is rather little—perhaps no—cultural inheritance outside the human/great ape clade. Cultural inheritance is a key part of the explanation of human evolution. But it is not a phenomenon of much general evolutionary significance. Genetic inheritance, on this view, is the core inheritance mechanism, supplemented (and perhaps even subverted) by meme-based inheritance in a few clades. This conception seems to fit not just the models of Boyd and Richerson (1985), but also the ideas of Dawkins (1989) and Dennett (1995). But it will miss most of the examples discussed in 1–5 earlier.

The alternative is to try to shoehorn a very heterogeneous range of examples into the category of cultural inheritance, treating, for example, the fidelity of seabirds to the nesting sites, the fidelity of phytophagous insects to the plant types on which they hatched, or the pheromone-mediated transmission of queen morphology in ants all as instances of cultural transmission. There is nothing catastrophically wrong with treating cultural transmission as a broad category. But it does have two disadvantages. First, 1–5 seem to form a very diverse set of examples, and it may be a mistake to choose a terminology that conflates their differences. Indeed, that will be the burden of the fourth section of this chapter. Second, the choice of this paradigm can make an informational conception of inheritance seem less problematic than it is. When a young hominid is explicitly taught the technique of knapping a handaxe, there clearly is a flow of information across generations. If teaching is not the transmission of information, nothing is. Other cases are far more problematic. The symbiotic fungus on which leafcutters depend is transmitted from one nest to the next, and through exquisite adaptation of the winged queens. But it is much less obvious that this is transmission of information across the generations.

These points lead us to a less grudging recognition of the variety of inheritance mechanisms. On the "extended replicator" conception, the replicators are that set of developmental resources that are adapted from the transmission of similarity across the generations. So it is a multiple inheritance model, but it takes the number of channels determining similarity to be an open empirical question. Moreover, the centrality of genetic transmission is not built into the model itself. It is an empirical question whether genetic transmission has deeper evolutionary roots—whether it originated earlier in life's history—than any other kind of replicator. After all, it is quite certain that the mechanisms that construct, copy, and translate DNA are one complex product of evolution. DNA transmission and use is certainly not the original route of cross-generational similarity. Indeed, on Maynard Smith and Szathmary's view, the construction of the DNA/protein transcription system is the third of seven major transformations in life's history (Maynard Smith and Szathmary 1999: 17). Equally, it is an empirical question whether genetic transmission is more widespread across the tree of life than any other set of replicators. It might be, for example, that cytoplasmic factors transmitted in mitotic cell divisions are as ancient and as widely used as DNA replication and transcription itself. My own view is that DNA-based transmission of similarity *is* of fundamental significance. But that is not built into the structure of the theory.

It is time to summarize the state of play. I began by describing a striking phenomenon: organisms engineer their own environments, often in ways that are patterned across generations. In doing so, they typically also engineer the environment of their offspring. This downstream engineering has both developmental and evolutionary consequences. Developmentally, downstream environmental engineering is one causal channel through which parent/offspring similarity is generated. This has evolutionary signifi-

cance in itself, because (as has been pointed out since Lehrman [1953]) selection-driven evolutionary change depends on the existence of such similarities but it is indifferent to their mechanism. So ecological inheritance seems to be a channel across the generations that at least supplements genetic inheritance. But downstream engineering has further evolutionary significance. By modifying their offspring's environment, organisms modify selective pressures on their own lineage. And hence we can expect to find evolutionary feedback loops of various kinds. Downstream ecological engineering causes changes in the stream of gene copying across generations, and those changes in turn modify ecological engineering.

I then sketched three different responses to this ecological, developmental, and engineering nexus. One downplays the evolutionary significance of the organism-environment boundary but does not rethink inheritance. Another rethinks inheritance mechanisms but not the organism-environment boundary. Developmental systems theory and niche construction, in somewhat different ways, make this boundary central both to action in the environment (interaction) and the establishment of similarity across the generations (replication). On this perspective there is a very intimate linkage between the two phenomena. The issue is: is this a bug or a feature? In the fourth section I shall be suggesting that very significant aspects of niche construction should *not* be treated as inheritance systems. So, having to date emphasized the similarities among 1–5— the elements of ecological inheritance—I want now to disaggregate, exposing some important differences.

Hoyle's World

Suppose you are a biological engineer working for Hoyle & Co., and hence you are in the business of supplying an empty planet with an indigenous biota. You plan to stock the planet with a seed of life and let evolution do its work. You want to produce a rich, complex, well-adapted, and varied biota—biota in many ways like life here. So you will design your seedling in ways that make selection and evolution effective. As we all know, one necessary condition of selection-driven evolution is heritability. For no adaptive change will take place unless the offspring of successful organisms are likely to resemble their parent(s) in the respects that make them successful. So you will need to design an inheritance mechanism. To put it a little tendentiously, you will need some mechanism to insure that the biological characteristics of the parent are replicated in its descendants. One possible mechanism would be to ensure similarity by the transmission from parent to offspring of a set of replicators, that is, a set of factors each of which makes some predictable causal contribution to the organism's biological organization. Thus if two organisms received similar replicator sets, they would develop a similar biological organization. What would the specifications for our replicator sets look like?

I shall argue that our inheritance system should meet three general specifications. First, it must somehow *block outlaws*. Complex living systems depend on the cooperation of many components. The division of labor and adaptive specialization seen in, for example, an ant nest depend on the linked reproductive fate of the ants within the nest. If some ants in the nest could reproduce independently of others, very likely circumstances would arise in which the reproductive interests of these individuals would not be identical to that of the nest, and defection would undermine cooperative integration. Similarly, if an organism is to be built by a team of replicators cooperating together, those replicators must have a shared evolutionary fate. For otherwise the temptation to defect will undermine cooperative organism building.

Second, the replication system should insure the *stable transmission* of phenotypes over the generations. Richard Dawkins never tires of re-minding us that evolutionary innovation depends on cumulative selection. "No adaptation without cumulation" should be written on his tombstone. He is right, of course. New complex structures are very unlikely to arise through a single mutational change. For if there are major changes to many aspects of an organism's phenotype, it is very likely that at least one will lead to catastrophe. Hence we need phenotype stability: biological organisation must be reliably rebuilt over many generations, not a few.

Third, selection depends on the *generation of variation*. If evolution is to build a disparate biota, or one characterized by adaptive complexity, the replication system must have the potential to generate a large number of distinct phenotypes. Given these desiderata, I suggest that a replication system should have the following characteristics.

Anti-Outlaw Conditions

(C1) Replicators should be transmitted vertically. Replicators are to be transmitted only to offspring and from parents.

(C2) Replicators should be transmitted simultaneously.

(C3) The transmission of the replicator set should not be biased. Either all an organism's replicators are transmitted to each descendant, or each replicator has an equal chance of being transmitted to each descendant.

(C4) Replicators should have a "ballistic" commitment to their biological role. Once some factor becomes adapted for its role as one of a set of replicators, there should be no turning back. Evolution-by-design should block any evolutionary escape from the replicating role.

Stability Conditions

(C5) The copy-fidelity of the generation of replicators from generation to generation should be high.

(C6) The replicator/organization map should be robust. Thus (1) the replicator/organization map should have redundancies built into it, and (2) to the extent that the causal channel from replicator to organization depends on context, both internal and external, that context should be stable and predictable.

Generation of Variation

(C7) The array of possible replicator sets should be very large; possibly even unbounded.

(C8) The effect of a replicator on the biological organization of its carrier should be well-behaved. That is, the replicator/organization map should be smooth. A small change in the replicator set should generate a small change in biological organization.

(C9) The generation of biological organization from the replicator set should be modular. The replicators as a whole should not generate the biological organization of the organism as a whole; rather, replicators, or small sets of replicators, should be designed so that they make a distinctive contribution to the generation of one or a few traits, and relatively little distinctive contribution to others.

Some of these criteria are self-explanatory, but other may need some defence. Let me begin with anti-outlaw criteria. As noted, the evolution of complex phenotypic structures is an evolutionary achievement that depends for its success on the suppression of defection. This idea was originally developed in the context of group selective explanations of altruism (Williams 1966; Sober and Wilson 1998), and it has subsequently been applied to the evolution of multicelled organisms (Buss 1987; Michod 1999). So the most evolvable inheritance systems are those that suppress outlaws. Outlaws, in turn, are replicators that go it alone. Hence the ban on evolutionary escape (C4) and the ban on biased transmission—one in, all in (C3). C1, the insistence on vertical transmis-

sion, has the same motivation, since one way a replicator can go it alone is through horizontal or oblique transmission.

One idea behind C2, the demand for simultaneous transmission, is that drip-feeding replicators from one organism to the next opens the door to transmission biases. For replicators that arrive first can filter those that cross later. Imagine, for example, that (a) the replicator set for an organism—a cockroachoid, perhaps—includes a set of bacterial symbionts that are critical to food digestion and which are passed from parent to descendant; (b) the organism is fittest if it has several different clones of symbiont; (c) symbionts compete to some extent within the roachoid, and hence each clone would be fitter if it were alone; (d) different clones are passed at different times. In that case, any mutation in an early passage clone that closed the door behind it would be favored in it, even though it reduced roachoid fitness. As it happens, these are not biologically unrealistic assumptions. There are arthropod species which standardly have a suite of symbiont passengers (perhaps because different bacterial species have different peak efficiencies with different foods) and competition between them within the host is a possibility (Frank 1996).

The stability criteria I take to be uncontroversial, though their application turns out to be less transparent than the criteria themselves. C5 demands high fidelity copying of the replicator set, and C6 demands high fidelity use of that set. So let's turn to the generation of variation. Selection depends on the existence of selectable variation. Hence we need a rich array of replicator packages (C7). Maynard Smith argues for the centrality of this criterion: he argues that a crucial transition in evolution is the shift from limited systems of replication (where the number of variants is smaller than the number of members of the evolving population) to unlimited systems where the number of variants is as large or larger than the population size. The argument links back to the importance of cumulative selection in adaptive change: new structures are built slowly, in small steps. Microeveolution requires a rich supply of variation, and that requires a rich array of replicator sets (Maynard Smith and Szathmary 1995; Maynard Smith and Szathmary 1999).[1]

But that is not all it requires. We also need to consider the conditions under which replicator packages are mapped onto selectable variation. I have been persuaded by Dawkins's argument (1982, in the final chapter) for the evolutionary importance of the life cycle. He argues that the cycle from single-cell to organism to singlle-cell is central to evolution. For an alteration to the replicator package that acts early, at the single-celled stage, can generate a genuinely novel structure; it can have an overall effect on the organism. Dawkins suggests that without a developmental bottleneck mutations can have only a local effect. I think the importance of the single cell stage can be overstated. Consider, for example, the caterpillar/butterfly transition, or any other point in organism life cycles where major reorganization takes place. Even so, I think Dawkins is essentially right. Hence the condition on simultaneous and hence early transfer, rather than drip-feeding replicators one by one. Because the whole replicator set is transferred early, alterations in the set have the potential to act early and have a global effect.[2]

C8, the requirement for a smooth replicator/ organization map, flows from the work of Kauffman (1993, 1995). Suppose a population of bald pigs are trapped on an island in a period of global cooling. Can selection save the population from extinction by engineering furry descendants? Only if slightly furrier pigs have a replicator set close in character to that of the bald pigs; only if marginally still furrier pigs have a replicator set close to that of the slightly furry ones, and so on. Hence the requirement for smooth mapping.

I am also impressed by the arguments of Muller, Wagner, Raff, and Dawkins on the significance of developmental modularity.[3] First, they see an important connection between continued evolutionary plasticity in a lineage and developmental modularity. For, as Wimsatt has shown, unless development is modular, phenotypes will become generatively entrenched (Wimsatt and Schank 1988). It is hard to change developmental sequences if the development of any characteristic is linked to the development of many characteristics. For a change is likely to ramify, having many effects on the developed phenotype, and some of these are nearly certain to be deleterious. Thus, to the extent that development is holistic, the more complex the organism, and the more it has been elaborated over evolutionary time, the less significant further change there can be in that lineage.

The point that adaptive change would be impossible if development were holistic has been made before. Lewontin, for example, has pointed out that such change requires traits to be "quasi-independent"; there are at least some developmental trajectories that allow one to be changed without affecting others (Lewontin 1978: 169). Modularity adds something to this picture: modules are (often with modification) reusable. Thus, Muller and Wagner claim:

The more we learn about molecular mechanisms of development in widely different organisms, the higher the number of conserved mechanisms that become known. Some of them do indicate homology of morphologically divergent characters.... Still others illustrate that highly conserved molecular mechanisms may be used in radically different development contexts, indicating that the machinery of development consists of modular units that become recombined during evolution. (Mueller and Wagner 1996: 11)

An important example is the invention of cell types. For the most part, animals do not differ from one another because they are built from different sorts of cells. Rather, they build different structures from similar cell toolkits. Evolution only had to discover the trick of making a certain type of cell once. This seems to be a quite general phenomenon. Other examples are the mechanisms for initiating eye and limb formation (Gilbert et al. 1996: 366). So evolutionary/developmental complexes once discovered can then be co-opted and used for other purposes. But that, of course, depends on developmental modularity.

Moreover, as Dawkins (1996) points out, there are links between modularity, redundancy, and evolvability. If the replicator/organization map is modular, there is a fair chance that introducing redundancy into the replicator set through duplication will result in the duplicated, hence redundant, structure in the organism's organization. Most of the time, of course, such extra structure will be deleterious and selection will dispose of the resultant replicator set. But not quite always, and then the new structures can form the basis of genuine evolutionary novelty. Gould and Dawkins do not agree on much, but they do agree on the importance of redundancy in scaffolding evolutionary change. And I am happy to join that consensus. So there are many reasons for thinking that the ideal inheritance system should satisfy C9.

I suggest that inheritance mediated by a replication system satisfying C1–C9 would be evolutionarily potent; it would be highly evolvable. It would meet the Hoyle conditions; it would be (as I shall say) a full Hoyle. Such a replicator set would not, of course, be sufficient to generate a disparate and adaptively complex biota. The world must be cooperative too. Adaptive regimes must remain stable for evolutionarily significant periods of time, for complex structure evolves only through cumulative selection. Moreover, "adaptive landscapes" must be continuous: small changes in phenotype should make only a small difference to fitness. Furry pigs will never evolve if the fitness of somewhat furry pigs is not intermediate between bald and fully furry ones (Lewontin 1978). It might also be the case that some major evolutionary transitions—perhaps the invention of the eukaryote cell—depends on a very lucky accident rather than being the predictable result of an appropriate replicator set and selection regime. Even so, the existence of

full Hoyle replication systems would be of special significance to evolution. Their existence and character would be a central part of the explanation of the biological character of our world.

It is time now to forgo the pleasures of a priori biology and return to actual mechanisms of inheritance. I shall argue that: (1) no actual inheritance system is a full Hoyle; (2) genetic inheritance has a high Hoyle score; (3) some of the cross-generational channels lumped into ecological inheritance are close to Hoyle standard, but many are not; and (4) meeting the Hoyle standard does not depend on taking the replicator set to carry information about the biological organization they will help build.

Gene-like Replication Systems

The Hoyle conditions fit no actual inheritance mechanisms perfectly. This is no surprise, inasmuch as satisfying some conditions militates against satisfying others. It is, for example, well known that fidelity trades off against variation. Griffiths and Gray (1994) argue that the delivery of some developmental resources must be precisely timed, and if they are right this militates against simultaneous transmission. But genetic replication does fit the conditions quite well. Gene transmission is vertical. It is simultaneous and early, hence giving variant replicators a good chance to act early in an organism's life cycle. It is not quite outlaw-proof, of course. Meiotic drive and sex ratio-distorting genes are outlaws, and hence transmission is not quite unbiased. The adaptation of genes to their role in replicator packages may not be quite an evolutionary sink either. That depends on where viruses come from. But genetic replication largely solves the problem of preventing individual replicators from going it alone.

Fidelity and richness are also uncontroversial. It is typically supposed that the fidelity of genetic replication is high, even in prokaryotic lineages that lack the error-correcting machinery of eukaryotes. But that depends on measuring fidelity by comparing the base sequence of a copy with the base sequence of its parent gene. Measuring fidelity that way, it is high. But if we take the replicator to include not just the base sequence but also the molecular machinery and relations to other sequences through which it exerts its phenotypic effects, then base sequence fidelity is not fidelity *simpliciter*. Successful replication would involve the replication of this unit of phenotypic action, not just the base sequence. Moreover, for sexual organisms our measure of fidelity will depend on the size of individual replicators. Long sequences are likely to be altered by recombination. So there are unanswered questions about the fidelity of genetic replication. Even so, on reasonable ways of settling these questions, genetic replication will probably count as a high-fidelity system.

There is a similar ambiguity with the respect to the richness of genetic replication. For though there is a truly vast number of different base sequences—as Dennett puts it, the Library of Mendel is huge (Dennett 1995)—it is far from obvious that that is the right way of counting replicators. Presumably, we should count in terms of phenotypic effect: Two sequences are different if they generate predictable differences in biological organization.[4] Even so, no one denies that genetic replication is rich. In context, and in concert with much else, it generates a splendid range of biological organizations and could generate many more.

Modularity and robustness are much more controversial. However, genetic replication is at least minimally modular. For the DNA/RNA/ protein transcription and translation system is an instance of a modular replicator/organization mapping, though it exemplifies modularity at a fine-grained and local scale. The extent to which we shall see a vindication of Wagner's hypothesis that there exists an array of gene-development–adaptation modular complexes, and that the existence of this array explains the continuing evolutionary plasticity of lineages, remains unknown. As far as I know, the extent

of coarse-grained modularity in development remains an open empirical question. There are similarly open questions about robustness. There seems to be plenty of redundancy built into genetic replication, at least with eukaryotes with paired genes. But the degree to which development is canalized remain to be settled.

In sum, it is clear that although genetic replication is not the full Hoyle, that system goes close to meeting the Hoyle conditions. Inheritance based on gene-like replication yields a highly evolvable biota. The same cannot be said for all that is packed into ecological inheritance.

It is somewhat ironic that when nongenetic inheritance is recognized at all, it is most commonly recognized in the guise of dual inheritance models in which cultural transmission supplements genetic inheritance. For cultural replication seems to me to have a low Hoyle score. It turns out that the conditions that allow cultural replication restrict its richness and effect quite sharply. Limited cultural inheritance is possible, but the conditions allowing the range of variation to increase also reduce fidelity. Perhaps imprinting by phytophagous insects on their plant targets, or the philopatry shown by sea birds to their nesting sites, are valid instances of nongenetic replication, but transmission of this type of behavioral similarity from parent to offspring will not generate cumulative change, for the system is not rich enough. Tomasello has pointed out that *cumulative cultural evolution* demands something like the capacity for true imitation. He argues that primate societies do not show cultural evolution because apes do not ape. They learn from others by social priming—their attention, and exploratory trial-and-error learning, is directed in part by what they see others do. So they can learn what to be interested in: what constitutes a resource or a danger. But they do not copy the action patterns of others. Hence insight—the discovery of a new way to extract nuts, fish for termites, or reach inaccessible food—will not be transmitted to other members of the group. In effect, without imitation, copy fidelity is too low to sustain a ratchet effect, and hence for cultural evolution to generate cumulative change (Tomasello forthcoming). There has been a recent documentation of the existence of a quite large set of behavioral traditions distinctive of particular chimpanzee communities (Whiten et al. 1999), but there is no evidence in this data either of cumulative cultural evolution or even of traditions that have persisted over many generations.

Yet once organisms have the cognitive sophistication to imitate, the robustness and modularity of cultural transmission will be eroded. For once cognitive sophistication sufficient for imitation has evolved, the connection between experience and behavior will be highly context-sensitive. Patterns of behavioral similarity will not be transmitted deeply through the generations. The causal influence of a replicator on biological organization is always context-dependent. It is always scaffolded by features both internal and external. Hence if the cross-generational stability condition is to be met, these contextual factors must be persistent, or reliably remade anew in each generation. These conditions may be met by a few elements in hominid cultures. Learning may be scaffolded, internally, by sensitive periods or externally by reliably reoccurrent and highly salient features of the environment. Language probably meets both these conditions. But many of the elements of human and hominid cultures probably do not meet the criteria that underwrite cross-generation stability. After all, the crux of the argument against behaviorism in psychology is that the effect of experience on behavior is highly variable. So if an organism's psychology is not captured by behaviorist models, cross-generation similarities of behavior will not be well explained by dual inheritance models. Yet if its psychology is captured by behaviorist models, its learning capacities are unlikely to be sophisticated enough to support imitative learning.

Set aside these problems. There is nothing in the system of cultural transmission that filters outlaws. Ideas and other cultural constructs spread horizontally and obliquely, not just vertically. They are not replicated and transmitted en masse (like a computer being loaded with its system disk) but are drip-fed. There is every opportunity for transmission biases. If this is a replication system at all, it is one made for banditry on a massive scale. So if ideas and the like are not typically viruses of the mind (as Dawkins takes religion to be) then that fact poses a serious problem for theories of cultural evolution. Memes should often be outlaws. If they are not usually bad for their carrier's fitness, this suggests to me that cultural transmission should not be seen as an inheritance mechanism at all.

So much for dual inheritance. The action of organisms in physically engineering their environment and that of their descendants seems not to fit the Hoyle criteria well. John Maynard Smith has sharply distinguished between limited and unlimited systems of heredity, and at most the transmission of physically engineered surroundings seems to be a limited system. But I think the problem is more pressing. Transmission is not vertical. Indeed, it is not even individual. It is diffuse. Groups of trees engineer their soil structures or a fire-prone understory; individual trees do not make their microenvironments for themselves and their descendants. In most cases groups of animals make warrens, trackways, track-and-bowl systems, beaver lodges, termite mounds, and other structures that are ultimately taken over by the next generation. If this is transmission at all, it is diffuse and development is holistic. Contrast the case of the continuously occupied beaver lodge with a case where there really does seem to be inheritance, transmission, and a life cycle: cases in which mated queens leave ant and termite nests to found a new nest and build a new structure. In such cases, there obviously is a new generation—an F2 nest—and a life cycle. But where the inherited system of resources is not tied to anything like a life cycle, there seems to be no moment of transmission at all. The next generation just gradually comes to occupy, use, and renovate the lodge as the previous generation dies out. So especially when generations overlap, we have here a resource modified by many organisms, for themselves and their descendants, and used by many. Beavers are major ecological engineers, and their engineering certainly has downstream effects, but these effects do not constitute an inheritance system. We might treat some of these diffuse cases as individual vertical transmission by taking the units in question to be groups. But the conditions that allow group selection are quite onerous. So this is not a general solution to the problem.

On the other hand, some extragenetic inheritance systems do fit the Hoyle picture well. Perhaps cytoplasmic factors in the egg that determine the basic positional layout of the early embryo do. The transmission of such factors may constitute a very important system, since it is both ancient and widespread. But my favorite examples concern the transmission of obligate symbionts. Because symbiosis is a common biological phenomenon, this category is important. It is not a minor quirk of a few clades. And here the fit with evolvability criteria seems to be good. The transmission of these symbionts is often early. It is typically unbiased. There is a considerable variety of evolutionary specificity in these biological relationships. Mycorrhizal associations between fungi and their associated plants are not very species specific. But symbiosis is often highly evolutionarily stable and mutually obligatory. The symbiotic microorganisms cannot survive alone, and nor can their partner. It is a mechanism with high fidelity. It is reliable, with often delicate adaptations to insure successful transmission. A specific species associates with a specific species, sometimes so much so that the species branching pattern of the one models that of the associated clade. However, because each member of the partner retains a good deal of metabolic and developmental integrity, I assume that development is both modular and robust. Many, perhaps most, changes in leafcutter morphology, physiology, and behavior would have no impact on the specific association with their codependent fungus.

Highly evolvable systems are rich: they have the capacity to produce many variants with which evolution can work. But though the principle is clear, applying it is not. Maynard Smith has argued that highly evolvable replication systems must be digital, for only such systems maintain fidelity levels high enough for cumulative selection (Maynard Smith 1996). However, the digital/analog distinction, and hence the fidelity of replication, can be assessed only in the context of replicator/reader systems. Our symbol reading system determines the fact that

7 7 7 8 **8** 8

contains only two symbol types. The same point applies to counting replicator variation. It is the system that uses genetic resources that determine whether two base sequences are of the same type or not. Does the leafcutter-fungus replication system count as a high fidelity/low variety system because it will only generate one kind of system: a symbiotic association between an ant and a fungus? Or will all the differences in ant and in fungus count as different variants? Suppose, for example, that the amount of fungus the queen takes on her founding flight has some relation to the probability of the fungus failing to grow, or on fungus growth rate. If so, this would add the possibility of heritable variation to the system. That variation might be digital if, say, these differences were governed by threshold effects. Or they might be analog if, say, growth rate varied smoothly with sample size.

Obviously, then, settling richness is both empirically and conceptually complex. Depending on how these issues shake out, symbiont transmission may count as a limited replication system or, more likely, a large family of limited systems. Even if it does, they may be extraordinarily important. There is no doubt that symbiosis is ecologically important, inasmuch as nitrogen-fixing bacteria in legumes and mycorrhizal associations with trees are fundamental to terrestrial plant ecology, as are coral reefs to the ecology of shallow tropical seas. But symbiosis may be of great evolutionary significance in the generation of novelty. It is now widely accepted that the eukaryotic cell is an evolutionarily frozen symbiosis (Dyer and Obar 1994).

In general, symbiotic inheritance is less outlaw proof than genetic inheritance. Although some symbiotic organisms have no future without their hosts, there are many cases where the association is less rigid than this (Thompson 1994). Even when association is wholly obligatory, the transmission of, say, symbiotic bacteria often involves a significant number of organisms, and that opens the door to within-organism competition that reduces host fitness. Strikingly, it turns out that there are instances of host adaptation to minimize the outlaw problem, by segregating some symbionts into a germline group destined for passage to the next generation, and a somatic group that will play a role in host metabolism. For example, fulgoroid planthoppers segregate one of their bacterial symbiont species into a large, differentiated and reproductively disabled form, and a small, undifferentiated form stored in a separate part of their body and destined for their eggs (Frank 1996). The upshot, though, is that while in some host/symbiont associations anti-outlaw provisions are in force, genetic replication seems better screened against outlaws.

In sum, then, some host/symbiont systems meet anti-outlaw conditions quite well, with vertical and early transmission. They meet stability conditions well, for replication fidelity is high, and the effect of symbiont on host phenotype is robust. They probably meet some of the conditions on the generation of variation, though I think many crucial details of the natural history of these associations remain unknown. So my best guess is that they have a high Hoyle score, though not as high as the genetic replication system.

I would like to extract two further points from these examples. First, I think there is an important heuristic moral to be drawn. Treating symbiont transmission as an inheritance system with a high Hoyle score suggests an empirical research program that would otherwise be invisible. Among the questions we should ask are:

1. Can we find anti-outlaw mechanisms? These might include: simultaneous transfer; host adaptations for blocking all but vertical transmission; host adaptations for limiting the number of individual organisms transferred; and host adaptations for evolutionary capture of symbionts by taking over the provision of critical metabolic resources to the associate.

2. Is there evidence about the range of variation? Does the host phenotype differ, if different combinations/quantities of symbionts are transferred? Do genetic differences in host or associate change the nature of their association?

3. The discovery of high Hoyle score replication systems on our world raises a profound evolutionary problem. How and why could evolution assemble such an inheritance mechanism? Replicators are adapted for their role of insuring that offspring are like their parents. That is no surprise, for most departures from similarity will be bad news. But a system that insures accurate replication across a generation is one thing, an evolvable replication system quite another. It is possible that a highly evolvable replication system might evolve in response to some local evolutionary demand. But my working assumption is that a high Hoyle score replication system is likely to be the result of some lineage level selection for evolvability. That is an idea we can test. Can we find evidence of increased evolvability by comparing symbiont rich clades with symbiont poor sister clades? We might measure by species richness and/or morphological and ecological diversity.

Second, meeting the Hoyle conditions does not depend on the flow of information across a generation. Indeed, the point of preformationist inheritance mechanisms of this kind is that you can *dispense* with information. Because you have a sample of the fungus that can be grown, you do not need information on how to make the fungus.

That, of course, leaves open the status of gene-based inheritance. We might reasonably conjecture that gene-based inheritance systems are information-based systems precisely because they are not based on sample-to-product inheritance. This is a thought that must be held for another time. However, I will suggest one way it might be developed.

We rightly think that perception involves a flow of information from distal events to the mind. For perceptual representation tracks the world despite great flux in the proximal channel between, say, perceived tiger and the tiger percept. Color constancy mechanisms, for example, seamlessly compensate for great variation in illumination conditions. So there is a robust relationship between distal source and internal registration, and only between them. Along similar lines, if there were a stable relationship between genetic structure and phenotypic structure despite variation in developmental route, we would rightly see the genetic structure as carrying information about the phenotype. On that way of reading the situation, the idea of genes as information carriers would depend on the existence of a certain type of canalization, not of genetic variation but of developmental variation. It would depend on the right type of robustness (C6). For now, though, I will leave the issue open. It is, however, important to see that if the argument of this section is right, gene-based inheritance is of special importance in explaining life's disparity whether the idea of the informational genome can be vindicated or not.

Recapitulation

Let me briefly summarize the state of play as it now is, if the argument of this chapter is right. I have (first) identified a cluster of characteristics of inheritance systems that support evolvability. Inheritance systems with most of these characteristics have a high Hoyle score. Second, I have argued that genetic inheritance does have a high

Hoyle score. Third, I claim that downstream environmental engineering is not in general an inheritance system or cluster of inheritance systems. Fourth, I argued that some elements lumped into environmental engineering are inheritance mechanisms with a high Hoyle score. And finally, I decoupled Hoyle scores from the idea that replicator sets carry information specifying phenotypes. For sample-based systems of inheritance are not information flow systems, but on certain empirical bets, genetic systems might be.[5]

Notes

1. Thus, for example, they argue: "One could argue that for replicators that are not modularly replicated, variants can arise only through 'macromutations'.... Hence what members of autocatalytic cycles as limited hereditary replicators lack is the ability to undergo microevolution: hereditary is almost always exact" (Szathmary and Maynard Smith 1997: 559).

2. Peter Godfrey-Smith has pointed out to me that early transfer does not entail the existence of a developmental bottleneck. Vegetatively reproducing plants presumably meet C2 through asexual cell division. So if a bottleneck were a necessary condition on a highly evolvable biota, even setting aside necessary features of the abiotic environment, evolvability may not be fully captured through a specification of the replication system. There remains, though, the possibility that it is captured indirectly. Maynard Smith has argued that the bottleneck is a consequence of the anti-outlaw conditions; it is an adaptation to ensure common fate (Maynard Smith 1988).

3. See Mueller and Wagner (1996), Wagner (1995), Wagner and Altenberg (1996), Raff (1996), and the discussion of "kaleidoscope embryology" in Dawkins (1996). I discuss modularity in development further in Sterelny (forthcoming), and the issue is also taken further in Brandon (1999).

4. Thus treating genes as difference-makers, in line with Sterelny and Kitcher (1988) and Sterelny and Griffiths (1999, chap. 4.3). They make a difference, of course, only relative to an appropriate genetic, cellular, and environmental context. But even given this qualification, the identification of gene types is far from uncontrover-

sial and straightforward; see Sterelny and Griffiths (1999, chap. 4.1–4.3).

5. Thanks to the participants in the ANU 1999 genetic information workshop, and especially to Russell Gray, Paul Griffiths, Susan Oyama, and James Maclaurin for their comments on an earlier version of this paper.

References

Boyd, R., and P. Richerson. (1985). *Culture and the Evolutionary Process.* Chicago: University of Chicago Press.

Brandon, R. N. (1999). The units of selection revisited: The modules of selection. *Biology and Philosophy* 14: 167–180.

Buss, L. (1987). *The Evolution of Individuality.* Princeton, NJ: Princeton University Press.

Caro, T., and M. Hauser. (1992). Teaching in non-human animals. *Quarterly Review of Biology* 67: 151–174.

Dawkins, R. (1982). *The Extended Phenotype.* Oxford: Oxford University Press.

Dawkins, R. (1989). *The Selfish Gene.* (1st ed. 1976). Oxford: Oxford University Press.

Dawkins, R. (1996). *Climbing Mount Improbable.* New York: W. W. Norton.

Dennett, D. C. (1995). *Darwin's Dangerous Idea.* New York: Simon and Schuster.

Dyer, B. D., and R. A. Obar. (1994). *Tracing the History of the Eukaryotic Cell: The Enigmatic Smile.* New York: Columbia University Press.

Frank, S. A. (1996). Host control of symbiont transmission: The separation of symbionts into germ and soma. *American Naturalist* 148: 1113–1124.

Gilbert, S. F., J. M. Opitz, and R. Raff. (1996). Resynthesising evolutionary and developmental biology. *Developmental Biology* 173: 357–372.

Gray, R. (1992). Death of the gene: Developmental systems strike back. In P. E. Griffiths (Ed.), *Trees of life: Essays in philosophy of biology,* pp. 165–209. Dordrecht: Kluwer Academic.

Griffiths, P. E., and R. Gray. (1994). Developmental systems and evolutionary explanation. *Journal of Philosophy* 91: 277–304.

Griffiths, P. E., and R. D. Gray. (1997). Replicator II: Judgement day. *Biology and Philosophy* 12: 471–492.

Jones, C., J. Lawton, and M. Shaclak. (1997). Positive and negative effects of organisms as physical ecosystems engineers. *Ecology* 78: 1946–1957.

Kauffman, S. A. (1993). *The Origins of Order: Self-organization and Selection in Evolution.* New York: Oxford University Press.

Kauffman, S. A. (1995). *At Home in the Universe.* New York: Oxford University Press.

Keller, L., and K. G. Ross. (1993). Phenotypic plasticity and "cultural transmission" in the fire ant, *Solenopsis invicta. Behavioural Ecology and Sociobiology* 33: 121–129.

Laland, K. N., Odling-Smee, F. J., and Feldman, M. W. (2000). Niche construction, biological evolution and cultural change. *Behavioral and Brain Sciences* 23: 131–175.

Lehrman, D. S. (1953). Critique of Konrad Lorenz's theory of instinctive behaviour. *Quarterly Review of Biology* 28(4): 337–363.

Lewontin, R. (1974). The analysis of variance and the analysis of causes. *American Journal of Human Genetics* 26: 400–411.

Lewontin, R. C. (1978). Adaptation. *Scientific American* 239: 156–169.

Lewontin, R. C. (1982). Organism and environment. In H. C. Plotkin (Ed.), *Learning, Development and Culture,* pp. 151–170. New York: Wiley.

Lewontin, R. C. (1983). The organism as the subject and object of evolution. *Scientia* 118: 65–82.

Lewontin, R. C. (1985). Adaptation. In R. Levins and R. Lewontin, *The Dialectical Biologist,* pp. 65–84. Cambridge: Harvard University Press.

Lewontin, R. C. (1991). *Biology as Ideology: The Doctrine of DNA.* New York: HarperCollins.

Maynard Smith, J. (1988). Evolutionary progress and the levels of selection. In M. Nitecki (Ed.), *Evolutionary Progress,* pp. 219–230. Chicago: University of Chicago Press.

Maynard Smith, J. (1996). Evolution—natural and artificial. In M. Boden (Ed.), *The Philosophy of Artificial Life,* pp. 173–178. Oxford: Oxford University Press.

Maynard Smith, J., and E. Szathmary. (1995). *The Major Transitions in Evolution.* New York: W. H. Freeman.

Maynard Smith, J., and E. Szathmary. (1999). *The Origins of Life: From the Birth of Life to the Origins of Language.* Oxford: Oxford University Press.

Michod, R. E. (1999). *Darwinian Dynamics: Evolutionary Transitions in Fitness and Individuality.* Princeton, NJ: Princeton University Press.

Mueller, G. B., and G. P. Wagner. (1996). Homology, hox genes and developmental integration. *American Zoologist* 36: 4–13.

Odling-Smee, F. J. (1994). Niche construction, evolution and culture. In T. Ingold (Ed.), *Companion Encyclopedia of Anthropology,* pp. 162–196. London: Routledge.

Odling-Smee, F. J., K. N. Laland, and M. W. Feldman. (1996). Niche construction. *American Naturalist* 147: 641–648.

Raff, R. (1996). *The Shape of Life: Genes, Development and the Evolution of Animal Form.* Chicago: University of Chicago Press.

Sober, E., and D. S. Wilson. (1998). *Unto Others: The Evolution of Altruism.* Cambridge, MA: Harvard University Press.

Sterelny, K. (forthcoming). Development, evolution and adaptation. *Philosophy of Science, (supplementary volume).*

Sterelny, K., and P. Griffiths. (1999). *Sex and Death: An Introduction to Philosophy of Biology.* Chicago: University of Chicago Press.

Sterelny, K., and P. Kitcher. (1988). The return of the gene. *Journal of Philosophy* 85: 339–360.

Sterelny, K., K. Smith, and M. Dickison. (1996). The extended replicator. *Biology and Philosophy* 11: 377–403.

Szathmary, E., and J. Maynard Smith. (1997). From replicators to reproducers: The first major transitions leading to life. *Journal of Theoretical Biology* 187: 555–571.

Thompson, J. N. (1994). *The Coevolutionary Process.* Chicago: University of Chicago Press.

Tomasello, M. (forthcoming). Two hypotheses about primate cognition. In C. Heyes and L. Huber (Eds.), *Evolution of Cognition.* Cambridge, MA: MIT Press.

Wagner, G. P. (1995). The biological role of homologues: A building block hypothesis. *Neues Jahrbuch für Geologie und Palaontologie* 195: 279–288.

Wagner, G. P., and L. Altenberg. (1996). Complex adaptations and the evolution of evolvability. *Evolution* 50: 967–976.

Whiten, A., J. Goodall, and W. C. McGrew. (1999). Culture in chimpanzees. *Nature* 399: 682–685.

Williams, G. C. (1966). *Adaptation and Natural Selection.* Princeton, NJ: Princeton University Press.

Wilson, D. S. (1997). Biological communities as functionally organized units. *Ecology* 78: 2018–2024.

Wimsatt, W. C., and J. C. Schank. (1988). Two constraints on the evolution of complex adaptations and the means of their avoidance. In M. H. Nitecki (Ed.), *Evolutionary Progress,* pp. 231–275. Chicago: University of Chicago Press.

24 Developmental Systems Theory and Ethics: Different Ways to Be Normative with Regard to Science

Cor van der Weele

Biological knowledge and biological technology have become immensely consequential. Genetics is already a central part of the development of biotechnology, and, with the Human Genome Project approaching its conclusion, will become ever more powerful. Because genetics and its social implications such as prenatal and presymptomatic genetic diagnosis touch upon life and death, health and illness, development and identity, it need not surprise that they generate great hopes and fears, and many normative questions.

DST is one approach to raise normative considerations. It is critical of an overemphasis on genetics in biology, with an important eye to the social consequences of such one-sidedness. But it does not involve ethical norms, against which to test, for example, applications of biotechnology.

What does DST have to do with ethics? In order to answer that question, this chapter will distinguish three different ways to be normative with regard to science. The three normative enterprises to be considered are ethics, science criticism, and STS (Science and Technology Studies).

Ethics and Science

Ethics belongs to philosophy. It is that part of philosophy that systematically reflects on morality, be it in a normative, a descriptive or a meta-sense, the latter involving analysis of moral language and reasoning, the former two involving the question how we ought to live. Ethical theory is a philosophical framework within which to reflect on moral actions, moral judgements, or moral character. Classical notions on which ethics is built are obligations (Kant), utility (utilitarianism), or virtues (Aristotle). Major efforts in ethics are directed to rendering these respective approaches coherent and convincing, discussing their differences or finding more convincing alternatives, all directed toward finding conceptual frameworks within which the question how we ought to live can be answered.

The concern that such theoretical fights are not necessarily fruitful for ethical practice has led Beauchamp and Childress (1979 and later editions) to design a medical ethics that could be helpful in practice. They left the goal that ethics should be founded by (a single) ethical theory behind; instead, they focused on common morality, the great advantage of which they felt was that agreement can be found there that is forever lacking on the level of theory. From common morality they abstracted common principles, which are general and universal rules and which form the core of their approach, that has been extremely influential. Justification for these principles is not a matter of giving them a theoretical foundation; justification is found through an approach called reflective equilibrium (Rawls 1971), in which coherence is sought between concrete judgements and more abstract ethical notions, such as principles. This looking for coherence is essentially an ongoing social affair that has no real endpoint. There is no absolute justification.

This approach certainly takes ethics away from abstract realms, bringing it in closer contact with the world. It remains a philosophical undertaking, however, in that the approach has mainly been developed, discussed and justified by philosophers.

Beauchamp and Childress focus on medical ethics, and the common morality they refer to is thus domain-specific; it is the morality associated with medical practice. For this domain, they presented four principles: autonomy, nonmalificence, benificence, and justice. With the help of the principles many practical dilemmas in biomedical ethics can be tackled, as Beauchamp and Childress show in their book. Automatic solutions are not the result, because the different principles can point in the direction of different solutions. But the approach does enable decision-making on the basis of argument concerning the application and balancing of relevant ethical principles. In other words, practical ethics is a matter of the ongoing application and elucidation of moral principles.

How does such an ethics relate to science?

It is not controversial that, like medical practice, scientific inquiry is subject to ethical evaluation; various principles are clearly relevant and applicable. In the case of scientific knowledge, domain specific principles include general rules specifying that science should extend the domain of knowledge, that the gathering of knowledge should be done in fair ways, and that (human) subjects of scientific research should be dealt with respectfully. Research on humans is subject to ethical evaluation in that fundamental human rights have to be respected. Increasingly, the interests of nonhuman animals are also subject to ethical consideration. Thus, ethical principles are relevant to scientific research and lead to restrictions on doing research.

When it comes to the content of knowledge, it is more problematic what ethics could have to say. Ethicists, by their training as moral philosophers, are not in a position to judge the specific contents of science, but they often do reflect on scientific knowledge in more general ways. For example, in controversial areas such as IQ research, the Human Genome Project, cloning, embryo research, or xenotransplantation, it can and has been argued that there should be restrictions on knowledge gathering because of potential worrisome social consequences. Scientists tend to resist such proposals. There is a strong internal normativity in science that says that knowledge is a worthwhile goal, irrespective of its content or of its consequences. Ethicists in their turn often feel that scientists claim an amount of freedom that is not appropriate given the enormous social impact of their work. However, there is no generally shared common morality from which to derive widely recognized principles to override the internal normativity of science. In this situation, many ethicists prefer not to push. Peter Singer (1996: 227–228), for example, writes that there are strong pragmatic reasons not to put ethical restrictions on the content of science: attempts to prohibit research will only produce suspicion and will almost always backfire.

The practical implications of scientific knowledge are a completely different matter. Many widely held general moral principles, all the principles of biomedical ethics among them, are relevant when it comes to social implementation, and ethics is productive and fruitful here. For example, the ethical studies surrounding the Human Genome Project are almost totally directed to a study of its implications, as is reflected in the name of the undertaking, ELSI, which stands for "Ethical, Legal and Social Implications." Apart from questions about research ethics (e.g., concerning privacy and safety of research subjects) the great majority of questions call for studies of the implications of the Genome Project for public health activities, genetic testing, conceptions of humanity, concepts of race and ethnicity, and so on. In all these areas, policies should be worked out in which society accommodates the challenges posed by genetic knowledge and technology in an ethically responsible way.

Thus, in dealing with science, ethics is fruitful with regard to implications and with regard to research ethics, while it has not been successful in dealing with the content of scientific knowledge.

Science Criticism

Precisely because it does not deal with the contents of scientific knowledge, Ruth Hubbard does not have faith in ELSI. The program, she writes in *Exploding the Gene Myth* (Hubbard and Wald 1993: 159), will not affect the decisions made in science. It will not question the current emphasis on genes as determining our development, health, and behavior, and it will not get in the way of science. On the contrary, the human genome managers will decide which ethical and legal questions will be asked.

Hubbard's approach to science is normative, but her domain is not ethics. She is one of those critical scientists who do address the contents of science.

Science criticism, as Proctor (1991: 232) uses the term in his book *Value-Free Science?* and as I shall use it here, rests on a rejection of the neutrality of science, by making a connection between the implications of science and content. A major theme in science criticism is that because scientific knowledge heavily influences social attention, it should be responsible knowledge. Exactly what this means may of course vary, since criticism in science is many-voiced, but a persistent theme within biology is criticism on genetic determinism. It is the main subject of Ruth Hubbard's normative approach. By focusing too heavily on genes, scientists draw our attention away from societal causes regarding health and disease. There must be a balance, she writes (Hubbard and Wald 1993: 61), between individual health care and public health measures. An excessive preoccupation with individual concerns and responsibilities is detrimental to health when it encourages us to neglect the social conditions that affect us all. It does not lead to necessary cleaning up of factories; instead, it leads to identification of suspectibilities of individual workers to various dangerous substances.

According to this line of reasoning, the present overemphasis on genes is represented in the Human Genome Project. Therefore, the American Council for Responsible Genetics (1990) has been warning for many years against the Human Genome Project, criticizing the reductionist view of causality that detracts from other biological processes and from social factors. The council has commented on the comparatively large amounts of money spent on mapping and studying genes, saying that "the genome project vastly exaggerates the importance of genes, especially at this time, when a deteriorating environment and economy make it increasingly difficult for most people to live healthful lives" (1990: 4).

The theme of the increasing geneticizing of health and disease is prominent in science criticism in biology, such as in the volume *Are Genes Us?* (1994), edited by Carl Cranor. For example, Evelyn Fox Keller argues in her paper in Cranor's book (1994: 95–97) that though some of the experimental studies in the context of the Human Genome Project are inevitably bound to undermine the simplistic notions on which they rely, the project certainly encourages the trend that genes are now often seen not only as causing disease, but as defining disease.

Against the overemphasis on genetics and its tendency for causal reductionism, most critical biologists stress an interactive, systemic nature of causality. Genes are players in a system. This is where DST fits in, in ways that are elaborated elsewhere in this book.

The content of science thus matters because of its context. Alan Garfinkel has argued that because of this context, science cannot be value free. I think his argumentation and analysis are congenial to many critical scientists. In chapter 5 of his book *Forms of Explanation*, called "The Ethics of Explanation," Garfinkel (1981) argues that science cannot give value-free explanations. His point of departure is Weber's view that science is like a map that can tell us how to get somewhere but that cannot tell us where we should go. This view is shared by Hempel and others who hold that a division of labor exists between value-free causal accounts, given by scientists, and value judgments, made by policy-makers and others. Garfinkel disagrees. If scientists are concerned with statements of the form "A causes B", he says, this implies by no means that science is value-free. The reason is that causal statements are always made in a social context. It can be wrong to make true causal statements. For example, it is morally wrong to say, "If you look in the attic, you'll find Anne Frank" in the presence of Nazi search parties, even if it is "merely" a causal statement. It is not the presence of value-words but the context that makes the statement non-neutral (Garfinkel 1981: 137).

Simple as it is, says Garfinkel, the point is often missed by scientists who want to forget about the context of their research, as in war industries. When the context of relevance is not so clear, and when applications of research are not in sight, it is more difficult to evaluate the situation; after all, "any fact may end up aiding some evil cause. So what are we to do?" (p. 138). But even in seemingly neutral situations, Garfinkel maintains, "A causes B" is not value-free. Why not? Because explanations are incomplete and this incompleteness is always potentially relevant. Ever since Mill, it has been clear that explanations typically mention only one or two causal factors as the cause of a phenomenon, giving all the other factors the status of background. Explanations unavoidably involve choices. The incompleteness of explanations becomes morally relevant when science becomes a guide for social choices. Any explanatory framework recognizes only certain alternatives and therefore guides you to specific solutions. The choices involved need certainly not be made for moral or political reasons: "The value ladenness is a fact about the explanation, not its proponents. It is value laden insofar as it insists ... that change come from this sector rather than that" (Garfinkel 1981: 141). All kinds of nonmoral mechanisms exist to explain the choices involved, such as tradition, available equipment, expertise, and so on.

The same need for choices arises with regard to methodological criteria; Levins (1966, 1968) as well as Van der Steen (1993a, 1993b) have argued that methodological trade-offs cannot be avoided because theories cannot satisfy all methodological criteria at the same time.

The unavoidability and consequentiality of choices is a core insistence in science criticism: scientific choices matter in social contexts. In many cases this is clear; many historical examples have been analyzed, by critical scientists and others, that show how socially relevant biological knowledge can be. The history of eugenics and of IQ-testing are among those examples.

But describing context involves again choices. Garfinkel specifies his social normativity in political terms, calling his view of reality Marxist. Lewontin too has been very explicit, seeing science as part of wider social ideology. Writing

with Rose and Kamin (Rose, Lewontin, and Kamin 1984), as well as with Levins (Levins and Lewontin 1985), he has argued that genetic determinism flourishes as part of a bourgeois, reductionist, Cartesian world view that does not encourage questions about the interdependence of organisms and their environment. A dialectical worldview is needed, of which a dialectical biology is an element.

Not all critical scientists present such analyses. Their normative context is often relatively unspecified, nor do they analyze in detail exactly how better knowledge will lead to a better society. And this cannot be surprising, since social or political normativity is a motivating element but not the primary domain of analysis for most critical scientists. With an eye on relations between science and society, science criticism focuses primarily on the *content* of science. As Susan Oyama writes in one of her papers: "Some of my reasons for working on the nature-nurture problem stem from concerns about publicly contested issues of, say, intelligence, race or sex, but most have to do with the kinds of distinctions that are made in the scientific work that draws on and feeds these larger disputes" (Oyama 2000, chap. 11). This balance applies clearly not only to her reasons but also to her critical efforts. Social concerns are there, but they are mostly in the background. This background is also present when she says, for example, that the analysis of the concept of nature is important in part because the concept is "ethically resonant" (Oyama 1999). This suggests that (some) concepts help shape social problems and solutions, but leaves it open exactly how this happens, and it does not involve explicit moral normativity.

In order to be in a position to address the precise and technical content of science, a high level of expertise is required. Thus, critical scientists are typically scientists themselves. Lewontin's expertise in population genetics enabled him to present a detailed critique of the neglect of gene/environment interaction in causal analyses in biology, and to insist that the proper object of study is the relation of genotype, phenotype, and environment, expressed in norms of reaction (Levins and Lewontin 1985: 114). Norms of reaction describe biological outcomes of given genotypes as functions of environmental variables. The argument is a piece of theoretical biology for which Lewontin's expertise as a population geneticist and theoretical biologist is the relevant expertise. Likewise, Oyama's criticism on determinist reasoning in *The Ontogeny of Information* (Oyama 1985/2000) is based on relevant expertise and gains its influence from being a clear and well informed expert analysis, and the same applies to Ruth Hubbard's argument in *Exploding the Gene Myth* (Hubbard and Wald 1993) that there is more to health and disease than meets the genetic eye.

That the content of science matters for society is an assumption that is shared by many, and much hope is often placed on a non-(genetically) reductionist biology. For example, in *The Biotech Century*, Jeremy Rifkin (1998) points out that a biology that talks in reductionist terms leads to very different kinds of biotechnological practices than a biology that takes a more systemic, integrative approach to nature. The former encourages a biotechnology that, for example, looks at genetically engineered plants not as parts of larger environments but as self-contained devices in isolation. Rifkin sees especially promising developments in the direction of a more integrative biology in developmental biology. In this field, he says, with reference to the work of Stuart Newman, a new middle ground (not a DST-style formulation) between nature and nurture is being found in which understanding of the subtle relationships between genotype and phenotype, and between environmental triggers and genetic expression, is growing. In developmental biology, the idea of the almighty master molecules is giving way to "a more sophisticated understanding of genes as integral components of more complex networks that make up both an organism and its environment" (Rifkin 1998: 156–157). This view is no doubt on the overoptimistic side for

developmental biology as a whole. If it were not, Rifkin would not have had to notice, as he does, that "still little research has been done, to date, on how genetic predispositions interact with toxic materials in the environment, the metabolizing of different foods, and lifestyle to effect genetic mutations and genotypic expressions" (Rifkin 1998: 229). But he is certainly right that developmental biology is a field in which systemic complexity is increasingly drawing attention, with DST as an important attention drawer.

In systemic approaches, the findings of the Human Genome Project are not the Holy Grail of biology. They do not have much meaning in themselves, and they have to be interpreted in a larger network of heterogeneous causal interactions. Such a view implies that the Human Genome Project is not, or at least should not be, just an everlasting gene-centered affair by definition. Much depends on what is done with the sequence information. If the next step is an enormous effort into *functional genomics* in which gene/environment interaction are the center of attention (through the study of norms of reaction and the molecular mechanisms underlying them), the Human Genome Project could be a step toward a systemic biology, though it could still be questioned whether it is a good or necessary step.

Rifkin hopes that a new approach in biology will make a difference for biotechnological practices, and this is suggestive of the ways in which science matters. First, its reigning paradigms matter, because they act as searchlights and bestowers of meaning. Second, its precise results matter, for example on how development is influenced by toxins and other environmental causes. Study of reaction norms is a valuable instrument for the latter. The information generated by the study of reaction norms has its restrictions from a DST point of view, as Russell Gray (1992) has pointed out: reaction norms do not look at processes but only at outcomes of development; therefore they are not sophisticated enough to generate a picture of development that "includes a range of codeveloping life-history trajectories over a range of codeveloping environments" (Gray 1992: 174). Nevertheless, in my view (Van der Weele 1999, chap. 5), norms of reaction do have an important place in the study of developmental systems. They can be seen as a first step, which can be followed by more detailed studies on life trajectories, or on the mechanisms of gene expression. (For another defense of norms of reaction, see Kitcher forthcoming.)

What can a more systemic biology accomplish? Is it going to change the world for the better? This assumption, apart from being too simplistic, is not always shared by those who are expected to benefit from a richer biology. For example, many homosexuals have not shown themselves too unhappy with deterministic genetic accounts of homosexuality, even if the accounts turned out to be very problematic from a scientific point of view, because such accounts make homosexuality seem less fraught with complex and highly moralized choices. Instead, it would be a matter of simple, unavoidable nature. Susan Oyama, when expressing her anxiety about the biological turn (Oyama 2000, chap. 10) is at the same time sensitive to the complexities of social processes, saying that while the disempowered may be unable to prevent a swing back to biology anyway, the discussion might encourage scrutiny of the whole issue, which could be a good thing. But given the dichotomy between insides and outsides on which this discussion rests, her preference is, here as elsewhere, to question this dichotomous conceptual foundation, rather than "trying to turn such a dangerously loaded dichotomy to one's advantage." In her view, the entanglement of biological arguments in beliefs about the fixity of nature has too often served to prejudge complex matters, while a reconstructed biology "does not so readily lend itself to these uses" (Oyama 2000, chap. 10).

Oyama's careful discussion rightly suggests that a more complex biology cannot accomplish all kinds of social work by itself. Let me take eugenics as a further example. A nonreductionist biology is often seen as a weapon against eugenics. Rifkin (1998) is one example among many. He refers to Beckwith when he says that "a more balanced presentation of the relationship between genetics and environment needs to be made in the public arena, lest we risk the new science becoming the handmaiden for a eugenics-based politics" (Rifkin 1998: 158). But here, too, the point can only be that a more complex biology "does not so readily lend itself to such uses." It would be much too strong to say that a more complex biology would or should prevent eugenic policies, for several reasons.

First, eugenic policies are feasible on the basis of good and rich biology, too—although presumably with less oversimplification. The fact that gene/environment interactions are important does not imply that the outcome of every genetic defect becomes completely unpredictable. Although it is important to acknowledge that the effects of genes depends on all kinds of other genes and environmental conditions, there are many cases in which these effects will be fairly predictable in a normal range of genetic and environmental conditions. Within DST it is stressed strongly that the contingency of causal influences in development does not imply unpredictability. On the contrary, in normal situations in which all the normal causal factors are there, outcomes of development are highly predictable. In other words, given a normal environment, the influence of chromosomal abnormalities and genetic mutations will to a certain degree be predictable in many cases. Thus, an extra chromosome 21 leads to Down syndrome, even though its precise manifestations are quite unpredictable. Whether a person will eventually develop Huntington's disease can also be predicted on the basis of genetics alone, in the environments we are familiar with. In short, predictions based on information about genes will not vanish completely on the basis of a sound biology.

Second, there is more to eugenics than biology. The old eugenics was formed by not only biological forces but by social and political forces as well. While eugenics once took the form of coercive measures, later proponents of eugenics emphasized free individual choice, arguing for a "democratic eugenics." Whether new forms of genetic testing, predicting, and prevention are to be called eugenic is a matter of definition. If the term eugenics is reserved for the old ways, in which race and coercion were central, they are not. If everything that is directed toward genetic health in future generations is to be seen as eugenics, they are. In his book *The Lives to Come* Philip Kitcher talks about "laissez-faire" eugenics with respect to genetic counseling and testing (Kitcher 1996: 196). Laissez-faire eugenics, in which everyone is to be her or his own eugenicist, attempts to honor individual freedom and choice. This is an attractive feature of the new eugenics. On the other hand, individual choices are not made within a vacuum, and disturbing or even disastrous scenarios are still feasible. Yet in this era of genetic knowledge, some form of eugenics is inescapable, says Kitcher, and it is important to find the safest options. I agree, and it seems clear that in order to find safe options it is necessary to consider the science that is made use of as well as the social and political conditions, the goals that the technologies are used for, and how they affect people's lives.

Appropriate Technology

How exactly do science and society interrelate? It is not surprising that the analyses of critical scientists regarding this question are often general or tentative, given that their primary target of criticism is science itself. A separate academic discipline exists, science and technology studies, STS (e.g., Bijker, Hughes, and Pinch 1987; Latour 1987), which is dedicated to detailed study of how science shapes society and is shaped by it, with a special emphasis on relations of power and knowledge. This is the third domain of normativity with regard to science that I will consider.

Like science criticism, STS is concerned with the content of science. Unlike science criticism, is

does not look at scientific disciplines from the point of view of the discipline itself, but from the point of view of social science: science is described as a social activity in a social context. For a long time, normative reflection was almost absent from large parts of STS, as the discipline emphasized that empirical understanding was its goal. Normative reflection was associated with *taking sides* and was explicitly avoided. Many STS-ers, like many critical scientists, might be in part motivated by normative social considerations, but they were careful not to let this show up in their work. The difference is that for STS-ers, this work is not, say, biology, but social studies of biology.

But the antinormative atmosphere of STS is lessening, partly through explicit reconsideration of normative issues. Thus, Hans Radder (1992, 1998a, 1998b) has argued that questionable normative consequences go unnoticed as the result of the absence of normative reflection. For example, actor-network theory only emphasizes the side of winners by being interested exclusively in how chains of power in networks are strengthened. The voice of the losers is not represented in actor-network theory; actors vanish from sight, even cease to be actors, as soon as they are unsuccessful in building chains. A more symmetrical treatment of all the relevant actor perspectives would be normatively desirable, according to Radder (1992: 162).

Radder himself has been practicing constructive normative reflection in his book *In and About the World* (Radder 1996). In this book, he presents a strategy for the development of "appropriate technology." Recommendations for an appropriate agricultural biotechnology, he thinks, require that we do not stay within the boundaries of common ethical approaches, since that would yield recommendations that are too general and unpractical. Instead, we have to make a detour through the complexities of modern technological systems and their realization in the living world. He proposes to ask questions concerning the desirability of the realization of technological systems, concerning the desirability of the choices resulting from this realization, and concerning the desirability of the conditions needed for the realization. In addition, we should ask whether we know enough to answer these questions in a sensible way. In order to answer those questions, an integration of knowledge and expertise in the fields of biotechnology, technology assessment, science and technology studies, and ethics is needed. In the process of implementation, active participation of the people involved is a procedural necessity.

Participation is a key element in the growing amount of normative proposals concerning the social implementation of science and technology. Participatory methods are increasingly developed and implemented, such as "Constructive Technology Assessment," to which Radder's proposal is close in atmosphere. It is a field in which the world of STS meets the world of policy studies and Technology Assessment (TA). Newer brands of TA stress the importance of participation by all the social groups somehow involved with or affected by the development of new technologies. Because participation is an area full of pitfalls and practical difficulties, much explorative and experimental work is being done in this area, in order to enable participatory strategies to be successful. Various instruments to encourage public participation are developed, such as models and instruments for participatory debate and strategy development. The underlying normativity is a democratic one, and the direction in which these efforts point are aptly called the democratization of technology development.

One important issue in finding fruitful methods of participation concerns the fundamental question which roles the various participants have, and can have, in such procedures. For example, with regard to scientific knowledge, what are laypeople assumed to contribute? One possibility is that they contribute *values*, with which they try to judge technological projects. On this view, the relevant knowledge involved in public debates comes from science, while the public is there for the input of public norms and values with the help of which normative decisions about technological proposals are to be made. Such a division of tasks reflects the division of work between science generating knowledge, and ethics (derived from common morality) generating norms. A different possibility is that laypeople also have things to say with regard to knowledge, for example because they possess relevant knowledge of their own. Brian Wynne (e.g., Wynne 1996), writing about risk assessment, argues for this view. The role of knowledge with regard to risks is to control them. It is often thought that while science aims for control over life circumstances and develops the knowledge needed for it, laypeople resist such control and have no relevant knowledge either. Wynne attacks his view. Laypeople do aim for control, but since in daily life, unlike in scientific experiments, there is no possibility to keep conditions under control, the kind of knowledge that is useful for control in daily life is much more local and contextual than scientific knowledge typically is (Wynne 1996: 70). Personal agency and responsibility play a large role in it, and exactly these are in danger when people become dependent on scientific knowledge. The fundamental risk that people run in our "risk society," according to Wynne, is a risk to their identity, through dependency upon expert systems. What is needed is a search for more legitimate, more contextual forms of knowledge in public domains, in which laypeople are involved (Wynne 1996: 78). Wynne's normative project, clearly, is the protection of human agency and control in a world full of expert systems.

With regard to genetics, there are senses in which the risks he mentions are far from farfetched. For example, information about genetic defects typically has the character of risk figures which are not only very hard to interpret but are also in themselves devoid of personal meaning. The possibilities for genetic analysis are growing rapidly. In the near future, DNA-chips (devices that contain many pieces of single-stranded DNA) will become available that enable rapid analysis of many parts of the genome at once. If such analysis is going to inform you that you have an increased (but not exactly known) risk to get diseases X, Y, and Z, this information may inherently suggest that you should take preventive measures. Whether you want to have this information and to become a more or less full-time risk evader is a question that is intimately related to the question of what constitutes personal control. The widespread assumption that more knowledge enables better choices may diminish the social space for such questions and causes people to become fearful and dependent in decision making. As yet this is an underrecognized problem.

Participatory normative projects, especially if critical questions about the sources of legitimate knowledge are involved, recall a theme of the previous section: good science does not automatically lead to a good society. Along Wynne's lines of thought, the authority of science is even a threat to human agency and control. This threat comes from scientific expertise in general and is not confined to only some kinds of such expertise. Yet at the same time, it may be possible to distinguish among more and less threatening forms of science. Scientific knowledge that promises to be uniquely authoritative in solving practical questions is more threatening to human agency and control than scientific knowledge as it is encouraged by DST, that actively acknowledges the complexity and multiple layers of the world. Such knowledge, again, "does not so easily lend itself to such uses." The latter kind of science will tend to be more hospitable to the importance of local knowledge as well as the importance of human agency in the determination of normatively desirable social outcomes. An important question for scientists as well as policymakers is whether or not they acknowledge the possibility that in the implementation of science and technology important knowledge does not only come from science but from other domains of human life as well. In other words, development of participatory social instruments that

can fruitfully deal with the problems of how to relate scientific and lay knowledges is a major challenge for the implementation of appropriate technology.

Integration?

I have considered three different normative undertakings with regard to science. The aim has not been to be complete as to possible approaches, but, by showing distinct possibilities, placing in perspective DST's position in the normative landscape.

Let me recapitulate and sketch some potential integrative developments. First, ethics and science criticism. The approach taken by ethics primarily comes from philosophy. It is characterized by close attention to a conceptually coherent normative framework. Ethics does not itself judge scientific knowledge; with regard to scientific knowledge, it is in an outsider's position. Science criticism, on the other hand, focuses on science. Its normativity is aimed at sound and responsible knowledge. It concentrates on methodology, metaphors, blind spots, missing facts, interpretation of data, problem definition, statistical evaluation, or other dimensions in which different choices can make social differences. An important motivation is that knowledge is an important factor in shaping society, and that sound knowledge is important to help shape a sound society.

The two approaches thus appear to be very dissimilar, the one stressing ethical norms, while taking scientific knowledge for granted, the other stressing sound science. As far as critical scientists talk about ethics it is either in a tentative form, by saying, for example, that science is "ethically resonant," or by referring to general political analysis and worldviews. The direct relation between those worldviews and biological knowledge is just as problematic as that between ethical norms and biological knowledge.

For some purposes, the differences between ethics and science criticism do not really seem to be a problem, because they go with a nice division of tasks: science criticism deals with science, while ethics deals with its moral implications. However, through this division of labor each undertaking has only very limited legitimacy, and a thorough analysis of the ethical resonance of knowledge cannot be undertaken. It would be interesting to see whether some rapprochement between scientific and ethical normativity is possible.

A potential area of rapprochement of these two approaches may be framed by focusing on the concept of attention. Science criticism centrally invokes the issue of selective attention, saying for example that genetic determinism draws undue attention to genetic causes and solutions for social problems. I have suggested elsewhere (Van der Weele 1999: 135) that being normative about strategies of directing scientific attention could be called an ethics of attention. Interestingly, an approach which by some has been called an ethics of attention does exist within ethics. Its source is in the work of Iris Murdoch (1991, first published in 1970), who refers to origins of her view in the work of Simone Weil. The context is not science, but the directing of personal attention. Starting from the view that our energy is naturally directed in selfish ways, Murdoch sees the essence of moral excellence as outward attention. Clarity of vision is the normative ideal: unsentimental, detached, unselfish, objective attention. The direction of such attention is outward, away from the self, toward the world, and the ability so to direct attention she calls love (Murdoch 1991: 66). Martha Nussbaum (1990) takes a much similar approach. In her many discussions of the novels of Henry James, it is richness of perception that is the primary condition of moral excellence. Following James, she writes that our moral task is to be "people on whom nothing is lost," who are "finely aware and richly responsible" (Nussbaum 1990: 148).

The work of Murdoch and Nussbaum is not tailor-made for thinking about selective attention in science. In the first place, it is personal, not scientific attention that is their object. Further, they do not deal with the problem that we cannot attend to everything at once; their writings at times suggest that we can. But what does make their work interesting with regard to knowledge is that for them, the core of morality is not in moral theory, principles, or rules, but in perception. Moving to a less personal sphere, it is not so big a step to substitute research, or research questions, for perception. It is interesting to consider the thought that for science, a fair and rich distribution of attention, however that could be judged, could be an ideal that comes close to Hubbard's implied ideal of a balance of attention. It is also in line with many of the emphases in DST. For example, in her chapter "The Conceptualization of Nature," Susan Oyama (2000) discusses choices that biologists face in their search for knowledge. Such choices include what to pay attention to and what to problematize, says Oyama. Referring to Donna Haraway, she argues for a biology of multiple and embodied perspectives. Though cast in different terms, this seems to me close in spirit to the ideal of a biology in which fair and rich distribution of attention is a central challenge.

Likewise, but more briefly: There are also clear points of contact between science criticism and STS. Though the first undertaking focuses primarily on scientific normativity and the second on social analysis in the construction of science and technology, both are concerned with contents of knowledge. The development of socially responsible technology requires a thorough analysis in the intermediate area of relations between various kinds of knowledge, be it lay knowledge in relation to scientific knowledge or knowledge from various scientific disciplines. The aim to use more inclusive and integrative knowledge in the construction of responsible technology could be fused with the aim to find more democratic ways of developing technology, as in participatory forms of technology assessment.

Further analysis and development of such integrative work is beyond this chapter. The main purpose here has been to distinguish and compare different normative approaches to science, in order to shed light on relations between DST and ethics. After what has gone before, the conclusion can only be that while DST is clearly *ethically resonant*, and though integration can be imagined, DST and ethics in their present forms are clearly distinguished as normative enterprises, or, if you prefer, are worlds apart.

References

Beauchamp, T. L., and J. F. Childress. (1994). *Principles of Biomedical Ethics.* (4th ed.). New York: Oxford University Press.

Bijker, W. E., T. P. Hughes, and T. J. Pinch. (Eds.) (1987). *The Social Construction of Technological Systems.* Cambridge, MA: MIT Press.

Council for Responsible Genetics (1990). *Position Paper on Genetic Discrimination.* Boston: Council for Responsible Genetics.

Cranor, C. F. (Ed.) (1994). *Are Genes Us? The Social Consequences of the New Genetics.* New Brunswick, NJ: Rutgers University Press.

Garfinkel, A. (1981). *Forms of Explanation.* New Haven: Yale University Press.

Hubbard, R., and E. Wald. (1993). *Exploding the Gene Myth: How Genetic Information Is Produced and Manipulated by Scientists, Physicians, Employers, Insurance Companies, Educators and Law Enforcers.* Boston: Beacon Press.

Keller, E. F. (1994). Master molecules. In C. F. Cranor (Ed.), *Are Genes Us? The Social Consequences of the New Genetics,* pp. 89–98. New Brunswick, NJ: Rutgers University Press.

Kitcher, P. (1996). *The Lives to Come: The Genetic Revolution and Human Possibilities.* New York: Simon and Schuster.

Kitcher, P. (forthcoming). Battling the undead: How (and how not) to resist genetic determinism. In R. Singh, C. Krimbas, J. Beatty, and D. Paul (Eds.), *Thinking about Evolution: Historical, Philosophical, and Political Perspectives.* Cambridge: Cambridge University Press.

Latour, B. (1987). *Science in Action.* Milton Keynes: Open University Press.

Levins, R. (1966). The strategy of model building in population biology. *American Scientist* 54: 421–431.

Levins, R. (1968). *Evolution in Changing Environments.* Princeton, NJ: Princeton University Press.

Levins, R., and R. Lewontin. (1985). *The Dialectical Biologist.* Cambridge, MA: Harvard University Press.

Murdoch, I. (1991). *The Sovereignty of Good.* London: Routledge.

Nussbaum, M. C. (1990). *Love's Knowledge: Essays on Philosophy and Literature.* New York: Oxford University Press.

Oyama, S. (1985). *The Ontogeny of Information: Developmental Systems and Evolution.* Cambridge: Cambridge University Press. (2d rev. ed., Durham, NC: Duke University Press, 2000.)

Oyama, S. (1999). The nurturing of natures. Unpublished manuscript.

Oyama, S. (2000). *Evolution's Eye: A Systems View of the Biology-Culture Divide.* Durham, NC: Duke University Press.

Proctor, R. N. (1991). *Value-Free Science? Purity and Power in Modern Knowledge.* Cambridge, MA: Harvard University Press.

Radder, H. (1992). Normative reflections on constructivist approaches to science and technology. *Social Studies of Science* 22: 141–173.

Radder, H. (1996). *In and About the World: Philosophical Studies of Science and Technology.* Albany: State University of New York Press.

Radder, H. (1998a). The politics of STS. *Social Studies of Science* 28: 325–333.

Radder, H. (1998b). Second thoughts on the politics of STS. *Social Studies of Science* 28: 344–348.

Rawls, J. (1971). *A Theory of Justice.* Cambridge, MA: Harvard University Press.

Rifkin, J. (1998). *The Biotech Century: Harnessing the Gene and Remaking the World.* London: Victor Gollancz.

Rose, S., R. C. Lewontin, and L. J. Kamin. (1984). *Not in Our Genes: Biology, Ideology and Human Nature.* Harmondsworth: Penguin.

Singer, P. (1996). Ethics and the limits of scientific freedom. *The Monist* 79: 218–229.

Van der Steen, W. J. (1993a). *A Practical Philosophy for the Life Sciences.* Albany: State University of New York Press.

Van der Steen, W. J. (1993b). Towards disciplinary disintegration in biology. *Biology and Philosophy* 8: 385–397.

Van der Weele, C. N. (1999). *Images of Development: Environmental Causes in Ontogeny.* Albany: State University of New York Press.

Wynne, B. (1996). May the sheep safely graze? A reflexive view of the expert-lay knowledge divide. In S. Lash, B. Szerszynski, and B. Wynne (Eds.), *Risk, Environment and Modernity: Towards a New Ecology.* London: SAGE Publications.

Index